Fishery Resources, Environment, and Conservation in the Mississippi and Yangtze (Changjiang) River Basins

Sponsors

The following sponsors supported the publication of this book:

The Nature Conservancy
Lower Mississippi River Conservation Committee
U.S. Fish and Wildlife Service
U.S. Geological Survey
University of Arkansas at Pine Bluff

Fishery Resources, Environment, and Conservation in the Mississippi and Yangtze (Changjiang) River Basins

Edited by

Yushun Chen

Aquaculture and Fisheries Center, University of Arkansas-Pine Bluff
1200 North University Drive, Pine Bluff, Arkansas 71601, USA

Duane C. Chapman

U.S. Geological Survey, Columbia Environmental Research Center
4200 New Haven Road, Columbia, Missouri 65201, USA

John R. Jackson

Department of Biological Sciences, Arkansas Tech University
1701 North Boulder Avenue, Russellville, Arkansas 72801, USA

Daqing Chen

Yangtze River Fisheries Research Institute, Chinese Academy of Fishery Sciences
No. 8, 1st Wudayuan Road, East Lake Hi-Tech Development Zone
Wuhan, Hubei 430223, P. R. China

Zhongjie Li

State Key Laboratory of Freshwater Ecology and Biotechnology, Institute of Hydrobiology
Chinese Academy of Sciences, 7 South Donghu Road, Wuhan 430072, P. R. China

K. Jack Killgore

U.S. Army Corps of Engineers, Engineer Research and Development Center
Environmental Laboratory
3909 Halls Ferry Road, Vicksburg, Mississippi 39180, USA

Quinton Phelps

Big Rivers and Wetlands Field Station, Missouri Department of Conservation
3815 East Jackson Boulevard, Jackson, Missouri 63755, USA

Michael A. Eggleton

Aquaculture/Fisheries Center, University of Arkansas at Pine Bluff
1200 North University Drive, Box 4912, Pine Bluff, Arkansas 71601, USA

American Fisheries Society Symposium 84

Proceedings of the Symposium "Fishery Resources and Environment of the Mississippi and Yangtze (Changjiang) River Basins: Common Challenges and Shared Perspectives"
Held in Little Rock, Arkansas, USA
September 11, 2013

American Fisheries Society
Bethesda, Maryland, USA
2016

Suggested citation formats follow.

Entire book

Chen, Y., D. C. Chapman, J. R. Jackson, D. Chen, Z. Li, K. J. Killgore, Q. Phelps, and M. A. Eggleton, editors. 2016. Fishery resources, environment, and conservation in the Mississippi and Yangtze (Changjiang) River basins. American Fisheries Society, Symposium 84, Bethesda, Maryland.

Chapter in book

Yuan, J., Y. Xia, Z. Li, Z. Yin, and J. Liu. 2016. Changes in fisheries resources in the Hanjiang River and Danjiangkou Reservoir, China. Pages 179–191 *in* Y. Chen, D. C. Chapman, J. R. Jackson, D. Chen, Z. Li, K. J. Killgore, Q. Phelps, and M. A. Eggleton, editors. Fishery resources, environment, and conservation in the Mississippi and Yangtze (Changjiang) River basins. American Fisheries Society, Symposium 84, Bethesda, Maryland.

Printed in the United States of America on acid-free paper.

Library of Congress Control Number 2016943174

ISBN 978-1-934874-44-8
ISSN 0892-2284

American Fisheries Society Web site: www.fisheries.org

American Fisheries Society
5410 Grosvenor Lane, Suite 110
Bethesda, Maryland 20814
USA

Contents

Preface

During the 143rd American Fisheries Society (AFS) annual meeting (September 8–12, 2013, Little Rock, Arkansas), the first Mississippi–Yangtze River basin symposium was presented with the session title "Fishery Resources and Environment of the Mississippi and Yangtze (Changjiang) River Basins: Common Challenges and Shared Perspectives." This book was developed from the presentations in the symposium.

The Mississippi and Yangtze (Changjiang) River basins, the largest basins of North America and Asia, serve as the principal navigational waterways and main water sources and play important economic, social, cultural, and ecological roles in the two continents. Maintaining healthy and productive fisheries and the integrity of aquatic ecosystems are extremely important for achieving sustainability in the two basins. Both basins share many taxa, humans exploit some of the same groups of fish, and both basins have experienced some of the same environmental challenges to their fisheries, which, to date, have not been comparatively analyzed. This book covers (1) fishery resources, geology, land use, hydrology, and environment; (2) endangered and invasive species, biodiversity and conservation; and (3) anthropogenic stressors, floodplains, and river restorations in the two basins.

We thank all of the authors, reviewers, and Aaron Lerner and Debby Lehman of the AFS publication division for their help and persistence in producing this volume. We hope you find this book useful.

Symbols and Abbreviations

The following symbols and abbreviations may be found in this book without definition. Also undefined are standard mathematical and statistical symbols given in most dictionaries.

A ampere
AC alternating current
Bq becquerel
C coulomb
°C degrees Celsius
cal calorie
cd candela
cm centimeter
Co. Company
Corp. Corporation
cov covariance
DC direct current; District of Columbia
D dextro (as a prefix)
d day
d dextrorotatory
df degrees of freedom
dL deciliter
E east
E expected value
e base of natural logarithm (2.71828...)
e.g. (exempli gratia) for example
eq equivalent
et al. (et alii) and others
etc. et cetera
eV electron volt
F filial generation; Farad
°F degrees Fahrenheit
fc footcandle (0.0929 lx)
ft foot (30.5 cm)
ft^3/s cubic feet per second ($0.0283\ m^3/s$)
g gram
G giga (10^9, as a prefix)
gal gallon (3.79 L)
Gy gray
h hour
ha hectare (2.47 acres)
hp horsepower (746 W)
Hz hertz
in inch (2.54 cm)
Inc. Incorporated
i.e. (id est) that is
IU international unit
J joule
K Kelvin (degrees above absolute zero)
k kilo (10^3, as a prefix)
kg kilogram
km kilometer
l levorotatory
L levo (as a prefix)
L liter (0.264 gal, 1.06 qt)
lb pound (0.454 kg, 454g)
lm lumen
log logarithm
Ltd. Limited
M mega (10^6, as a prefix); molar (as a suffix or by itself)
m meter (as a suffix or by itself); milli (10^{-3}, as a prefix)
mi mile (1.61 km)
min minute
mol mole
N normal (for chemistry); north (for geography); newton
N sample size
NS not significant
n ploidy; nanno (10^{-9}, as a prefix)
o ortho (as a chemical prefix)
oz ounce (28.4 g)
P probability
p para (as a chemical prefix)
p pico (10^{-12}, as a prefix)
Pa pascal
pH negative log of hydrogen ion activity
ppm parts per million
qt quart (0.946 L)
R multiple correlation or regression coefficient
r simple correlation or regression coefficient
rad radian

S	siemens (for electrical conductance); south (for geography)
SD	standard deviation
SE	standard error
s	second
T	tesla
tris	tris(hydroxymethyl)-aminomethane (a buffer)
UK	United Kingdom
U.S.	United States (adjective)
USA	United States of America (noun)
V	volt
V, Var	variance (population)
var	variance (sample)
W	watt (for power); west (for geography)
Wb	weber
yd	yard (0.914 m, 91.4 cm)
α	probability of type I error (false rejection of null hypothesis)
β	probability of type II error (false acceptance of null hypothesis)
Ω	ohm
μ	micro (10^{-6}, as a prefix)
′	minute (angular)
″	second (angular)
°	degree (temperature as a prefix, angular as a suffix)
%	per cent (per hundred)
‰	per mille (per thousand)

Part 1 Overview of the Two Basins

American Fisheries Society Symposium 84:3–34, 2016

The Mississippi River: A Place for Fish

Harold L. Schramm, Jr.*
*U.S. Geological Survey, Mississippi Cooperative Fish and Wildlife Research Unit
Mississippi State University, Mail Stop 9691, Mississippi State, Mississippi 39762, USA*

Brian S. Ickes
*U.S. Geological Survey, Upper Midwest Environmental Sciences Center
2630 Fanta Reed Road, La Crosse, Wisconsin 54603, USA*

Abstract.—The Mississippi River flows 3,734 km from its source at Lake Itasca, Minnesota to its outlet at the Gulf of Mexico. Along its course, it collects water from portions of two Canadian provinces and 41% of the conterminous United States. Although greatly altered for navigation and flood control throughout much of its length, the Mississippi River remains an important fishery resource that provides habitat for 188 species of fishes and recreational and commercial fishing opportunities. The objectives of this chapter are to describe the contemporary fisheries habitat throughout the Mississippi River, identify how management to achieve human benefits influences the fishes and their habitats, and summarize efforts to conserve and enhance fish habitat. The 826-km headwater reach is entirely in Minnesota and remains largely unaltered. The reaches that extend 1,059 km from St. Anthony Falls, Minnesota to above the confluence with the Missouri River near St. Louis, Missouri have been altered by impoundment that has affected floodplain function, increased sedimentation of backwaters, and homogenized the formerly diverse aquatic habitats. After the confluence with the Missouri River, the Mississippi River flows freely for 1,849 km to the Gulf of Mexico. The alterations of the free-flowing reaches of greatest significance to the fisheries resource are reducing the duration and height of the flood pulse as a consequence of shortening the river channel, disconnection of the river from its historic and present floodplain, and loss of secondary channel–island complexes. Engineering features to improve commercial navigation have also added habitat and, when wisely manipulated, can be used to rehabilitate habitat. Some aspects of water quality have improved, but legacy chemicals and nutrient-laden inflows and sediments remain problems. Although true restoration in the sense of restoring all environmental conditions to an unaltered state is unlikely, the future value of the Mississippi River as a fisheries resource will depend on actively maintaining diverse and accessible aquatic habitats to support food webs and water quality suitable for fishes.

Introduction

The Mississippi River has been and remains one of the great defining natural resources of North America. From the 1600s to the 1800s, the Mississippi River served as a pathway for exploration of a young nation and as a trade corridor. In the 1800s, the fertile soils created by the same geologic forces that formed the Mississippi River were tilled for agriculture and generated the need for reliable navigation and flood control that would change the ecological structure and function of America's greatest river in the present and probably in the future. The role of the Mississippi River in the social fabric of America is divulged in numerous books, including

* Corresponding author: hschramm@usgs.gov

Mark Twain's *Life on the Mississippi* (Twain 1883), John Barry's *Rising Tide* (Barry 1997), John Anfinson's *The River We Have Wrought* (Anfinson 2005), and Calvin Fremling's *Immortal River* (Fremling 2005). Considerably less attention has been directed at the life in the Mississippi River, although there have been notable contributions in the technical literature (e.g., Eckblad 1986; Jahn and Anderson 1986; Fremling et al. 1989; Baker et al. 1991; Schramm 2004). This chapter seeks to advance and update those efforts.

The recent form of the Mississippi River was established after the Wisconsin glaciation 10,000–14,000 years ago. The Mississippi River basin, which includes the Mississippi River, the Missouri River and its tributaries, and the Ohio River and its tributaries, as well as other large rivers such as the Arkansas and Illinois rivers, collects water from 4.8 million km^2 that spans portions of two Canadian provinces and 41% of the conterminous United States (Figure 1). The predominant land use of the basin is cropland (58%), followed by range and barren lands (21%) and woodlands (18%; Goolsby 2000). The Mississippi River extends 3,734 km from its headwaters at Lake Itasca, Minnesota to the Gulf of Mexico south of New Orleans, Louisiana. The upper 895 km of the Mississippi River courses through Minnesota, and the lower 491 km is entirely in Louisiana; in between, the river forms the borders between Minnesota, Wisconsin, Iowa, Illinois, Missouri, Kentucky, Arkansas, Tennessee, Mississippi, and Louisiana. The mean annual discharge of the lower Mississippi River (i.e., including flows from the Missouri and Ohio rivers) is 16,908 m^3/s (Poore et al. 2001), and

Figure 1. The watershed of the Mississippi River includes two Canadian provinces and 41% of the conterminous United States. Image credit: U.S. Geological Survey, Upper Midwest Environmental Sciences Center.

the Mississippi River ranks seventh among the world's rivers, ordered by discharge (www.waterencyclopedia.com/Re-St/Rivers-Major-World.html).

The objectives of this chapter are to describe the contemporary fisheries habitat throughout the Mississippi River, identify how management to achieve human benefits influences the fishes and their habitats, and summarize efforts to conserve and enhance fish habitat. Many factors affect fish and fish habitat, but climate, geomorphology, and hydrology are the master variables that determine the structure and function of the Mississippi River floodplain ecosystem. These master variables can have independent effects, but they also interact; climate, for example affects hydrology that in turn affects geomorphology, and geomorphology affects hydrology.

The Mississippi River flows north to south through 18 degrees of latitude, and the climate grades from cool temperate at its origin (median frost-free days = 113 at Lake Itasca, Minnesota) to subtropical at its terminus(median frost-free days = 300 at New Orleans, Louisiana). The river provides a home to 188 species of fishes in 31 families that include several coolwater fishes such as esocids and percids, but most are warmwater fishes such as cyprinids, catostomids, ictalurids, and centrarchids (Schramm et al. 2016, this volume).

While climate is a rather continuous gradient, geomorphology and hydrology change incrementally and sometimes rather abruptly as a consequence of the interaction of geology and confluence of seventh-, eighth-, and ninth-order rivers, such as the Arkansas, Missouri, and Ohio rivers, with the Mississippi River. Imposed on this template are engineered alterations, themselves influenced by geomorphology, to modify the river to accommodate human needs. As a result of these natural processes and human activities, the ecology of the river and the fish habitat it provides are best understood when the river is divided into five ecological reaches[1](Figure 2): the headwaters; the impounded segment, which is further divided into upper impounded and lower impounded reaches due to geomorphic differences; and the free-flowing segment, which is divided into upper free-flowing and lower free-flowing reaches as a result of the substantial increase in discharge from the confluence of the Ohio River at the upstream end of the lower free-flowing reach.

Geomorphology

Headwaters Reach

The headwaters reach extends 826 km from its source at Lake Itasca to Coon Rapids, Minnesota (Figure 2). Six dams have created relatively small eutrophic and mesotrophic multipurpose reservoirs that provide water supply, flood control, hydropower, and recreational use, and five low-head dams have been constructed on natural lakes to stabilize water levels, but most of the reach is free flowing. The sinuous channel has a sand and gravel substrate and is flanked by wetlands and forested riparian zones throughout much of the reach (Figure 3); the riparian zone remains largely unaltered except for short segments in several small cities (Eric Altena, Minnesota Department of Natural Resources, personal communication).

---

[1] The Mississippi River Commission Act of 1879 established the Mississippi River Commission (MRC) and delegated to this body the authority for the improvement of the Mississippi River from its headwaters to the Head of the Passes near its mouth. The MRC recognizes three reaches of the Mississippi River based on engineering solutions to achieve the purposes of navigation and flood control: headwaters, the upper Mississippi River system (upper St. Anthony Falls to the mouth of the Ohio River), and the lower Mississippi River. This chapter recognizes five reaches of the Mississippi River delineated by geomorphic and hydrologic conditions that affect fish habitat. The headwaters reach of this chapter coincides with the headwaters of the MRC. The upper impounded, lower impounded, and upper free-flowing reaches of this chapter coincide with the upper Mississippi River (also referred to as the upper Mississippi River system) of the MRC. The lower free-flowing reach in this chapter coincides with the lower Mississippi River of the MRC.

Figure 2. The 3,734-km Mississippi River varies from the headwaters at Lake Itasca to its outlet into the Gulf of Mexico. The river has been variously divided for political reasons. The five reaches identified here largely reflect geomorphic changes.

Impounded Reaches

The impounded reaches extend 1,059 km from upper St. Anthony Falls in Minneapolis, Minnesota to Lock and Dam 26 (Melvin Price locks and dam), approximately 8 km upstream of the confluence with the Missouri River near St. Louis, Missouri, traversing a north–south axis of flow through 10° of latitude. The historic river is a braided channel with glacial outwash substrate, primarily sand and gravel; however, some areas of exposed bedrock are present in the lower impounded reach at Keokuk, Iowa and in the segment bordered by Missouri. The river and its floodplain are flanked by steep bluffs of sandstone, limestone, and dolomite throughout most of the impounded reach. Throughout both impounded reaches, surface water response to impoundment created a riverine upper part of the pool (the segment of river between two consecutive locks and dams), a transitional middle section of flooded braided channels and backwaters, and a larger open impounded lake-like area in the lower part of each navigation pool (Theiling and Burant 2013), but impoundment for navigation has had differing effects in the two reaches. The upper (Pools 1–13, 533 km) and lower (Pools 14–26, 526 km) impounded reaches differ in the width of the floodplain (Table 1); the floodplain ranges from 1.6 to 3.2 km wide and totals approximately 201,000 ha (only 3% is disconnected by levees) throughout the upper impounded reach, while the lower impounded reach averages 8 km wide and totals approximately 407,000 ha (50% disconnected by levees; USACE 2000).

Figure 3. The headwaters reach begins at Lake Itasca, Minnesota. Throughout much of its course, it flows through wetlands and forests, and the riparian zone remains largely undeveloped. Photograph credits: Doug Ohman, Pioneer Photography.

The impounded reaches of the Mississippi have a long history of development that has greatly influenced the geomorphology, hydrology, and ecology of the river (Sparks 1995; USGS 1999; Anfinson 2005; Johnson and Hagerty 2008; Theiling and Burant 2013). In recognition of extensive and multiple-use values, the federal government legislated the impounded reaches as a nationally significant ecosystem and a nationally significant commercial navigation system (Upper Mississippi River Management Act of 1986), which has facilitated research and monitoring of fish and fish habitat. This

Table 1. Floodplain areas for reaches of the Mississippi River. Floodplain area is the area of the historic floodplain, disconnected floodplain is the portion of the historic floodplain that is now separated from the river by levees, permanently inundated floodplain is the area of the floodplain that is aquatic habitat throughout most or all of the year, and active floodplain is the area seasonally inundated in most years. Data from Delaney and Craig (1997), Schramm et al. (1999), and USACE (2000).

	Headwaters	Upper impounded	Lower impounded	Upper free-flowing	Lower free-flowing
Floodplain area (ha)	132,740	200,728	407,123	268,312	10,117,361
Disconnected floodplain (%)	<0.01	3	50	82	93
Permanently inundated floodplain (ha)	0	83,923	60,263	27,084	51,154
Active floodplain (ha)	132,740	197,716	202,711	125,678	600,690

has led to more information about the status of fish and fisheries and also better understanding of actions needed to sustain them (USGS 1999; Johnson and Hagerty 2008; Ickes et al. 2005a).

Alterations in the impounded reaches have been incremental over time and centered on making the river navigable for commercial shipping and flood control to allow development of the floodplain for agriculture and urban areas. Physical, chemical, and ecological changes associated with these alterations are detailed and synthesized in several governmental reports (Scientific Assessment and Strategy Team 1994; USGS 1999; Johnson and Hagerty 2008). The most important and notable changes relative to Mississippi River fishes are (1) impoundment for navigation, (2) levee development for flood control and floodplain agricultural production, and (3) in-channel engineering to maintain a navigable channel (see Box 1).

Navigation-related channel modifications in this reach began in 1823 with snag removal and sandbar dredging, progressed to the construction of channel-training structures by 1873, and culminated in the construction of 29 dams and ship locks, most constructed in the 1930s (Fremling and Claflin 1984; Anfinson 2005; Johnson and Hagerty 2008; Figure 4). The locks and dams raised and stabilized water levels during low-discharge periods, with these effects most apparent in the lower impounded areas of each navigation pool. This had the immediate consequence of permanently inundating much of the active floodplain[2] of the lower third of each navigation pool (Table 1).

Since impoundment (1930s to present), sediment deposition has increased in contiguous off-channel areas, impairing these critical habitats for fishes, and waves in the lower lake-like portions of the pools have reduced islands and associated littoral zones, thereby reducing habitat diversity (Knights et al. 1995; Raibley et al. 1997; Johnson et al. 1998; Ickes et al. 2005a). Additionally, the locks and dams are semipermanent barriers to fish passage throughout this reach, and passage opportunities vary as a function of species, seasonal changes in the hydrograph, and local dam operations (Ickes et al. 2001; Knights et al. 2002; Zigler et al. 2003; Wilcox et al. 2004; Tripp et al. 2013).

The floodplain in the impounded reach, except for a 420-km reach between the Chippewa River confluence in Wisconsin and Rock Island, Illinois that was protected as a federal fish and wildlife refuge in 1924 (Upper Mississippi River National Wildlife and Fish Refuge), has been extensively developed for agriculture. This has had two major effects on impounded reach floodplains. First, levee construction, more prominent in the lower impounded reach, to facilitate agricultural production on the rich alluvial floodplain soils disconnected a substantial portion of the historic floodplain from the river (Table 1). Second, increases in the area tilled for agriculture and increased fertilizer application throughout the entire upper basin (Midwestern United States) has resulted in increased sediment and nutrient loads delivered to the river (DeHaan1998; Goolsby et al. 2000; Houser and Richardson 2010). Continuing sediment deposition in low-flow, off-channel, and impounded habitats within the navigation pools has had a homogenizing effect on water depth and substrate composition in these pooled areas, reducing habitat quality and quantity for fishes and other fauna. High-nutrient loads from agricultural nonpoint sources often result in eutrophic water quality conditions and substantial diel swings in dissolved oxygen in off-channel habitats (Houser and Richardson 2010).

The impounded reaches now exist as a complex mosaic of private and public lands managed to satisfy agriculture, urban, navigation, commercial, recreation, and natural resource conservation interests. Complex ownership patterns, multijurisdictional management authorities, and the multiuse nature of the impounded reaches present substantial challenges to natural resource

[2] That portion of the historic floodplain that continues to be seasonally inundated in some years. Depending on location, the active floodplain is bounded by man-made levees.

Box 1. Engineered structures for navigation and flood control

Dikes, also known as wing dams or wing dikes, are used to direct the flow of water. Originally made of wooden pilings, dikes are now built with large quarried rock (see figure below). Typically, the dikes are rooted or anchored to the shore (a trench is

Formerly built from pilings (top) and now made of large quarried rock (middle), dikes are used to direct water flow to maintain alignment and depth of the navigation channel. Dikes are usually constructed as dike fields containing several dikes. Where possible, dikes are now being modified with notches (bottom) to allow some water to pass through the dike at lower river stages to maintain flow through the dike field or a secondary channel. Photograph credits: Moore (1972), U.S. Army Corps of Engineers, and H. L. Schramm.

(Box continues)

Box 1. Continued

dug into the alluvial material landward of the river bank and filled with quarried rock) and extend into the river perpendicular to the thalweg; however, various configurations and placements of dikes (e.g., lateral dikes, L dikes) are used to achieve specific results. Commonly, dikes are constructed in groups of three or more (dike fields), with each successive downstream dike extending further into the channel. Dikes and dike fields are often installed upstream of secondary channels to serve the dual purpose of preventing the river from changing channels and to direct the energy of the flow to scour the primary navigation channel (see figure below). More than 230 and 617 linear km of dikes have been constructed in the upper and lower free-flowing reaches, respectively (Hutoff et al. 2013). Dikes are also used in the impounded reaches, in addition to the dams, as a means to divert flow from side channels, increase flow velocity in the main navigation channel, and promote bed scour within each navigation pool to ensure a 2.74-m-deep commercial navigation channel. Nearly all dikes in the impounded zone are submersed at normal pool elevations.

Bendway weirs are underwater rock dikes placed on the channel bottom (not connected to shore) on the outside bend of the river. In contrast to dikes built from shore to direct flow of the main channel away from shore and toward the channel, bendway weirs can be used to direct flows toward the shallow, convex shore (point bar) to widen a channel or straighten a bend to facilitate navigation.

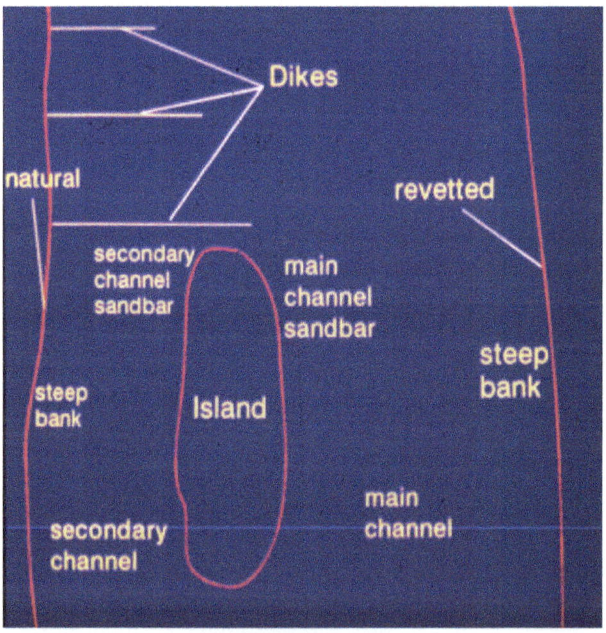

Dikes direct the flow to the navigation channel, and revetted banks forestall erosion on the outside of river bends. Together, these engineering structures maintain channel alignment and facilitate scouring of the navigation channel.

(Box continues)

Box 1. Continued

Revetments are materials applied to the river banks to forestall bank erosion and maintain channel alignment. Historically made from willow *Salix* spp. bundles (small trees and branches woven into a mat) anchored to the shore, revetments are now made of large quarried rock (impounded reaches and upper free-flowing reach) or articulated concrete mattress (formed 30 cm × 70 cm × 7 cm concrete slabs interconnected by malleable stainless steel wire) overlain with large quarried rock (lower free-flowing reach). The articulated concrete mattress is placed from the top of the bank to the toe of the channel, and the quarried rock is placed on top of the mattress from the top of the bank to several meters below the low-water level (see figure below). Revetments have been installed on approximately 1,750 km river bank in the free-flowing reaches.

Revetments made of quarried rock or connected concrete slabs overlaid with quarried rock are installed on outside (concave) river bends and other locations to forestall bank erosion. Revetments made of willow bundles were replace by quarried rock. Revetments today in the lower free-flowing reach consist of articulated concrete mattress overlain with large quarried rock (bottom). Photograph credits: U.S. Army Corps of Engineers and H. L. Schramm.

(Box continues)

Box 1. Continued

Levees are earthen structures installed on the floodplain to protect properties from rising river waters (see figure below). The levees were originally engineered to contain the 100-year flood (a flow that, based on historic data, has a 1% chance of occurring in any given year), but levees continue to be raised to contain the project design flood, a worst-case scenario that is a flood discharge that exceeds the 500-year flood. Construction of levees requires large quantities of soil excavated from the land adjacent to the levees. The resulting borrow pits are relatively large and deep and, although filled by sedimentation over time, can provide permanent-water lakes on the active floodplain that support diverse and abundant fish assemblages (Cobb et al. 1984; Miranda and Killgore 2013).

Levees have been installed throughout the length of the impounded and free-flowing reaches to contain the Mississippi River within a relatively narrow portion of its historic floodplain. Photograph credit: H. L. Schramm.

managers throughout the basin (Ickes et al. 2005b; Garvey et al. 2010).

Free-Flowing Reaches

The free-flowing reaches of the Mississippi River begin 8 km upstream from the confluence of the Missouri River at the tailwaters of Lock and Dam 26 and extend 1,849 km to the Gulf of Mexico. As the name implies, no dams impede the flow of water or fragment the river, but the free-flowing reaches have been extensively altered to create and maintain a commercial navigation channel by installation of wing dikes to direct water flows and revetments to forestall bank erosion (see Modifications for navigation and flood control). Throughout these reaches, the river is a meandering channel with a predominantly sand substrate and isolated areas of gravel (Figure 5).

The upper and lower free-flowing reaches differ in floodplain width, river width, and discharge. The floodplain of the upper free-flowing reach is less than 80 km wide and terminates at natural bluffs in many areas. The lower free-flowing reach begins at the confluence of the Ohio River, which contributes 45% of the mean annual discharge passing the lower free-flowing reach (Meade 1995). Mean annual discharge is 5,380 m^3/s at St. Louis, Missouri and increases to 16,990 m^3/s at Vicksburg (Poore et al. 2001); the highest recorded discharge at Vicksburg is 61,140 m^3/s (Biedenharn et al. 2000). The greatly in-

Locks and dams built for navigation and flood control have converted the Mississippi River from Minneapolis, Minnesota to St. Louis, Missouri into a series of impoundments that have permanently inundated much of the historic floodplain. Photograph credits: Upper Mississippi River Restoration Program.

creased discharge is carried in a wider and generally deeper channel, and the river has a broader historic floodplain that averages almost 100 km wide.

The free-flowing reach has a lower gradient than the impounded reach, and dams are not needed to maintain sufficient depth for commercial navigation. Dikes are used to train the channel so the energy of the current can be used to maintain a deeper and more open channel and, in conjunction with revetments to forestall erosion, to maintain channel alignment. Installation of dikes and revetments is now more than 90% complete throughout the free-flowing reach, but inspection and maintenance of these structures is a perpetual process. Dredging shallow areas and removing snags (large trees) were the principal forms of channel maintenance in the 19th century, and dredging is still used as necessary during extremely low river stages or where channel training structures fail to maintain sufficient channel depth for commercial navigation. The dredged material is usually disposed into the channel to be carried downstream by the river flow.

As in the impounded reach, the fertile Mississippi alluvial valley has been extensively developed for agriculture. Much of this land is within the historic Mississippi River floodplain, and private levees have been constructed since agricultural development of the area began in the 19th century. Following the catastrophic flood of 1927, the U.S. Army Corps of Engineers Mississippi River and Tributaries Project initiated the construction of federal mainline levees throughout the length of the unimpounded river. The extent and magnitude of levees increase downstream as the floodplain widens, high bluffs become less frequent, and human use of the floodplain increases. The emplacement and construction of the levees were mostly

Figure 5. The free-flowing reach winds through a broad alluvial valley. The river is flanked by a broad floodplain throughout much of the reach, but the levees are set adjacent to the river, which is almost entirely main channel habitat, in much of the heavily industrialized lower 335 km. Photograph credits: U.S. Army Corps of Engineers, Lower Mississippi Conservation Committee, and H. L. Schramm.

completed by 1935, but the levees have been incrementally raised (by government and private parties) over time in response to large flood risks (Smith and Winkley 1996; Pinter 2005). As in the impounded reach, the levees disconnect the river from much of the historic floodplain (Table 1). However, in the free-flowing reach, the levees are set back from the river throughout most of its length to accommodate floodwater conveyance, and a substantial area of active floodplain remains. The area of active floodplain varies along the length of the lower free-flowing reach, ranging from an average of 44 ha/km in the heavily industrialized lower 375 km where the levees frequently are immediately adjacent to the river channel to more than 1,800 ha/km in the middle portion (Schramm et al. 1999).

While many changes to the Mississippi River occurred after the flood of 1927, the major alteration in the free-flowing reach was cutoffs, also called neck cutoffs (Biedenharn et al. 2000). This process involves digging a new channel across the narrow neck of a long river bend (meander loop or oxbow) to create a shorter, straighter path for the river and leaving the cutoff bend as a secondary or abandoned channel or an oxbow lake. Sixteen neck cutoffs (14 were engineered, 2 were natural) shortened the river by 244 km during 1929–1942 (Biedenharn and Watson 1997). The shortening and straightening of the river facilitated navigation and reduced flooding by increasing downstream conveyance. However, this drastic change—a 30% shortening of the river within the impacted reach (originally 885 km from near Memphis, Tennessee to near Natchez, Mississippi)—occurred in less than 13 years. In comparison, 14–17 natural cutoffs occurred per hundred years before 1884 (Biedenharn and Watson 1997). This drastic and rapid shortening altered the natural sinuosity that triggered channel realignment and altered the slope, which triggered a massive slope recovery process now known as "head cutting" (a migrating channel degradation upstream, channel aggradation downstream of the alteration; Winkley and Brooks 1980). The head cutting is diminished but still lingering (Biedenharn and Watson 1997; Biedenharn et al. 2000; Pinter et al. 2010).

Hydrology

While the hydrograph of the headwaters remains largely unchanged except in the relatively short sections affected by the six impoundments, the hydrographs of the impounded and free-flowing reaches have been substantially altered. In the impounded reaches, changes in the hydrograph are a culmination of many factors, including climate, basin-scale land use and tributary alterations that affect runoff and delivery, and channel-training structures that influence conveyance (Pinter et al. 2008, 2010), but locks and dams placed to raise the minimum stage to provide sufficient depth for navigation during low-flow seasons exert the dominant influence on the hydrograph (Figure 6). The hydrologic effect of the dams varies among and within navigation pools (Johnson and Hagerty 2008), but the significant change is a river that now permanently inundates much of the historic floodplain in the middle and lower portions of each navigation pool (Table 1). The dams have five principal ecological consequences: (1) reduced allochothonous production in the permanently inundated portion of the floodplain that would be expected to result from cyclic flooding of the floodplain (cf. Junk et al. 1989; Bayley 1995); (2) a vast shallow-water zone lateral to the main stem that is vulnerable to sedimentation and lacking the replenishing flows from the river; (3) accumulation of fine-particle sediments that provide poor rooting substrate for aquatic plants and that are easily resuspended by wave action, leading to turbid and light-poor environments for aquatic plant growth; (4) geomorphic homogenization due to sedimentation of off-channel (i.e., lateral to the channel) areas and loss of islands from wave-driven erosion in the lake-like lower reach of navigation pools; and (5) restricted longitudinal connectivity during times of the year when the dams are closed to maintain sufficient depth for navigation.

Most navigation dams have their greatest hydrologic effect during low flows. At high flows (e.g., spring floods), gates are opened and the river flows freely. Thus, the locks and dams raise (relative to preimpoundment conditions) and stabilize water levels within each navigation pool during low flows in the summer and fall, but have minor consequences on flows and river stages during high flows. The exceptions to this are Locks and Dams 25 and 26 that are managed as midpoint control dams (Flinn et al. 2008). This hydrologic management technique attempts to maintain sufficient channel depth for commercial navigation at low to medium flows based upon the water-level elevation at the mid-point of each pool, which results in greater water elevation variability above the downstream dam than is experienced at the other locks and dams. Midpoint control permits the rapid conveyance of floodwaters out of these lower two pools during high flows, which usually occur in the spring, but results in a dewatering or drawdown in the impounded and contiguous off-channel backwater zones. Dewatering occurs rapidly and has both positive and negative impacts on local fish communities in impounded and off-channel habitats (Flinn et al. 2008).These dewatering events occur in response to high flows and, thus, vary annually in response to flood magnitude.

Although lacking the dramatic and obvious effects of dams, the hydrology of the free-flowing river has also been affected by several engineering activities. In the upper free-flowing reach, alterations of the hydrologic regime are largely attributed to wing dikes and bendway weirs (Pinter et al. 2010). In the lower free-flowing river, the lower and briefer contemporary flood pulse (Figure 6) is largely the result of the 16 cutoffs that occurred during 1929–1942 (Pinter et al. 2010). The altered flood pulse had an unmeasured but potentially large adverse effect on the river floodplain ecosystem as the flood pulse magnitude and duration were greatly reduced (Schramm 2004), and the floodplain inundation ended in most years before the water temperature reached the threshold for growth and reproduction of the warmwater fish assemblage (Schramm and Eggleton 2006).While eliminating flooding to avoid negative social and economic impacts is a legislatively mandated goal of present river management, a rise in the river stage is essential to connect the river to permanent or semipermanent floodplain waters and to make the primary and secondary production on the floodplain available to the river fishes (Junk et al. 1989; Bayley 1995).

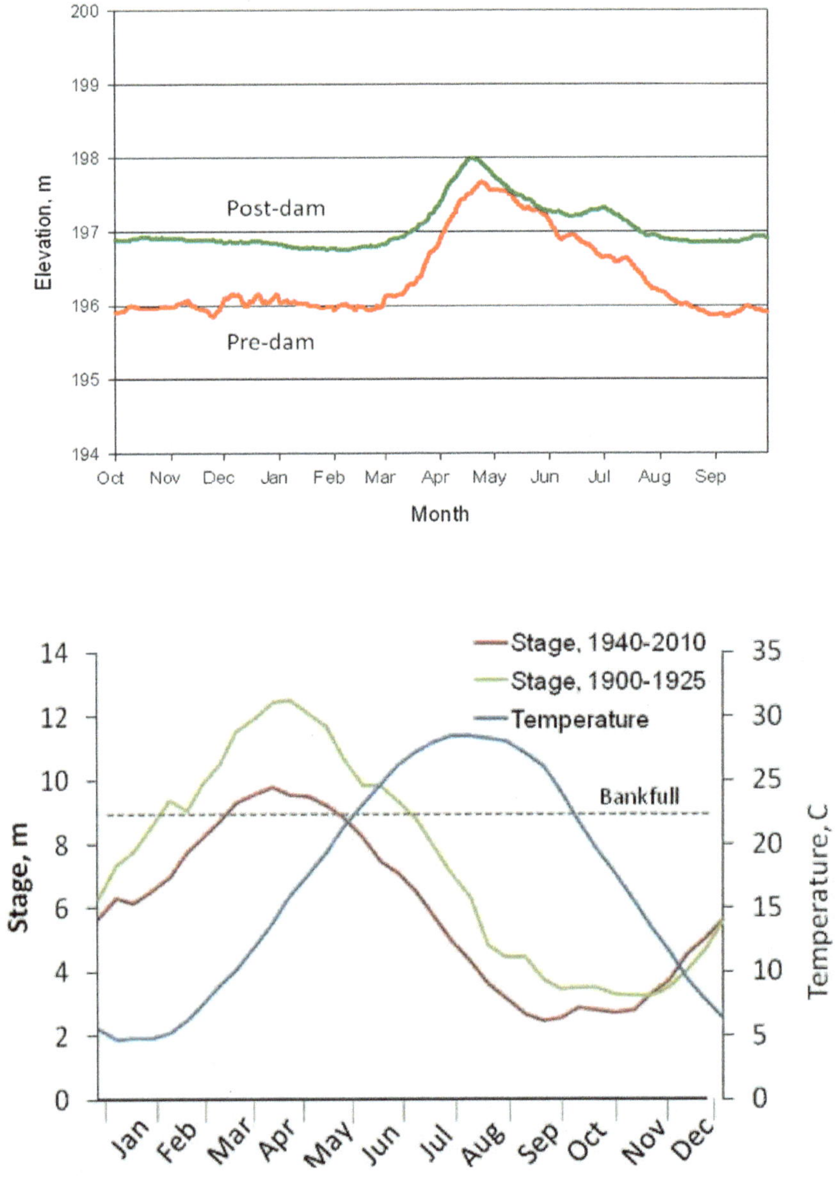

Figure 6. Mean annual hydrographs for the impounded reach (Pool 6, upper panel) and the free-flowing reach at Vicksburg, Mississippi (lower panel).

Channel incision (degradation), in part resulting from the cutoffs in 1929–1942 (Biedenharn and Watson 1997; Biedenharn et al. 2000; Pinter et al. 2010), also impacts connection of the main-stem river with the floodplain. Today, greater discharges are needed for the river to rise above the banks and inundate the floodplain than before major river alterations. Further, some floodplain lakes and abandoned channels that remained connected to the main stem above a certain low river stage and discharge are now disconnected (perched above river elevation) at the same discharge and suffer greater reduction in depth and surface area as water drains from these lakes.

Fish Habitat

Geomorphology and hydrology interact to create and maintain fish habitat in unaltered rivers. Conversely, geomorphic and hydrologic modifications by humans often impair and reduce habitat quantity and quality in river systems. However, despite extensive alterations that have undoubtedly changed habitat, the Mississippi River still provides diverse habitat for fishes (Table 2). Habitats in the unaltered headwaters are typical of first-, second-, and third-order lowland streams with riverine and lateral lakes. Alterations in the impounded and free-flowing reaches have added habitats. Dikes and dike fields can be considered a net gain in habitat in that main channel habitat is converted to a heterogeneous channel border habitat. At the microspatial scale, rock dikes add surface area of hard substrate and complex habitat with numerous interstices. Dikes also create heterogeneous environments at larger spatial scale. Current velocity generally is decreased upstream of the dikes, and at higher river stages areas of fast current velocity and turbulence occur at dike tips and over and immediately downstream of the dikes. Deep scours usually form immediately downstream of dikes, and sandbars build upstream of dikes. The revetted banks also increase the area of hard, rough substrate and provide a more stable environment compared to steep natural banks in high current velocity areas. These added man-made habitats are, in some cases, used differently by fish assemblages. In the upper impounded reach, Madejczyk et al. (1998) found fish biomass greater along wing dikes than barren shorelines, and wing dikes had greater species richness than other shoreline habitat types. The diversity and abundance of fishes associated with dikes were attributed to the proximity of these structures to the main channel and the diverse range of depth and current velocity provided. Madejczyk et al. (1998) also suggest that added rocky substrate in the forms of wing dikes

Table 2. Mississippi River fish habitats. Impounded reach habitats are modified from Rasmussen (1979) and Wilcox (1993), and free-flowing reach habitats are modified from Cobb and Clark (1981) and Baker et al. (1991). Terminology follows Armantrout (1998).

Headwaters reach	Impounded reach	Free-flowing reach
Main channel	Main channel	Main channel
Riffles	Dikes and dike fields	Dikes and dike fields
Pools	Steep natural banks	Steep natural banks
Natural banks	Revetted banks	Revetted banks
Sandbars	Sandbars	Sandbars (inside bends)
Side channels	Side channels	Island/secondary channel complexes
Natural lakes (glacial and fluvial)	Sloughs	Connected abandoned channels
Impoundments	Islands (natural and created)	Permanent floodplain lakes
Tailwaters	Lake (impoundment)	Ephemeral floodplain lakes
	Tailwaters	Borrow pits
	Floodplain	

and revetted banks may be important in the pools of the impounded reaches where sedimentation postimpoundment has covered essential rock and gravel substrates needed by some species. Pennington et al. (1983) found only minor differences in species composition and generally similar abundance of fishes between natural and revetted banks in the lower free-flowing reach, but found some species appeared to prefer one bank type over the other. In the lower impounded reach, Farabee (1986) found fish more abundant on revetted banks than natural banks but, like Pennington et al. (1983), found fish assemblages to be generally similar. However, revetted banks replace steep natural banks that contribute large woody material to the river and provide habitat extensively used by burrowing mayflies and caddisflies (Beckett et al. 1983; Bingham et al. 1988), catfishes (Driscoll et al. 1999), and probably other fauna and appear to be a preferred habitat for the federally endangered Pallid Sturgeon *Scaphirhynchus albus* (Herrala et al. 2014).

Locks and dams in the impounded reach create lentic habitats while lotic habitat persists in dam tailwaters and the upper reaches of each pool. Impoundment and subsequent elevation and stabilization of water level has created extensive backwater and reduced-velocity habitats lateral to the main channel. In the upper impounded reaches, the lentic conditions favored establishment of aquatic macrophytes that added structural and biological diversity.

Although diverse habitats are present (USACE 2000), habitat quality (USGS 1999; Johnson and Hagerty 2008), connection to the main channel (Sparks 1995; Ickes et al. 2005a; Miranda 2005; Garvey et al. 2010), and natural persistence and rejuvenation of habitats (McHenry et al. 1984; USGS 1999; Johnson and Hagerty 2008) are important management issues. Numerous native fish species depend on backwater habitat (shallow, reduced, or no current velocity; aquatic or inundated terrestrial vegetation) during some part of their life cycle (Sheaffer and Nickum 1986; Koel 2004; O'Hara et al. 2007; Schramm et al. 2016), and primary and secondary production in these backwater habitats fuel the riverine food web (e.g., Eckblad et al. 1984; Junk et al. 1989). Backwater habitat conditions are provided by side channels, sloughs, abandoned channels, floodplain lakes, and, in the headwaters and free-flowing reaches, large areas of the active floodplain when seasonally inundated. However, sedimentation and poor water quality are reducing the suitability of these habitats to fish, especially in the impounded reaches (Simons et al. 1975; Theiling 1999; Ickes and Houser 2008). Further, backwater habitats are of limited value to the river floodplain ecosystem if they are not connected to the river due to blocked connections resulting from sediment accretion (Simons et al. 1975; Theiling 1999), infrequent and insufficient fluctuations in river stage (Ickes et al. 2005b), or channel incision (Biedenharn and Watson 1997). Isolation of backwater habitats affects fish access but also affects habitat suitability. Crites et al. (2012) attributed declines in fish relative abundance, species richness, and species diversity to adverse temperature and water quality conditions in a side channel isolated from river flows during low-water stages (early summer through late winter) in the upper impounded reach.

Islands are important habitat because they provide shallow-water sandbar habitat and, in the impounded reach, areas where littoral vegetation can grow (Johnson and Jennings 1998; Knights et al. 2008). In the free-flowing reaches, islands separate the main channel from secondary channels that usually have reduced flow and current velocity and often contain steep natural banks. Thus, the fisheries value of islands is better understood by considering the entire island and secondary channel complex (Knights et al. 2008). Compared to the main channel, habitat conditions change substantially in the secondary channel as river stage changes, making secondary channels desirable for riverine fishes and for those preferring reduced-current conditions at different points in time (Driscoll et al. 1999; Barko and Herzog 2003). Dike fields placed upstream of secondary channels to divert flow to the main channel reduce the flow through the secondary channel. The reduced flow is a potential benefit that increases habitat diversity (e.g., riverine habitat with reduced current velocity and stable steep natural banks); however, the reduced flow also results in sediment accretion such that, over time, flow through the secondary channel ceases and many secondary channels become dry during low-water stages.

Habitat Conservation and Rehabilitation

Some efforts are being made to conserve and rehabilitate (sensu Gore and Shields 1995[3]) fish habitat. The programs and process of conserving and rehabilitating fish habitats throughout the Mississippi River are described in Benjamin et al. (2016, this volume). Here, we summarize several recent and ongoing rehabilitation efforts.

The headwaters reach is entirely within the state of Minnesota and is managed as all other state waters. In the impounded reach, the need for habitat rehabilitation is recognized, and this need has been actuated through a series of partnership ventures among state and federal agencies, academic institutions, and nongovernmental organizations. For example, the Upper Mississippi River Conservation Committee, representing the states of Minnesota, Wisconsin, Iowa, Illinois, and Missouri, has more than a 50-year history in planning and supporting conservation, monitoring, and rehabilitation activities, largely by playing a coordinating role among science, management, and public stakeholder organizations on the upper Mississippi River. In 1986, the U.S. Congress implemented a "balancing uses" paradigm on the upper Mississippi River through the Water Resources Development Act of 1986, recognizing the upper Mississippi River as both a nationally significant commercial navigation system and nationally significant ecosystem. This led to the development and implementation of the Upper Mississippi River Restoration Program with two primary elements: (1) a program of river rehabilitation, known as the Habitat Restoration and Enhancement Project element, and (2) system-wide monitoring of key ecosystem attributes, known as the Long Term Resources Monitoring element. While this program has amassed extensive data that have contributed to great advancement in river and fish ecology and has benefitted 41,278 ha of habitat (Karen Hagerty, U.S. Army Corps of Engineers, Rock Island District, personal communication), additional habitat rehabilitation could be beneficial.

Recent experimental management in limited areas in the upper impounded reach has sought to temporarily reset the minimum allowed pool elevation to a lower elevation during the summer growing season, with the constraint that discharges remain sufficient to not impair commercial navigation. Known as a summer drawdown, this hydrologic management technique tries to emulate the preimpoundment hydrograph by dewatering large portions of the impounded zone to promote sediment drying and compaction and growth of emergent annual vegetation, thereby improving habitat for fishes and other fauna. Preliminary assessments indicate that these drawdowns enhance marsh vegetation, have positive effects on fish spawning and nursery areas, and have no adverse effects on fish populations, water quality, or contaminant bioavailability (Kenow et al. 2015).

The ecological value of backwater habitats for fishes has been long recognized (Knights et al. 2008), and limited efforts are underway in isolated sites in the impounded reaches to rehabilitate and manage these habitats to benefit fish populations (Figure 7; Ickes et al. 2005a; Knights et al. 2008). The benefits of natural alluvial island habitat to riverine biota are well documented in other

[3] Restoration of the Mississippi River, particularly the impounded and free-flowing reaches, will require restoring the natural flow regime (Poff et al. 1997), an action that is not realistically possible. We have purposely referred to beneficial modifications of individual sites as rehabilitation.

Figure 7. Habitat rehabilitation activities in the impounded reaches include dredging channels to increase fresh water flows into backwater areas (upper left), using dredged material to build island in the lower portions of navigation pools (middle left), and building chevron dikes that provide deepwater, low-current habitat (bottom left). In the free-flowing reach, notching dikes (upper right) allows water to flow through secondary channels, converting shallow secondary channels that dewater during low river stages (middle right) to diverse island–secondary channel complexes (bottom right).Photograph credits: Upper Mississippi River Restoration Program and Jan Hoover, U.S. Army Corps of Engineers, Engineering Research and Development Center.

large rivers (Thorp 1992; Gunnell and Petts 2002; Hintz et al. 2014). Many islands have been lost in the lake-like lower portions of navigation pools due primarily to wave erosion (LaRoe et al. 1995), and efforts are underway to use channel dredge materials to rebuild these islands. Preliminary assessments suggest that these islands provide desirable aquatic habitat (e.g., development of aquatic vegetation) and are used by fishes, but rigorous evaluations of fisheries benefits of island and channel dike construction efforts are rare (Gutreuter 2004; Remo et al. 2013). Additional rehabilitation projects are anticipated. The Water Resources and Development Act of 2007 authorized US$1.72

$\times 10^9$ for ecosystem restoration in navigable portions of the upper and lower impounded reaches[4]. To date, annual appropriations have been $19–33 million.

Notable habitat conservation and rehabilitation activities have been initiated in the upper free-flowing reach (Figure 7). The Middle Mississippi River National Wildlife Refuge was established in 2000. This refuge, which consists of seven disjunct tracts of land totaling 2,800 ha adjacent to the river, will expand the area and ecological functionality of the floodplain and backwater habitats and return agricultural lands to natural floodplain forests. Site-specific projects have enhanced riverine habitat in the upper free-flowing reach. Removal or notching of closing structures (dikes or other engineered structures placed to prevent water flow through secondary channels, sloughs, and other waters lateral to the main channel) and other dikes to allow water to flow through secondary channels and side channels has been done to improve the connectivity of these habitats with the river. Further, habitat is being improved within side channels by placement of hard points (small rock dikes) and woody debris to increase habitat diversity and by dredging to deepen and improve connection with the main channel. Seven side channels were rehabilitated from 1997 to 2012 in the upper free-flowing reach (T. Keevin, U.S. Army Corps of Engineers, St. Louis District, personal communication). Dike notching has been used to build islands in dike fields by creating moderate current velocity channels adjacent to the riverbank, essentially creating miniature secondary channel–island complexes. The added diversity of depth and current velocity in the created habitat supports diverse species and multiple life stages; the secondary channel habitat is extensively used by lentic species that would otherwise be restricted to the limited backwater habitat (Allen 2010). Chevron dikes—U-shaped rock structures with the open end facing downstream (Figure 7)—have been installed in several locations in the upper free-flowing reach, with the dual benefit of directing flow to maintain the navigation channel and creating deepwater, reduced-current habitat needed by some fish species, particularly during the winter. These structures have created the intended habitat conditions and increased habitat diversity (Remo et al. 2013). Schneider (2012) found mixed responses of the fish assemblage to chevron dike installation; species diversity and abundance of large-bodied riverine fishes and lentic/backwater species increased, but some taxa, notably benthic riverine chubs *Macrhybopsis* and *Hybopsis*, decreased.

Unlike the impounded and upper free-flowing reaches where recognition of extensive and multiple values led to passage of Upper Mississippi River Management Act of 1986, which in turn facilitated research and monitoring to support fishery conservation and enhancement, the limited and undocumented fishery use in the lower free-flowing reach has precluded any mandate for active fisheries and habitat management. However, fisheries and water quality agencies in the states bordering the lower free-flowing reach (Arkansas, Kentucky, Louisiana, Mississippi, Missouri, and Tennessee), organized under coordination of the Lower Mississippi River Conservation Committee in 1992, have identified and prioritized habitat rehabilitation projects. Their recommendations for habitat rehabilitation and water quality monitoring and management are reemphasized in the "Lower Mississippi River Resource Assessment" (USACE 2015a) presently under final public review and the "Assessment of Natural Resource Habitat Needs Final Report" (USACE 2015b). Completed and planned activities have focused on restoring flow in secondary channels, which is usually accomplished by notching dikes (Figure 7) to increase flow and scour the secondary channel and thereby maintain the island–secondary channel complex. Killgore et al. (2012) have developed a quantitative method for prioritizing rehabilitation in the more than 100

[4] Some of these funds will be spent in the lower Illinois River and the upper free-flowing reach of the Mississippi River as these river segments are legislatively part of the upper Mississippi River system.

secondary channel–island complexes in the lower free-flowing reach. Fourteen secondary channels totaling 91 km were rehabilitated from 2006 to 2013. Rigorous evaluation of these secondary-channel rehabilitation efforts is needed to provide guidance through an adaptive management framework for future habitat restoration.

Rehabilitation efforts will remain constrained by management for flood control and navigation (Ickes and Houser 2008) and affected by fluctuations in river discharge and stage within and among years. Most efforts to accomplish rehabilitation require working with the river and often use the river; for example, to scour a channel after a dike is notched or to accomplish a summer drawdown. The river works in predictable ways, but discharge in the river is unpredictable. Theiling et al. (2014) describe the difficulties, limitations, and frustrations of habitat rehabilitation efforts and their sufficient evaluation in the upper and lower impounded reaches and upper free-flowing reaches; these same barriers undoubtedly apply to the lower free-flowing reach

Water Quality

Water quality in the Mississippi River is a culmination of habitat alteration, urbanization, industrialization, watershed development for agriculture and current agricultural practices, and legal requirements played out on a multijurisdictional stage. Further, some of the activities that are the most detrimental to Mississippi River water quality originate hundreds of kilometers distant from the river. A comprehensive review of water quality in the Mississippi River is available in a National Research Council (2008) report, and additional information about reach-specific water quality issues are provided for the impounded and the upper free-flowing reaches (Weiner and Sandheinrich 2010) and for the lower free-flowing reach (LMRCC 2014); much of what follows is gleaned from those works and references therein. Here, we note only those water quality issues of concern to fish and fisheries in the Mississippi River and its floodplain (e.g., we do not treat Gulf of Mexico hypoxia).

It is generally true that water quality in the Mississippi River has improved since 1972 with the passage of the Clean Water Act (CWA); however, the validity of this statement is largely restricted to point-source pollutants and nonpoint-source agriculture-related chemicals that are now banned from use. Adverse short-term and acute impacts of toxic chemicals are rare and temporary, but legacy chemicals (e.g., chlorinated hydrocarbons, chlorinated aromatic compounds, and metals) remain in the sediments of the Mississippi River and tributaries and can adversely affect fish reproduction (e.g., Koch 2005; Koch et al. 2006). Further, endocrine disrupting chemicals also have adverse effects on fish reproduction (e.g., Harshbarger et al. 2000; Bistodeau et al. 2006; Hinck et al. 2009), and both the input of these chemicals and the understanding of their effects can be expected to increase. Nitrate load increased about 10-fold from predevelopment conditions (Goolsby and Battaglin 2001) and 300% from the 1950s to 1990s (Goolsby et al. 2000). This has significant consequences for the Gulf of Mexico, and Gulf hypoxia has been extensively investigated. A less known but significant concern of the elevated nitrate load, especially when coupled with increased phosphorus loadings, for Mississippi River fisheries is the accelerated phytoplankton and macrophyte production in river lakes and backwaters (Houser and Richardson 2010).

Sediments have positive and negative effects on the Mississippi River. Sediments cause loss of habitats in off-channel zones, carry adsorbed nutrients and unwanted chemicals, and hinder growth and expansion of desirable aquatic vegetation. Sediments benefit channel stability in the free-flowing reach by reducing channel degradation and maintaining and building coastal wetlands and barrier islands in the deltaic outlet of the Mississippi River.

The Mississippi River provides drinking water for millions of people in cities along its route, aquatic recreation, and fish for harvest via recreational and commercial fishing, so water quality is a significant concern, yet water quality assessment efforts are infrequent and insufficient and do not monitor the entire river. This is partly a consequence of the multijurisdictional nature of the river and that the river is viewed principally as a navigation resource. The insufficient attention to water quality also can be attributed to the fact that except for the headwaters reach that lies entirely within Minnesota, Mississippi River water quality is a culmination of activities upstream in the Mississippi River and in its tributaries, so poor water quality cannot be remedied locally. Only 8 of the 11 states bordering the Mississippi River monitor water quality, and water quality information is especially scarce in the lower free-flowing reach. Some states bordering the impounded reaches list the Mississippi River as impaired and have fish consumption advisories in place. However, inconsistencies in determining impairment and contaminant levels necessary to trigger fish consumption advisories render such decisions rather uninformative in terms of health and safety of using the water for recreation or eating the fish. For example, the same segment of the Mississippi River is listed as impaired by Illinois and Wisconsin but not by Iowa. The states' consumption advisories in place are primarily for mercury and polychlorinated biphenyl chemical compounds.

Fisheries Benefits to Man

Direct benefits of the Mississippi River fisheries resource, measured in economic units, are miniscule compared to the economic benefits of transportation and agricultural production, as well as avoiding the economic losses from flooding, that are provided by maintaining navigation and flood control. Commercial fish harvest is not allowed in the headwaters reach. Commercial fish-harvest data have been compiled by the Upper Mississippi River Conservation Committee since 1953. Estimated total commercial fish harvest in the impounded reach and upper free-flowing reach has never exceeded 6,000 metric tons per year and has steadily declined since the mid-1990s (Figure 8). The decline is attributed to dwindling markets and diminished profits to commercial fishers and is not a consequence of overfishing (Michelle Marron, Wisconsin Department of Natural Resources, personal communication). Estimates of commercial harvest in the lower free-flowing reach are limited. The most comprehensive assessment of commercial landings from the lower free-flowing reach is for Kentucky and reflects partial harvest (fish from this portion of the river can be landed in other states) from only 85 km of the Mississippi River. The recent increase is from harvest of nonnative bigheaded carps, primarily Silver Carp *Hypophthalmichthys molitrix*.

The headwaters reach provides significant recreational fishing opportunities. Angler use in a 97-km section of the lower headwaters reach was 9.8 angler-hours per ha during a 4.5-month fishing season (Altena 2008). Although recent estimates are not available, fishing effort in the impounded reach was 18–64 angler-hours per ha per year, and total recreational use was valued at \$1.0–1.2 × 10^9 per year and supported 18,000 jobs 20 or more years ago (Fleener 1975; Ackelson 1979; Watson and Hawkinson 1979; Farabee 1993; USACE 1994). In comparison, recreational fishing effort in U.S. reservoirs averages 88 angler-hours per ha per year (Miranda 1999). Neither fishing effort nor recreational value has been estimated for the lower free-flowing reach, but recreational fishing effort is very low (H. L. Schramm, personal observation).

Issues and Needs for a Healthy and Productive Fisheries Resource

All Reaches

Maintaining biodiversity.—Loss of aquatic flora and fauna certainly is not unique to the Mississippi River, but altered hydrology, including impoundment, and instream habitat impairments and loss significantly challenge persistence and productivity of the native fauna. Schramm et al. (2016) report that 188

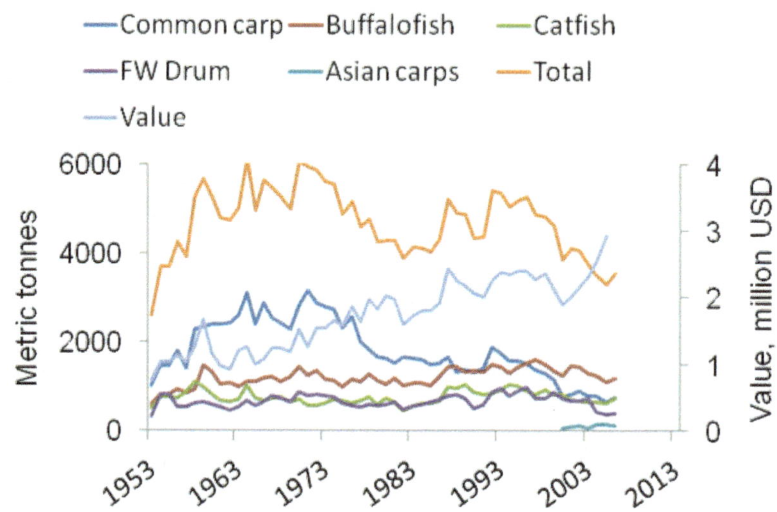

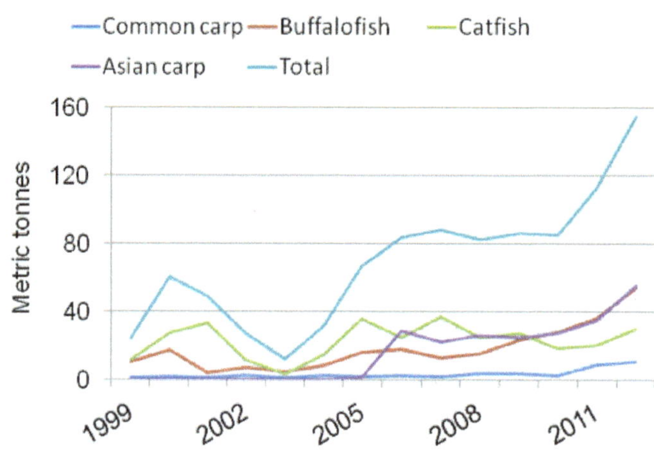

Figure 8. Commercial fish harvest in the upper and lower impounded reaches and the upper free-flowing reach (upper panel) and in the Kentucky waters of the lower free-flowing reach (lower panel). Data provided by M. Marron, Wisconsin Department of Natural Resources, for the impounded and upper free-flowing reaches; data for Kentucky provided by D. Bunnell, Kentucky Department of Fish and Wildlife Resources.

fish species persist in the Mississippi River, but five species have been extirpated from individual reaches of the Mississippi River during the past 50 years. We have noted that the Mississippi River still provides diverse fisheries habitat and that alterations to the river have added habitat, but it is also apparent that some habitats are degrading and being lost. Although one of the five reach-specific extirpations (Alabama Shad *Alosa alabamae* from the lower impounded reach) is probably a consequence of blocked migration, clear reasons for the other extirpations are not apparent (Schramm et al. 2016). It is possible that the extirpations are an indicator of long-term habitat degradation and loss. The establishment of nonnative species (see below), which is favored by altered hydrology and geomorphic conditions (Poff et al. 1997), further threatens persistence of native species.

Although substantial information exists about the ichthyofauna of the Mississippi

River (Schramm et al. 2016), knowledge about trends in species occurrence and abundance would benefit from contemporary, comprehensive assessments in all reaches. These assessments would help monitor trends in abundance of rare fishes, but they would also be useful for monitoring trends in the distribution and abundance of nonnative fishes.

Managing nonnative species.—Fed by rivers east of the Rocky Mountains and west of the Appalachian Mountains, the Mississippi River is a potential home to fishes introduced throughout much of the United States. As a navigation corridor connected to international commerce via the Great Lakes and the Gulf of Mexico, the ecosystem is repeatedly challenged with freshwater invaders from around the globe (Ickes 2008). Notable examples of recently established nonnative aquatic species include Silver Carp, Bighead Carp *Hypophthalmichthys nobilis*, and dreissenid mussels; additional species have been established in particular reaches (e.g., European faucet snail [also known as mud bythnia] *Bythnia tentaculata*; Schramm et al. 2016). White Perch *Morone americana*, Round Goby *Neogobius melanostomus*, and European Ruffe *Gymnocephalus cernua* established in nearby, connected waters are potential invaders. The Silver Carp and Bighead Carp that have now colonized the impounded and free-flowing reaches of the Mississippi River escaped from warmwater aquaculture facilities sited on tributaries to the Mississippi River; these invasive species have attained high biomass in some areas of the Mississippi River (Chapman and Hoff 2011) and pose significant threats to native biota and freshwater ecosystems (Kolar et al. 2007). Methods to eradicate established populations and prevent further introductions do not currently exist.

Impounded Reach

Habitat loss.—Impoundment has contributed to island loss, sedimentation of secondary channels and backwaters, and an overall homogenization of habitat (e.g., loss of islands, less variation in depth and current velocity). Koel (2004) found greater fish species richness and diversity in river reaches with a greater diversity of contiguous habitats (connected channels, side channels, backwater lakes, and marshes) than reaches with less habitat diversity and demonstrated the importance of secondary channels and backwaters connected to the channel to preserve fish species richness and diversity. Roach et al. (2009) found that trophic position (food-chain length) increased along a lateral connectivity gradient and attributed the increase in trophic position to a greater diversity of forage organisms. We suggest that existing habitat diversity be maintained and degraded and lost habitats rehabilitated. Effective actions could include island construction in the lake portions of navigation pools to increase littoral habitat and deepening and restoring flow and connection to secondary channel and backwater habitats to improve water quality and provide winter habitat for fishes.

Water quality.—Good water quality is fundamental to good fisheries habitat. Reduction of point-source pollutants and endocrine disruptive chemicals will benefit fish populations and also ensure that fish are safe for human consumption. Excessive nutrients and turbidity affect primary production and directly or indirectly create unsuitable water quality conditions for fish. Curbing the input of nutrients and turbidity-causing sediments is a watershed issue (although commercial navigation and recreational boats contribute to turbidity), but remedies to the effects of accumulating nutrients and sediments remain a challenge to fisheries resource managers.

Navigation.—The Mississippi River and Tributaries project that provides flood protection and navigation for the entire Mississippi River has an approximately 2.0×10^{11} annual economic impact (www.bigrivercoalition.org/), and cargo conveyed on the Mississippi River is a substantial part of that impact. Navigation directly and indirectly affects fisheries resources. Direct effects include mortality resulting from entrainment (Bartell and Campbell 2000; Killgore et al. 2001, 2011; Gutreuter et al. 2006; Miranda and Killgore 2013) and wave-induced stranding (Adams et al. 1999). The direct effects are possibly greater in the impounded reach where the navigation channel is often shallower and narrower. Indirect effects of navigation include development of the riverbanks by the infrastructure that supports shipping and the risks of accidental release of toxic substances.

Free-Flowing Reaches

Altered hydrology.—Hydrology is a foundational element of river floodplain ecosystem structure and function (Poff et al. 1997). The altered hydrology reduces the spatial and temporal extent of floodplain inundation. Further, the briefer flood pulse decouples the hydrologic and thermal conditions (Junk et al. 1989; Schramm and Eggleton 2006).Thus, the altered hydrologic regime is a systemic problem that affects the spawning, recruitment, and growth of fishes dependent on the floodplain and directly or indirectly affects the entire river floodplain ecosystem through nutrient cycling (Schramm et al. 2009, 2015) and cascading trophic interactions (Eckblad et al. 1984). Restoring an ecologically functional hydrologic regime is not compatible with current management for flood prevention and navigation. It should be noted that channel degradation (incision) has a similar effect to a lower hydrograph in that greater discharge and higher river stages are needed to connect the river to the floodplain and floodplain water bodies (Poff et al. 1997). Thus, floodplain inundation is even less frequent and of shorter duration than expected from the altered hydrograph in portions of the free-flowing reach where channel degradation has occurred.

Loss of in-river habitat.—Loss of in-river habitat threatens biodiversity and fisheries productivity. Of particular concern are losses of secondary channel–island complexes and steep natural banks. Efforts are ongoing to conserve and rehabilitate these habitats, and these efforts may warrant expansion. It is noteworthy that engineering activities to conserve and rehabilitate secondary channel–island complexes are often as simple as putting a notch in one or several existing dikes (or building a new dike with a notch).

Loss of active floodplain and river-floodplain connection.—Seasonal inundation of the floodplain and connection between the river and the floodplain contribute to fish production (Junk et al. 1989; Barko et al. 2006; Phelps et al. 2014). Tall and continuous levees have severed the river from more than 82% of the historic floodplain, and the altered hydrograph has reduced inundation of the active floodplain and connection of floodplain waters with the river. In addition, floodplain inundation and connection have been reduced in part of the lower free-flowing reach because the river channel is incising in response to the change in bed slope triggered by 16 neck cutoffs.

Failure to recognize that a functional floodplain still exists and can be managed to achieve ecological benefits perpetuates the problems associated with loss of floodplain. It is true that the active floodplain is less than 18% (18% in the upper free-flowing and 7% in the lower free-flowing reaches, Table 1) of the historic floodplain area. The flood pulse concept (Junk et al. 1989) predicts that restoring connection to the vast historic floodplain would increase fish production (assuming the historic hydrologic regime were also restored) as well as other ecological services (Schramm et al. 2009, 2015). Further, some managers and resource conservation organizations have forecast reduced economic losses from floods if the levees were removed and the river was allowed to reclaim its natural floodplain (Interagency Floodplain Management Review Committee 1994). The debate about floodplain management is between tear the levees down (termed "the nonstructural solution") and build the levees higher (termed "the structural solution") and does not consider management of the present active floodplain. Regardless of debate position on levees, we suggest that there is a fundamental flaw in reasoning about whether and how to manage the floodplain. The entirety of the historic floodplain was not inundated every year; although records are not available, massive floods are infrequent events, and the entire floodplain was infrequently inundated. Any model that predicts losses or benefits from flood protection should consider flood magnitude and frequency. Assuming for simplicity and for lack of appropriate models that ecological benefits and adverse economic impacts are linearly related to area of floodplain inundated, the levees have neither decreased ecological benefits by 82% nor increased economic benefits by 82% on an annual basis, but rather by a much smaller percentage for both. Lost in the debate amid the large area gained for human development or lost for ecological services is the 647,000 ha of active floodplain that remains (Schramm et al. 1999). The merits of managing the remaining active floodplain warrant consideration, as suggested by the use of a serendipitously connected expanse of floodplain by fishes during the 2011 flood (Phelps et al. 2014), and are included as priority opportunities for rehabilitation in the Lower Mississippi River Resources Assessment (USACE 2015a).

Connectivity between the river channel and floodplain water bodies also affects the structure and function of fish assemblages in floodplain lakes. Miranda (2005) found connectivity, quantified as likelihood of connection with the river, affected the structure of fish assemblages in oxbow lakes of the Yazoo River, a tributary of the Mississippi River with a similar fish assemblage.

Lack of public use, awareness, and value.—A final issue that affects future management of the free-flowing reach, particularly the lower portion, is the lack of the general public's use and familiarity with the river (A. Rodgers, U.S. Fish and Wildlife Service, personal communication). Use requires access. Access is limited and difficult to provide in a system with 12 m annual river-stage fluctuations and where good access sites exist that can be reached without travel across the active floodplain. The river provides abundant fishing opportunities and can be safely navigated with appropriate equipment and caution (LMRCC 2013). In an effort to increase awareness and use, and thereby increase peoples' value of the lower free-flowing reach, the LMRCC has produced a guide that provides information about access and safe use and fishing tips (LMRCC 2013), and the Lower Mississippi River Resources Assessment (USACE 2015a) ranks provision of access as a high priority.

Conclusions

The Mississippi River can be thought of as multiple rivers, not just ecologically but also socially and economically. The river has been altered, and these alterations affect the geomorphology and hydrology that, in turn, determine structure and function of a river floodplain ecosystem of which fish are a part. While the immediate and near-term consequences of these alterations can be severe, possibly more severe are the long-term consequences of a river floodplain ecosystem in which the river is fixed in space and no longer creates new habitat (Petts 1989; Baker et al. 1991; Gore and Shields 1995). Habitat loss due to sedimentation is a natural consequence of aging of alluvial rivers, but unaltered and unconstrained rivers are also continually creating new habitat while old aquatic habitats are lost to sedimentation and channel realignment. Lacking creation of new habitats, net loss of aquatic habitat can be expected. Such appears to be the fate of the Mississippi River, at least the reaches downstream of upper St. Anthony Falls that are managed for navigation. Thus, needs and opportunities abound for conserving and rehabilitating the impounded and free-flowing reaches. Managing the fisheries resources and the ecosystem that sustains them can be done within the opportunities and constraints of a dynamic and unpredictable system with an altered hydrologic regime managed for navigation and flood control. Many gaps in knowledge remain to be filled to fully understand the structure and function of the Mississippi River. Hopefully, river and fisheries scientists will be called upon to fill those gaps, but as river and fisheries scientists move forward they should remain aware that inferences will be based on information obtained from a geomorphically and hydrologically altered system that may be persisting in a changing climatic regime.

Acknowledgments

We thank Bob Anderson, Charles Camillo, and Jack Smith, U.S. Army Corps of Engineers, for providing data not available in the published literature. Angelina Rodgers, U.S. Fish and Wildlife Service and Lower Mississippi River Conservation Committee Coordinator, and Tom Keevin, U.S. Army Corps of Engineers, also diligently assisted efforts to obtain information and reports not available in the published literature. David Biedenharn, U.S. Army Corps of Engineers, retired, patiently explained channel dynamic processes. Charles Theiling, U.S. Army Corps of Engineers, provided habitat areas for Table 1 for impounded and the upper free-flowing reaches. Input from an anonymous reviewer helped broaden and strengthen the content of this chapter. Funding was provided by U.S. Geological Survey. Last, we thank numerous river rats, including commercial fishers, fisheries biologists, and engineers, for the wisdom they have shared over many years.

References

Ackelson, M. C. 1979. Recreational use survey, Pool 9, Mississippi River. Upper Mississippi River Conservation Committee, Rock Island, Illinois.

Adams, S. R., T. M. Keevin, K. J. Killgore, and J. J. Hoover. 1999. Stranding potential of young fishes subjected to simulated vessel-induced drawdown. Transactions of the American Fisheries Society 128:1230–1234.

Allen, T. C. 2010.Middle Mississippi River islands: historical distribution, restoration planning, and biological importance. Doctoral dissertation. University of Missouri, St. Louis.

Altena, E. 2008. Mississippi River creel survey from St. Cloud to Coon Rapids, May 12 to September 30, 2007. Minnesota Department of Natural Resources, Completion Report, Little Falls, Minnesota.

Anfinson, J. O. 2005. The river we have wrought: a history of the upper Mississippi. University of Minnesota Press, Minneapolis.

Armantrout, N. B. 1998. Glossary of aquatic habitat inventory terminology. American Fisheries Society, Bethesda, Maryland.

Baker, J. A., K. J. Killgore, and R. L. Kasul. 1991. Aquatic habitats and fish communities in the lower Mississippi River. Reviews in Aquatic Sciences 3:313–356.

Barko, V. A., and D. P. Herzog. 2003. Relationships among side channels, fish assemblages, and environmental gradients in the unimpounded upper Mississippi River. Journal of Freshwater Ecology 18:377–382.

Barko, V. A., D. P. Herzog, and M. T. O'Connell. 2006. Response of fishes to floodplain connectivity during and following a 500-year flood event in the unimpounded upper Mississippi River. Wetlands 26:244–257.

Barry, J. M. 1997. Rising tide. Simon & Schuster, New York.

Bartell, S. M., and K. R. Campbell.2000. Ecological risk assessment of the effects of the incremental increase of commercial navigation traffic 25, 50, 75, and 100 percent increase of 1992 baseline traffic on fish. Interim report for the Upper Mississippi River, Illinois Waterway System Navigation Study. U.S. Army Corps of Engineers, Vicksburg, Mississippi.

Bayley, P. B. 1995. Understanding large river-floodplain ecosystems. BioScience 45:153–158.

Beckett, D. C., C. R. Bingham, L. G. Sanders, D. B. Mathis, and E. M. McLemore. 1983. Benthic macroinvertebrates of selected aquatic habitats of the lower Mississippi River. U.S. Army Corps of Engineers, Waterways Experiment Station, Environmental Laboratory, Report E-83-10, Vicksburg, Mississippi.

Biedenharn, D. S., C. R. Thorne, and C. C. Watson. 2000. Recent morphological evolution of the lower Mississippi River. Geomorphology 34:227–249.

Biedenharn, D. S., and C. C. Watson. 1997. Stage adjustment in the lower Mississippi River, USA. Regulated Rivers: Research and Management 13:517–536.

Bingham, C. R., C. H. Pennington, and R. E. Coleman. 1988. An ecological investigation of revetted and natural bank habitats in the lower Mississippi River. U.S. Army Corps of Engineers, Waterways Experiment Station, Environmental Laboratory, Report E-87-5, Vicksburg, Mississippi.

Bistodeau, T. J., L. B. Barber, S. E. Bartell, R. A. Cediel, K. J., Grove, J. Klaustermeier, C. Woodard, K. E. Lee, and H. L., Schoenfuss, 2006. Larval exposure to environmentally relevant mixtures of alkylphenolethoxylates reduces reproductive competence in male Fathead Minnows. Aquatic Toxicology 79: 268–277.

Chapman, D. C., and M. H. Hoff, editors. 2011. Invasive Asian carps in North America. American Fisheries Society, Symposium 74, Bethesda, Maryland.

Cobb, S. P., and J. R. Clark.1981. Aquatic habitat studies on the lower Mississippi River, river mile 480 to 530. Report 2: aquatic habitat mapping. U.S. Army Corps of Engineers, Waterways Experiment Station, Miscellaneous Paper E-80-1.Vicksburg, Mississippi.

Cobb, S. P., C. H. Pennington, J. A. Baker, and J. E. Scott. 1984. Fishery and ecological investigations of main stem levee borrow pits along the lower Mississippi River. U.S. Army Corps of Engineers, Mississippi River Commission, Vicksburg, Mississippi.

Crites, J. W. Q. E. Phelps, K. N. S. McCain, D. P. Herzog, and R. A. Hrabik. 2012. An investigation of fish community and water quality compositions in an isolated side channel of the upper Mississippi River. Journal of Freshwater Ecology 27:19–29.

DeHaan, H. C. 1998. Large river sediment transport and deposition: an annotated bibliography. U.S. Geological Survey, Environmental Management Technical Center, Long Term Resource Monitoring Program Technical Report 98-T002, Onalaska, Wisconsin.

Delaney, R. L., and M. R. Craig. 1997. Longitudinal changes in Mississippi River floodplain structure. U.S. Geological Survey, Upper Mississippi River Long Term Resource Monitoring Program, Project Status Report PSR 97-02, Onalaska, Wisconsin.

Driscoll, M. T., W. R. Davis, and H. L. Schramm, Jr. 1999. Relative abundance of catfishes in main channel and secondary channel habitats in the lower Mississippi River. Pages 231–238 *in* E. R. Irwin, W. A. Hubert, C. F. Rabeni, H. L. Schramm, Jr., and T. Coon, editors. Catfish 2000: proceedings of the first international ictalurid symposium. American Fisheries Society, Symposium 24, Bethesda, Maryland.

Eckblad, J. W. 1986. The ecology of pools 11–13 of the upper Mississippi River: a community profile. U.S. Fish and Wildlife Service Biological Report 85(7.8).

Eckblad, J. W., C. S. Volden, and L. S. Weilgart. 1984. Allochthonous drift from backwaters to the main channel of the Mississippi River. American Midland Naturalist 111:16–22.

Farabee, G. B. 1986. Fish species associated with revetted and natural main channel border habitats in Pool 24 of the upper Mississippi River. North American Journal of Fisheries Management 6:504–508.

Farabee, G. B. 1993. Recreational use of Pool 24, upper Mississippi River. Missouri Department of Conservation, Jefferson City.

Fleener, G. G. 1975. Recreational use of Pool 21, Mississippi River. Missouri Department of Conservation, Jefferson City.

Flinn, M. B., S. A. Adams, M. R. Whiles, and J. E. Garvey. 2008. Biological responses to contrasting hydrology in backwaters of the upper Mississippi River Navigation Pool 25. Environmental Management 41:468–486.

Fremling, C. R. 2005. Immortal river: the upper Mississippi in ancient and modern times. University of Wisconsin Press, Madison.

Fremling, C. R., and T. O. Claflin. 1984. Ecological history of the upper Mississippi River. Pages 5–24 *in* J. G. Wiener, R. V. Anderson, and D. R. McConville, editors. Contaminants in the upper Mississippi River. Butterworth Publishers, Boston.

Fremling, C. R., J. L. Rasmussen, R. E. Sparks, S. P. Cobb, C. F. Bryan, and T. O. Claflin. 1989. Mississippi River fisheries: a case history. Pages 147–158 *in* D. P. Dodge, editor. Proceedings of the international large rivers symposium. Canadian Journal of Fisheries and Aquatic Sciences Special Publication 106.

Garvey, J., B. S. Ickes, and S. Zigler. 2010. Challenges in merging fisheries research and management: the upper Mississippi River experience. Hydrobiologia 640:125–144.

Goolsby, D. A. 2000. Mississippi basin nitrogen flux believed to cause Gulf hypoxia. Eos, Transactions of the American Geophysical Union 82:321–332.

Goolsby, D. A., and W. A. Battaglin. 2001. Long-term changes in concentrations and flux of nitrogen in the Mississippi River basin, USA. Hydrological Processes 15:1209–1226.

Goolsby, D. A., W. A. Battaglin, B. T. Aulenbach, and R. P. Hooper.2000. Nitrogen flux and sources in the Mississippi River basin. Science of the Total Environment 248:75–86.

Gore, J. A., and F. D. Shields. 1995. Can large rivers be restored? BioScience 45:142–152.

Gunnell, A. M., and G. E. Petts. 2002. Island-dominated landscapes of large floodplain rivers, a European perspective. Freshwater Biology 47:581–600.

Gutreuter, S. 2004. Challenging the assumption of habitat limitation: an example from centrarchid fishes over an intermediate spatial scale. River Research and Applications 20:413–425.

Gutreuter, S., J. M. Vallazza, and B. C. Knights. 2006. Persistent disturbance by commercial navigation alters the relative abundance of channel-dwelling fishes in a large river. Canadian Journal of Fisheries and Aquatic Sciences 63:2418–2433.

Harshbarger, J. C., M. J. Coffey, and M. Y. Young. 2000. Intersexes in Mississippi River Shovelnose Sturgeon sampled below Saint Louis, Missouri, USA. Marine Environmental Research 50:247–250.

Herrala, J. R., P. T. Kroboth, N. M. Kuntz, and H. L. Schramm, Jr. 2014. Habitat use and selection by adult Pallid Sturgeon in the lower Mississippi River. Transactions of the American Fisheries Society 143:153–163.

Hinck, J. E., V. S. Blazer, C. J. Schmitt, D. M. Papoulias, and D. E. Tillitt. 2009. Widespread occurrence of intersex in black basses (*Micropterus* spp.) from U.S. rivers, 1995–2004. Aquatic Toxicology 95:60–70.

Hintz, W. D., A. P Porreca, J. E. Garvey, Q. E. Phelps, S. J. Tripp, R. A. Hrabik, and D. P. Herzog. 2014. Abiotic attributes surrounding alluvial islands generate critical fish habitat. River Research and Applications 31:1218–1226.

Houser, J. N., and W. B. Richardson. 2010. Nitrogen and phosphorus in the upper Mississippi River: transport, processing, and effects on the river ecosystem. Hydrobiologia 640:71–88.

Hutoff, F., N. Pinter, and J. W. F. Remo.2013. Theoretical analysis of wing dike impact on river flood stages. Journal of Hydraulic Engineering 139:550–556.

Ickes, B. S. 2008. Fisheries indicators: nonnative fishes. Pages 73–75 *in* B. L. Johnson and K. H. Hagerty, editors. Status and trends of selected resources of the upper Mississippi River System. U.S. Geological Survey, Upper Midwest Environmental Sciences Center, Long Term Resource Monitoring Program Technical Report 2008-T002, La Crosse, Wisconsin.

Ickes, B. S., M. C. Bowler, A. D. Bartels, D. J. Kirby, S. DeLain, J. H. Chick, V. A. Barko, K. S. Irons, and M. A. Pegg. 2005a. Multiyear synthesis of the fish component from 1993 to 2002 for the Long Term Resource Monitoring Program. U.S. Geological Survey, Upper Midwest Environmental Sciences Center, Long Term Resource Monitoring Program Technical Report 2005-T005, La Crosse, Wisconsin.

Ickes, B. S., and J. N. Houser. 2008. Introduction. Pages 11–22 *in* B. L. Johnson and K. H. Hagerty, editors. Status and trends of selected resources of the upper Mississippi River System. U.S. Geological Survey, Upper Midwest Environmental Sciences Center, Long Term Resource Monitoring Program Technical Report 2008-T002, La Crosse, Wisconsin.

Ickes, B. S., J. Vallazza, J. Kalas, and B. Knights. 2005b. River floodplain connectivity and lateral fish passage: a literature review. U.S. Geological Survey Contract Completion Report submitted to U.S. Fish and Wildlife Service, Quincy, Illinois.

Ickes, B. S., J. H. Wlosinski, B. C. Knights, and S. J. Zigler. 2001. Fish passage through dams in large temperate floodplain rivers: an annotated bibliography. U.S. Geological Survey, Upper Midwest Environmental Sciences Center, La Crosse, Wisconsin. Available: www.umesc.usgs.gov/ltrmp_fish/fish_passage_biblio.html. (April 2014).

Interagency Floodplain Management Review Committee. 1994. Sharing the challenge: floodplain management into the 21st century. U.S. Government Printing Office, Washington, D.C.

Jahn, L. A., and R. V. Anderson. 1986. The ecology of Pools 19 and 20, upper Mississippi River: a community profile. U.S. Fish and Wildlife Service Biological Report 85(7.6).

Johnson, B. L., and K. H. Hagerty, editors. 2008. Status and trends of selected resources of the upper Mississippi River system. U.S. Geological Survey, Upper Midwest Environmental Sciences Center, Long Term Resource Monitoring Program Technical Report 2008-T002, La Crosse, Wisconsin.

Johnson, B. L., and C. A. Jennings. 1998. Habitat associations of small fishes around islands in the upper Mississippi River. North American Journal of Fisheries Management 18:327–336.

Johnson, B. L., B. C. Knights, J. W. Barko, R. F. Gaugush, D. M. Soballe, and W. F. James. 1998. Estimating flow rates to optimize winter habitat for centrarchid fish in Mississippi River (USA) backwaters. Regulated Rivers: Research and Management 14:499–510.

Junk, W. J., P. B. Bayley, and R. E. Sparks.1989. The flood pulse concept in river-floodplain ecosystems. Pages 110–127 *in* D. P. Dodge, editor. Proceedings of the international large rivers symposium. Canadian Journal of Fisheries and Aquatic Sciences Special Publication 106.

Kenow, K. P., G. L. Benjamin, T. W. Schlagenhaft, R. A. Nissen, M. Stefanski, G. J. Wege, S. A. Jutila, and T. W. Newton. 2015. Process, policy, and implementation of pool-wide drawdowns on the upper Mississippi River: a promising approach for ecological restoration of large impounded rivers. River Research and Application, DOI: 10.1002/rra.2857.

Killgore, K. J., J. J. Hoover, B. R. Lewis, and R. Nassar. 2012. Ranking secondary channels for restoration using an index approach. U.S. Army Corps of Engineers, Engineer Research and Development Center, ERDC TN-EMRRP-ER-15, Vicksburg, Mississippi.

Killgore, K. J., S. T. Maynord, M. D. Chan, and R. M. Morgan, I. I. 2001. Evaluation of propeller-induced mortality on early life stages of selected fish species. North American Journal of Fisheries Management 21:947–955.

Killgore, K. J., L. E. Miranda, C. E. Murphy, D. M. Wolff, J. J. Hoover, T. M. Keevin, S. T. Maynord, and M. A. Cornish. 2011. Fish entrainment rates through towboat propellers in the upper Mississippi and Illinois rivers. Transactions of the American Fisheries Society 140:570–581.

Knights, B. C., B. S. Ickes, and J. N. Houser. 2008. Fish Assemblages in off-channel areas of the upper Mississippi and Illinois rivers: implications for habitat restoration at management-relevant scales. U.S. Army Corps of Engineers, Long Term Resource Monitoring Program Completion Report 2007APE07, Rock Island, Illinois.

Knights, B. C., B. L. Johnson, and M. B. Sandheinrich. 1995. Responses of Bluegills and Black Crappies to dissolved oxygen, temperature, and current in backwater lakes of the upper Mississippi River during winter. North American Journal of Fisheries Management 15:390–399.

Knights, B. C., J. M. Vallazza, S. J. Zigler, and M. R. Dewey. 2002. Habitat and movement of Lake Sturgeon in the upper Mississippi River system, USA. Transactions of the American Fisheries Society 131:507–522.

Koch, B. T. 2005. Organochlorine compounds in middle Mississippi River Shovelnose Sturgeon (*Scaphirhynchus platorynchus*): bioaccumulation and reproductive consequences. Master's thesis. Southern Illinois University, Carbondale.

Koch, B. T., J. E. Garvey, J. You, and M. J. Lydy. 2006. Elevated organochlorines in the brain-hypothalamic-pituitary complex of intersexual Shovelnose Sturgeon. Environmental toxicology and Chemistry 25:1689–1697.

Koel, T. M. 2004. Spatial variation in fish species richness of the upper Mississippi River system. Transactions of the American Fisheries Society 133:984–1003.

Kolar, C. S., D. C. Chapman, W. R. Courtenay, C. M. Housel, J. D. Williams, and D. P. Jennings. 2007. Bigheaded carps: a biological synopsis and environmental risk assessment. American Fisheries Society, Special Publication 33, Bethesda, Maryland.

LaRoe, E. T., G. S. Farris, C. E. Puckett, P. D. Doran, and M. J. Mac, editors. 1995. Our living resources: a report to the nation on the distribution, abundance, and health of U.S. plants, animals, and ecosystems. U.S. Department of the Interior, National Biological Service, Washington, D.C.

LMRCC (Lower Mississippi River Conservation Committee). 2013. Fishing the lower Mississippi River. Lower Mississippi River Conservation Committee, Vicksburg, Mississippi. Available: www.lmrcc.org/wp-content/uploads/2013/09/LMR-Fishing-Guide-Version2.pdf. (January 2016).

LMRCC (Lower Mississippi River Conservation Committee). 2014. Summary of available water quality assessment of the lower Mississippi River. Lower Mississippi River Conservation Committee, Vicksburg, Mississippi.

Madejczyk, J. C., N. D. Mundahl, and R, M. Lehtinen. 1998. Fish assemblages of natural and artificial habitats within the channel border of the upper Mississippi River. American Midland Naturalist 139:296–310.

McHenry, J. R., J. C. Ritchie, C. C. Cooper, and J. Verdon. 1984. Recent rates of sedimentation in the Mississippi River. Pages 99–117 *in* J. G. Wiener, R. V. Anderson, and D. R. McConville, editors. Contaminants in the upper Mississippi River. Butterworth Publishers, Boston.

Meade, R. H. 1995. Geology, hydrology, sediments, and engineering of the Mississippi River. Pages 13–30 *in* R. H. Meade, editor. Contaminants in the Mississippi River, 1987–1992. U.S. Geological Survey Circular 1133.

Miranda, L. E. 1999. A typology of fisheries in large reservoirs of the United States. North American Journal of Fisheries Management 19:536–550.

Miranda, L. E. 2005. Fish assemblages in oxbow lakes relative to connectivity with the Mississippi River. Transactions of the American Fisheries Society 134:1480–1489.

Miranda, L. E., and K. J. Killgore. 2013. Entrainment of Shovelnose Sturgeon by towboat navigation in the upper Mississippi River. Journal of Applied Ichthyology 29:316-322.

Moore, N. 1972. Improvement of the lower Mississippi River and tributaries, 1931-1972. Mississippi River Commission, Vicksburg, Mississippi.

National Research Council. 2008. Mississippi River Water Quality and the Clean Water Act: progress, challenges, and opportunities. The National Academies Press, Washington, D.C. Available: www.nap.edu/catalog/12051/mississippi-river-water-quality-and-the-clean-water-act-progress. (February 2016).

O'Hara, M., B. S. Ickes, E. Gittinger, S. DeLain, T. Dukerschein, M. Pegg, and J. Kalas. 2007. Development of a life history database for upper Mississippi River fishes. U.S. Geological Survey, Upper Midwest Environmental Sciences Center, Long Term Resource Monitoring Program 2007-T001, La Crosse, Wisconsin.

Pennington, C. H., J. A. Baker, and M. E. Potter.1983. Fish populations along natural and revetted banks on the lower Mississippi River. North American Journal of Fisheries Management 3:204–211.

Petts, G. E. 1989. Historical analysis of fluvial hydrosystems. Pages 1–18 *in* G. E. Petts, editor. Historical change of large alluvial rivers: Western Europe. Wiley, Chichester, U.K.

Phelps, Q. E., S. J. Tripp, D. P. Herzog, and J. E. Garvey. 2014. Temporary connectivity: the relative benefits of large river floodplain inundation in the lower Mississippi River. Restoration Ecology 23:53–56.

Pinter, N. 2005. Policy forum: one step forward, two steps back on U.S. floodplains. Science 308:207–208.

Pinter, N., A. A. Jemberie, J. W. F. Remo, R. A. Heine, and B. S. Ickes. 2008. Flood trends and river engineering on the Mississippi River system. Geophysical Research Letters 35:L23404.

Pinter, N., A. A. Jemberie, J. W. F. Remo, R. A. Heine, and B. S. Ickes. 2010. Cumulative impacts of river engineering, Mississippi and lower Missouri rivers. River Research and Applications 26:546–571.

Poff, N. L., J. D. Allan, M. D. Bain, J. L. Karr, K. L. Prestegaard, B. D. Richter, R. E. Sparks, and J. C. Stromberg. 1997. The natural flow regime. BioScience 47:769–784.

Poore, R. Z., J. Darling, H. J. Dowsett, H. J., and L. Wright. 2001. Variations in river flow to the Gulf of Mexico: implications for paleoenvironmental studies of Gulf of Mexico marine sediments. U.S. Geological Survey Bulletin 2187.

Raibley, P. T., K. S. Irons, T. M. O. Hara, K. D. Blodgett, and R. E. Sparks. 1997. Winter habitats used by Largemouth Bass in the Illinois River, a large river-floodplain ecosystem. North American Journal of Fisheries Management 17:401–412.

Rasmussen, J. L. 1979. UMRCC fisheries compendium, 2nd edition. Upper Mississippi River Conservation Committee, Rock Island, Illinois.

Remo, J. W. F., A. Khanal, and N. Pinter.2013. Assessment of chevron dikes for the enhancement of physical-aquatic habitat within the middle Mississippi River, USA. Journal of Hydrology 501:146–162.

Roach, K. A., J. H. Thorp, and M. D. DeLong. 2009. Influence of lateral gradients of hydrologic connectivity on trophic positions of fishes in the upper Mississippi River. Freshwater Biology 54:607–620.

Schneider, B. M. 2012. Changes in fish use and habitat diversity associated with placement of three chevron dikes in the middle Mississippi River. Master's thesis. Southern Illinois University Edwardsville.

Schramm, H. L., Jr. 2004. Status and management of fisheries in the Mississippi River. Pages 301–333 *in* R. Welcomme and T. Petr, editors. Proceedings of the second international symposium on the management of large rivers for fisheries, volume 1. Food and Agriculture Organization of the United Nations, Regional Office for Asia and the Pacific, RAP Publication 2004/16, Bangkok, Thailand.

Schramm, H. L., Jr., M. S. Cox, T. E. Tietjen, and A. W. Ezell.2009. Nutrient dynamics in the lower Mississippi River floodplain: comparing present and historic hydrologic conditions. Wetlands 29:476–487.

Schramm, H. L., Jr., and M. A. Eggleton.2006. Applicability of the flood-pulse concept in a temperate floodplain river ecosystem: thermal and temporal components. River Research and Application 22:543–553.

Schramm H. L., Jr., M. A. Eggleton, and R. B. Minnis. 1999. Spatial analysis of floodplain habitat critical to lower Mississippi River fishes. Mississippi Cooperative Fish and Wildlife Research Unit, Starkville.

Schramm, H. L, Jr., J. T. Hatch, R. A. Hrabik, and W. T. Slack. 2016. Fishes of the Mississippi River. Pages 53–77 *in* Y. Chen, D. Chapman, J. Jackson, and D. Chen, editors. Fishery resources, environment, and conservation in the Mississippi and Yangtze (Changjiang) River basins. American Fisheries Society, Symposium 84, Bethesda, Maryland.

Schramm, H. L. Jr., W. B. Richardson, and B. C. Knights. 2015. Managing the Mississippi River floodplain: achieving ecological benefits requires more than hydrological connection to the river. Pages 171–202 *in* P. F. Hudson and H. Middlekoop, editors. Geomorphology and integrated floodplain management. Springer, New York.

Scientific Assessment and Strategy Team.1994. Science for floodplain management into the 21st century. Part 5: Report of the Interagency Floodplain Management Review Committee to the Administration Floodplain Management Task Force. Interagency Floodplain Management Review Committee, Washington, D.C.

Sheaffer, W. A., and J. G. Nickum. 1986. Backwater areas as nursery habitats for fishes in Pool 13 of the upper Mississippi River. Hydrobiologia 136:131–139.

Simons, D. B., M. A. Stevens, P. F. Lagasse, S. A. Schumm, and Y. H. Chen. 1975. Environmental inventory and assessment of navigation Pools 24, 25, 26, upper Mississippi and lower Illinois rivers: a geomorphic study. U.S. Army Corps of Engineers, St. Louis District, St. Louis.

Smith, L. M., and B. R. Winkley. 1996. The response of the lower Mississippi River to river engineering. Engineering Geology 45:433–455.

Sparks, R. E. 1995. Need for ecosystem management of large rivers and their floodplains. BioScience 45:168–182.

Theiling, C. 1999. River geomorphology and floodplain habitats. Pages 4-1–4-21 *in* U.S. Geological Survey. Ecological status and trends of the upper Mississippi River system 1998: a report of the Long Term Resource Monitoring Program element. U.S. Geological Survey, Long Term Resource Monitoring Program 99-T001, La Crosse, Wisconsin.

Theiling, C. H., and J. T. Burant. 2013. Flood inundation mapping for integrated floodplain management: upper Mississippi River system. River Research and Applications 29:961–978.

Theiling, C. H., J. A. Janvrin, and J. Hendrickson. 2014. Upper Mississippi River restoration: implementation, monitoring, and learning since 1986. Restoration Ecology 23:157–166.

Thorp, J. H. 1992. Linkage between island and benthos in the Ohio River, with implications for riverine management. Canadian Journal of Fisheries and Aquatic Sciences 49:1873–1882.

Tripp, S., R. Brooks, D. Herzog, and J. Garvey.2013. Patterns of fish passage in the upper Mississippi River. River Research and Applications 30:1056–1064.

Twain, M. 1883. Life on the Mississippi. James R. Osgood & Company, Boston.

Upper Mississippi River Management Act of 1986. U.S. Code, volume 33, section 652.

USACE (U.S. Army Corps of Engineers). 1994. Economic impacts of recreation on the upper Mississippi River system. USACE, St. Paul District, St. Paul, Minnesota.

USACE (U.S. Army Corps of Engineers). 2000. Upper Mississippi River system habitat needs assessment. USACE, St. Louis District, Summary Report, St. Louis, Missouri. Available: www.umesc.usgs.gov/data_library/related_study_efforts/hna/11414_bro.pdf. (December 2015).

USACE (U.S. Army Corps of Engineers). 2015a. Lower Mississippi River resource assessment. USACE, Memphis District, Final Assessment Report, Memphis, Tennessee. Available: www.mvm.usace.army.mil/Portals/51/docs/missions/projects/LMRRA/LMRRA%20Public%20Review%20Draft%202.pdf. (December 2015).

USACE (U.S. Army Corps of Engineers). 2015b. Lower Mississippi River resource assessment: assessment of natural resource habitat needs. Final report. Available: www.mvm.usace.army.mil/Portals/51/docs/missions/projects/LMRRA/LMRRA%20Habitat%20final%20report.pdf. (December 2015).

USGS (U.S. Geological Survey). 1999. Ecological status and trends of the upper Mississippi River system 1998: a report of the Long Term Resource Monitoring Program. U.S. Geological Survey, Upper Midwest Environmental Sciences Center, Long Term Resource Monitoring Program 99-T001, La Crosse, Wisconsin.

Watson, L. D., and B. W. Hawkinson.1979.Recreational use of Pool 5, Mississippi River. Minnesota Department of Natural Resources, Investigational Report No. 362, St. Paul.

Weiner, J. G., and M. B. Sandheinrich. 2010. Contaminants in the upper Mississippi River: historic trends, responses to regulatory controls, and emerging concerns. Hydrobiologia 640:49–70.

Wilcox, D. B. 1993. An aquatic habitat classification system for the upper Mississippi River system. U.S. Fish and Wildlife Service, Environmental Management Technical Center, EMTC93-T003, Onalaska, Wisconsin.

Wilcox, D. B., E. L. Stefanik, D. E. Kelner, M. A. Cornish, D. J. Johnson, I. J. Hodgins, S. J. Zigler, and B. L. Johnson. 2004. Improving fish passage through navigation dams on the upper Mississippi River system. ENV Report 54, Upper Mississippi River - Illinois Waterway System Navigation Study. U.S. Army Engineer District, Rock Island, Illinois.

Winkley, B. R., and J. H. Brooks. 1980. Navigation and flood control through geometric alignment. U.S. Army Corps of Engineers, Vicksburg District, Hydraulics Branch, Potamology Section, Potamology Investigation Report 300-3, Vicksburg, Mississippi.

Zigler, S. J., M. R. Dewey, B. C. Knights, A. L. Runstrom, and M. T. Steingraeber. 2003. Movement and habitat use by radio-tagged Paddlefish in the upper Mississippi River and tributaries. North American Journal of Fisheries Management 23:189–205.

American Fisheries Society Symposium 84:35–52, 2016

Status and Development of the Yangtze River: An Ecological View

Huatang Deng, Jiawen Ba, and Daqing Chen*
Yangtze River Fisheries Research Institute, Chinese Academy of Fishery Sciences
No. 8, 1st Wudayuan Road, East Lake Hi-Tech Development Zone
Wuhan, Hubei 430223, P. R. China

Abstract.—The Yangtze River is the largest river in China, with a drainage area encompassing a land area of 1.8 × 10^6 km^2. The river main stem flows approximately 6,300 km from west to east through 11 provinces, eventually emptying into the East China Sea. The total basin area of the Yangtze River provides more than 40% of China's gross domestic product, sustains nearly 500 million people, and plays a vital role in the overall socioeconomic development of China. Because of longitudinal differences in geomorphology, the Yangtze River is often divided into upper, middle, and lower reaches for the purposes of study and description. In general, the Yangtze River basin has a warm climate and abundant rainfall. The subtropical monsoon climate accounts for 70–80% of its annual rainfall, which produces predictable annual flood events during summers of each year. The Yangtze River ecosystem also has diverse aquatic habitats that support high biodiversity. The Yangtze River main stem contains more than 3,000 tributaries and approximately 4,000 lakes that provide important habitats for thousands of aquatic species, which include 378 fish species. Yet, the ecological communities of the Yangtze River have been dramatically impacted in recent decades due to high degrees of anthropogenic activities associated with the economic development of China. Significant biodiversity declines occurring throughout the basin have been largely attributed to hydropower dam construction, overexploitation of major fisheries, water pollution, and habitat fragmentation. In addition, widespread disconnection of the Yangtze River main stem from its floodplain lakes, lake reclamation projects, invasions by exotic species, and other serious ecological problems threaten the survival of the remaining aquatic resources in the basin. To offset the adverse effects of these threats, a number of nature reserves have been established. Additionally, protection and restoration measures, including spring fishing closures, artificial population enhancement strategies, and complete habitat restoration projects, have been implemented throughout the basin. Although these projects have yielded some encouraging results to date, they are still considered inadequate. It is recommended that an integrated strategy that includes cross-functional communication mechanisms, more comprehensive monitoring, ecological rehabilitation, and government legislation is still needed. This will insure that utilization and development in the Yangtze River ecosystem are sustainable and balanced with conservation and ecological needs.

* Corresponding author: chdq@yfi.ac.cn

Introduction

The Yangtze River (Chang Jiang or "Long River") is one of the cradles of Chinese civilization and a lifeline for the sustainable development of China's economy and society. The Yangtze River basin is home to more than 500 million people, which is one-third of the population of China. The basin produces about 34% of China's total grain output and generates as much as 40% of the nation's gross domestic product. The Yangtze River basin enjoys a warm climate and abundant rainfall and contains thousands of tributaries and lakes that support high levels of biodiversity throughout the basin. However, numerous environmental issues are threatening the Yangtze River, such that the general health of the ecosystem has begun to attract greater attention in recent years from the Chinese government and public alike. This article provides an overview of the current status of the Yangtze River ecosystem, discusses primary threats to biodiversity, and highlights the achievements of recent ecological conservation efforts in the basin. We also present recommendations that should ensure that utilization and development in the Yangtze River ecosystem are sustainable and balanced with conservation and ecological needs.

Status of the Natural Environment of the Yangtze River

The Yangtze River is by far the largest river in China. The river is approximately 6,400 km long and drains a basin area of 1.8 million km^2, which encompasses 19% of total area of China (Figure 1). In terms of length and discharge, the Yangtze River is the third longest river in the world and contains the fifth greatest discharge (Zeng 1990; Chen et al. 2001a, 2001b). The river originates from the Tuotuo River, which is on the southwestern side of a glacier in the Geladandong Mountains of the Tibetan Plateau. The Yangtze River main stem flows through 11 provinces, including Qinghai, Tibet, Sichuan, Yunnan, Chongqing, Hubei, Hunan, Jiangxi, Anhui, Jiangsu, and Shanghai, before emptying into the East China Sea at Chongming Island. The Yangtze River's tributaries extend into eight more provinces, including Gansu, Shanxi,

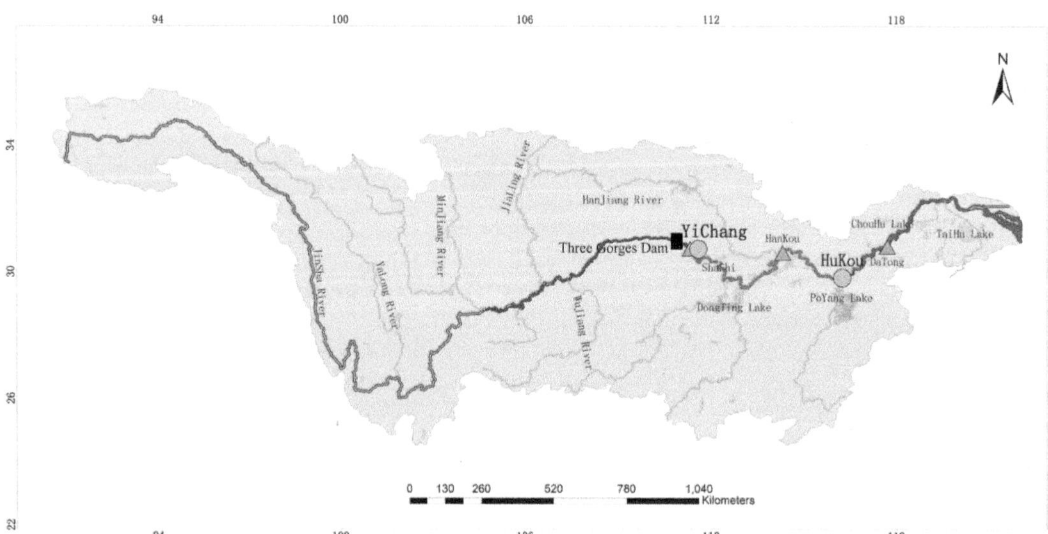

Figure 1. The Yangtze River drainage basin. The major tributaries, interior lakes, and the site of Three Gorges Dam are indicated. The cities of Yichang and Hukou are the two dividing points for the three reaches of the river. Gray triangles indicate the positions of three key hydrological gauging stations at Yichang, Hankou, and Datong, which represent the upper, middle, and lower Yangtze River, respectively.

Table 1. Basic descriptive information for each the upper, middle, and lower reaches of the Yangtze River. Data compiled from Zeng (1990), Chen et al. (2001), and Yang et al. (2008).

Measure	Upper basin	Middle basin	Lower basin
Length (km)	4,500	950	930
Basin area (km²)	1,000,000	680,000	120,000
Channel			
width (km)	0.5–1.5	1–2	2-4
depth (m)	5–20	6-15	10–20
slope (10-5)	10–40	2–3	0.5–1.0
Annual temperature (°C)	-4–16	14–16	16-18
Annual precipitation (mm)	200–1,100	>1,100	>1,100
Major tributaries/drainages	Yalong River Mingjiang River Jialing River Wujiang River	Dongting Lake drainage Hanjiang River Poyang Lake drainage	Taihu Lake Chaohu Lake

Guizhou, Henan, Zhejiang, Guangxi, Fujian, and Guangdong (Zeng 1990).

Topography

A variety of geomorphologies occur longitudinally from the upstream to the downstream reaches of the Yangtze River. The river flows across three major morphologic steps from its source on the Tibetan Plateau to the eastern coast, constituting a total elevation change of 5,400 m. The Tibetan Plateau in southwestern China constitutes the first step, with an average elevation of more than 4,000 m above mean sea level (msl). The second step ranges from about 500–2,000 m above msl in elevation and is comprised of high mountains and mountain-girt basins with a southwest to northeast orientation. The third step contains elevations ranging from 0 to 500 m above msl and constitutes the more lowland, fluvial reach of the Yangtze River in eastern China (Yang et al. 2007).

The Yangtze River is generally divided into three distinct reaches for classification and study (Table 1; Figure 1). The upper reach is the largest and extends from the river's source on the Tibetan Plateau downstream to Yichang, which is located at the lower end of the Three Gorges Reservoir. The upper reach is 4,504 km in length with a drainage area of 1.0×10^6 km². In this reach, the Yangtze flows primarily across mountainous regions with steep channel slopes ($1.0–4.0 \times 10^{-4}$) and swift currents (Zeng 1990; Chen et al. 2001b). The Yangtze River middle reach extends from Yichang downstream to Hukou and contains a connection with Poyang Lake in the downstream portion of the reach. This reach is 950 km in length and has a total drainage area of 6.8×10^5 km². Low-relief topography with channel slopes of $2.0–3.0 \times 10^{-5}$ and a meandering channel typify this reach in central China (Tong and Han 1982; Chen et al. 2001b). The lower reach of the Yangtze River flows the final 930 km from Hukou to an estuary that discharges into the East China Sea. This lower reach has a total drainage area of 1.2×10^5 km² and meanders among plains and hills. This reach contains much braiding of the Yangtze River channel, which is associated with a low gradient of about $0.5–1.0 \times 10^{-5}$ (Chen et al. 2001b; Yang et al. 2007). According to the topographical step patterns mentioned above, the middle and lower reaches of the Yangtze River are located on the third and lowest step.

Water system

The Yangtze River is well known for its well-developed water systems that contain more than 3,000 tributaries and approximately 4,000 lakes (Zeng 1990). Floodplain lakes connected with the Yangtze River main stem and its tributaries form elaborate riverine-lacustrine networks throughout the basin. There are 49 tributaries with drainage areas greater than 1.0×10^4 km^2, whereas drainage areas of the Jialing, Hanjiang, Minjiang, and Yalong rivers each exceed 1.0×10^5 km^2 (Figure 1). The Yangtze River also contains diverse lake clusters that differ greatly in size, shape, and morphometry, including those on the Qinghai–Tibet Plateau and the Yunnan–Guizhou Plateau in the upper basin and many located in the middle and lower basin. The middle and lower basin of the Yangtze River contains the largest concentration of freshwater lakes in China. In this portion of the Yangtze River basin, there are 1,685 natural lakes with individual surface areas greater than 1 km^2, which comprise 63% of all such lakes in China. Furthermore, these lakes have a total surface area of 62,686 km^2, which equates to 77% of the total surface area of all such lakes in China (Yang et al. 2011). In fact, the middle and lower Yangtze River basin contain the Poyang, Dongting, Taihu, and Chaohu lakes, which are four of China's five largest freshwater lakes (Figure 1).

Climatic and hydrological variation

The Yangtze River basin contains a mostly warm and humid climate. The majority of the basin is dominated by a subtropical monsoon climate with four distinct seasons. Mean annual temperature varies greatly across the different Yangtze River reaches due to its enormous elevational gradient from west to east (Table 1). Mean annual temperature in most areas of the middle and lower reaches ranges from 16°C to 18°C. Because of topography and elevational differences, mean annual temperature differences in the upper reach are large, varying from 16°C in the Sichuan subbasin to –4°C in the headwaters upstream of Yushu. In general, mean annual temperature in the highest step of the Yangtze River basin is less than 10°C (Yang et al. 2008). Over the past 50 years, mean annual temperatures in the Yangtze River basin have steadily increased, with rapid increases occurring since 2000 (Su et al. 2006; Yang et al. 2008).

The Yangtze River basin receives fairly abundant precipitation, with a mean annual precipitation of 1,127 mm/year. Because of its large area and monsoon climate, basin precipitation exhibits strong spatial and temporal variability (Table 1; Yang et al. 2006). Mean annual precipitation in the basin ranges from 270 to 500 mm/year in the upper basin compared to 1,600–1,900 mm/year in the middle and lower basin (Gemmer et al. 2008). On an annual basis, the Yangtze River basin experiences cold-dry winters and warm-wet summers, with 70–80% of the annual precipitation occurring during the summer seasons (Yu et al. 2009). Annual precipitation has increased markedly in the Yangtze River basin during the past few decades. For example, summer precipitation totals have increased by 112 mm during the past 40 years in the middle and lower basin. Concurrently, the proportional contribution of summer precipitation to the annual precipitation total has increased marginally from 38% in the 1960s to 42% in the 1990s (Su et al. 2006). Similarly, storm frequencies have increased in recent decades, especially with respect to extreme events containing greater flooding (Su et al. 2006).

Runoff in the Yangtze River basin is recharged primarily by precipitation. However, precipitation patterns are variable and runoff exhibits high within-year variation. For example, hydrologic gauging stations in the upper, middle, and lower Yangtze River basin indicated that summer and fall runoff accounted for 78%, 71%, and 68% of annual runoff, respectively (Yang et al. 2008). Because the Yangtze River basin receives monsoon-related precipitation, annual floods occur during summers. Historically, the Yangtze River has been known for its large floods and has four provinces (Hunan, Hubei, Jiangxi, and Anhui) located in areas highly at risk to damaging floods (Yang et al. 2008; Yu et al. 2009). Since the 1950s, major floods have occurred more frequently and tend to have greater water levels (means and peaks) for longer durations compared to earlier periods. For example, a series of floods in 1998 produced by monsoon-related precipitation in the middle basin lasted 2.5 months from early July through mid-September. The recurrence interval for these floods was calculated to be about 60 years (Yu et al. 2009).

Water discharge from the Yangtze River is approximately 920 km^3/year, which ranks fifth among the world's rivers. In terms of sediment, the Yangtze River discharges approximately 4.8×10^8 metric tons per year into the East China Sea, which is the fourth largest sediment load in the world (Chen et al. 1988; Milliman and Syvitski 1992). Although less than 50% of the river's discharge derives from the upper basin, sediment loads in the Yangtze River are much greater in the upper basin compared to the middle and lower basin. Over the past few decades, numerous dams and reforestation projects initiated along the Yangtze River have had little impact on water discharge. However, these projects have drastically altered annual and seasonal sediment loads (Figure 2; Xu and Milliman 2009). Prior to the construction of the Three Gorges Dam, the reduction of the Yangtze River's sediment load occurred in two phases between 1950 and 2002. First, the closure of the Danjiangkou Reservoir on the Hanjiang River tributary in 1968 significantly reduced sediment loads. Second, the installation of numerous dams and water and soil conservation measures in the Jialing River tributary subbasin after 1985 also reduced sediment loads to the Yangtze River (Figure 2; Yang et al. 2006). Following impoundment of the Three Gorges Reservoir on the Yangtze River main stem in 2003, 60% of the sediment load entering the reservoir became trapped during flood seasons. Because downstream channel erosion has not yet counteracted the sediment-trapping process of the Three Gorges Reservoir, sediment deposited into the Yangtze River estuary downstream is expected to continue decreasing for the near future (Xu and Milliman 2009).

Serving function

The Yangtze River is an enormous water resource, equaling about 9.96×10^{11} m^3 of available water. However, the ratio of water development to utilization is only 17% for the basin, which is below average for China. The full development potential of the Yangtze River's water resources has been pursued only in recent decades. The Yangtze River has high hydropower potential due to abundant water quantity coupled with the 5,400-m elevational drop in the basin. Although 95% of the elevational drop is isolated in the upper reach, the theoretical hydropower that could be developed is still on the order of 2.78×10^8 kW, which accounts for 40% of the total hydropower that can be developed in China. The degree of hydropower development is currently relatively low in the Yangtze River basin. Through 2001, 2,441 hydropower stations had been completed or were under construction. These stations combined have an installed capacity of 6.97×10^7 kW, which equates to only 27% of that total power technically feasible for the Yangtze River basin.

The Yangtze River is known as the "Golden Waterway" and is considered to be the most significant inland waterway in China. The river functions as the key transportation artery that connects the central, southeast, and eastern regions of China. The Yangtze River basin contains more than 3,600 rivers that could potentially support commercial navigation. By the end of 2002, river-training activities in the Yangtze River basin had produced a total of 66,000 km of navigable river courses, which included 2,628 km in the Yangtze River main stem. However, this figure accounts for only about 50% of the potential total for the entire navigable river courses in China. In light of the increase in navigable waterways throughout the Yangtze River basin, commercial shipping has grown rapidly. The commercial freight volume of the Yangtze River now exceeds the Mississippi and Rhine rivers and has become the world's busiest inland commercial waterway.

Aquatic biodiversity

The Yangtze River basin supports diverse and abundant aquatic resources due to ample water resources and abundant natural habitats. Unofficially, there are more than 1,100 species of aquatic biota resident in the Yangtze River; in addition, there are numer-

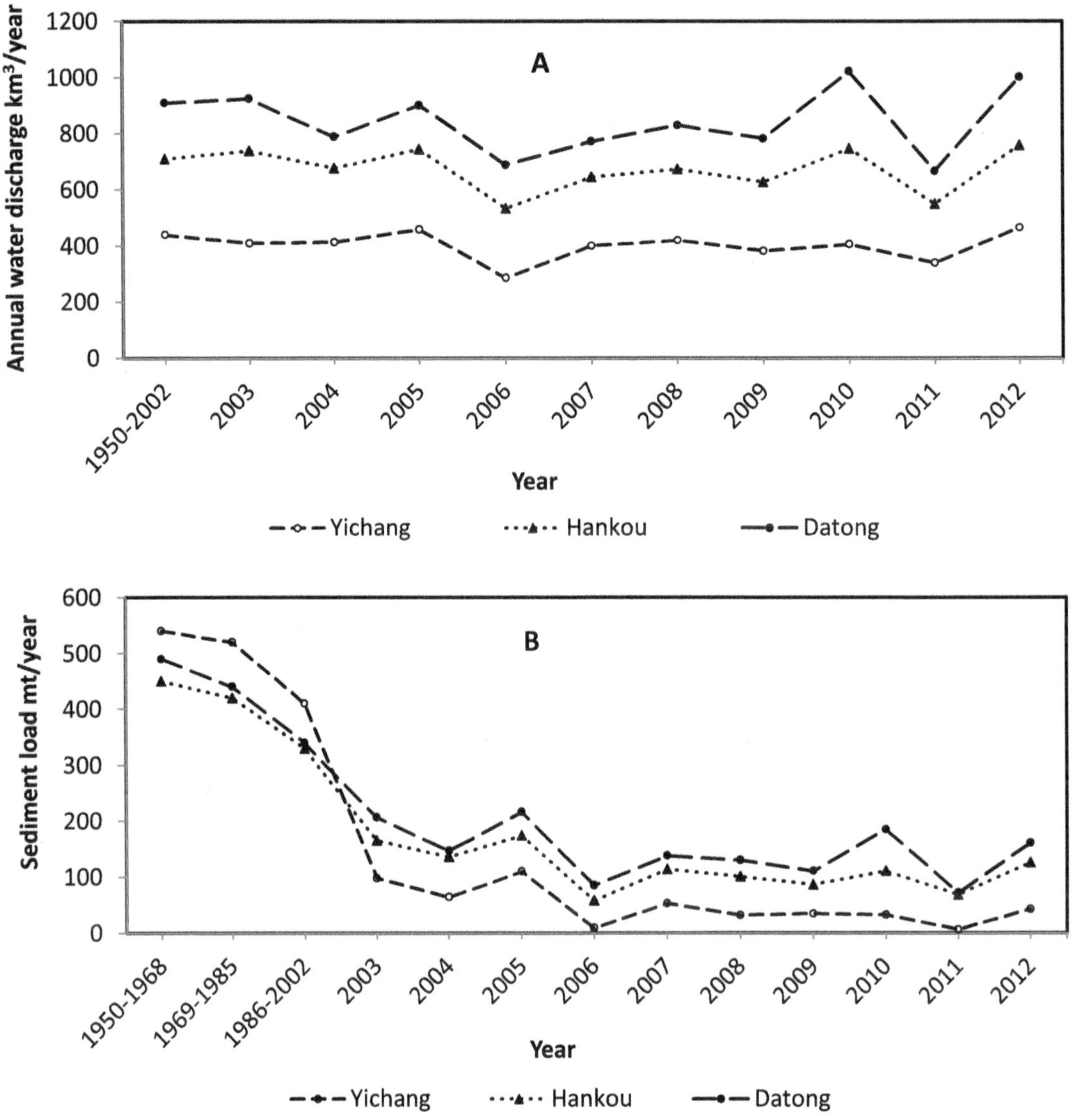

Figure 2. Temporal variations of mean **(A)** annual water discharge and **(B)** sediment loads at three key gauging stations at Yichang (upper basin), Hankou (middle basin), and Datong (lower basin). Water discharges during 1950–2002 and sediment loads for 1950–1968, 1969–1985, and 1985–2002 were generated from averages across these years, respectively. Data were cited from Yang et al. (2006) and the *River Sediment Bulletin of China* (2003–2012), published by the Ministry of Hydrology.

ous species of migratory animals that inhabit the basin seasonally. For example, in Poyang Lake of the lower basin, 310 species of birds have been documented, of which 129 species are strictly aquatic species. Amphibians in the Yangtze River basin account for 34% of the endemic species in China, with several of these species having very limited distributions within only one or a few subbasins. The Yangtze River dolphin or baiji *Lipotes vexillifer*, which inhabits only the middle and lower reaches of the river, is the most endangered cetacean species in the world (Zhang et al. 2003). Similarly, the Yangtze finless porpoise *Neophocaena phocaenoides asiaeorientalisis* represents the world's only freshwater species of porpoise. This species is endemic to the Yangtze River basin but is presently confined to fragmented habitats largely in the Poyang and Dongting lakes (Zhao et al. 2008).

There are several relict species that exist in the Yangtze River, though most have declined and could soon face extinction. The Chinese alligator *Alligator sinensis* is a class I protected species in China and is critically endangered in the wild, though up to 10,000 individuals are held in captivity around the world (Thorbjarnarson and Wang 1999). Presently, there is limited potential for reintroduction of this species due to high degrees of habitat loss and fragmentation to the species' native habitats in the Yangtze River lower basin (Dudgeon 2010). Additionally, the Chinese giant salamander *Andrias davidianus* is the world's largest amphibian and was formerly widely distributed throughout the basin in reaches that were 200–1,000 m above msl. Although the Chinese giant salamander is listed as a class II protected species in China (Wang et al. 2004), wild populations have declined substantially since the 1950s. Population declines have been largely attributed to pollution, habitat loss and fragmentation, and hunting. Since the 1980s, up to 14 nature preserves have been established to, in part, protect and conserve Chinese giant salamanders. The species also is now widely produced on private farms.

Fish biodiversity and fisheries resources

The Yangtze River basin contains the greatest diversity of freshwater fishes in China (Chen et al. 2009). A total of 378 fish species distributed among 14 orders, 32 families, and 144 genera have been documented from the basin, of which 338 are freshwater species. The Yangtze River fish fauna contributes almost one-third of the total number of freshwater fish species known from China (Chen et al. 2004; Yu et al. 2005; Chen et al. 2009). Included in this fauna are 11 migratory fish species, as well as 29 fish species that inhabit only brackish waters in the downstream estuary (Yu et al. 2005). Of the 378 resident fish species throughout the Yangtze River basin, 162 species or 43% of the entire fauna are endemic to the basin (Yu et al. 2005; Chen et al. 2009). In particular, the Yangtze River upper basin contains 112 of the endemic fish species, which represent nearly 70% of the basin's endemic fauna (Fan et al. 2006).

The Yangtze River contains environmental conditions that are ideal for supporting various commercial fisheries, and is considered the cradle of freshwater fisheries production in China (Chen et al. 2009). Fisheries yields in the Yangtze River basin account for about 65% of the total freshwater fisheries production for all of China (Chen et al. 2002). Many economically important fisheries thrive in the Yangtze River basin. Primary among these fisheries are the four major Chinese carps (i.e., Black Carp *Mylopharyngodon piceus*, Grass Carp *Ctenopharyngodon idella*, Silver Carp *Hypophthalmichthys molitrix*, and Bighead Carp *H. nobilis*). However, other significant commercial fisheries exist in the Yangtze River for Bronze Gudgeon *Coreius heterodon*, Largemouth Bronze Gudgeon *C. guichenoti*, Reeves Shad *Macrura reevesi*, Japanese Grenadier Anchovy *Coilia ectenes*, Osbeck's Grenadier Anchovy *C. mystus*, and Japanese Eel *Anguilla japonica* (Chen et al. 2009). During 1949–2000, fishery catches in the Yangtze River ranged from 100,000–400,000 metric tons/year, and averaged 250,000 metric tons/year (Chen et al. 2004). The river also has provided sources of high-quality broodstock that have fueled the development of large commercial industries for many species. For example, the broodstock sources of the four Chinese carps that are widely cultivated throughout China originally came from the Yangtze River basin. The Yangtze River fish fauna has been shaped by many fish–environment relationships that have developed over evolutionary time scales. Some fish species have evolved special adaptations and have become highly selective for specific conditions common in the basin (Yang et al. 2008). Many fishes in the Yangtze River basin begin seasonal migrations within the system triggered by seasonal water level fluctuations (Fu et al. 2003). However, as the Yangtze River economy develops and human activities increase, the basin is faced with a series of environmental problems that affect fisheries resources and the river environment (Chen et al. 2009).

Overexploitation of mostly unregulated fisheries and environmental degradation resulting from many different anthropogenic activities have impacted fisheries production through the Yangtze River basin. These impacts have decreased the amount and quality of many riverine habitats, and in the case of hydroelectric dams, block fish migration pathways; severe decreases in the quality of many fisheries are directly attributable to these impacts (Chen et al. 2009). Several recent fisheries characteristics support this contention. First, the numbers of fish species classified as endangered continue to increase. China's *Red Data Book of Endangered Animals* lists 25 species of fishes (Yue and Chen 1998). In fact, three of China's largest freshwater fishes, the Chinese Sturgeon *Acipenser sinensis*, Yangtze Sturgeon *A. dabryanus*, and Chinese Paddlefish *Psephurus gladius*, are presently all class I protected species in China (Cui and Wang 2006). Coincident with the increases in human activities that fuel economic development throughout the Yangtze River basin, these and other rare fish species continue to face high risks for extinction. Second, there have been dramatic decreases in the commercial catches of several major fishes. Annual commercial fish catches have decreased gradually, though steadily, following the historical peak of 427,000 metric tons in 1954. Although an average of 320,000 metric tons/year was captured during the 1950s, current commercial catches are approximately 60,000 metric tons/year, less than one-fifth that in previous decades (Figure 3). Furthermore, the commercial sup-

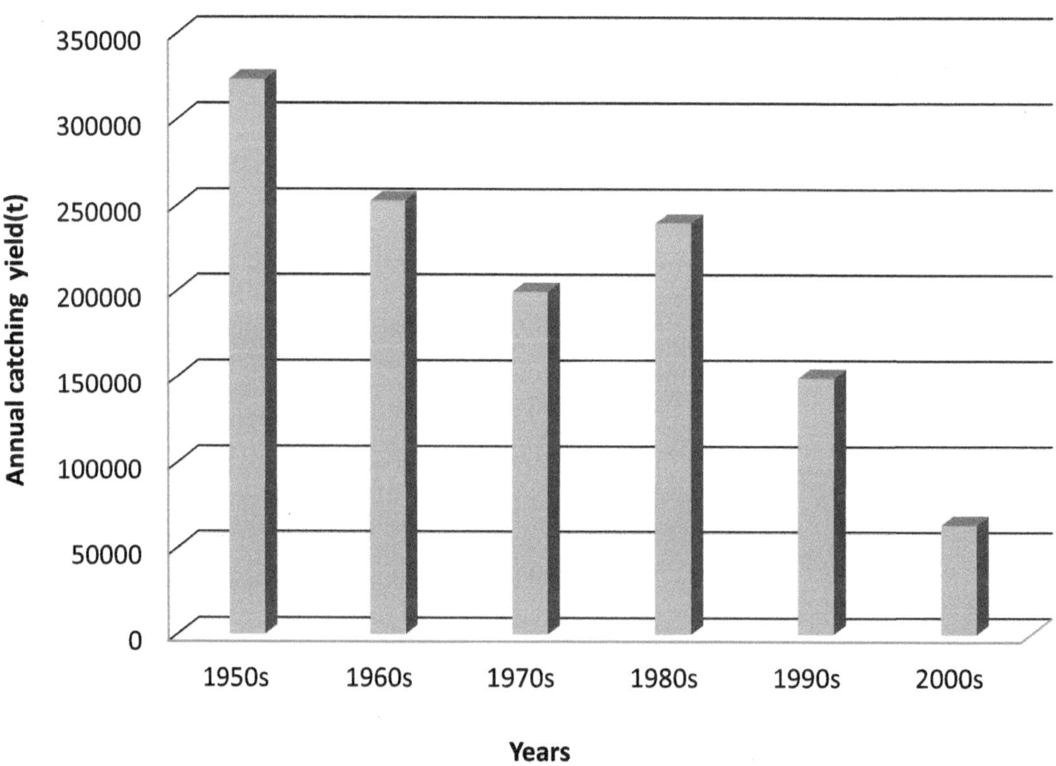

Figure 3. Changes in fish catches by decade in the Yangtze River basin. Data are from Zeng (1990) and the *Bulletin on the Ecological and Environmental Monitoring Results of the Three Gorges Project* (1997–2010). t = metric tons.

ply of fry for the four major Chinese carps has declined to less than 1.0×10^9 in recent years compared to 3.0×10^{10} in previous decades. Third, other commercial aquatic species such as the Chinese mitten-handed crab *Eriocheir sinensis* and Japanese Eel were no longer caught during spawning seasons in the Yangtze River basin due to environmental degradation and habitat losses. Finally, commercial fish catches from the Yangtze River indicate that the community structure contains decreased abundances of long-lived, large-bodied fishes coinciding with sharp increases in the abundances of short-lived, small-bodied fishes of less commercial value (Fu et al. 2003).

Main Threats of Economic Development and its Influences on the Ecosystem

Construction of hydropower dams

The first hydropower facility constructed in the Yangtze River basin was the Shilongba Dam hydropower station built in the upper basin on the Tanglangchuan River in 1910. In the century since then, numerous hydropower plants have been built throughout the Yangtze River basin. Of the 551 hydropower plants with a dam height greater than 60 m and capacity exceeding 100,000 kW that have been built worldwide, 48 are located in the Yangtze River basin. In fact, the Yangtze River basin contains the largest concentration of such hydropower projects in the world. However, hydropower development in the upper basin of the Yangtze River is less 20% of its full potential. Additionally, in the Yangtze River headwaters upstream of Yibin where the elevational gradients are greater, hydropower development is less than 5% of its full potential. In order to meet the increasing energy demands needed to fuel economic development in China, hydropower development is occurring rapidly throughout Yangtze River basin.

In the 1980s, planning began for six large hydropower projects on the Yangtze River. Gezhouba Dam was completed in 1988, and was the first dam built on the Yangtze River main stem (Figure 4). The Gezhouba Dam has blocked spawning migrations for fishes, degraded spawning habitats, and fragmented populations of both large sturgeons (Chinese Sturgeon and Yangtze Sturgeon) and the Chinese Paddlefish, which are all protected species in the Yangtze River (Wei et al. 1997; Chen et al. 2004). The abundance of Chinese Sturgeon has greatly declined during the past 30 years, despite that spawning habitats were re-established below Gezhouba Dam in recent years (Wei et al. 1997). The reproduction of Yangtze Sturgeon and Chinese Paddlefish is now restricted to the upper basin of the Yangtze River, with the latter species on the verge of extinction (Dudgeon 2010).

The well-known Three Gorges Dam (TGD) is 38 km upstream from Gezhouba Dam and was completed in 2006 (Figure 4). The dam is 185 m high and impounds Three Gorges Reservoir (TGR), which is 600 km in length long with a total storage capacity of approximately 3.93×10^{10} m^3. The reservoir operation alters hydrologic conditions in the Yangtze River, which affects the life cycles and habitats of resident fishes. It is suspected that flow regulation in TGR and riverbed erosion processes due to sediment-starved flows also may alter some downstream fluvial processes. These alterations are expected to somewhat destabilize channel meandering and possibly impact braided channels, which represent vital habitat for several endangered or otherwise rare species (Wang et al. 2005). As a result, one-quarter of these 40+ fish species associated with these habitats are presently exhibiting population declines, and may possibly vanish from the system in the near future (Chen et al. 2009).

The impact of dams on aquatic wildlife is a complex process that occurs gradually over long-term time scales. Moreover, if cascade dams are constructed as proposed in the upper Yangtze River main stem, it is expected they will impact the overall ecosystem in a cumulative fashion. At present, long-term comprehensive studies are needed to better understand the cumulative effects of these dams on the Yangtze River ecosystem.

Figure 4. Dams and reservoirs along the Yangtze River main stem: **(A)** Gezhouba Dam, which was built in 1981 and is 38 km downstream of Three Gorges Dam; **(B)** Three Gorges Dam (TGD); and **(C)** Three Gorges Reservoir (TGR) at Wushan, which is approximately 200 km upstream of TGD (This reach of TGR is highly constricted and characterized by an incised channel and steep littoral zones.); and **(D)** Three Gorges Reservoir (TGR) at Fengdu, which is approximately 400 km upstream of TGD (In this reach, the river channel becomes wider and less steep.).

Fishery overexploitation

Overexploitation of commercial fisheries in the Yangtze River is a serious issue, and is, in part, responsible for the net decline in aquatic biodiversity. According to monitoring data, there are more than 136,000 fishing vessels and 160 types of fishing gears operated throughout the Yangtze River basin (Figure 5; Chen et al. 2004). At present, commercial fish catches are believed to have exceeded the Yangtze River's carrying capacity; thus, further exploitation of these fisheries should not be possible. The depletion of the main economically important fish stocks has become detectable, with fish community structures tending towards smaller sizes and younger age structures (Liu et al. 2005). In fact, many commercial fisheries are exhibiting characteristics of growth overfishing. Chen et al. (2009) reported that the exploitation coefficients of the 10 most economically important fishes, which included the four Chinese carps, all indicated overexploitation (i.e., coefficients ≥0.7). In addition, destructive fishing gears and methods have been adopted to increase capture efficiencies and increase profits. Specifically, the illegal use of explosives, poisons, and electricity has lead to disastrous consequences for some fisheries resources. Small-mesh benthic fyke nets are utilized almost universally in lakes and streams of the lower and middle Yangtze River, and these gears have impacted juvenile abundances for many species. For instance, the juveniles of the major economically important fish species mentioned above accounted for 62% of the catches from small-mesh benthic fyke nets used near some villages in the region of Poyang Lake (Chen et al. 2004).

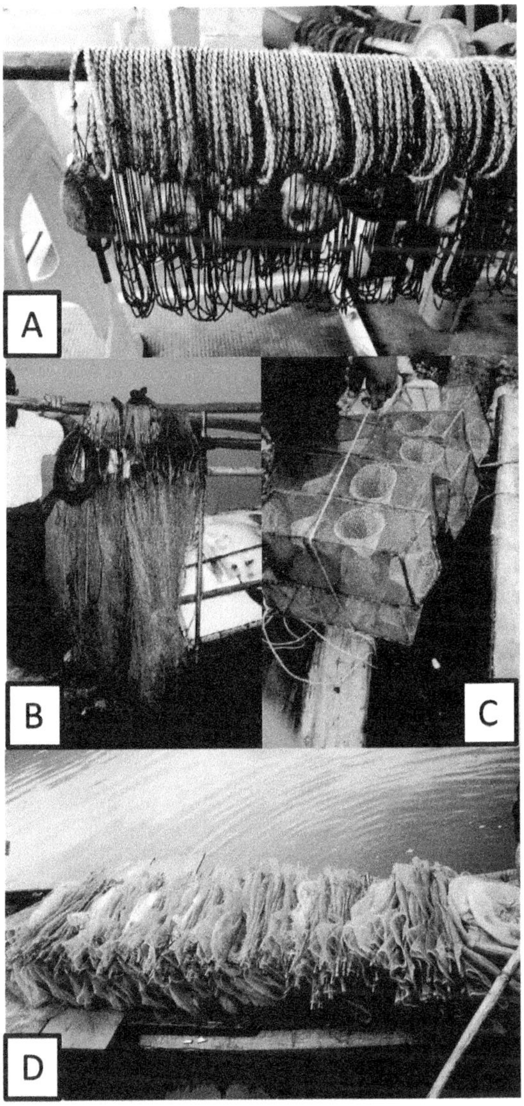

Figure 5. Common types of fish collection gears used in the Yangtze River basin: **(A)** rolling hooks, **(B)** trammel nets, **(C)** shrimp traps, and **(D)** small-mesh benthic fyke nets.

The overexploitation of fisheries resources also is related to intensive aquaculture practices being conducted in the Yangtze River basin. With aquaculture being increasingly practiced in the basin, stocked fishes have seriously affected biodiversity in some areas (Fang et al. 2006). For example, stocked Grass Carp have overgrazed aquatic vegetation in the middle basin of the Yangtze River. One submersed vascular plant species (*Potamogeton maackianus*) has completely vanished in Donghu Lake near the city of Wuhan (Chen 1980).

Water pollution

Rapid development of agriculture and industry, and widespread urbanization throughout the Yangtze River basin has polluted surface waters in recent years. According to preliminary data, 34% of the discharge of polluted waters in China occurs in the Yangtze River basin alone. In 2005, the annual pollution discharge to the Yangtze River reached 2.96×10^{10} metric tons, with most of the wastewater discharged directly into the river without any treatment. Although the water quality of the Yangtze River main stem is still generally acceptable, pollution levels are severe in the vicinities of large cities along the main stem and in tributaries and floodplain lakes near those cities. Pollution belts along the Yangtze River have increased to more than 600 km of the main stem, with an increasing trend through time. In particular, the Minjiang, Tuojiang, Xiangjiang, and Huangpu river tributaries are heavily polluted, with severe eutrophication occurring in Chenghai, Dianchi, Gantang, Caohu, Westlake, Taihu, and Dianshan lakes. In other areas of the Yangtze River basin, overuse of agricultural pesticides and fertilizers also have resulted in local areas of poor water quality from nonpoint source pollution.

Construction of large dams along the Yangtze River alters the natural hydrological conditions, which has, in some cases, exacerbated the water pollution effects. Overall, lower current velocities and decreased runoff to the Yangtze River main stem due to reservoir sluice operations have diminished purification capacities of the adjacent floodplain habitats, and thus, increased water pollution effects (Figure 6). For example, most of the embayments of TGR are classified as eutrophic, and contain serious algal blooms, largely composed of dinoflagellates *Peridiniopsis* spp. and the diatom *Cyclotella* sp., especially during the spring seasons (Cai and Hu 2006). Because TGR has more than 1,200 km of coastline and 0.1 km^2 hydrofluctuation areas, preservation

Figure 6. Threats for the aquatic resources of the Yangtze River basin: **(A)** fishing boats alongside the bank, **(B)** wastes in backwater reservoir, **(C)** great numbers of juvenile fishes harvested; and **(D)** *Neosalanx taihuensis*, which is a species of icefish and an exotic invasive captured in Three Gorges Reservoir.

of water quality in TGR is of paramount importance to the ecological health of the entire Yangtze River ecosystem.

Other common pollutants in the Yangtze River include heavy metals, oils, hydrocarbons, and excessive nutrients, which are directly discharged into the rivers and affect the entire river ecosystem. Eutrophication and the accumulation of toxic compounds have known disruptive effects on the physiology and reproduction of many fishes (Fan et al. 2006). Strengthening government regulatory control over pollution sources is of vital importance to improving water quality throughout the Yangtze River ecosystem.

River–lake disconnection and lake reclamation

There are numerous lakes in the Yangtze River basin with a total surface area exceeding 15,200 km^2, with 90% of these lakes located in the middle and lower reaches of the basin. Historically, these lakes were naturally connected to the Yangtze River main stem, and were subject to annual flooding from the river. Since the 1950s, large numbers of engineering projects have occurred, many of which involved construction of large sluices along both banks of the Yangtze River. These projects were aimed at flood control, agricultural irrigation, land reclamation, and to regulate water levels for control of schistosomiasis transmission (Wang and Jiang 1998). As a result of these projects, nearly all the Yangtze River's floodplain lakes became artificially disconnected from the river main stem, with the exception of the two largest lakes—Dongting and Poyang (Fu et al. 2003). Concurrently, disconnection occurred not only between the river main stem and the lakes, but also between lakes in close proximity to each other. Constructed sluice gates blocked exchanges of fishes between the river main stem and lakes; thus, habitats and populations within the Yangtze River basin were successively fragmented and aquatic biodiversity was significantly reduced, especially with respect to migratory fishes (Xie and Chen 1999; Chen et al. 2009). In fact, since the 1950s, the number of fish species in disconnected floodplain lakes has decreased from about 80 species to 52 species at present (Chen et al. 2009). At the same time, the proportional abundances of migratory fish species in these lakes has decreased from 50% to less than 30% (Chen et al. 2009).

The middle basin of the Yangtze River has experienced extensive land-use changes and large land-reclamation projects over the past few decades (Fang et al. 2005; Zhao et al. 2005). The spread of urbanization and increases in land reclamation for other more economically beneficial uses have reduced the overall areas of natural wetlands along the Yangtze River. The surface area of the Jianghan group of lakes has decreased 62% from 8,304 to 3,210 km^2, with nearly three-quarters of the natural wetlands disappearing during this time. Similarly, the surface area of Dongting Lake has decreased by 49% during the past 80 years, from 4,955 km^2 in the 1930s to 2,518 km^2 in 1998. This decrease in lake surface area was largely attributed to rapidly a growing human population in the vicinity of the lake, which involved extensive land reclamation that eliminated many natural wetlands and dewatered some areas (Zhao et al. 2005).

Biological invasions

Biological invasions or introduction of nonindigenous species is another potential threat to biodiversity throughout the Yangtze River basin. In many plateau lakes in the upper Yangtze River basin, the invasion by exotic fish species could become a very severe problem. For example, the number of native fish species in Dianchi Lake, located in the upper Yangtze River basin, has declined from 25 in the 1940s to five by the 2000s. The fish community structure in Dianchi Lake has become more similar to those in the middle and lower Yangtze River basin due to species introductions, some of which were on purpose while others were not (Zhou 2000; Xie et al. 2001). Recently, a total of 23 invasive fish species have been reported in TGR and the upper basin of the Yangtze River, some populations of which are still in the initial stages of expansion (Figure 6; Ba and Chen 2012). Meanwhile, there are several exotic fish species, such as Channel Catfish *Ictalurus punctatus*, Russian Sturgeon *Acipenser gueldenstaedti*, Amur Sturgeon *A. schrenki*, and a hybrid sturgeon widely cultivated in the aquaculture industry. These species also may pose potential threats for invasion throughout the basin.

Invasive species will threaten the native fishes through competition for habitat and forage, leading to greatly reduced numbers of native species, and possibly the local extinction of some (Zhang et al. 1997). The threat of further invasions of exotic fishes to the Yangtze River ecosystem will be constant and permanent (Yu et al. 2005), with no clear technical solutions for eliminating them.

Current Measures of Ecological Conservation

By the end of the 20th century, the environmental deterioration resulting from widespread development of the Yangtze River basin has attracted great attention from local and national governments and the Chinese public. In 1994, the Chinese government endorsed the sustainable development strategy as a national priority and policy. However, numerous constructive measures have been adopted with strong conservation-minded goals, and some achievements have been attained in recent decades.

Creation of natural reserves

As of 2008, there were 827 natural reserves of various types established throughout the Yangtze River basin. In fact, the total area of these reserves accounts for 20% of the whole Yangtze River valley, which exceeds the national average for other basins. A network of natural reserve of various types with strong ecological functions have taken shape throughout the basin (Li 2011). Natural reserves are designed to protect all essential terrestrial and aquatic or marine environments, including forests, steppe, marine and fresh waters, wetlands, and coastal habitats, all of which support numerous species of fishes, birds, amphibians, reptiles, and other wildlife. Among these reserves, four state-level aquatic natural reserves and about 100 state-level aquatic germplasm reserves have been established throughout the Yangtze River basin. These reserves are explicitly for the protection of aquatic species, especially endemics, and their habitats. The National Reserve of the Upper Yangtze for Endangered and Endemic Fishes was created in 2000 and expanded in 2005. This reserve is presently the only state-level fisheries reserve in the upper Yangtze River basin, and is now the largest aquatic reserve in China. The reserve encompasses an area of 331.7 km^2, and largely protects three widely known endangered fishes (*Psephurus gladius, Acipenser dabryanus,* and *Myxocyprinus asiaticus*), and 67 other endemic fishes and their habitats (Fan et al. 2006).

Closed fishing seasons

In 2002, a pilot project involving closed seasons for commercial fishing was implemented by China's Ministry of Agriculture for the middle and lower basin of the Yangtze River. This effort was further extended to the whole Yangtze River basin in 2003 after the approval of the China's State Council. The scope of the closed season included the Yangtze River main stem from Deqin County, Yunnan downstream to the estuary. Selected reaches of important tributaries, including Poyang and Dongting lakes, also were closed to commercial fishing. The new regulation stipulated that commercial fishing was not allowed annually from February 1 through April 30 upstream of Gezhouba Dam, and from April 1 through June 30 downstream of Gezhouba Dam. Concurrently, provincial and local governments enforced the regulation and implemented punishments for violators. These departments are mandated to strictly enforce laws against illegal fishing activities, including electrofishing, poisoning, and use of explosives, in order to maintain fishery production levels for the Yangtze River basin. Although monitoring does not yet suggest whole-scale recovery of the fisheries resources, the implementation and enforcement of the closed commercial fishing season has, to some degree, raised the visibility of fisheries conservation and restoration within the Chinese public (Chen et al. 2009).

Artificial enhancement of fisheries

In order to enhance natural reproduction and stock production, local governments and relevant fisheries authorities along the Yangtze River have engaged in artificial fish spawning and stocking programs aimed at supplementing wild populations. Artificial enhancement includes establishment of artificial spawning conditions and habitats at hatcheries, and the release of the produced juveniles, and sometimes mature adults, into lakes and rivers. The species released varies spatially in the Yangtze River but includes *Spinibarbus sinensis, Onychostoma sima, Sinila beorendahli, Schizothorax prenanti, Silurus meridionalis, Pelteobagrus vachelli,* and *Leiocassis longirostris* in the upper basin and *Megalobrama amblycephala, Plagiognathops microlepis, Xenocypris davidi, Squaliobar buscurriculus, Pelteobagrus fulvidraco, Siniperca chuatsi, Channa asiatica, C. maculata, Culter alburnus,* and *Hemisalanx prognathous* in the lower basin (Chen et al. 2012). These activities have promoted fishery resources conservation in the Yangtze River basin. For example, more than 14 spawning and fry nursery bases have been built in the TGR region of the river alone, with 130 million juvenile fishes released into the reservoir during 2007–2010. Since the commencement of fish stocking activities in 2006, the annual average natural fish yield in TGR has gradually increased by 10–20% per year. According to incomplete statistics, almost 4.5 million Chinese Sturgeon juveniles have been released into the Yangtze River basin during 1984–2005. Similarly, 720,000 Chinese Suckers *Myxocyprinus asiaticus* also were released during 2005–2007 (Zhang et al. 2008). These demonstration activities play vital roles in educating the public about the importance of environmental protection within the context of fishery sustainability.

Lake restoration projects

The frequent floods, decreased biodiversity, degraded ecosystem conditions, and an awareness of the importance of natural wetlands stimulated the Chinese government to prohibit land reclamation projects along the central Yangtze River in the late 1990s (Wang 1998). Following unprecedented catastrophic flooding in 1998, the Chinese government implemented lake restoration projects, with emphasis on the floodplain of the central Yangtze River. Measures such as reconversion of agricultural lands to lakes, leveling protective embankments for floodwater discharge, resettlement of human populations to new settlement areas have been successfully implemented. Consequently, during the period of 1978–1998, the number of lakes increased from 250 to 258 (3.2% increase), and the total area of coverage increased from 1,839.1 to 2,144.4 km^2 (16.6% increase) in the Jianghan River floodplain alone (Fang et al. 2005). Moreover, a study of land conversion in the Honghu Lake region during 1987–1998 indicated that the watered area nearly doubled from 457.5 km^2 in 1987 to 854.1 km^2 in 1998 (Zhao et al. 2003).

Prospects and Recommendations for Sustainable Basin Development

The Yangtze River basin is one of the most important regions in China, and it has the greatest capacity to positively affect the nation's economy and population. Despite the immense ecological value and biodiversity of the Yangtze River ecosystem, there are mounting economic, social, and political pressures to continue further development throughout of the basin. This development would largely entail rapidly increasing levels of industrialization coupled with widespread urbanization, and the environmental degradation that accompanies such development. In fact, the continued development of the entire nation of China is being linked to the capacity for sustainable development within the Yangtze River basin. However, despite the obvious economic advantages of development in the Yangtze River basin, balancing the ecosystem's ecological value and biodiversity with economic development that is environmentally sustainable should be of paramount importance to China.

A healthy Yangtze River ecosystem should balance conservation and environmental concerns with satisfying human demands associated with economic development. Ecologically, the river should continue to support its high levels of biodiversity and commercial fishery yields that are sustainable. Large-scale conservation planning and actions at the appropriate scales are needed to improve the Yangtze River ecosystem in order to maintain an overall healthy river ecosystem. Some of the recommendations presented herein, though far from comprehensive, may produce positive results through carefully planned conservation-minded programs. If implemented, these programs should ensure that a healthy Yangtze River ecosystem over the long-term will be attainable eventually for future generations.

Establish cross-functional communication mechanisms

The administration and conservation of the Yangtze River involves a number of government agencies, local authorities, institutes, universities, and various stakeholder groups. Overall programming, policies, and management aimed at river conservation are developed by departments of water, environment, forestry, agriculture, and other agencies, and implemented jointly in concert with local governments and the relevant river basin authorities. At the present time, establishing cross-functional communication mechanisms among these different authorities for developing a successful integrated river basin management plan for the Yangtze River is essential for the future.

Strengthen comprehensive monitoring operation

Continued monitoring throughout the Yangtze River basin will provide constantly updated data sets that will be used towards the management and conservation of the ecosystem. However, in order to achieve this goal, it is necessary to enhance the current programs to further improve monitoring of the entire Yangtze River basin. These enhancements should include remote automated monitoring stations linked to ground stations by satellite, which would provide nearly real-time data for the different locations in the river.

Enhance ecological rehabilitation

The Yangtze River is a complex ecosystem composed of a network of water systems that support multiple uses. These water systems contain wide varieties of environments and habitats that support an enormous number of wildlife species and, thus, biodiversity. Restoring the integrity of the water system network is the necessary first step towards the ecological rehabilitation of the entire Yangtze River basin. Additionally, ecological engineering measures should be taken to restore the habitat suitability of important waters for fisheries such as the spawning grounds, rearing habitats, and feeding grounds for economically important species. In some newer systems such as reservoirs, fisheries are being developed using sound ecological principles (i.e., ecological fishery) such that the fisheries are sustainable over the long-term.

Reinforce legislation and improve conservation management

Legislation and legitimate regulation are key to promoting river administration and improving river management. At present, further enforcement of relevant laws and regulations are needed to ensure promotion and implementation of an integrated Yangtze River basin management plan.

Acknowledgments

This work was funded by the Special Fund for Agro-Scientific Research in the Public Interest (201303056-5;200903048).

References

Ba, J., and D. Chen. 2012. Invasive fishes in Three Gorges Reservoir area and preliminary study on effects of fish invasion owing to impoundment. Journal of Lake Sciences 24:185–189. (In Chinese.)

Cai, Q.H., and Z.Y. Hu. 2006. Studies on eutrophication problem and control strategy in the Three Gorges Reservoir. Acta Hydrobiologica Sinica 30:7–11. (In Chinese.)

Chen, D., X. Duan, S. Liu, and W. Shi. 2004. Status and management of the fisheries resources of the Yangtze River. Pages 173–182 *in* R. Welcomme and T. Petr, editors. Proceedings of the second international symposium on the management of large rivers for fisheries, Volume 1. Food and Agriculture Organization of the United Nations, Regional Office for Asia and the Pacific, Bangkok, Thailand.

Chen, D., S. Li, and K. Wang. 2012. Enhancement and conservation of inland fisheries resources in China. Environmental Biology of Fishes 93:531–545.

Chen, D., F. Xiong, K. Wang, and Y. Chang. 2009. Status of research on Yangtze fish biology and fisheries. Environmental Biology of Fishes 85:337–357.

Chen, D. Q., S. P. Liu, and X. B. Duan. 2002. A preliminary study of the fisheries biology of main commercial fishes in the middle and upper reaches of the Yangtze River. Acta Hydrobiologica Sinica 26:622–626. (In Chinese.)

Chen, H. D. 1980. Structure and dynamics of macrophyte communities in Lake Donghu, Wuhan. Oceanologia et Limnologia Sinica 11:275–83.

Chen, J. Y., H. T. Shen, and C. X. Yun. 1988. Processes of dynamics and geomorphology of the Changjiang estuary. Shanghai Scientific and Technical Publishers, Shanghai. (In Chinese.)

Chen, Z., J. F. Li, and H. T. Shen. 2001b. Yangtze River, China: historical analysis of discharge variability and sediment flux Geomorphology 41:77–91.

Chen, Z., L. Z. Yu, and A. Gupta. 2001a. Yangtze River, China: an introduction. Geomorphology 41:73–75.

Cui, H., and L. Wang. 2006. A brief discussion on the legal protection of fishery resources along the Yangtze River. Resources and Environment in the Yangtze Basin 15:58–60. (In Chinese.)

Dudgeon, D. 2010. Requiem for a river: extinctions, climate change and the last of the Yangtze. Aquatic Conservation: Marine and Freshwater Ecosystems 20:127–131.

Fan, X., Q. Wei, J. Chang, J., H. Rosenthal, J. He, D. Chen, L. Shen, H. Du, and D. Yang. 2006. A review on conservation issues in the upper Yangtze River: a last chance for a big challenge: can Chinese Paddlefish (*Psephurus gladius*), Dabry´s Sturgeon (*Acipenser dabryanus*) and other fish species still be saved? Journal of Applied Ichthyology 22:32–39.

Fang, J., Z. Wang, S. Zhao, Y. Li, Z. Tang, D. Yu, L. Ni, H. Liu, P. Xie, L. Da, Z. Li, and C. Zheng. 2006. Biodiversity changes in the lakes of the central Yangtze. Frontiers in Ecology and the Environment 4:369–377.

Fang, J. Y., S. Rao, and S. Q. Zhao. 2005. Human-induced long-term changes in the lakes of the Jianghan plain, central Yangtze. Frontiers in Ecology and the Environment 3:186–192.

Fu, C., J. Wu, J. Chen, Q. Wu, and G. Lei. 2003. Freshwater fish biodiversity in the Yangtze River basin of China: patterns, threats and conservation. Biodiversity and Conservation 12:1649–1685.

Gemmer, M., T. Jiang, B. Su, and Z. W. Kundzewicz. 2008. Seasonal precipitation changes in the wet season and their influence on flood/drought hazards in the Yangtze River basin, China. Quaternary International 186:12–21.

Li, H.Q. 2011. Construction status quo and eco-protection of natural conservation areas in the Changjiang River basin. Resources and Environment in the Yangtze Basin 20:150–155. (In Chinese.)

Liu, S. P., X. B. Duan, D. Q. Chen, F. C. Liao, and W. J. Chen. 2005. Studies on status of fishery resources in the middle reach of the Yangtze River. Acta Hydrobiologica Sinica 29:708–711. (In Chinese.)

Milliman, J. D., and J. P. M. Syvitski. 1992. Geomorphic/tectonic control of sediment discharge to the ocean: the importance of small mountainous rivers. Journal of Geology 100:525–544.

Su, B. D., T. Jiang, and W. B. Jin. 2006. Recent trends in observed temperature and precipitation extremes in the Yangtze River basin, China Theoretical and Applied Climatology 83:139–151.

Thorbjarnarson, J., and X. Wang. 1999. The conservation status of the Chinese alligator. Oryx 33:152–159.

Tong, Z.J. and Q.W. Han. 1982. River channel morphology and evolution below Danjiangkou Dam, Hangjiang River. Hydrological Bureau of Changjiang River Planning Office, Wuhan, China. (In Chinese.)

Wang, K. Y. 1998. The management and exploitation of Dongting Lake. Hunan People's Publishing House, Changsha, China. (In Chinese.)

Wang, S., and J. Jiang. 1998. History of Chinese Lakes. Science Press, Beijing. (In Chinese.)

Wang, X. M., K. J. Zhang, Z. H. Wang, Y. Z. Ding, W. Wu, and S. Huang. 2004. The decline of the Chinese giant salamander *Andrias davidianus* and implications for its conservation. Oryx 38:197–202.

Wang, Z., J. H. W. Lee, and D. Cheng. 2005. Impacts of the TGP project on the Yangtze River ecology and management strategies. International Journal of River Basin Management 3:237–246.

Wei, Q., F. Ke, J. Zhang, P. Zhuang, J. Lou, R. Zhou, and W. Wang. 1997. Biology, fisheries, and conservation of sturgeons and paddlefish in China. Environmental Biology of Fishes 48:241–255.

Xie, P., and Y. Chen. 1999. Threats to biodiversity in Chinese inland waters. Ambio 28:674–681.

Xie, Y., Z. Li, W.P. Gregg, and D. Li. 2001. Invasive species in China: an overview. Biodiversity and Conversation 10:1317–1341.

Xu, K., and J. D. Milliman. 2009. Seasonal variations of sediment discharge from the Yangtze River before and after impoundment of the Three Gorges Dam. Geomorphology 104:276–283.

Yang, G., L. Weng, and L. Li. 2008. Yangtze conservation and development report, 2007. Science Press, Beijing.

Yang, G., C. Zhu, and Z. Jiang. 2011. Yangtze conservation and development report, 2011. Yangtze Press, Wuhan, China. (In Chinese.)

Yang, G. F., Z. Chen, F. L. Yu, Z. H. Wang, Y. W. Zhao, and Z. Q. Wang. 2007. Yangtze River of

China hydromorphological study: sediment rating parameters and implications. Geomorphology 85:166–175.

Yang, Z., H. Wang, Y. Saito, J. D. Milliman, K. Xu, S. Qiao, and G. Shi. 2006. Dam impacts on the Changjiang (Yangtze) River sediment discharge to the sea: the past 55 years and after the Three Gorges Dam. Water Resources Research [online serial] 42:W04407.

Yu, F., Z. Chen, X. Ren, and Y. G. Yang. 2009. Analysis of historical floods on the Yangtze River, China: characteristics and explanations Geomorphology 113:210–216.

Yu, X. D., T. H. Luo, and H. Z. Zhou. 2005. A research on the large-scale pattern for the species diversity of fishes in the Yangtze basin. Biodiversity Science 13:473–495. [In Chinese.]

Yue, P., and Y. Chen. 1998. China red data book of endangered animals: pisces. Science Press, Beijing. (In Chinese.)

Zeng, X. C. 1990. Fishery resources of the Yangtze River basin. Marine Press, Beijing. (In Chinese.)

Zhang, G., W. Cao, and Y. Chen. 1997. Effects of fish stocking on lake ecosystems in China. Acta Hydrobiologica Sinica 21:271–280. (In Chinese.)

Zhang, H., Q. W. Wei, D. G. Yang, H. Du, H. J. Zhang, and X. H. Chen. 2008. Analysis of river habitat diversity based on current velocity gradients: example from the core reach of Chinese Sturgeon Nature Reserve of the Yangtze River in Yichang, Hubei Province. Chinese Journal of Ecology 27:667–674. (In Chinese.)

Zhang, X., D. Wang, R. Liu, Z. Wei, Y. Hua, Y. Wang, Z. Chen, and L. Wang. 2003. The Yangtze River dolphin or baiji (*Lipotes vexillifer*): population status and conservation issues in the Yangtze River, China. Aquatic Conservation: Marine and Freshwater Ecosystems 13:51–64.

Zhao, S., J. Fang, S. Miao, B. Gu, S. Tao, C. Peng, and Z. Tang. 2005. The 7-decade degradation of a large freshwater lake in central Yangtze River, China. Environmental Science and Technology 39:431–36.

Zhao, S. Q., J. Y. Fang, W. Ji, and Z. Y. Tang. 2003. Lake restoration from impoldering: impact of land conversion on riparian landscape in Honghu Lake area, central Yangtze. Agriculture, Ecosystems and Environment 95:111–118.

Zhao, X., J. Barlow, B. L. Taylor, R. L. Pitman, K. Wang, Z. Wei, B. S. Stewart, S. T. Turvey, T. Akamatsu, R. R. Reeves, and D. Wang. 2008. Abundance and conservation status of the Yangtze finless porpoise in the Yangtze River, China. Biological Conservation 141:3006–3018.

Zhou, W. 2000. A preliminary study on endangerment mechanism of freshwater fish species in wetland ecosystem of Yunnan. Biodiversity Science 8:163–168.

American Fisheries Society Symposium 84:53–77, 2016

Fishes of the Mississippi River

Harold L. Schramm, Jr.*
U.S. Geological Survey, Mississippi Cooperative Fish and Wildlife Research Unit
Mississippi State University, Mail Stop 9691, Mississippi State, Mississippi 39762, USA

Jay T. Hatch
James Ford Bell Museum of Natural History, University of Minnesota
10 Church Street SE, Minneapolis, Minnesota 55455, USA

Robert A. Hrabik
Missouri Department of Conservation
2206 West St. Joseph Street, Perryville, Missouri 63775, USA

William T. Slack
U.S. Army Engineer Research and Development Center, Waterways Experiment Station EEA
3909 Halls Ferry Road, Vicksburg, Mississippi 39180, USA

Abstract.—This chapter provides a listing of fishes known to be present in the Mississippi River from the headwaters at Lake Itasca, Minnesota to the Gulf of Mexico terminus. A total of 188 species are presently known from the Mississippi River, including 3 diadromous species and 17 nonnative species that have established self-sustaining populations in one or more reaches of the Mississippi River. Species are classified into three relative abundance categories and noted as residents, peripherals, introduced (established nonnatives), or strays (introduced but not established). The diversity of fishes varies longitudinally with 78 species in the reach from the headwaters to St. Anthony Falls; 113 and 105 species in the upper (Upper St. Anthony Lock and Dam to Pool 13) and lower (Pools 14–26) impounded reaches, respectively; and 121 and 136 species in the upper (Missouri River confluence to Ohio River confluence) and lower (Ohio River to the Gulf of Mexico outlet) free-flowing reaches, respectively. Although the composition of the Mississippi River fish fauna has changed little despite more than 80 years of habitat alteration, the extirpation of five species in individual reaches of the river may be a forewarning of a system losing resiliency and indicate the need for habitat conservation and rehabilitation to conserve the biodiversity of North America's largest river.

Introduction

The Mississippi River courses 3,700 km from its source at Lake Itasca, Minnesota, USA to its outlet approximately 160 km downstream of New Orleans, Louisiana. Although highly altered geomorphologically and hydrologically throughout much of its length downstream of the headwaters (the upper 824 km from Lake Itasca, Minnesota to St. Anthony Falls, Minnesota), the Mississippi River still provides diverse fish habitat (Schramm and Ickes 2016, this volume). As would be expected of a river with a watershed that stretches from the Appalachian Mountains in the east to the Rocky Mountains in the west, the Mississippi River is home to a diverse fish fauna. Of the approximately 1,200 species of freshwater fishes known from the inland waters

* Corresponding author: hschramm@usgs.gov

of continental North and Central America (Walsh et al. 2009), 188 are part-time or full-time residents of the Mississippi River (excluding marine fishes of occasional occurrence near the mouth). A comprehensive assessment of the Mississippi River compiled by Fremling et al. (1989) listed 193 species in 27 families. Schramm (2004) updated the Fremling et al. (1989) list to include fishes of known or expected occurrence based on more recent compilations by Baker et al. (1991), Pitlo et al. (1995), and Warren et al. (2000). Excluding strays and marine species restricted to the lower 150 km, Schramm (2004) reported 140 resident species. This chapter presents an assessment of the current ichthyofauna of the Mississippi River.

Methods

The assessment reported in this paper began with the lists of Mississippi River fishes provided by Fremling et al. (1989) and Schramm (2004). Data used to describe the fishes in the Minnesota portion of the Mississippi River (headwaters and upper impounded reaches) were obtained from Eddy and Underhill (1974), Phillips et al. (1982), Hatch et al. (2003), Dickson and Tomelleri (2008), Schmidt and Proulx (2009), collections of the James Ford Bell Museum of Natural History, and the Fishes of Minnesota Mapper (www.dnr.state.mn.us/maps/fom/index.html). Data used to describe fish distribution and abundance in Mississippi River reaches along Missouri's border were obtained from the Missouri Department of Conservation Fish Community and Heritage databases, Pflieger (1997), and the U.S. Geological Survey Long Term Resource Monitoring Program database (www.umesc.usgs.gov/data_library/fisheries/graphical/fish_front.html). In addition, data assembled from databases maintained by the Illinois Department of Natural Resources, Illinois Natural History Survey, U.S. Army Corps of Engineers, U.S. Fish and Wildlife Service, and U.S. Geological Survey used to describe fish distribution in Missouri in preparation for the third edition of *Fishes of Missouri* were also consulted. Data for fish distribution in the upper and lower free-flowing reaches were obtained from Etnier and Starnes (1993), Lee et al. (1980), Mississippi Museum of Natural Science Ichthyology Collection records, Robison and Buchanan (1988), and Ross (2001). Records were updated with current published information, state and regional databases, museum records, examination of museum collections, recent collection efforts of the authors, and the personal knowledge of a team of reviewers comprised of eight additional ichthyologists and fisheries biologists with extensive experience working on the Mississippi River (listed in Acknowledgments). Nomenclature follows *Common and Scientific Names of Fishes from the United States, Canada, and Mexico, 7th Edition* (Page et al. 2013). Our assessment of Mississippi River fishes is for the Mississippi River floodplain ecosystem (i.e., the mainstem river and the contiguous floodplain). We provide separate lists of species for five reaches of the Mississippi River as described in Schramm and Ickes (2016): headwaters (Lake Itasca, Minnesota to Upper St. Anthony Falls Lock and Dam, Minnesota), upper impounded (Lock and Dam Pools 1–13), lower impounded (Lock and Dam Pools 14–26), upper free-flowing (Missouri River confluence to Ohio River confluence), and lower free-flowing reaches (Ohio River confluence to Gulf of Mexico; Table 1).

Relative abundance of species is meaningful in many ways, including assessing changes in abundance spatially (e.g., among river reaches, upstream and downstream of distribution barriers) and temporally (e.g., since alteration, since establishment of a nonnative species). We assigned relative abundance categories to species by reach following modification of the five-tier abundance scale (abundant, common, occasional, uncommon, and rare) used by Fremling et al. (1989) and adopted by Schramm (2004). Assigning abundance categories, however, is laden with problems. Quantitative estimates of fish abundance are not available for all fishes or for all reaches. Furthermore, even with standardized sampling as conducted in the upper and lower impounded reaches and the

Table 1. Relative abundance and backwater and fluvial dependence of fishes in the headwaters, impounded, and free-flowing reaches of the Mississippi River. Abundance categories are abundant (A; usually present in relatively large numbers in samples collected in appropriate habitats and with appropriate sampling gear), common (C; usually present in moderate numbers in samples collected in appropriate habitats and with appropriate sampling gear), uncommon (U; infrequently or rarely collected in samples collected with appropriate sampling gear in any habitat), and extirpated (E; absent for at least 50 years from collection efforts deemed sufficient to detect the species). Parenthetical designations indicate diadromous (D), introduced (I; a species not native to this reach of the Mississippi River that has established a self-sustaining population), peripheral (P; a species native to the Mississippi River basin that periodically enters this reach of the river from established populations in tributaries or other waters naturally connected to the Mississippi River but does not usually occupy the Mississippi River during one or more phases of its life cycle), or stray (S; an introduced species that is not established in the reach). Species lacking parenthetical designation are considered native and resident to the Mississippi River. Backwater-dependent fish are those that require backwater/lentic habitat for one or more life stages of life functions, and fluvial-dependent fish are those that require lotic habitat for one or more life stages or life functions.

Family Species	Reach Headwaters	 Upper impounded	 Lower impounded	 Upper free-flowing	 Lower free-flowing	Backwater dependent	Fluvial dependent
Petromyzontidae							
Chestnut Lamprey *Ichthyomyzon castaneus*		U	C (P)	C (P)	C (P)		Yes
Silver Lamprey *I. unicuspis*		U	C (P)	U (P)	U (P)		Yes
American Brook Lamprey *Lethenteron appendix*		U (P)					Yes
Carcharhinidae							
Bull Shark *Carcharhinus leucas*				U (P)			
Acipenseridae							
Lake Sturgeon *Acipenser fulvescens*	U (S)	U	U	U	U		Yes
Gulf Sturgeon *A. oxyrinchus desotoi*					U (D)		Yes
Pallid Sturgeon *Scaphirhynchus albus*			U	U	U		Yes
Shovelnose Sturgeon *S. platorynchus*		C	A	C	A		Yes
Polyodontidae							
Paddlefish *Polyodon spathula*		U	C	C	C		Yes
Lepisosteidae							
Alligator Gar *Atractosteus spatula*			E	U	U	Yes	
Spotted Gar *Lepisosteus oculatus*[a]		U	U	U	U	Yes	
Longnose Gar *L. osseus*		U	C	C	C	Yes	
Shortnose Gar *L. platostomus*	U (P)	C	C	C	C	Yes	
Amiidae							
Bowfin *Amia calva*	U	C	U	U	U	Yes	

Table 1. Continued.

Family Species	Reach: Headwaters	Upper impounded	Lower impounded	Upper free-flowing	Lower free-flowing	Backwater dependent	Fluvial dependent
Hiodontidae							
Goldeye *Hiodon alosoides*		U	U	C	C		
Mooneye *H. tergisus*	U (S)	U	C	U	U		Yes
Anguillidae							
American Eel *Anguilla rostrata*	U (D)	U (D)	U (D)	U(D)	U (D)		
Clupeidae							
Alabama Shad *Alosa alabamae*			E	U (D)	U (D)		Yes
Skipjack Herring *A. chrysochloris*		U	U	C	C		Yes
Gizzard Shad *Dorosoma cepedianum*	U (S)	A	A	A	A	Yes	
Threadfin Shad *D. petenense*				U	U	Yes	
Cyprinidae							
Central Stoneroller *Campostoma anomalum*	U (P)	U (P)	U (P)	U (P)	U (P)		Yes
Goldfish *Carassius auratus*		U (I/P)	U (I)	U (I)	U (I)	Yes	
Northern Redbelly Dace *Chrosomus eos*	U					Yes	
Finescale Dace *C. neogaeus*	U (P)					Yes	
Grass Carp *Ctenopharyngodon idella*	U (S)	U (S)	U (I)	U (I)	U (I)		Yes
Bluntface Shiner *Cyprinella camura*[b]					U (P)		
Red Shiner *C. lutrensis*		U (P)	C	C	U		
Spotfin Shiner *C. spiloptera*	C	A	C	U			Yes
Blacktail Shiner *C. venusta*			U (P)	U	U		Yes
Steelcolor Shiner *C. whipplei*					U (P)		Yes
Common Carp *Cyprinus carpio*	U (I)	A (I)	A (I)	C (I)	U (I)	Yes	
Gravel Chub *Erimystax x-punctatus*				U			
Western Silvery Minnow *Hybognathus argyritis*				U	U		Yes
Brassy Minnow *H. hankinsoni*	U	U (P)				Yes	
Cypress Minnow *H. hayi*					U	Yes	
Mississippi Silvery Minnow *H. nuchalis*		U	U	C	C		Yes
Plains Minnow *H. placitus*				U	U		Yes

Table 1. Continued.

Family Species	Reach Headwaters	 Upper impounded	 Lower impounded	 Upper free-flowing	 Lower free-flowing	Backwater dependent	Fluvial dependent
Cyprinidae (continued)							
Pallid Shiner *Hybopsis amnis*		U	U	E	U	Yes	
Clear Chub *H. winchelli*					U (P)		
Silver Carp *Hypophthalmichthys molitrix*		U (S)	C (I)	C (I)	C (I)		Yes
Bighead Carp *H. nobilis*		U (S)	U (I)	U (I)	U (I)		Yes
Striped Shiner *Luxilus chrysocephalus*[c]				U(P)	U (P)		Yes
Common Shiner *L. cornutus*	C	U (P)					Yes
Ribbon Shiner *Lythrurus fumeus*[t]					U(P)	Yes	
Redfin Shiner *L. umbratilis*[b]			U (P)	U (P)			
Sturgeon Chub *Macrhybopsis gelida*				U	U		Yes
Shoal Chub *M. hyostoma*		U	C	A	C		Yes
Sicklefin Chub *M. meeki*				U	U		Yes
Silver Chub *M. storeriana*		C	C	C	C		Yes
Hornyhead Chub *Nocomis biguttatus*	U	U (P)					Yes
Golden Shiner *Notemigonus crysoleucas*	U	C	U	U (P)	U (P)	Yes	
Pugnose Shiner *Notropis anogenus*	U						
Emerald Shiner *N. atherinoides*	U	A	A	A	A		Yes
River Shiner *N. blennius*		C	C	C	C		Yes
Bigeye Shiner *N. boops*			U (P)	U (P)	U (P)		Yes
Silverjaw Minnow *N. buccatus*				U (P)			
Ghost Shiner *N. buchanani*		E	U	U	U		Yes
Bigmouth Shiner *N. dorsalis*	U	U (P)	U	U (P)			Yes
Blackchin Shiner *N. heterodon*	U					Yes	
Blacknose Shiner *N. heterolepis*	U					Yes	
Spottail Shiner *N. hudsonius*	U	C	C	U (P)	U (P)		Yes
Longnose Shiner *N. longirostris*					U (P)		Yes
Taillight Shiner *N. maculatus*					U (P)	Yes	
Ozark Minnow *N. nubilus*				U (P)			Yes

Table 1. Continued.

Family Species	Reach: Headwaters	Reach: Upper impounded	Reach: Lower impounded	Reach: Upper free-flowing	Reach: Lower free-flowing	Backwater dependent	Fluvial dependent
Cyprinidae (continued)							
Chub Shiner *N. potteri*					U (P)		Yes
Silverband Shiner *N. shumardi*			U	U	C		Yes
Sand Shiner *N. stramineus*	U	U (P)	U (P)	U (P)			Yes
Weed Shiner *N. texanus*		C			U	Yes	
Mimic Shiner *N. volucellus*[d]	U	U	U	U	U		Yes
Channel Shiner *N. wickliffi*[d]	U (S)	A	A	A	A		Yes
Pugnose Minnow *Opsopoeodus emiliae*		U	U	U (P)	U (P)	Yes	
Suckermouth Minnow *Phenacobius mirabilis*[e]		U(P)	U (P)	U (P)	U (P)		Yes
Bluntnose Minnow *Pimephales notatus*	U	U	U	U (P)	U (P)		Yes
Fathead Minnow *P. promelas*	U	U	U	U (P)	U (P)		
Bullhead Minnow *P. vigilax*	U (I)	C	C	C	C		Yes
Flathead Chub *Platygobio gracilis*				U	U		Yes
Blacknose Dace *Rhinichthys atratulus*	U	U (P)					Yes
Longnose Dace *R. cataractae*	U	U (P)					Yes
Creek Chub *Semotilus atromaculatus*	U	U (P)	U (P)	U (P)	U (P)		Yes
Catostomidae							
River Carpsucker *Carpiodes carpio*	U (S)	C	A	A	A		Yes
Quillback *C. cyprinus*	U (S)	C	C	U	U		Yes
Highfin Carpsucker *C. velifer*		U	U	U	U		Yes
Longnose Sucker *Catostomus catostomus*			U (S)				
White Sucker *C. commersonii*	C	U	U	U			Yes
Blue Sucker *Cycleptus elongatus*		U	U	U	U		Yes
Northern Hog Sucker *Hypentelium nigricans*	U (I)	U		U (P)			Yes
Smallmouth Buffalo *Ictiobus bubalus*		C	A	A	A	Yes	
Bigmouth Buffalo *I. cyprinellus*	U	U	C	C	C	Yes	

Table 1. Continued.

Family Species	Reach					Backwater dependent	Fluvial dependent
	Headwaters	Upper impounded	Lower impounded	Upper free-flowing	Lower free-flowing		
Catostomidae (continued)							
Black Buffalo *I. niger*		U	U	U	U	Yes	
Spotted Sucker *Minytrema melanops*		A	U		U		Yes
Silver Redhorse *Moxostoma anisurum*	U	C	U	U (P)			Yes
River Redhorse *M. carinatum*		C		U (P)			Yes
Golden Redhorse *M. erythrurum*		U	U	U (P)	U (P)		Yes
Shorthead Redhorse *M. macrolepidotum*[f]	C	A	C	U	U		Yes
Greater Redhorse *M. valenciennesi*	U	U					Yes
Ictaluridae							
White Catfish *Ameiurus catus*				U (S)			
Black Bullhead *A. melas*	U	U	U	U	U	Yes	
Yellow Bullhead *A. natalis*	U	U	U	U	U	Yes	
Brown Bullhead *A. nebulosus*	U	U	E		U (P)	Yes	
Blue Catfish *Ictalurus furcatus*			U	A	A		
Channel Catfish *I. punctatus*	U (I)	A	A	A	C		
Mountain Madtom *Noturus eleutherus*					U (P)		Yes
Stonecat *N. flavus*	U (S)	U	U	U	U		Yes
Piebald Madtom *N. gladiator*[g]					U (P)		Yes
Tadpole Madtom *N. gyrinus*	U	U	U	U	U	Yes	
Freckled Madtom *N. nocturnus*			U	U	U		Yes
Flathead Catfish *Pylodictis olivaris*	U (S)	C	A	A	A		
Osmeridae							
Rainbow Smelt *Osmerus mordax*	U (I)	U (S)		U (S)	U (S)		
Salmonidae							
Cisco *Coregonus artedi*	C						
Lake Whitefish *C. clupeaformis*	U						
Rainbow Trout *Oncorhynchus mykiss*		U (I/P)	U (I/P)		U (S)		Yes
Brown Trout *Salmo trutta*		U (I/P)	U (I/P)				
Brook Trout *Salvelinus fontinalis*	U (I/P)						

Table 1. Continued.

Family Species	Reach: Headwaters	Upper impounded	Lower impounded	Upper free-flowing	Lower free-flowing	Backwater dependent	Fluvial dependent
Esocidae							
Grass Pickerel *Esox americanus vermiculatus*		U	U		U (P)	Yes	
Northern Pike *E. lucius*	C	C	U	U (P)		Yes	
Muskellunge *E. masquinongy*	U	U (P)	U (P)	U (S)		Yes	
Chain Pickerel *E. niger*					U(P)	Yes	
Central Mudminnow *Umbra limi*	U	U	U			Yes	
Percopsidae							
Trout-perch *Percopsis omiscomaycus*	U	U	U	U (P)			Yes
Aphredoderidae							
Pirate Perch *Aphredoderus sayanus*		U	U		U(P)	Yes	
Gadidae							
Burbot *Lota lota*	U	U	U (P)				Yes
Mugilidae							
Striped Mullet *Mugil cephalus*				U (P)	U (P)		Yes
Atherinopsidae							
Brook Silverside *Labidesthes sicculus*	U	U	U	U	U		Yes
Mississippi Silverside *Menidia audens*				U	C		Yes
Belonidae							
Atlantic Needlefish *Strongylura marina*					U (P)		
Fundulidae							
Golden Topminnow *Fundulus chrysotus*					U (P)	Yes	
Banded Killifish *F. diaphanus*[h]	U	U(P)				Yes	
Starhead Topminnow *F. dispar*		U (P)	U		U (P)	Yes	
Blackstripe Topminnow *F. notatus*			U	U	U (P)	Yes	
Blackspotted Topminnow *F. olivaceus*				U (P)	U (P)	Yes	
Rainwater Killifish *Lucania parva*					U (P)	Yes	

Table 1. Continued.

Family Species	Reach					Backwater dependent	Fluvial dependent
	Headwaters	Upper impounded	Lower impounded	Upper free-flowing	Lower free-flowing		
Poeciliidae							
Western Mosquitofish *Gambusia affinis*		U (I)	U	C	U	Yes	
Least Killifish *Heterandria formosa*					U (P)	Yes	
Gasterosteidae							
Brook Stickleback *Culaea inconstans*	U	U (P)				Yes	
Ninespine Stickleback *Pungitius pungitius*	U					Yes	
Syngnathidae							
Gulf Pipefish *Syngnathus scovelli*					U (P)		
Cottidae							
Mottled Sculpin *Cottus bairdi*	U						Yes
Moronidae							
White Bass *Morone chrysops*	U(S)	A	C	C	C		Yes
Yellow Bass *M. mississippiensis*		U	U	U	U		
Striped Bass *M. saxatilis*				U (S)	U (S)		Yes
Centrarchidae							
Shadow Bass *Ambloplites ariommus*				U (P)		Yes	
Rock Bass *A. rupestris*	C	C	U			Yes	
Flier *Centrarchus macropterus*					U (P)	Yes	
Green Sunfish *Lepomis cyanellus*	U	U	C	U	U	Yes	
Pumpkinseed *L. gibbosus*	C	U	U			Yes	
Warmouth *L. gulosus*		U	U	U	U	Yes	
Orangespotted Sunfish *L. humilis*	U (I)	C	C	C	U	Yes	
Bluegill *L. macrochirus*	C	A	A	C	C	Yes	
Dollar Sunfish *L. marginatus*					U (P)	Yes	
Longear Sunfish *L. megalotis*			U	U (P)	U	Yes	
Redear Sunfish *L. microlophus*					U	Yes	
Redspotted Sunfish *L. miniatus*					U	Yes	
Bantam Sunfish *L. symmetricus*					U	Yes	
Smallmouth Bass *Micropterus dolomieu*	C	C	U	U (P)			

Table 1. Continued.

Family Species	Reach Headwaters	Upper impounded	Lower impounded	Upper free-flowing	Lower free-flowing	Backwater dependent	Fluvial dependent
Centrarchidae (continued)							
Spotted Bass *M. punctulatus*				U (P)	U (P)	Yes	
Largemouth Bass *M. salmoides*	C	A	A	U	U	Yes	
White Crappie *Pomoxis annularis*	U (I/P)	C	A	C	C	Yes	
Black Crappie *P. nigromaculatus*	C	A	A	U	U	Yes	
Percidae							
Western Sand Darter *Ammocrypta clara*		U	U	U	U		Yes
Crystal Darter *Crystallaria asprella*		U	U	U	U		Yes
Mud Darter *Etheostoma asprigene*		C	U	U	U	Yes	
Rainbow Darter *E. caeruleum*		U (P)		U (P)	U (P)		Yes
Bluntnose Darter *E. chlorosoma*		U	U (P)	U (P)	U (P)	Yes	
Meramec Saddled Darter *E. erythrozonum*				U (P)			Yes
Iowa Darter *E. exile*	U	U (P)				Yes	
Fantail Darter *E. flabellare*		U (P)		U (P)			Yes
Swamp Darter *E. fusiforme*					U (P)	Yes	
Slough Darter *E. gracile*					U	Yes	
Harlequin Darter *E. histrio*					U		Yes
Least Darter *E. microperca*	U (P)					Yes	
Johnny Darter *E. nigrum*	C	C	C	U (P)	U (P)		Yes
Goldstripe Darter *E. parvipinne*					U (P)		
Cypress Darter *E. proeliare*					U (P)	Yes	
Orangethroat Darter *E. spectabile*				U (P)			Yes
Banded Darter *E. zonale*		U					Yes
Yellow Perch *Perca flavescens*	C	A	U		U (P)	Yes	
Logperch *Percina caprodes*	U	C	U	U (P)	U		
Gilt Darter *P. evides*				U (P)			Yes
Blackside Darter *P. maculata*[i]	U (I)	U					Yes
Slenderhead Darter *P. phoxocephala*		U	U	U (P)			Yes
Dusky Darter *P. sciera*				U (P)	U (P)		
River Darter *P. shumardi*		U	U	U	U		Yes
Saddleback Darter *P. vigil*					U (P)		Yes

Table 1. Continued.

Family Species	Reach					Backwater dependent	Fluvial dependent
	Headwaters	Upper impounded	Lower impounded	Upper free-flowing	Lower free-flowing		
Percidae (continued)							
Sauger *Sander canadensis*		C	C	U	U		Yes
Walleye *S. vitreus*	C	C	C	U	U		Yes
Sciaenidae							
Freshwater Drum *Aplodinotus grunniens*	U (I)	A	A	A	A		Yes
Elassomatidae							
Banded Pygmy Sunfish *Elassoma zonatum*					U	Yes	
Total species	78	113	105	121	136		
Total peripheral species	6	22	14	40	47		
Total diadromous species	1	1	1	2	3		
Total introduced resident (not peripheral) species	10	2	5	5	5		
Total strays	10	4	1	4	3		
Total resident native species extirpated	0	1	3	1	0		
Resident fluvial-dependent species	26	45	45	44	45		
Resident backwater-dependent species	27	34	35	25	33		
Resident fluvial-dependent species (native only)	22	45	42	42	42		
Resident backwater-dependent species (native only)	25	32	33	23	31		

[a] Spotted Gar, upper impounded reach: a specimen collected in Iowa was verified as a Spotted Gar. Subsequently, Sipiorski (2011) demonstrated that hybrids between Longnose Gar and Shortnose Gar produce a gar with distinct spots on the head and snout that could be misidentified as a Spotted Gar.
[b] Bluntface Shiner, Ribbon Shiner, and Redfin Shiner: no records of recent collection since Warren et al. (2000), but these species would be peripheral if collected in the Mississippi River; thus, we do not note as extirpated.
[c] Striped Shiner, lower free-flowing reach: no records of collection since 1978 (Fremling et al. 1989), but this species would be a peripheral if collected in the Mississippi River; thus, we do not note as extirpated.
[d] The taxonomy of Channel Shiner and Mimic Shiner is uncertain. The joint committee on names of fishes (American Fisheries Society and American Society of Ichthyologists and Herpetologists) has retained Channel Shiner in *Common and Scientific Names of Fishes from the United States, Canada, and Mexico* (Page et al. 2013). Page and Burr (2011) did not recognize Channel Shiner, citing inconsistent descriptions of populations across its range. Based on Hrabik (1997), we recognize Channel Shiner as a valid taxon and have included it here, but recent and preliminary (unpublished) genetic studies

Table 1. Continued.

have suggested that populations in the upper Mississippi River are highly introgressed with Mimic Shiner and possibly other species.

[e] Suckermouth Minnow, upper impounded reach: last valid collection record is 1953; not listed as extirpated because considered peripheral.

[f] Pealip Redhorse *Moxostoma pisolabrum* was elevated from synonymy with Shorthead Redhorse and it was suggested that *M. pisolabrum* and possible hybrids occur in the upper and lower free-flowing Mississippi River (Smith 1979). We recognize that Shorthead Redhorse are present in the upper and lower free-flowing reach, but data are not available to assess abundance of Pealip Redhorse at this time. We list abundance only for Shorthead Redhorse, aware that some of these individuals may be Pealip Redhorse in the free-flowing reaches.

[g] Northern Madtom *Noturus stigmosus* was reported as a stray by Fremling et al. 1989) and present by Warren et al. (2000) in the free-flowing reach but not collected; thus, we do not note as extirpated. Northern Madtom from Tennessee is now the Piebald Madtom (Thomas and Burr 2004). Piebald Madtom has been captured rarely in the Mississippi River near the mouth of the Hatchie River (H. W. Robison, Southern Arkansas University, retired, personal communication).

[h] Banded Killifish, upper impounded reach: no collection records since 1960, but this species would be a peripheral if collected in the Mississippi River; thus, we do not note as extirpated.

upper free-flowing reach as part of the Long Term Resource Monitoring Program (www.umesc.usgs.gov/ltrmp.html), it is not possible to compare abundance of fishes collected with different gears. In addition, abundance is expected to differ across trophic-level and fish-size spectra. For example, 10 individuals of a small cyprinid per 100 m^2 seine haul might be a relatively low abundance for that taxon, whereas four large catfish per 100 m^2 electrofishing or trawl transect might be considered a relatively high abundance for that taxon, size-class, and trophic level. In other words, abundance must be judged relative to expected abundance for a life stage and trophic level. Also, few, if any, species in the Mississippi River occur equally in all habitats, so habitat must be considered. Given the need for subjective categorization of abundance, we have adopted a simpler, three-tier relative abundance scale: abundant (usually present in relatively large numbers in samples collected in appropriate habitats [i.e., where the species and life stage is expected to live] and with appropriate sampling gear [i.e., a gear that effectively captures the species]), common (usually present in moderate numbers in samples collected in appropriate habitats and with appropriate sampling gear), and uncommon (infrequently or rarely collected in samples collected with appropriate sampling gear in any habitat; Table 1).

Although we have attempted to provide informative and comparable estimates of abundance of the fishes, a problem remains. The relative abundances we have assigned Mississippi River fishes are more or less a reflection of catch rates with standard or typically used large-river sampling gears used to assess fish populations and monitor fish assemblages (e.g., boat-mounted electrofishing, hoop netting, gill netting, seining, etc.). However, some taxa may be more abundant than generally known if researchers and managers were to employ nontraditional sampling methodologies. Sampling with benthic trawls (e.g., the Mini-Missouri trawl; Herzog et al. 2005, 2009) to capture small-bodied benthic fishes is one such example, and use of this method is becoming more common. Although not a recommended sampling method, greater abundances of madtoms (e.g., Stonecat and Freckled Madtom), Banded Pygmy Sunfish, and darters (e.g., Harlequin Darter) have been obtained by sampling rock wing dikes with sodium cyanide than by conventional sampling gears (D. Etnier, University of Tennessee, retired, personal communication). Further, rock baskets set in seemingly homogeneous sandy flats overnight are sometimes replete with Freckled Madtoms (R. A. Hrabik, personal observation), suggesting that madtoms emerge from cover and roam areas typically not thought to be preferred habitat for that taxonomic group. As a last example, trotlines have been found to effectively capture river sturgeons *Scaphirhynchus* spp., including the endangered Pallid Sturgeon (Killgore et al. 2007), and catch rate of river sturgeon is similar to catch rate of catfishes *Ictalurus* spp. (DeVries et al. 2015), particularly Blue Catfish, which are considered abundant based on catches with electrofishing and hoop nets. We have considered catch rate information from nontraditional gears in our assignment of abundance categories as known. We also note that our reach-specific assigned abundance categories were subject to review by the team of reviewers, and all suggested changes in abundance were fully considered.

We classified individual species as diadromous, peripheral (a species native to the Mississippi River basin that periodically enters a particular reach of the river from established populations in tributaries or other waters naturally connected to the Mississippi River but does not usually occupy the Mississippi River during one or more phases of its life cycle), introduced (a species not native to this reach of the Mississippi River that has established a self-sustaining population), or stray (an introduced species that is not established in the reach). Note that, as abundance categories, these classifications are reach specific. These classifications were based on interpretation of published information and the authors' and reviewers' personal observations and knowledge.

We have included in Table 1 only those

species that are supported by valid collection records in the past 50 years. Similarly, species noted as extirpated are only those considered residents (i.e., not peripheral) and of known (validated) former occurrence that have not been collected for at least 50 years. A shorter time period for judging continued presence certainly would be desirable, but in light of the lack of frequent and standardized sampling riverwide, 50 years was selected to provide reasonable certainty of extirpation.

Fishes were classified as fluvial dependent or backwater dependent if available life history information or the authors' personal knowledge were sufficient to make reliable judgments. Fluvial-dependent fishes are those that require flowing water (lotic conditions) for one or more life stages to complete their life cycles and are equivalent to rheolimnophilic (Poddubny and Galat 1995) and rheophilic B (Schiemer and Waidbacher 1992) fishes. Backwater-dependent fishes are those that require standing or nonflowing water (limnetic conditions) for one or more life stages to complete their life cycle. Fishes lacking fluvial- or backwater-dependent classification should not be inferred to be habitat generalists (i.e., able to complete their life cycle in limnetic or lotic conditions), as information for many of these species was insufficient to classify as either fluvial- or backwater-dependent.

Results

A total of 188 species of fishes in 31 families are known to occur in the Mississippi River (Table 1). Forty-four of the 188 species were considered peripherals in all reaches where they occur.

Thirteen species are considered strays in one or more reaches where they occur. White Catfish is considered a stray in the only reach where collected (upper free-flowing reach). Rainbow Smelt is a stray in the impounded and free-flowing reaches. The source of the Rainbow Smelt is presumed to be Missouri River impoundments where this nonnative species has become established, but it also could have originated from Lake Michigan via the Illinois River (Burr and Mayden 1980). Rainbow Smelt most likely entered the headwaters reach through human transfer and is now established in that reach (C. Kavanaugh, Minnesota Department of Natural Resources, personal communication). Rainbow Trout is considered peripheral in the impounded reaches where introduced populations have become established in tributary streams, but is considered a stray in the lower free-flowing reach where it likely becomes displaced from stocked streams within the basin. Striped Bass are listed as strays in the upper and lower free-flowing reaches. It is possible that some Striped Bass are migratory Gulf Coast Striped Bass, as this fish was known to be common in Mississippi coastal rivers (McIlwain 1967); however, Long et al. (2013) do not show Gulf Coast Striped Bass west of the Pontchartrain Basin, Louisiana, and this race is considered extirpated from most Gulf Coast streams. The Striped Bass reported from the upper and lower free-flowing reaches are probably Atlantic-strain fish that escaped from numerous reservoirs in the Mississippi River basin where this fish has been stocked. It is possible that these fish are now completing their life cycle within the Mississippi River and have established a population. Larval Striped Bass have been collected (Gallagher and Conner 1980; Schramm and Pennington 1981), but neither the strain nor recruitment has been assessed. Thus, we have classified Striped Bass as strays. Several species considered strays in the headwaters reach are indigenous in most or all other reaches: Lake Sturgeon, Mooneye, Gizzard Shad, Channel Shiner, River Carpsucker, Quillback, Stonecat, Flathead Catfish, and White Bass. These species are among a total of 13 that either have been introduced or have moved beyond the historic St. Anthony Falls barrier and into the headwaters since 1995 (Hatch et al. 2003) but for which we have no evidence of reproduction. Lower St. Anthony Falls Lock and Dam (completed in 1956) and Upper St. Anthony Falls Lock (completed in 1963; dam was completed in 1951; U.S. Army Corps of Engineers

2015) provided a means of passage through the St. Anthony Falls barrier.

Diadromous species include catadromous American Eel that migrate as far upriver as the headwaters, anadromous Gulf Sturgeon in the lower free-flowing reach, and Alabama Shad in the upper and lower free-flowing reaches. Coker (1930) reported the Alabama Shad from the lower impounded reach near Keokuk, Iowa, but this species has not been reported since; we regard Alabama Shad as extirpated in the lower impounded reach. Striped Bass are anadromous, but as noted above, individuals in the free-flowing reaches are probably strays, or progeny of strays, from tributary reservoirs.

Thirteen native species and one native species complex are resident in all five reaches: Bowfin, Emerald Shiner, Mimic Shiner/Channel Shiner complex (see Table 1, footnote d), Bigmouth Buffalo, Shorthead Redhorse, Black Bullhead, Yellow Bullhead, Tadpole Madtom, Brook Silverside, Green Sunfish, Bluegill, Largemouth Bass, Black Crappie, and Walleye. Twenty-two additional species occur in all reaches but are either peripheral, introduced, or strays in one or more reaches: Lake Sturgeon, Shortnose Gar, Mooneye, Gizzard Shad, Central Stoneroller, Grass Carp, Common Carp, Golden Shiner, Spottail Shiner, Bluntnose Minnow, Fathead Minnow, Bullhead Minnow, Creek Chub, River Carpsucker, Quillback, Channel Catfish, Flathead Catfish, White Bass, Orangespotted Sunfish, White Crappie, Johnny Darter, and Freshwater Drum.

Seventeen nonnative species have established self-sustaining populations in one or more reaches. Only the Common Carp is established in all reaches. Four species, all cyprinids, are established in the lower three reaches: Goldfish, Grass Carp, Silver Carp, and Bighead Carp. The latter two fishes escaped confinement, and reproducing populations have been documented in the lower three reaches. While the rapid increases in abundance of the Silver Carp and Bighead Carp support establishment of these fishes in the three lower reaches, the evidence for establishment is not as clear for Grass Carp. Grass Carp spawning has been documented in the lower free-flowing reach (Conner et al. 1980) and observed in tributaries to the upper free-flowing reach (R. Hrabik, personal observation), but juveniles and adults are seldom collected. Grass Carp are extensively used for aquatic vegetation control in public and private waters, so it is possible that the occasional collections of large Grass Carp are a result of escapement and not natural reproduction and recruitment. Nevertheless, based on documented spawning and collection of young of year, Grass Carp are considered established in the lower impounded and free-flowing reaches. Grass Carp have been collected in the headwaters and upper impounded reaches, but available information does not support self-sustaining populations in these two reaches. Similarly, Silver Carp and Bighead Carp are present in the upper impounded reach, but populations at this time are not considered self-sustaining.

The high incidence of introduced species (10) in the headwaters reach is mostly a result of species historically downstream of St. Anthony Falls, a natural distribution barrier, gaining access to the headwaters when navigation locks were completed in 1963 (Fremling et al. 1989). A dam at Coon Rapids, Minnesota, for a time appeared to block further upstream migration, but at least three of the upstream-invading species — Northern Hog Sucker, Orangespotted Sunfish, and Blackside Darter — have moved well beyond the dam. Channel Catfish have been intentionally stocked above the dam since 1963 and now occur in much of the lower half of the reach. The introduced Brook Trout and White Crappie are considered peripheral.

The diversity of fishes varies longitudinally. The headwaters reach is home to 78 confirmed fish species; of the 71 resident species (i.e., not diadromous or peripheral), 10 are introduced and 10 are strays (Table 1). The upper impounded reach has 113 confirmed species; the 90 species considered residents includes 2 introduced species (2 additional introduced species are peripheral) and 4 stray species. Fish species richness declines to 105 confirmed species in the lower impounded reach; 90 species are resident, 6 of which are introduced. Fish species richness increases in the upper and lower free-flowing reach. A total of 121 species are confirmed present in the upper free-flowing reach; the 79 resident species include 5 introduced species and 4 stray species. Greatest species richness occurs in the lower free-flowing reach where 136 confirmed species are present; however, only 86 of these species are resident, including 5 introduced species and 3 stray species. Despite longitudinal variation in the total number of species, the number of resident fishes is relatively similar among reaches, except for the headwaters.

Of 53 resident species (i.e., not diadromous or peripheral) in the headwaters reach for which sufficient information was available to assign habitat dependence, 26 (22 native) were fluvial-dependent and 27 (25 native) were backwater-dependent (Table 1). The majority of the resident species were fluvial-dependent in the impounded and free-flowing reaches. Interpretation of the frequency of fluvial- and backwater-dependent fishes is constrained by fishes of unknown habitat dependency (some of which may be able to complete their life cycle in lotic or lentic conditions), but the slight prevalence of native backwater-dependent species in the largely free-flowing headwaters reach and the prevalence of fluvial-dependent species in the impounded reaches seem counterintuitive. However, the headwaters reach has six reservoirs and flows through nine glacial lakes, two of which have low-head dams to stabilize water levels primarily for recreational and shoreline development benefits, in addition to a largely unaltered floodplain (Schramm and Ickes 2016, this volume). The navigation channel and much of the channel borders in the impounded reaches provide lotic conditions for fluvial-dependent fishes in these reaches. Thus, expansive lentic habitats exist in the headwaters reach and lotic conditions persist in the impounded reaches of the river. The upper and lower free-flowing reaches both provide expansive lotic habitat for fluvial species as well as backwater habitat in connected lateral channels and active floodplain habitat (i.e., the batture or the portion of the levee-constrained floodplain that seasonally floods during many years). The diverse habitats in the batture of the lower free-flowing reach may also provide suitable conditions for some of the species classified as peripheral. Most of the fish were subjectively classified as peripheral based on their usual association with small-stream ecosystems. A challenge in classifying a taxon as resident or peripheral in the Mississippi River is that the river floodplain ecosystem includes the channel, channel border, a variety of side channels and abandoned channels, and the floodplain that also contains a diversity of habitats. As a result, some fishes classified as peripheral may, indeed, be resident to the Mississippi River floodplain ecosystem, and this possible error would be more likely in the free-flowing reaches that have both relatively expansive floodplains and a variety of channel border and lateral channels.

Changes from Previous Listings

Lists of fish fauna for the entire Mississippi River have been compiled by Fremling et al. (1989) and Schramm (2004). Both lists, which draw heavily from other published species lists, include species of known and hypothetical occurrence. Here, we identify differences between the present and the two previous lists as well as lists of fishes in the lower two-thirds of the lower free-flowing reach compiled by Warren et al. (2000). Deletions of a species from one or more reaches are based on a lack of existing specimens or records that, in our professional judgment, have been validated by a competent fish biologist. Note that "stray" is used by Fremling et al. (1989) to denote peripheral or the result of an inland stocking (cf., a nonnative species that is not established as used by us).

Chestnut Lamprey has been added as peripheral in the lower free-flowing reach.

Southern Brook Lamprey *Ichthyomyzon gagei*, included as a possible stray in Fremling et al. (1989) in the lower free-flowing reach, has been deleted.

American Brook Lamprey has been added as peripheral in the upper impounded reach.

Bull Shark, listed as lacking recent collection records by Fremling et al. (1989), is confirmed as present. Although known to be euryhaline and survive relatively long periods of time in freshwater, reproduction in freshwater appears unlikely (Montoya and Thorson 1982); thus, we consider it peripheral. As the source of this species is the Gulf of Mexico, it must traverse the lower free-flowing reach to reach the upper free-flowing reach. Nevertheless, valid collections from the lower free-flowing reach are not available, and thus, we do not include this taxon in the lower free-flowing reach.

Lake Sturgeon has been added in the lower free-flowing reach.

Spotted Gar has been added to the upper impounded reach; however, it is possible that Spotted Gar in this reach may be hybrids between Longnose Gar and Shortnose Gar (see Table 1, footnote a). Longnose Gar, listed as present in headwater lakes by Fremling et al. (1989), has been deleted from the headwaters reach.

Shortnose Gar, listed as lacking recent collection records in the headwaters reach by Fremling et al. (1989), is present but considered peripheral.

Goldeye has been deleted from the headwaters reach.

Mooneye has been added as a stray in the headwaters reach.

American Eel, listed by Fremling et al. (1989) as lacking recent collection records in the headwaters, has been confirmed as present.

Gizzard Shad, considered an abundant resident in the headwaters by Fremling et al. (1989), has been listed as an uncommon stray. Gizzard Shad have not been collected in extensive surveys in the headwaters reach upstream of Coon Rapids Dam (E. Altena, Minnesota Department of Natural Resources, personal communication).

Central Stoneroller, listed as a resident in the headwaters reach by Fremling et al. (1989), is considered peripheral.

Goldfish, considered introduced in the headwaters reach by Fremling et al. (1989), is not present in this reach.

Southern Redbelly Dace *Chrosomus erythrogaster* was listed by Fremling et al. (1989) as a possible stray in the upper impounded reach. We found no collection records of this species since 1978. Because this species would probably be peripheral if it occurred, we do not list as extirpated in the upper impounded reach. Warren et al. (2000) included Southern Redbelly Dace in the lower free-flowing reach, possibly because of occurrence in a few tributaries, but we found no collection records for the Mississippi River; we do not list this taxon in the lower free-flowing reach. Southern Redbelly Dace has been deleted from the list.

Grass Carp was listed by Fremling et al. (1989) as an established nonnative in the upper impounded and upper and lower free-flowing reaches and as a stray in the lower impounded reach. Evidence for establishment of Grass Carp anywhere in the Mississippi River is equivocal. We list Grass Carp as established in the lower impounded and the upper and lower free-flowing reaches based on observations of a range of sizes of this taxon, and we regard Grass Carp as strays in the headwaters and upper impounded reaches.

Spotfin Shiner, included as a stray in the upper free-flowing reach by Fremling et al. (1989), is considered resident in this reach.

Blacktail Shiner has been added as a peripheral in the lower impounded reach. Listed as a stray by Fremling et al. (1989) in the upper free-flowing reach, we consider this taxon resident in the upper free-flowing reach.

Gravel Chub, included by Warren et al. (2000) and Schramm (2004) in the free-flowing reaches, has been retained in the upper free-flowing reach as peripheral but deleted from the lower free-flowing reach.

Western Silvery Minnow was included as present in the upper and lower free-flowing reaches and as a probable stray in the lower impounded reach by Fremling et al. (1989).We found no evidence to support occurrence of this taxon in the lower impounded reach and have deleted it. Warren et al. (2000) listed Western Silvery Minnow in the lower free-flowing reach, and we found valid collection records (1983) in both the upper and lower free-flowing reaches. We have retained this taxon in the free-flowing reaches.

Mississippi Silvery Minnow has been deleted from the headwaters reach.

Pallid Shiner was listed as lacking recent collection records in the lower impounded reach and not listed in the upper free-flowing reach by Fremling et al. (1989). Pallid Shiner was listed only for the impounded reaches by Schramm (2004). This species is confirmed as present in the lower impounded reach and, based on valid historic collection records, considered extirpated in the upper free-flowing reach.

Silver Carp and Bighead Carp are new additions to the fish fauna since 1989.

Common Shiner, listed as resident in the upper and lower impounded reaches by Fremling et al. (1989) and Schramm (2004), is considered peripheral in the upper impounded reach and has been deleted from the lower impounded reach.

Redfin Shiner was included as a resident in the upper impounded reach and as a stray in the lower impounded and lower free-flowing reaches by Fremling et al. (1989); we retained this taxon as peripheral in the lower impounded reach and deleted it from the upper impounded and the lower free-flowing reaches.

Sturgeon Chub has been deleted from the lower impounded reach but added to the lower free-flowing reach.

Shoal Chub replaces Speckled Chub *Macrhybopsis* (also *Hybopsis*) *aestivalis* as a result of elevation of subspecies to species status (Eisenhour 2004) but does not represent an addition to the impounded or free-flowing reaches.

Bluehead Chub *Nocomis leptocephalus*, noted as a stray in the lower free-flowing reach by Fremling et al. (1989), has been deleted.

Pugnose Shiner, listed by Fremling et al. (1989) as a stray in the upper impounded reach, has been added in the headwaters reach and deleted from the upper impounded reach.

Silverjaw Minnow, listed as a stray in the lower free-flowing reaches by Fremling et al. (1989), has been deleted.

Ghost Shiner, listed as lacking recent collection records in the lower impounded reach by Fremling et al. (1989), is verified as present.

Ironcolor Shiner *Notropis chalybaeus*, noted as a stray in the lower free-flowing reach by Fremling et al. (1989), has been deleted.

Blackchin Shiner has been deleted from upper impounded reach. Although collections have been reported in the past 20 years, specimens have not been verified and records are not accepted.

Blacknose Shiner has been deleted from upper impounded reach.

Spottail Shiner, considered present in the free-flowing reaches by Fremling et al. (1989) and Warren et al. (2000), has not been recently collected in the lower free-flowing reach. This species was collected at the mouth of the Ohio River in 1963 (Fishnet2, www.fishnet2.net). We consider this species peripheral in the free-flowing reaches.

Ozark Minnow has been deleted from the upper impounded reach.

Chub Shiner, listed as lacking recent collection records by Fremling et al. (1989), is confirmed as present and considered a peripheral.

Rosyface Shiner *Notropis rubellus* was included by Fremling et al. (1989). This species is now Carmine Shiner *N. percobromus*, a result of removing Carmine Shiner from synonymy with Rosyface Shiner (Wood et al. 2002). However, the only record of Carmine Shiner is a lone record from an upper impounded reach tributary (Little Maquoketa River, Iowa). Thus, Carmine Shiner has been deleted.

Sand Shiner, considered resident in the upper impounded, lower impounded, and upper free-flowing reaches by Fremling et al. (1989), is considered peripheral in these reaches.

Pugnose Minnow, listed as lacking recent collection records in the upper free-flowing reach by Fremling et al. (1989), is verified as present.

Suckermouth Minnow has been added in the lower free-flowing reach but is considered peripheral. The last valid collection of this taxon in the upper impounded reach was in 1953; however, the taxon is not considered extirpated in this reach because it would be peripheral if present.

Longnose Dace has been added to the headwaters reach as resident and to upper impounded reach as peripheral.

Pearl Dace *Semotilus margarita* (now Northern Pearl Dace *Margariscus nachtriebi*) has been deleted.

Longnose Sucker has been added in the lower impounded reach. Although more than 500 km from its known range (Lee et al. 1980), the record is supported by a voucher specimen. We have included the taxon as a stray.

White Sucker, listed as probable stray in the lower impounded and upper free-flowing reach by Fremling et al. (1989), is confirmed as present and considered resident in both reaches.

Creek Chubsucker *Erimyzon oblongus* west of the Appalachian Mountains has been replaced by Western Creek Chubsucker *E. claviformis* (Bailey et al. 2004). This species and Lake Chubsucker *E. sucetta* have been deleted; there are no historic records of these taxa in the Mississippi River, so they are not considered extirpated.

Northern Hog Sucker, listed as a stray in the lower impounded, upper free-flowing, and lower free-flowing reaches by Fremling et al. (1989), has been retained as peripheral in the upper free-flowing reach but deleted from the lower impounded and lower free-flowing reaches.

River Redhorse has been added to the lower impounded reach. Warren et al. (2000) listed this taxon as questionable for the lower free-flowing reach. We did not find records to verify this taxon present in the lower free-flowing reach.

Golden Redhorse, listed as resident by Fremling et al. (1989) in the upper free-flowing reach, is considered peripheral.

Pealip Redhorse *Moxostoma pisolabrum* was formerly considered a subspecies of Shorthead Redhorse. Nelson et al. (2004) elevated it to full species rank based on Harris et al. (2002). Page et al. (2013) retained it as a full species, but this is controversial (see Table 1, footnote f). Warren et al. (2000) listed Pealip Redhorse in the lower free-flowing reach. We retained Shorthead Redhorse and recognize that Pealip Redhorse may also be present.

Blacktail Redhorse *Moxostoma poecilurum*, noted as a stray in the upper and lower free-flowing reaches by Fremling et al. (1989), has been deleted.

White Catfish is listed as a stray in the upper impounded and upper free-flowing reaches by Fremling et al. (1989). We have retained this taxon as a stray for the upper free-flowing reach but deleted it from the upper impounded reach.

Brown Bullhead is considered extirpated in the lower impounded reach.

Blue Catfish was listed by Fremling et al. (1989) as lacking recent collection records in the upper impounded reach. We could not verify the presence of this taxon in the upper impounded reach and have deleted it.

Speckled Madtom *Noturus leptacanthus*, noted as a stray in the lower free-flowing reach by Fremling et al. (1989), has been deleted.

Freckled Madtom has been added to the upper free-flowing reach; the taxon is confirmed present and considered resident in the lower free-flowing reach.

Northern Madtom *Noturus stigmosus* was

listed in Fremling et al. (1989) and Schramm (2004). The Northern Madtom from Tennessee is now the Piebald Madtom (see footnote g, Table 1). Northern Madtom has been deleted and Piebald Madtom added.

Rainbow Smelt has been added as introduced to the headwaters and as a stray in the upper impounded reach.

Cisco is present in headwaters reach lakes; this taxon has been retained in the headwaters reach and deleted from the impounded reaches.

Rainbow Trout has been added as an introduced peripheral in the lower impounded reach as a consequence of stocking in Iowa streams tributary to the Mississippi River.

Brown Trout has been added as an introduced peripheral in the lower impounded reach as a consequence of stocking in Iowa streams tributary to the Mississippi River.

Grass Pickerel, listed as a stray in the lower impounded reach by Fremling et al. (1989), is considered resident.

Muskellunge has been added and considered peripheral in the lower impounded reach and added as a stray in the upper free-flowing reach.

Central Mudminnow, listed as a stray in the upper impounded reach and not listed in the lower impounded reach by Fremling et al. (1989), is considered resident in both reaches.

Burbot, included by Fremling et al. (1989) in the upper free-flowing reach, has been deleted from that reach.

Northern Studfish *Fundulus catenatus*, noted as a stray in the upper and lower free-flowing reaches by Fremling et al. (1989), has been deleted.

Banded Killifish was not included by Fremling et al. (1989) but was listed by Schramm (2004); this taxon has been confirmed for the upper impounded reach and is considered peripheral.

Starhead Topminnow was removed from synonymy with *F. nottii* (now known as the Bayou Topminnow) after Fremling et al. (1989) noted it as a stray in the lower impounded and lower free-flowing reaches. It has been retained in the lower impounded reach as uncommon and in the lower free-flowing reach as peripheral.

White Perch *Morone americana* was listed as a possible occurrence in the lower free-flowing reach by Warren et al. (2000). We found no evidence to retain this nonnative species in this reach.

Striped Bass was listed as a resident of the free-flowing reaches by Fremling et al. (1989). As discussed previously, we consider this species a stray.

Shadow Bass, listed as a stray in the lower free-flowing reach by Fremling et al. (1989), has been deleted from this reach but added as peripheral in the upper free-flowing reach.

Flier, listed as a stray in the upper and lower free-flowing reaches by Fremling et al. (1989), is retained as peripheral in the lower free-flowing reach but deleted from the upper free-flowing reach.

Longear Sunfish, listed as a stray in the lower impounded reach by Fremling et al. (1989), is considered resident.

Redear Sunfish, listed as a stray in the lower impounded and upper free-flowing reaches by Fremling et al. (1989), has been deleted.

Spotted Sunfish *Lepomis punctatus* was included as a stray by Fremling et al. (1989). Redspotted Sunfish, formerly considered a subspecies of Spotted Sunfish (Warren 1992), replaces Spotted Sunfish and is considered a resident in the lower free-flowing reach.

Bantam Sunfish, listed as a stray by Fremling et al. (1989) and Schramm (2004), is considered resident in the lower free-flowing reach.

Smallmouth Bass, noted as a stray in the upper and lower free-flowing reaches by Fremling et al. (1989) and as present in the lower free-flowing reaches by Warren et al. (2000), is retained as peripheral in the upper free-flowing reach and deleted from the lower free-flowing reach.

Western Sand Darter, noted by Fremling et al. (1989) as lacking recent collection records in the upper free-flowing reach, is confirmed as present.

Scaly Sand Darter *Ammocrypta vivax*, noted as a stray in the lower free-flowing reach by Fremling et al. (1989), has been deleted.

Crystal Darter, not listed from the lower impounded and upper free-flowing reaches and listed as a stray in the lower free-flowing reach in Fremling et al. (1989), has been added as a resident for these reaches.

Rainbow Darter, listed as a resident in upper impounded and lower free-flowing reaches by Fremling et al. (1989), is considered peripheral in these reaches.

Bluntnose Darter, noted as lacking recent collection records in the upper impounded reach by Fremling et al. (1989), is confirmed as present in the upper impounded reach and has been added as peripheral in the lower impounded reach.

Iowa Darter has been added to the headwaters reach.

Fantail Darter, listed as a stray in the lower impounded reach by Fremling et al. (1989), has been deleted from this reach.

Harlequin Darter has been added to the lower free-flowing reach.

Orangethroat Darter, listed as a stay in the lower impounded reach by Fremling et al. (1989), has been deleted.

Banded Darter, listed as a stray in the lower free-flowing reach by Fremling et al. (1989), has been deleted from this reach. Recent Long Term Resource and Monitoring Program data indicate consistent presence of this taxon in Pool 8 of the upper impounded reach, and this taxon has been included as a resident in this reach.

Redfin Darter *Etheostoma whipplei*, noted as a stray in the lower free-flowing reach by Fremling et al. (1989), would now be Redspot Darter *E. artesiae* (Piller et al. 2001). The former has been has been deleted.

Logperch is considered peripheral in the upper free-flowing reach.

Gilt Darter was noted as lacking recent collection records in the upper impounded reach by Fremling et al. (1989) and Schramm (2004). The old record was based on a misidentified Blackside Darter (Hubbs 1945). The species has been deleted from this reach and added as peripheral in the upper free-flowing reach.

Blackside Darter, formerly listed as a probable stray in the upper and lower impounded reaches and the lower free-flowing reach by Fremling et al. (1989), has been confirmed as resident (introduced) in the headwaters and upper impounded reaches but is not present in other reaches.

Stargazing Darter *Percina uranidea*, listed for the lower free-flowing reach by Fremling et al. (1989) but not by Warren et al. (2000), has been deleted.

Banded Pygmy Sunfish, considered a stray in the lower free-flowing reach by Fremling et al. (1989), is considered resident.

Of the numerous marine fishes included by Fremling et al. (1989) in the lower reaches of the Mississippi River, we have retained only Atlantic Needlefish and Gulf Pipefish, both of which are considered to have resident inland populations (Douglas 1974).We have retained Bull Shark and Striped Mullet, which have verified occurrences far inland in the free-flowing reaches, but consider them peripheral.

Noteworthy are species now considered extirpated. Alligator Gar, migratory Alabama Shad, and Brown Bullhead are probably extirpated in the lower impounded reach. Ghost Shiner is considered extirpated in the upper impounded reach. Pallid Shiner is considered extirpated in the upper free-flowing reach.

Discussion

The Mississippi River differs geomorphologically and hydrologically along its length, and paralleling Schramm and Ickes (2016), we have recognized five distinctly and incrementally different reaches. Important for future conservation of the Mississippi River ichthyofauna in each reach is the conservation and rehabilitation of both fluvial (main channel and channel border) and backwater (sloughs, abandoned channels, floodplain lakes, and floodplains) habitats (see Schramm and Ickes 2016 for habitat definitions) in order to maintain the overall fish diversity. Of particular concern is the loss of suitable backwater habitats in the impounded reaches due to sedimentation and lack of flow in the now permanently inundated historic floodplain, as well as homogenization of habitats in the flowing-water channel (Schramm and Ickes 2016). In the free-flowing reaches, reduced flood pulses, thermal-hydrologic asynchrony (i.e., the flood pulse inundates the floodplain before the spawning and growing seasons of floodplain-dependent species; Junk et al. 1989; Schramm and Eggleton 2006), and channel incision that results in draining and disconnection of floodplain water bodies may jeopardize backwater-dependent species. Similarly, the loss of main channel border habitats (Schramm and Ickes 2016) may adversely affect fluvial-dependent fishes. Future conservation of the biodiversity and ecological function of the Mississippi River floodplain ecosystem will benefit from habitat rehabilitation to address these concerns.

The presence of 188 species and general persistence of species listed in previous comprehensive assessments (Fremling et al. 1989; Schramm 2004) suggest that the composition of the Mississippi River fish assemblage has changed little despite more than 80 years of extensive habitat alterations (Schramm and Ickes 2016 and references therein). However, the extirpation of five species—Alligator Gar, Alabama Shad, Pallid Shiner, Ghost Shiner, and Brown Bullhead—in reaches where they historically occurred may warn of future losses of biodiversity. Loss of anadromous Alabama Shad from the impounded reach can be attributed to locks and dams, but reasons for extirpation of the other taxa are not apparent. Assessments of habitat use and preferences, as well as environmental tolerances, of these taxa will help elucidate preferred or acceptable environmental conditions in the Mississippi River. If and when such information is assembled, it needs to be interpreted in the light of a river with altered habitats; in other words, measurements obtained in an altered system may provide information about conditions tolerated but not preferred conditions. Further, while the conspicuous habitat changes that we see today—locks and dams, levees, and most dikes and revetments—occurred during a relatively brief period from the late 1920s through the 1940s (Schramm and Ickes 2016), other potentially drastic, but now invisible, changes preceded these environmental alterations. For example, riparian forests were harvested to provide fuel during the steamboat era in the 19th century (Hurley 1997). The river widened considerably and resembled the lower Missouri River. Installation of dikes and revetments returned the river to widths similar to those before the tree harvest (Turner 2014). Are changes in species composition and abundance a result of contemporary alterations or the result of recovery from a previous, and possibly more severely, altered state?

Lack of rigorous, standardized, and comprehensive assessments has constrained judgment of extirpation, and it is possible that other taxa also are no longer present in the Mississippi River. Further, changes in abundance are extremely difficult to monitor and restrict our assessment of change to presence and absence. Large-river fishes are adapted to survive and persist under dynamic environmental conditions. This characteristic would be expected to make the Mississippi River fish assemblage relatively resilient to habitat alterations, yet some species are no longer extant in different reaches. The cause or causes of these extirpations are not evident and warrant investigation to guide management. Lacking specific guidance, maintaining habitat diversity should be the management mantra because it can only benefit the fish assemblage. Alterations of the river to achieve social and economic benefits (i.e., navigation and flood control) can, and over time probably will, homogenize habitats, but these alterations can, when wisely planned, also be used to conserve and increase habitat heterogeneity in the Mississippi River (Barko et al. 2004; Benjamin et al. 2016, this volume; Schramm and Ickes 2016).

Maintaining diverse river habitats, both spatially and temporally, will be critical to ensuring persistence of the diverse Mississippi River fish fauna.

Acknowledgments

Konrad Schmidt provided valuable input and guidance throughout this project. Previous drafts of this manuscript were reviewed by Melvin Bowler, Tom Buchanan, Dave Herzog, Brent Knights, John Lyons, John Olson, and Bobby Reed, and Table 1 reflects their collective wisdom. Funding for this research was provided by Mississippi Department of Wildlife, Fisheries and Parks, Mississippi State University, Missouri Department of Conservation, U.S. Army Engineer Research and Development Center (Mississippi River Geomorphology and Potamology Program), and U.S. Geological Survey.

References

Bailey, R. M., W. C. Latta, and G. R. Smith. 2004. An atlas of Michigan fishes with keys and illustrations for their identification. Miscellaneous Publication Number 192. Museum of Zoology, University of Michigan.

Baker, J. A., K. J. Killgore, and R. L. Kasul. 1991. Aquatic habitats and fish communities in the lower Mississippi River. Reviews in Aquatic Sciences 3:313–356.

Barko, V. A., D. P. Herzog, R. A. Hrabik, and J. S. Scheibe. 2004. Relationship among fish assemblages and main-channel border physical habitats in the unimpounded upper Mississippi River. Transactions of the American Fisheries Society 133:371–384.

Benjamin, G. L., A. J. Rodgers, and K. J. Killgore. 2016. Mississippi River ecosystem restoration: the past forty-plus years. Pages 311–350 *in* Y. Chen, D. C. Chapman, J. R. Jackson, D. Chen, Z. Li, K. J. Killgore, Q. E. Phelps, and M. A. Eggleton, editors. Fishery resources, environment, and conservation in the Mississippi and Yangtze (Changjiang) river basins. American Fisheries Society, Symposium 84, Bethesda, Maryland.

Burr, B. M., and R. L. L. Mayden. 1980. Dispersal of Rainbow Smelt, *Osmerus mordax* (Pisces: Osmeridae), into the upper Mississippi River. American Midland Naturalist 104:198–201.

Coker, R. E. 1930. Studies of common fishes of the Mississippi River at Keokuk, 1930. Bureau of Fisheries Document 1072. U.S. Government Printing Office, Washington, D.C.

Conner, J. V., R. P. Gallagher, and M. F. Chatry. 1980. Larval evidence for natural reproduction of the Grass Carp (*Ctenopharyngodon idella*) in the lower Mississippi River. Pages 1–19 *in* L. A. Fuiman, editor. Proceedings of the fourth larval fish conference. U.S. Fish and Wildlife Service, National Power Plant Team, FWS/OBS/43, Ann Arbor, Michigan.

DeVries, R. J., D. Hann, and H. L. Schramm, Jr. 2015. Increasing capture efficiency of Pallid Sturgeon *Scaphirhynchus albus* (Forbes and Richardson, 1905) and the reliability of catch rate estimates. Journal of Applied Ichthyology 31:603–608.

Dickson, T. J., and J. R. Tomelleri. 2008. The great Minnesota fish book. University of Minnesota Press, Minneapolis.

Douglas, N. H. 1974. Freshwater fishes of Louisiana. Claitor's Publishing Division, Baton Rouge, Louisiana.

Eddy, S., and J. C. Underhill. 1974. Northern fishes with special reference to the upper Mississippi Valley. University of Minnesota Press, Minneapolis.

Eisenhour, D. J. 2004. Systematics, variation, and speciation of the *Macrhybopsis aestivalis* complex west of the Mississippi River. Bulletin of the Alabama Museum of Natural History 23:9–47.

Etnier, D. A., and W. C. Starnes. 1993. The fishes of Tennessee. The University of Tennessee Press, Knoxville.

Fremling, C. R., J. L. Rasmussen, R. E. Sparks, S. P. Cobb, C. F. Bryan, and T. O. Claflin. 1989. Mississippi River fisheries. Pages 309–351 *in* D. P. Dodge, editor. Proceedings of the international large rivers symposium. Fisheries and Oceans Canada, Ottawa.

Gallagher, R. P., and J. V. Conner. 1980. Spatiotemporal distribution of ichthyoplankton in the lower Mississippi River, Louisiana. Pages 101–115 *in* L. A. Fuiman, editor. Proceedings of the fourth larval fish conference. U.S. Fish and Wildlife Service, National Power Plant Team, FWS/OBS/43, Ann Arbor, Michigan.

Harris, P. M., R. L. Mayden, H. S. Espinosa Perez, and F. G. Leon. 2002. Phylogenetic relationships of *Moxostoma* and *Scartomyzon* (Catostomidae) based on mitochondrial cytochrome *b* sequence data. Journal of Fish Biology 61:1433–1452.

Hatch, J. T., K. P. Schmidt, D. P. Siems, J. C. Underhill, R. A. Bellig, and R. A. Baker. 2003. A new distributional checklist of Minnesota fishes with comments on historical occurrence. Journal of the Minnesota Academy of Science 67:1–17.

Herzog, D. P., V. A. Barko, J. Scheibe, R. A. Hrabik, and D. E. Ostendorf. 2005. Efficacy of a benthic trawl for sampling small-bodied fishes in large river systems. North American Journal of Fisheries Management 25:594–603.

Herzog, D. P., D. E. Ostendorf, R. A. Hrabik, and V. A. Barko. 2009. The mini-Missouri trawl: a useful methodology for sampling small-bodied fishes in small and large river systems. Journal of Freshwater Ecology 24:103–108.

Hrabik, R.A. 1997. Meristic and morphometric variation of some populations of *Notropis volucellus* and *Notropis wickliffi* in the upper Mississippi River basin. Master's thesis. Southeast Missouri State University, Cape Girardeau.

Hubbs, C. L. 1945. Corrected distributional records for Minnesota fishes. Copeia 1945:13–22.

Hurley, A., editor. 1997. Common fields: an environmental history of St. Louis. Missouri Historical Society Press, St. Louis.

Junk, W. J., P. B. Bayley, and R. E. Sparks. 1989. The flood pulse concept in river-floodplain ecosystems. Pages 110–127 *in* D. P. Dodge, editor. Proceedings of the international large rivers symposium. Canadian Special Publication of Fisheries and Aquatic Sciences 106.

Killgore, K. J., J. J. Hoover, S. G. George, B. R. Lewis, C. E. Murphy, and W. E. Lancaster. 2007. Distribution, relative abundance and movements of Pallid Sturgeon in the free-flowing Mississippi River. Journal of Applied Ichthyology 23:476–483.

Lee, D. S., C. R. Gilbert, C. H. Hocutt, R. E. Jenkins, D. E. McAllister, and J. R. Stauffer. 1980. Atlas of North American freshwater fishes. North Carolina State Museum of Natural History, Publication 1980-12, Raleigh.

Long, E. A., C. L. Mesing, K. J. Herrington, R. R. Weller, and I. I. Wirgin. 2013. Restoration of Gulf Striped Bass: lessons and management implications. Pages 25–63 *in* J. S. Bulak, C. C. Coutant, and J. A. Rice, editors. Biology and management of inland Striped Bass and hybrid Striped Bass. American Fisheries Society, Symposium 80, Bethesda, Maryland.

McIlwain, T. D. 1967. Distribution of the Striped Bass, *Roccus saxatilis* (Walbaum), in Mississippi waters. Proceedings of the 21st Annual Conference Southeastern Association of Fish and Wildlife Agencies 21:254–257.

Montoya, R. V., and T. B. Thorson. 1982. The Bull Shark (*Carcharhinus leucas*) and Largetooth Sawfish (*Pristis perotteti*) in Lake Bayano, a tropical man-made impoundment in Panama. Environmental Biology of Fishes 7:341–347.

Nelson, J. S., E. J. Crossman, H. Espinoza-Pérez, L. T. Findley, C. R. Gilbert, R. N. Lea, and J. D. Williams. 2004. Common and scientific names of fishes from the United States, Canada, and Mexico. American Fisheries Society, Special Publication 29, Bethdsda, Maryland.

Page, L. M., and B. M. Burr. 2011. Peterson field guide to freshwater fishes of North America North of Mexico, 2nd edition. Houghton, Mifflin, and Harcourt Publishers, Boston.

Page, L. M., H. Espinosa-Pérez, L. T. Findley, C. R. Gilbert, R. N. Lea, N. E. Mandrak, R. L. Mayden, and J. S. Nelson. 2013. Common and scientific names of fishes from the United States, Canada, and Mexico, 7th edition. American Fisheries Society, Special Publication 34, Bethesda, Maryland.

Pflieger, W. L. 1997. Fishes of Missouri. Missouri Department of Conservation, Jefferson City.

Phillips, G. L., W. D. Schmid, and J. C. Underhill. 1982. Fishes of the Minnesota region. University of Minnesota Press, Minneapolis.

Piller, K. R., H.L. Bart, Jr., and C.A. Walser. 2001. Morphological variation of the Redfin Darter, *Etheostoma whipplei*, with comments on the status of the subspecific populations. Copeia 2001:802–807.

Pitlo, J., Jr., A. vanVooren, and J. L. Rasmussen. 1995. Distribution and relative abundance of upper Mississippi River fishes. Upper Mississippi River Conservation Committee, Rock Island, Illinois.

Poddubny, A. G., and D. L. Galat. 1995. Habitat associations of upper Volga River fishes: effects of reservoirs. Regulated Rivers: Research and Management 11:67–84.

Robison, H. W., and T. M. Buchanan. 1988. Fishes of Arkansas. University of Arkansas Press, Fayetteville.

Ross, S. T. 2001. Inland fishes of Mississippi. University Press of Mississippi, Jackson.

Schiemer, F., and H. Waidbacher. 1992. Strategies for conservation of a Danubian fish fauna. Pages 363–382 *in* P. J. Boon, P. Calow, and G. E. Petts, editors. River conservation and management. Wiley, London.

Schmidt, K., and N. Proulx. 2009. Status and critical habitat of rare fish species in the Mississippi River from the Coon Rapids Dam to the Iowa border. Final report submitted to the Minnesota Department of Natural Resources, State Wildlife Grants Program. Available: http://files.dnr.state.mn.us/eco/nongame/projects/consgrant_reports/2009/2009_schmidt_proulx2.pdf. (September 2015).

Schramm, H. L., Jr. 2004. Status and management of fisheries in the Mississippi River. Pages 301–333 *in* R. Welcomme and T. Petr, editors. Proceedings of the second international symposium on the management of large rivers for fisheries, volume 1. Food and Agriculture Organization of the United Nations, Regional Office for Asia and the Pacific, RAP Publication 2004/16, Bangkok, Thailand.

Schramm, H. L., Jr., and M. A. Eggleton. 2006. Applicability of the flood-pulse concept in a temperate floodplain river ecosystem: thermal and temporal components. River Research and Application 22:543–553.

Schramm, H. L., Jr., and B. S. Ickes. 2016. The Mississippi River: a place for fish. Pages 3-34 *in* Y. Chen, D. C. Chapman, J. R. Jackson, D. Chen, Z. Li, K. J. Killgore, Q. E. Phelps, and M. A. Eggleton, editors. Fishery resources, environment, and conservation in the Mississippi and Yangtze (Changjiang) River basins. American Fisheries Society, Symposium 84, Bethesda, Maryland.

Schramm, H. L., Jr., and C. H. Pennington. 1981. Aquatic habitat studies on the lower Mississippi River, river mile 480-530: report 6. Larval fish studies—pilot report. U.S. Army Waterways Experiment Station, Miscellaneous Paper E-80-1, Vicksburg, Mississippi.

Sipiorski, J. T. 2011. The world according to gars: the molecular systematics and comparative phylogeography of living gars (Actinopterygii: Lepisosteidae). Doctoral dissertation. Southern Illinois University, Carbondale.

Smith, P. W. 1979. The fishes of Illinois. University of Illinois Press, Urbana.

Thomas, M. R., and B. M. Burr. 2004. *Noturus gladiator*, a new species of madtom (Siluriformes: Ictaluridae) from Coastal Plain streams of Tennessee and Mississippi. Ichthyological Exploration of Freshwaters 15:351–368.

Turner, W. 2014. Voices of Missouri's rivers. Missouri Department of Conservation, Jefferson City.

U.S. Army Corps of Engineers. 2015. Upper Mississippi River locks and dams 2015. Available: www.mvr.usace.army.mil/Portals/48/docs/CC/FactSheets/MISS/UpperMississippiRiverLocksandDams2015.pdf. (October 2015).

Walsh, S. J., H. L. Jelks, and N. M. Burkhead. 2009. The decline of North American freshwater fishes. American Institute of Biological Sciences. Available: www.actionbioscience.org/biodiversity/walsh.html. (September 2015).

Warren, M. L., Jr. 1992. Variation of the Spotted Sunfish, *Lepomis punctatus* (Centrarchidae): meristics, morphometrics, pigmentation and species limits. Bulletin of the Alabama Museum of Natural History 12:1–47.

Warren M. L. Jr., B. M. Burr, S. J. Walsh, H. L. Bart, Jr., R. C. Cashner, D. A. Etnier, B. J. Freeman, B. R. Kuhajda, R. L. Mayden, H. W. Robison, S. T. Ross, and W. C. Starnes. 2000. Diversity, distribution, and conservation status of the native freshwater fishes of the southern United States. Fisheries 25(10):7–28.

Wood, R. M., R. L. Mayden, R. H. Matson, B. R. Kuhajda, and S. R. Layman. 2002. Systematics and biogeography of the *Notropis rubellis* species group (Teleostei: Cyprinidae). Bulletin of the Alabama Museum of Natural History 22:37–80.

American Fisheries Society Symposium 84:79–90, 2016

Development and Reform of Lake Commercial Fisheries in the Yangtze River Basin, China

Yuxi Lian and Jiashou Liu
State Key Laboratory of Freshwater Ecology and Biotechnology, Institute of Hydrobiology Chinese Academy of Sciences, 7 South Donghu Road, Wuhan 430072, P. R. China

Yushun Chen
State Key Laboratory of Freshwater Ecology and Biotechnology, Institute of Hydrobiology Chinese Academy of Sciences, 7 South Donghu Road, Wuhan 430072, P. R. China
and
Aquaculture and Fisheries Center, University of Arkansas-Pine Bluff 1200 North University Drive, Pine Bluff, Arkansas 71601, USA

Jing Yuan and Zhongjie Li*
State Key Laboratory of Freshwater Ecology and Biotechnology, Institute of Hydrobiology Chinese Academy of Sciences, 7 South Donghu Road, Wuhan 430072, P. R. China

Abstract.—Lakes are very important resources for fisheries production in China. The total area of lakes for commercial fisheries in China reaches 1.02 million ha, which accounts for 18% of the total freshwater aquaculture area. China has gained rich experience developing lake commercial fisheries over the past 60 years. In the 1950s, lake fisheries were primarily focused on the capture of natural aquatic animal species. With the success of the artificial reproduction of the four domestic carps (Silver Carp *Hypophthalmichthys molitrix*, Bighead Carp *H. nobilis* (also known as *Aristichthys nobilis*), Grass Carp *Ctenopharyngodon idella*, and Black Carp *Mylopharyngodon piceus*), stock-enhanced fisheries became the main production method in the 1960s and 1970s. During this period, large-scale resources investigations all over China were conducted, some stocking and management theories that were based on primary production developed, and fish production potentials for lakes were determined. The united fishing method was created during this period, which significantly increased the capture efficiency in large lakes. In the 1980s, semi-intensive and intensive aquaculture methods, including application of fertilizers and artificial feeds, cage culture, and pen culture, were applied to lake and reservoir fisheries, which substantially increased fishery production and also the income of fishermen. However, intensive aquaculture in lakes has caused serious environmental problems, such as ecosystem degradation, exhaustion of natural fisheries resources, decreased biodiversity, and increased eutrophication. Sustainable development of fisheries in lakes of the Yangtze River basin has been facing unprecedented challenges, both to the environment and to human society. More and more attention has focused on the balance between fisheries development and environmental protection in the past two decades. Ecofishery is a possible solution to this potential conflict. As a strategy for lake fisheries reform, it is suggested that use of natural food resources in lakes should be more efficient than present, and lake fisheries should be developed based on ecosystem restorations.

* Corresponding author: zhongjie@ihb.ac.cn

Introduction

The freshwater commercial fisheries of China play an important role in the world food fish supply, accounting for 68% of the total annual freshwater fisheries production of the world (Jia 2000). In China, about 97% of Chinese food fish is provided by freshwater fisheries and 20 million people are engaged in work related to freshwater fisheries (FAO 2014). China's freshwater fisheries mainly include those from rivers, ponds, lakes, and reservoirs. The category of lake fisheries in China is quite extensive; traditional capture of naturally produced fish, stock-enhanced fishery, and cage and pen culture compromise lake fisheries. Lake fisheries have always played an important role in Chinese traditional fisheries. For instance, in the 1950s, annual fishery capture from lakes was ~60,000–80,000 metric tons, which contributed to more than 80% of total freshwater fishery production in China (Ning et al. 1996). By 2010, lake fisheries production was 1,537,000 metric tons, which was approximately 20-fold greater than the 1950s harvest. However, this figure accounted for only 18.4% of the total freshwater fishery production due to concurrent increases in pond and reservoir fisheries (Figure 1; Bureau of Fishery, China Ministry of Agriculture 2011). For example, reservoir production was 13,600 metric tons in 1957 and accounted for only 2.4% of the total freshwater fishery production, but in 2013, reservoir production reached 3,536,581 metric tons and accounted for 12.6% of the total freshwater fishery production.

The important position of Chinese lake fisheries is determined by characteristics of Chinese lakes. Total lake area in China is vast. According to fishery statistics in 2005 and 2006, total lake area (not including ponds and reservoirs) was 8,141,400 ha and accounted for more than 37% of the total inland water area in China (Yang et al. 2010). China's lakes are numerous; there are more than 22,000 lakes with an area of less than 1 km^2 and about 2,693 lakes with an area of more than 1 km^2 (Liu and He 1992; Ma et al. 2011). Moreover, most of the small lakes (i.e., with an area <1 km^2) are shallow (Table 1) and located primarily in the eastern plains. These lakes are good for commercial fisheries because of the pleasant climate, rich fish resources, and abundant natural food. More than 99% of the lakes in China are located in the Qinghai–Tibet Plateau, the Eastern Plain, Inner Mongolia and Xinjiang, the Northeastern Plain, and the Yunnan–Guizhou Plateau (Figure 2; Wang and Dou 1998).

The lakes located in the Qinghai–Tibet Plateau and Inner Mongolia–Xinjiang are not

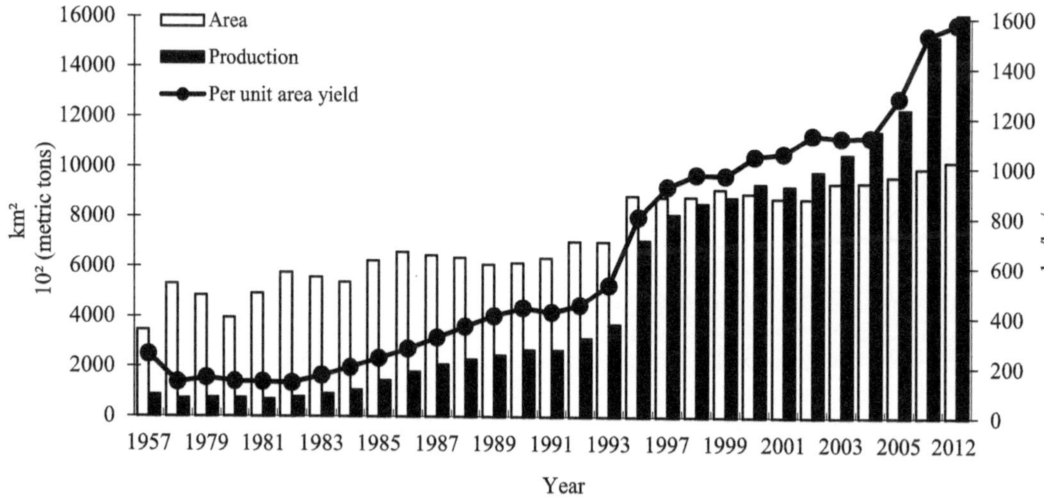

Figure 1. Annual fishery area and yield of lakes in China.

Table 1. Percent composition of Chinese lakes by number and area.

Category	<1 km^2	1-50 km^2	50-500 km^2	>500 km^2	Total
Number (%)	88.55	10.52	0.82	0.11	100
Area (%)	3.30	16.96	32.61	47.23	100

suitable for commercial fisheries because of low temperature and high salinity. The most developed area of lake fisheries in China is the Eastern Plain. This area is characterized as having a typical floodplain and contains the most productive commercial lake fisheries in China due to the high abundance of water (average annual runoff reaches 9.56 × 10^{11} m^3) and a climate conducive to fish production (average annual temperature is ~16–18°C; Changjiang Water Resources Commission 2007). The five largest freshwater lakes, Poyang, Dongting, Taihu, Hongze, and Chaohu, are all located in this region (Yang et al. 2004).

Development of Commercial Fisheries in the Yangtze Basin Lakes

Within the Eastern Plain, the Yangtze River basin accounts for the most developed lake fisheries of the region, providing 90% of China's lake fishery production (Liu and He 1992). Many lakes are located in this region, with more than 4,000 having surface areas greater than 0.5 km^2. Total lake area of the Yangtze basin lakes accounts for 31% of the total lake area in China (Wang and Dou 1998). There are about 370 anadromous, catadromous, and lacustrine fish species in the Yangtze River system (Fishery Resources Investigation Panel of the Yangtze River System 1990).

Commercial fisheries in the Yangtze River basin lakes have a long history. A taxation authority for commercial fisheries was established in the Yangtze River basin about 700 years ago during the Yuan Dynasty (Cong and Li 1993). Commercial fisheries in the Yangtze basin lakes have experienced substantial development over the past 60 years in total fishery area, overall production, fish species stocked, and fishing efficiency (Figure 3). In general, the development of com-

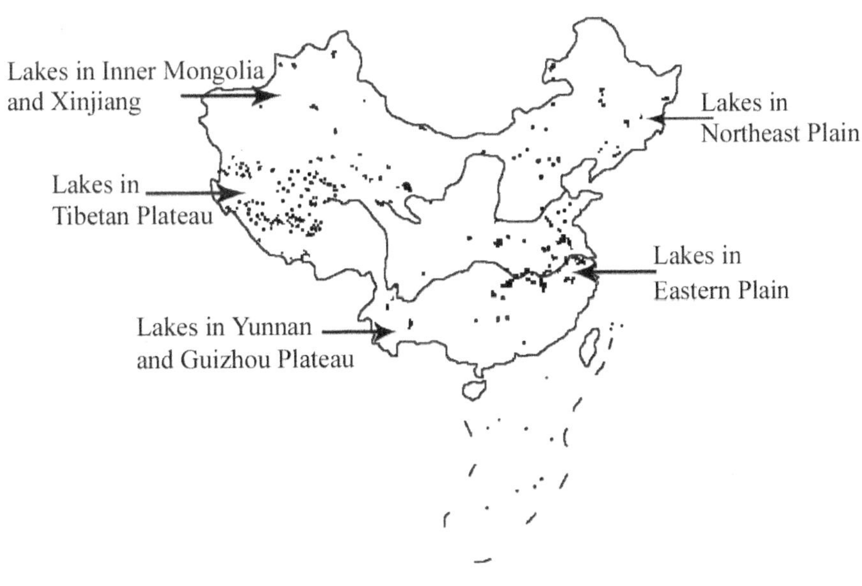

Figure 2. Distribution of lakes in China.

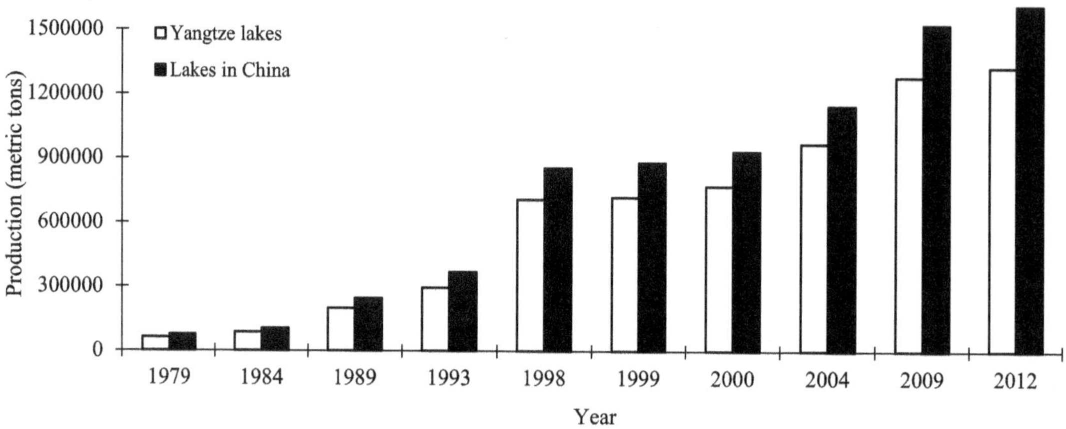

Figure 3. Commercial fishery production in Chinese lakes.

mercial fisheries in the Yangtze basin lakes can be divided into four stages.

Capture fishery stage

The first stage, before the 1960s, can be called the capture fishery stage. During this stage, the production of Yangtze basin lake fisheries was low and mainly depended on capture of naturally produced fish. Lake fisheries in this period were primarily family-run businesses and not well organized, so the utilization efficiency of lake fishery resources was very low (Zhang et al. 1997). In 1957, the Yangtze basin lake fishery area was 35,507 ha, total production was 39,948 metric tons, and yield was 1,122 kg/ha (Bureau of Fishery, China Ministry of Agriculture 1945–1975).

Stocking-enhanced fishery stage

The second stage, from the 1960s to 1970s, can be called the stocking-enhanced fishery stage. Fisheries in the Yangtze basin lakes began to implement artificial stocking. The success of artificial reproduction technology developed in this period promoted artificial stocking. Some small-to-medium lakes began to be stocked, and natural resources were used more effectively during this stage. Moreover, improved organization in lake fisheries production increased capture efficiency. But due to lack of efficient management, the development of Yangtze basin lake fisheries in this period was slow, with inconsistent harvest and a slowly increasing capture rate from year to year. Fishery management was in its infancy and did not take full advantage of natural resources. In addition, the main fishery species in these lakes were primarily limited to Common Carp *Cyprinus carpio* and the four domestic carps (Silver Carp *Hypophthalmichthys molitrix,* Bighead Carp *H. nobilis* [also known as *Aristichthys nobilis*], Grass Carp *Ctenopharyngodon idella,* and Black Carp *Mylopharyngodon piceus*) (Bureau of Fishery, China Ministry of Agriculture 1979). The Yangtze basin lake fisheries were at a low level of development in the 1970s. In 1979, total Yangtze basin lake fishery area was 303,600 ha and total fishery production was 62,265 metric tons. Productivity was 205 kg/ha, even lower than in 1957 (Bureau of Fishery, China Ministry of Agriculture 1979).

Lakes connected to the Yangtze River have the ability to modulate and store flood water during flood seasons. However, many lakes in the Yangtze River basin were reclaimed to land for food production in the 1960s and 1970s. For example, Dongting Lake, which used to be the largest freshwater lake in China, was decreased in area from 4,350 km^2 in the 1950s to 2,345 km^2 in the middle of the 1980s due to land reclamation and sediment deposition. In the 1950s, the area of lakes in the middle reach of the Yangtze River basin was 8,303.7 km^2, but only 3,210.2 km^2 remained in the 1990s (Yang et al. 2010). Reclamation of land from lakes not only reduced available fishery areas, but also weakened the capacity of lakes for flood control (Dong et al. 1999; Yang et al. 2010).

However, fisheries in the Yangtze basin lakes obtained some breakthrough progress at this stage. First, the success of artificial reproduction technologies of several economically important fishes, such as Silver Carp, provided enough fry for stocking, which reduced the previous reliance on collection of wild fry (Liu and He 1992). Second, great progress was made in freshwater fishing techniques. For example, the united fishing method—driving, blocking, gill netting, and trapping—was created during this stage and increased capture efficiency in large-water fisheries (Li and Xu 1995). For instance, more than 80% of target-size cyprinids can sometimes be caught in shallow lakes. Third, fishery production from small lakes also experienced substantial advancements. For example, the Institute of Hydrobiology, Chinese Academy of Sciences conducted extensive resource investigations and production techniques in Donghu Lake resulting in a productivity increase from 124 kg/ha in 1971 to 427 kg/ha in 1978, with an annual increasing rate of 23.5% (Liu 1980).

Aquaculture-based fisheries stage

The third stage, from the 1980s to 1990s, tended toward aquaculture-based fisheries. The Yangtze basin lake fisheries in this period developed rapidly, with intensive and semi-intensive aquaculture models started in the 1980s. These lake aquaculture models increased fishery production and income of fishermen substantially but also caused serious environmental problems.

Intensive and semi-intensive aquaculture in the Yangtze basin lakes started to boom in the 1980s and included fertilization, artificial feeds, cage culture, and pen culture, which targeted macrophyte-dominated lakes (Zeng 1995). After 1986, managers of Yangtze basin lake fisheries began to adjust management methods based on recent experience. Commercial fisheries of phytoplankton-dominated lakes still play an important role of lake fisheries, but fishery managers have changed their focus towards further developing fisheries in medium to large macrophyte-dominated lakes (Liu and Huang 2008), thus moving lake fisheries further towards intensive and large-scale production. At this stage, lake fisheries were shifted to a high-cost-with-high-output mode based on large artificial inputs (e.g., feed and fertilizers) instead of the traditional semiclosed cycle mode, which depends on the natural productivity of the resource. In 1999, Yangtze basin lake area in commercial fisheries was 551,484 ha, total production was 720,000 metric tons, and yield was 1,304 kg/ha (Bureau of Fishery, China Ministry of Agriculture 1999). The intensive and semi-intensive commercial fishery models increased the overall fishery production and fishermen's income but also caused serious environmental problems. For example, pen culture led to habitat fragmentation, excessive stocking of herbivorous fishes turned many macrophyte-dominated lakes into phytoplankton-dominated lakes, feeding and fertilization increased nutriment loading, and some endemic species disappeared from the lakes because of interspecific competition. All of these combined effects led to a destruction of ecosystem integrity, depletion of natural fishery resources, eutrophication, and decrease of biodiversity (Li et al. 1998; Cao et al. 2007).

In the 1990s, fishery markets demanded diverse fishery products of higher quality. Lake fisheries began to diversify and fishery products were no longer limited to the four domestic carps. Seventeen special high-value aquatic species gradually began to play an important role in lake fisheries, including Mandarin Fish *Siniperca chuatsi*, Southern Catfish *Silurus soldatovi meridionalis*, Chinese Longsnout Catfish *Leiocassis longirostris*, Salangid Icefish *Neosalanx taihuensis*, Chinese mitten crab *Eriocheir sinensis*, and Chinese softshell turtle *Pelodiscus sinensis* (Hu et al. 1999).

Modern sustainable ecological fishery stage

In the fourth stage, from the 1990s to the present, the intensive production of fish in

Yangtze basin lakes has shifted towards more environmentally sustainable practices due to social pressures and the need for a sustainable ecological commercial fishery as well as an emerging recreational fishery. The idea of the ecological fishery is to use natural resources in lakes following ecological and restoration principles with the aim of achieving a high and stable fishery yield as well as minimizing undesirable environmental effects by stocking and managing harvest and production appropriately (Jia 1995; Li 2004).

Entering the 21st century, the goals of Yangtze basin lake fishery management were no longer limited to maximization of production and are now shifting to ecologically sustainable fisheries. Ecological fisheries take full consideration of aquatic environments and the importance of biodiversity and are attentive to harmonious development of fisheries and the environment. The goal is to maintain and use lake fisheries while retaining other ecological structure and functions (Wen et al. 2002; Jin et al. 2003). This approach led to another rapid development of fisheries in the Yangtze basin lakes. From 2000 to 2012, Yangtze basin lake commercial fisheries encompassed 653,867 ha, produced 1,337,909 metric tons annually, and yielded 1,740 kg/ha (Bureau of Fishery, China Ministry of Agriculture 2012).

Recreational fishing in the Yangtze basin lakes is just beginning. The main uses are nonconsumptive sightseeing and fishing-watching. Tourists can enjoy the beautiful scenery and unique culture of fishing and share the joy of harvesting. The annual fishing in winter is for consumption. There is also angling for recreation only. However, recreational anglers are a small proportion of this segment (Chai et al. 2010).

At present, lake fisheries modes vary with lake areas. For most large lakes (e.g., area greater than 66,667 ha), a combined strategy entailing protection and stocking, and supplemented by aquaculture, has been adopted. Management of large lakes depends on the natural productivity of the lake. Large lakes seldom are highest in productivity; the average production of such lakes is 45–63 kg/ha. For example, Poyang Lake and Dongting Lake are examples of lakes in this category (Ning et al. 1996). Lakes with an area between 6,667 and 66,667 ha are typically characterized by good water quality and are rich in natural food. They mainly depend on natural reproduction but are supplemented by some stocking. Some of these lakes have higher fishery productivity. For example, Honghu and Liangzi lakes have production between 110 and 600 kg/ha. Medium-size lakes with an area between 333 and 6,667 ha have comprehensive cultivation systems with supporting equipment and management technologies. Medium-size lakes combine stocking and feeding but mainly depend on feeding, targeting high fish production between 375 and 750 kg/ha or even higher, such as Donghu Lake and Niushan Lake. Small lakes (e.g., area <333 ha) adopt intensive culture in the whole lake using fertilization, feeding, and comprehensive management. Most of these lakes have very high fish production, exceeding 750 kg/ha (Wu et al. 2000).

Different pen-culture forms are also adopted along lake shores and vary by shoreline bathymetry (Wu et al. 1991). In general, the bay and shore areas suitable for pen culture account for about 10% of a lake. These areas are easy to divide and managed for low or high intensity aquaculture (Figure 4). One aquaculture method using a pen and low levee was developed for lakes with substantial changes in depth over time. In this method, the levee retains water when the water level drops but will not prevent the water from exchanging when the water level rises. In another method, fences are set up at shallow shores with a stable current velocity to allow for flowing water aquaculture. Effluents from shoreline pen culture naturally mix into the pelagic zone, further enhancing productivity. Small patches of pen culture are adopted in some shallow lakes (Wu et al. 1991).

Problems of Fisheries in the Yangtze Basin Lakes

Since the 1950s, lake fisheries in the Yangtze River basin have changed from mainly

Figure 4. Pen culture in a Yangtze basin lake.

depending on natural reproduction to artificial stocking. Lake fishery production has increased substantially and productivity of some medium to small lakes can reach 3,000 kg/ha (Figure 5). In the meantime, the demand for aquatic products has turned from quantity to quality, which led to diversified aquatic products and improved product quality. However, a series of problems in Yangtze basin lake fisheries began to emerge at the same time.

Decrease in lake area

The area of Yangtze basin lakes decreased rapidly because of land reclamation and sedimentation (Figure 6). Reclamation of land from lakes not only weakened the capacity for flood control of lakes and rivers, but also reduced waters available for fisheries. Since 1950, almost 100 lakes in the Yangtze River basin have disappeared entirely, and more than 542 km^2 of lake area has been lost (Yang et al. 2010).

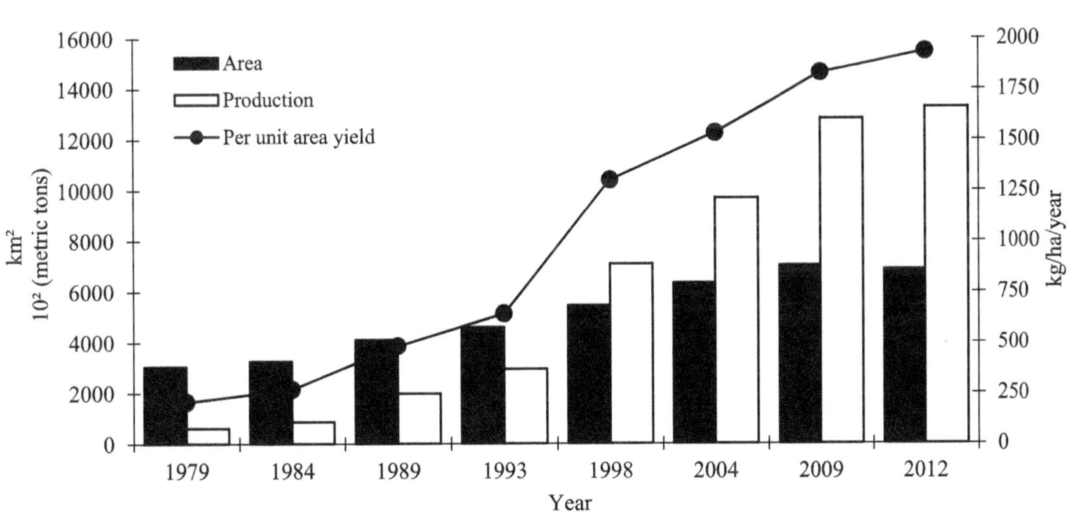

Figure 5. Fishery development in the Yangtze basin lakes (1979–2012).

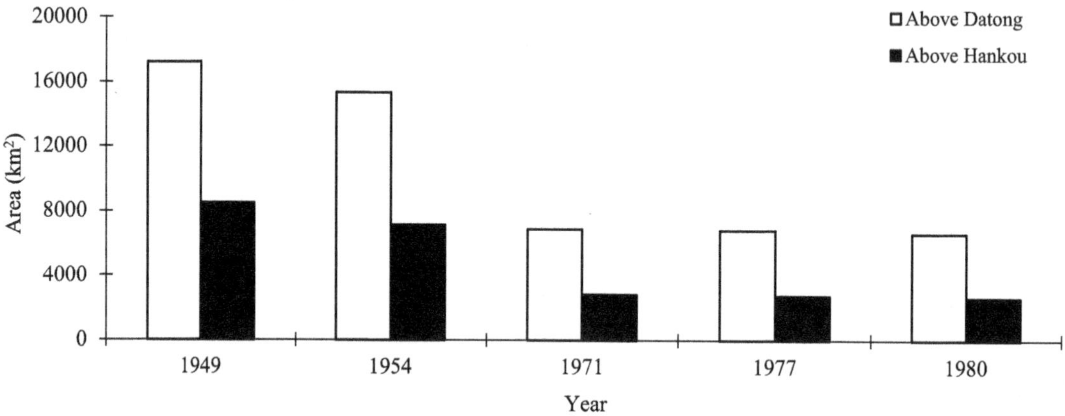

Figure 6. Lake area decrease in the middle reach of the Yangtze River since 1949. "Above Datong" includes the reach from Datong to the upper reaches of the Yangtze River; "Above Hankou" includes the reach from Hankou to the upper reaches of the Yangtze River (Yang et al. 2010).

Habitat fragmentation

Habitat fragmentation by water engineering projects has affected fisheries in the Yangtze basin lakes by blocking fish migration. Some fish species migrate between the Yangtze River and associated lakes for the purposes of spawning, feeding, or overwintering. Fish migration routes were blocked by dams and locks in the Yangtze River basin, which seriously affected recruitment and biodiversity in lakes. For example, in Honghu Lake, from the 1980s to 1990s, total fish species decreased from 74 to 57 because of habitat fragmentation and other reasons (Yang et al. 2010). Habitat fragmentation also blocked the intraspecific gene exchange and degraded the natural germplasm. Shi et al. (2009) provides an example of this relationship affecting Lake Anchovy Coilia ectenes in Taihu Lake.

Overstocking and overfishing

Chinese mitten crabs and Grass Carp were overstocked in many lakes, which shifted macrophyte-dominated lakes to phytoplankton-dominated lakes. Traditional stocking fisheries paid attention only to production and ignored protection of the aquatic environment, which seriously damaged lake ecosystems (Jin et al. 2003). Overfishing caused extreme decreases in fish resources and biodiversity that could not recover for a long time (Wu et al. 2000). Overfishing also resulted in catches of smaller fish, sometimes before they were mature. Moreover, illegal fishing gears and methods were used to harvest juvenile fishes, resulting in inadequate recruitment of fish capable of spawning (Zhang et al. 1996).

Water pollution and eutrophication

Economic development in the Yangtze River basin coincided with increases in industrial wastewater, domestic sewage, agricultural chemicals, chemical fertilizers, and waste feeds discharged into lakes (Zhang and Mei 1996; Liang et al. 2009). This water pollution resulted in an increase in fish diseases, which stimulated drug misuse in aquaculture and drug residues in aquatic products (Yang et al. 1999; Rico et al. 2012). Water pollution did not only lower the quality of aquatic products, but also caused a series of undesirable ecosystem effects, such as eutrophication, decrease in water transparency, and damage to aquatic plant resources. One striking example is Taihu Lake, in the Yangtze River basin near Shanghai (Figure 7). Taihu Lake was previously characterized by its abundant aquatic plants and good water quality, but in recent years cyanobacteria blooms have occurred frequently and aquatic plants have declined because of the

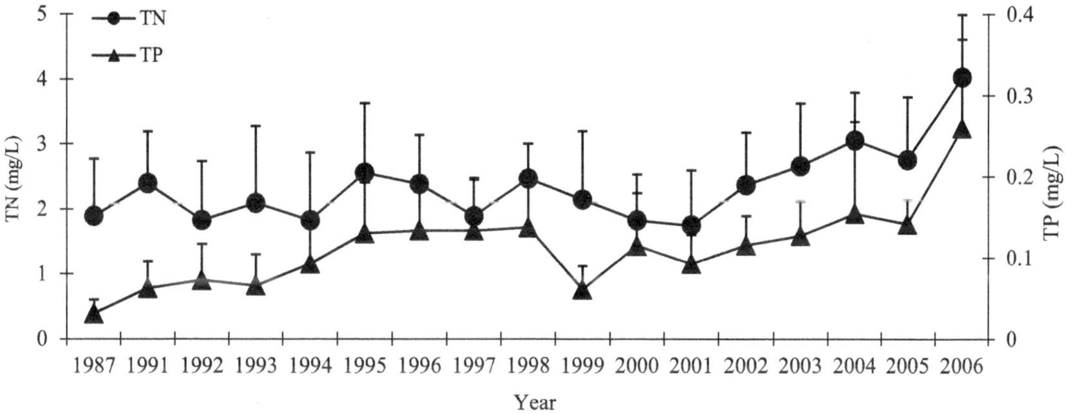

Figure 7. Variation of total nitrogen (TN) and total phosphorus (TP) in Lake Taihu during summer from 1987 to 2006 (Fan et al. 1997; Zhu 2008).

large amounts of industrial and agricultural wastewater discharges (Liu et al. 2007; Zhu 2008).

Lack of advanced fishery technology and management

The Yangtze River basin is the most developed area for lake fisheries in China. However, there remain many technology and management issues in lake fisheries: inadequate fishery facilities, lack of comprehensive and scientifically driven lake management, high labor input and cost, and low resource utilization efficiency. At present, the main cultured species are still the filter-feeding Silver Carp and Bighead Carp. However, past and current overfishing as well as poor fisheries and land management have created fisheries with low abundances of larger-bodied carnivores. As a result, many of the fisheries resources in these lakes have become dominated by undesirable, small-bodied fishes of low commercial value. (Xie et al. 2000; Zhang et al. 2008).

Reform Strategies for Fisheries in the Yangtze Basin Lakes

The general trend of lake-fishery management in China is as follows: large- and medium-sized lakes are mainly managed for natural propagation and stocking, and with some intensive cage and pen culture, depending on the type of water. Stocked species are not only limited to filter-feeding and benthic fishes, but also some carnivorous fishes, and there is a change from primary dependence on stocking towards protection and enhancement of natural resources and productivity. Small lakes are primarily under intensive culture to achieve high yields (Zhang et al. 1997). Lake fishery management should be developed into a comprehensive form and take full advantage of materials and energy within the lake ecosystem. It should gradually develop towards ecologically sustainable methods and attend to development and protection of natural resources, especially the protection of fish resources and environment. The management of lake fisheries should work towards multi-leveled, three-dimensional development and target both economy and ecology (Zhang 1992; Hu et al. 1999).

To improve fisheries of the Yangtze basin lakes, there are several tasks to consider:

1. Restoration and optimization of fishery resources and community structure. Establish breeding reserves to protect spawning grounds and larvae feeding grounds for key economic fish species. Spawning grounds should be protected and fishing should be forbidden during the breeding season. Lake sluice gates should be opened to the river during the breeding season so that

the juveniles spawned in the river can access lake nursery areas, which can improve the lake fish community and diversity. Efforts should be made to prevent the unintentional introductions of undesirable and potentially harmful species during stocking (Chen and Zhang 1986; Zeng 1995).

2. Improvement of aquaculture technology. Intensive culture technologies such as cage and pen culture should be developed and improved to take advantage of existing resource in both large and small lakes (Liu and He 1992). However, ecological principles should be followed to develop these intensive culture technologies. Further, biomanipulation may be applied in some lakes. For instance, carnivorous fishes must be managed and controlled to prevent excess predation on stocked fingerlings (Zeng 1995).
3. Optimize capture strategies. Reasonable stocking with a suitable capture strategy can ensure a healthy cycle in fishery ecosystems (Jin et al. 2003). Bycatch reduction can reduce damage to other fishes in the lake. In addition, standards for minimum harvest size and catch limits should be established in all commercial fishery lakes. Overfishing and harmful fishing methods should be strictly forbidden. Capture frequencies should be regulated to ensure that the fishermen can obtain higher profits (Wu et al. 2000; Wen et al. 2002).
4. Optimize lake fishery development modes. Fish processing should be further developed to improve the value of aquatic products and increase employment (Luo et al. 2005). Recreational fisheries should be another development angle for lake fisheries (Hu et al. 1999). Recreational fisheries are a relatively new fishery form in China that integrates many benefits, such as fishing, tourism, entertainment, and exercise. This new fishery form not only reduces environment pressure compared to that caused by the traditional fishery, but also brings greater economic and social benefits.

Conclusions

Fisheries in the Yangtze basin lakes have made great contributions to the freshwater fisheries of China, but there is great potential for further development and reformation. Many lakes remain that are not used or are not properly managed for fishery development. The current fishery production of the Yangtze basin lakes is 1,576 kg/ha annually, which can be further improved by adopting advanced aquaculture technologies and proper management. However, there are many problems related to lake fisheries in the Yangtze River basin, such as low economic benefits, huge resource consumption, environmental pollution, and water quality degradation. Therefore, coordinating the development of lake fisheries with drinking water supplies, tourism, and environmental conservation will be a primary goal of future Yangtze basin lake fishery development and reform.

References

Bureau of Fishery, China Ministry of Agriculture. 1949–1975. China fishery statistical yearbook. China Agriculture Press, Beijing. (In Chinese.)

Bureau of Fishery, China Ministry of Agriculture. 1979. China fishery statistical yearbook. China Agriculture Press, Beijing. (In Chinese.)

Bureau of Fishery, China Ministry of Agriculture. 1999. China fishery statistical yearbook. China Agriculture Press, Beijing. (In Chinese.)

Bureau of Fishery, China Ministry of Agriculture. 2011. China fishery statistical yearbook. China Agriculture Press, Beijing. (In Chinese.)

Bureau of Fishery, China Ministry of Agriculture. 2012. China fishery statistical yearbook. China Agriculture Press, Beijing. (In Chinese.)

Cao, L., W. Wang, Y. Yang, C. Yang, Z. Yuan, S. Xiong, and J. Diana. 2007. Environmental impact of aquaculture and countermeasures to aquaculture pollution in China. Environmental Science and Pollution Research International 14:452–462.

Chai, S., H. Dai, and H. Wang. 2010. A study on classification of recreational fishery resources and evaluation of its development. Journal of Ocean University of China 2010(1):32–37. (In Chinese.)

Changjiang Water Resources Commission. 2007. Changjiang zhi. [Knowledge of Changjiang.]

Encyclopedia of China Publishing House, Beijing. (In Chinese.)

Chen, G., and M. Zhang. 1986. Lake resources and part of the lake fisheries in China. Freshwater Fisheries 3:43–44. (In Chinese.)

Cong, Z., and T. Li., editors. 1993. History of fisheries in China. China Science and Technology Press, Beijing. (In Chinese.)

Dong, H., Q. Ming, and F. Shi. 1999. Impacts of reclamation on the flood regime in Poyang Lake and countermeasures. Journal of Lake Sciences 11(1):20–27. (In Chinese.)

Fan, C., J. Ji, and H. Chen. 1997. The current eutrophication status and trend of Lake Taihu and its comprehensive control countermeasure. Shanghai Environmental Science 16(8):4–7. (In Chinese.)

FAO (Food and Agriculture Organization of the United Nations). 2014. National aquaculture sector overview: China. Available: www.fao.org/fishery/countrysector/naso_china/en. (April 2014).

Fishery Resources Investigation Panel of the Yangtze River System.1990. The fishery resources of the Yangtze River system. China Ocean Press, Beijing. (In Chinese.)

Hu, C., J. Liu, and Z. Li. 1999. Sustainable utilization and protection of large water fishery resources. Reservoir Fisheries 19(5):50–53. (In Chinese.)

Jia, J. 1995. Ecological fishery and fishery ecology. Freshwater fisheries 25(4):29–32. (In Chinese.)

Jia, J. 2000. The freshwater fisheries of China in the 21st century. Freshwater Fisheries 30(1):3–6. (In Chinese.)

Jin, G., Z. Li, and P. Xie. 2003. The ecological foundation of sustainable lake fisheries and an example. Journal of Lake Sciences 15(1):69–75. (In Chinese.)

Li, M. 2004. A primary study on problems of ecological fishery. Modern Fisheries Information 19(11):5–8. (In Chinese.)

Li, S., and S. Xu., editors. 1995. Culture and capture of fish in Chinese reservoirs. International Development Research Centre, Ottawa.

Liang, W., Y. Cai, and L. Fang. 2009. Pollution in Taihu Lake China: causal chain and policy options analyses. Frontiers of Earth Science in China 3:437–444.

Liu, H., and Z. Huang, editors. 2008. Aquaculture in China. Science Press, Beijing. (In Chinese.)

Liu, J. 1980. A general report on increasing the fishery production of Lake Donghu. Oceanologia et Limnologia Sinica 11(2):185–188. (In Chinese.)

Liu, J., and B. He, editors. 1992. Freshwater fish aquaculture in China. Science Press, Beijing. (In Chinese.)

Liu, W., W. Hu, Y. Chen, X. Gu, Y. Chen, and J. Ji. 2007. Temporal and spatial variation of aquatic macrophytes in west Taihu Lake. Acta Ecologica Sinica 27(1):159–170. (In Chinese.)

Luo, X., G. Zou, G. Pan, and G. Yang. 2005. The status and development trends of freshwater fisheries in China. Journal of Yangtze University 2(5):98–102. (In Chinese.)

Ma, R., G. Yang, H. Duang, J. Jiang, S. Wang, X. Feng, A. Li, F. Kong, B. Xue, and J. Wu. 2011. Number, area, and spatial distributions of Chinese lakes. Science China Earth Sciences 41:394–401.

Ning, Z., K. Wei, and G. Chen. 1996. Development and reform of lake and reservoir fisheries in China. China Agriculture Press, Beijing. (In Chinese.)

Rico, A., K. Satapornvanit, M. M. Haque, J. Min, P. T. Nguyen, T. C. Telfer, and P. J. van den Brink. 2012. Use of chemicals and biological products in Asian aquaculture and their potential environmental risks: a critical review. Reviews in Aquaculture 4(2):75–93.

Shi, W., M. Zhang, K. Liu, D. Xu, and J. Duan. 2009. Stress of hydraulic engineering on fisheries in the lower reaches of the Yangtze River and compensation. Journal of Lake Sciences 21(1):10–20. (In Chinese.)

Wang, S., and H. Dou, editors. 1998. A directory of lakes in China. Science Press, Beijing. (In Chinese.)

Wen, Z., Z. Huang, and H. Liu. 2002. An ecological view: the sustainable development of lake fisheries in China. Reservoir Fisheries 22(4):1–3. (In Chinese.)

Wu, L., C. Liu, D. Zhang, and G. Liu. 2000. Problems and possible solutions of development and utilization of lake and reservoir fish resources in China. Journal of Jilin Agriculture University 22(4):100–104. (In Chinese.)

Wu, W., L. Jin, and N. Yang. 1991. The development trends of lake fisheries in China. Modern Fisheries Information 6(99):6–9. (In Chinese.)

Xie, S., Y. Cui, and Z. Li. 2000. Ecological studies on lake fisheries on piscivorous fishes: theory and methods. Acta Hydrobiologica Sinica 24(1):72–81.

Yang, H., X. Zhong, J. Han, L. Zhao. 1999. The effects of water pollution on fisheries of the Changjiang estuary. Journal of Fishery Sciences of China 6(5):78–82. (In Chinese.)

Yang, G., J. Jiang, S. Wang, Z. Liu, R. Wu, and P. Pu. 2004. The lake science development in China. Journal of Geographical Sciences 14(supplement):38–44.

Yang, G., R. Ma, L. Zhang, J. Jiang, S. Yao, M. Zhang, and H. Zeng. 2010. Lake status, major problems and protection strategy in China. Journal of Lake Sciences 22(1):85–92. (In Chinese.)

Zeng, Y. 1995. Development of comprehensive fishery technology about large-medium shallow plant-type lakes in China. Journal of Lake Sciences 7(1):85–92. (In Chinese.)

Zhang, G., W. Cao, and Y. Chen. 1997. Effects of fish stocking on lake ecosystems in China. Acta Hydrobilogica Sinica 21(3):271–280. (In Chinese.)

Zhang, Z. S., and Z. P. Mei. 1996. Effects of human activities on the ecological changes of lakes in China. GeoJournal 40(1–2):17–24.

Zhang, Y. 1992. Advances on aquaculture and propagation technology of lake and reservoir in China. Journal of Fisheries of China 16(2):179–187. (In Chinese.)

Zhang, T., Z. Li, and Q. Guo. 2008. Investigations on fishes and fishery of four lakes along the middle and lower basins of the Changjiang River. Acta Hydrobiologica Sinica 32(2):167–177.

Zhu, G. 2008. Eutrophic status and causing factors for a large, shallow and subtropical Lake Taihu, China. Journal of Lake Sciences 20(1):21–26. (In Chinese.)

Part 2 Endangered and Invasive Species, Biodiversity, and Conservation

American Fisheries Society Symposium 84:93–112, 2016

Paddlefishes and Sturgeons of the Yangtze and Mississippi Rivers: Status, Biology, and Management

QUINTON E. PHELPS*
Big Rivers and Wetlands Field Station, Missouri Department of Conservation
3815 East Jackson Boulevard, Jackson, Missouri 63755, USA

KELLY BAERWALDT
U.S. Fish and Wildlife Service, Midwest Region
1511 47th Avenue, Moline, Illinois 61265, USA

DAQING CHEN, HAO DU, LI SHEN, CHENYOU WANG, QIWEI WEI, AND HUI ZHANG
Yangtze River Fisheries Research Institute, Chinese Academy of Fishery Sciences
No.8, 1st Wudayuan Road, East Lake Hi-Tech Development Zone
Wuhan, Hubei 430223, P. R. China

JAN JEFFREY HOOVER
U.S. Army Engineer Research and Development Center
3909 Halls Ferry Road, Vicksburg, Mississippi 39180, USA

Abstract.—Paddlefishes (Polyodontidae) and sturgeons (Acipenseridae) are taxa of large, ancient fish shared by the Yangtze and Mississippi River basins. In the Yangtze, native Chinese Paddlefish (also known as Chinese Swordfish) *Psephurus gladius* are functionally extinct and Chinese Sturgeon *Acipenser sinensis* and Dabry's Sturgeon *A. dabryanus* are maintained through stocking. In the Mississippi River, native Paddlefish *Polyodon spathula* are secure, Pallid Sturgeon *Scaphirhynchus albus* are endangered, Shovelnose Sturgeon *S. platorynchus* are threatened, and Lake Sturgeon *A. fulvescens* are locally imperiled, although large robust populations of all these species exist in portions of their ranges. Paddlefishes and sturgeons of both rivers are habitat specialists, mature late in life (>7 years), and do not spawn every year (1–3-year intervals). With the exception of the Chinese Sturgeon, populations of all these species can be managed in both rivers through a combination of habitat preservation, habitat restoration, flow regulation, stocking of hatchery-reared fish, and regulation of harvest.

Common Challenges and Similar Faunas

The Yangtze and Mississippi basins are the largest river basins of Asia and North America and share similar environmental challenges. Both rivers are extensive, complex alluvial systems serving as the primary river arteries for China and the United States and for urban centers in the central portions of both countries. Both are impounded in their upper reaches and free-flowing to the sea in their lower reaches. These rivers provide navigational highways for the transportation of goods, water sources for consumption and irrigation, and fisheries for commercial and

* Corresponding author: quinton.phelps@mdc.mo.gov

recreational harvesters. As a result, the Yangtze and Mississippi rivers have essential economic, social, cultural, and ecological roles on both a national and global scale.

Because of connections between North American and Eurasia that ended in the Cretaceous period and because of their similar latitudes and geomorphology, the Yangtze and Mississippi rivers are inhabited by some of the same native animal groups (albeit different species), including alligators, softshell turtles, suckers, and several fish taxa.

Paddlefishes and sturgeons are large (5–500+ kg) fish shared by the Yangtze and Mississippi rivers. These megafishes are living fossils, with differing trophic classifications (i.e., piscivores and benthivores). Additionally, they share a high value as important natural resources and cultural icons. These groups, although taxonomically and ecologically distinct, share many characteristics in addition to their large size. Within a basin, they are monotypic to oligotypic: 1–3 species per group. They are nonperciform, lacking the spiny fins of perch-like fishes. They are adapted to life in turbid, fast-flowing environments; they spawn in swift main-stem open-channel habitats; and their reproduction is cued by seasonal changes in river flow. These last characteristics have made them sensitive to anthropogenic changes in hydrology. Native populations of both groups exhibited population declines in the 20th century, some of which have been precipitous and, in at least one case, probably irreversible.

Paddlefishes

Paddlefishes (class Actinopterygii, subclass Chondrostei, order Acipenseriformes, family Polyodontidae) are living fossils. Of the half-dozen species known, only two are extant. The extinct species, found in late Cretaceous deposits 70–75 million years old (Grande and Bemis 1991) or from more recent Eocene deposits 34–56 million years old (Cope 1883, 1885, 1886) are similar in form to each other and to extant species. All have small, inconspicuous teeth and a prominent, elongate snout (rostrum) with an internal network of stellate ossifications. Additionally, the caudal fin has prominent notched scales called fulcra. Extant species of paddlefish are superficially shark-like, with a cartilaginous skeleton, mostly naked skin, some denticle-like scales, and a heterocercal tail. Electroreceptors appear as spots and cover the surface of the rostrum, face, and operculum. Barbels are tiny and infrequently observed.

Yangtze River Paddlefish

Status.—The Chinese Paddlefish (also known as Chinese Swordfish) *Psephurus gladius* (Figure 1) is one of two extant paddlefish species. It is characterized by its wedge-shaped rostrum (along with several osteological features), distinguishing it from other fossil and living paddlefishes (Grande and Bemis 1991, 1998; Bemis and Kynard 1997) and inspiring common names like "elephant fish," "sword fish," and "beaked sturgeon" (Ping 1931; Liu et al. 1995). Unlike the more frequently depicted Paddlefish (also known as American Paddlefish) *Polyodon spathula*, the Chinese Paddlefish has a more cuneiform rostrum, a protrusible mouth, toothier jaws, shorter gill rakers, and more prominent fulcra along the upper lobe of the caudal fin (Tchang 1928; Wu 1930; Liu and Zeng 1988; Liu et al. 1995). This fish is mainly distributed in the Yangtze River basin (including the brackish water of the mouth)

Figure 1. Chinese Paddlefish of the Yangtze River. Photo credit: Q. Wei.

but was also recorded in the East China Sea, the Yellow Sea, and other rivers connected to those seas, such as the Qiantangjiang and Yongjiang rivers (YARSG 1988; Wei et al. 1997).

Within the Yangtze River, Chinese Paddlefish were commonly found in the mainstem channel. They may also be infrequently found in its tributaries, including the Minjiang, Tuojiang, and Jialinjiang rivers and Dongting and Poyang lakes (Liu and Zeng 1988). Some historical records exist for other basins, including the Hai, Yellow, and Huai rivers and the Bohai Sea (Gao et al. 2009). Construction of the Gezhouba Dam (closed in 1981) and Three Gorges Dam (closed in 2003) created two isolated subpopulations: one in the upper basin with access to known spawning grounds and the other in the middle and lower basin with uncertain access to spawning grounds (Liu et al. 1995). Principal threats to the species are hydroelectric projects (which block migration and alter habitats), overfishing, pollution, navigation, and dredging (Wei et al. 1997; Gao et al. 2009).

Chinese Paddlefish population declines began in the mid-20th century (YARSG 1988; Wei et al. 1997), and a substantial loss of genetic diversity occurred between the 1950s to the 1990s (Wu 2005). In the 1970s, the annual harvest of Chinese Paddlefish in the entire Yangtze River was about 25 metric tons. In the upper reach, above Chongqing, the catch was about 5 metric tons, with much of the catch consisting of large individuals; below Chongqing, the catch was greater, but most of the fish were smaller individuals (YARSG 1988). There were also substantial catches in the middle and lower Yangtze reaches at that time. However, population declines of Chinese Paddlefish in the 1980s were pronounced. Data for 1981–2002 from the upper, middle, and lower sections of the Yangtze River indicated that large numbers of young fish were observed in 1983 (587) and 1985 (819) below the Gezhouba Dam (Zhu and Yu 1987; Wei et al. 1997; Chen 2007; Gao et al. 2009; Zhang et al. 2009). However, only about 60 fish were observed for the entire period between 1986 and 1993 (Wei et al. 1997). Only two adult specimens (both females) were recorded in 2002 and 2003, despite extensive continued sampling. Among multiple stressors, overfishing in the 1970s and habitat degradation beginning in the 1980s are regarded as the main reasons for the decline (Wei et al. 1997).

Biology.—Unlike the filter-feeding Paddlefish of North America, the Chinese Paddlefish feeds principally on decapods and fishes, including pelagic minnows and benthic catfishes and loaches, with some prey of substantial size (Zhu and Yu 1987; Liu et al. 1995). Despite its ability to migrate long distances, the species reportedly lays under banks in still water and sucks prey into its large mouth (Williams 1883). It is presumed to be a strong swimmer and is believed to be diadromous (Ping 1931) with estuarine adults historically migrating more than 4,000 km to reach spawning grounds in the Jinsha River near Yibin (Liu and Zeng 1988; Liu et al. 1995).

Even though detailed biological investigations are now impossible because the Chinese Paddlefish is so rare, and presumed extinct, much is already known of its life history. In the Yangtze River, Chinese Paddlefish spawn in early spring in deep, swift water over firm substrates in the headwaters of the river (Liu and Zeng 1988). Males may outnumber females 2:1 and spawned eggs are large (2.3–3.8 mm diameter) and numerous, with fecundities of 178,000–786,000 per female (Liu et al. 1995; Wei et al. 1997). Within 30 d, fry can reach 150 mm, and by 60 d, 285 mm (Zhu and Yu 1987). The young drift downriver as they develop, with progressively larger young occurring further downstream (Liu et al. 1995; Wei et al. 1997). Juvenile growth during the first year is rapid (to 75 cm total length [TL]), and after that, they grow 15–20 cm/year (Ma et al. 1996). Mature males are 5–10 years old and mature females 6–16, with females only slightly larger than males. Their maximum size is unknown: traditional sources indicate a maximum length of 7 m (e.g., Ping 1931), but more recent data have indicated a maximum size of 363 cm (Q. Wei, unpublished data). Prior to 1980, mature individuals only occurred in the upper reaches of the Yangtze River, juveniles (4 cm TL) occurred in the Jiangan-Fengdu section of the upper Yangtze River, and fingerlings (75–250 mm TL, 5–95 g) occurred in the Xupu-Chongming section of the lower Yangtze River (YARSG 1988). Juveniles ranging 80–530 mm TL were found on the eastern tidal beach of Chongming Island (i.e., the Yangtze estuary) from 1983 to 1985. The Sichuan Fisheries Research Institute (1980) reported a spawning ground near Yibin as evidenced by the capture of a female with running eggs on April 6, 1980. The spawning site was described as 500 m in length, consisting of sand in the upper part and gravel or cobble in the lower part. Ages of spawning individuals ranged from 8 to 12 years, with egg diameters ranging from 2.8 to 3.8 mm. Juveniles were last recorded in the lower reaches of the river in 1995 (Chen 2007). Artificial reproduction of Chinese Paddlefish has not been successful.

Management.—Chinese Paddlefish, especially the young, are reportedly delicate in flavor and the species was "much sought by the people for food" and could be found in fish markets (Ping 1931). Although the fish was rare, it was still overexploited for meat through the 1970s (YARSG 1988). Fishing for Chinese Paddlefish was prohibited in 1983 by the national government, and since 1988, it was listed as a first-class state-protected animal in China. Chinese Paddlefish were readily captured in a variety of gear, including small-mesh nets, traps, and fyke nets, with 95% of the 231 records for 1980–2003 reported as fisheries bycatch (Wei et al. 1997; Chen 2007). Any kind of utilization of this fish is illegal. Currently, this species is considered to be on the verge of extinction or already extinct (Zhang et al. 2009).

It was hoped that Chinese Paddlefish populations would persist both above and below the completed Gezhouba and Three Gorges dams. The upper river population would have access to stable spawning grounds; the middle-lower river population would have access to limited (transient) spawning grounds and would be supplemented by downstream migrants passing over the dams (Wei et al. 1997; Gao et al. 2009). In addition, reservoirs would provide rearing habitat for Chinese Paddlefish and their prey, and both populations could be enhanced by stocking. Spawning and rearing preserves were proposed and, in some cases, established (e.g., Yichang and Chongming), but recent construction of the Xiangjiaba Dam inundated the principal upstream spawning areas for the species (Wei et al. 1997; Fan et al. 2006). Artificial propagation efforts, initially encouraging, have failed (Gao et al. 2009). More than 70% of citizens polled, however, indicate a willingness to pay approximately US$0.50 to $160 per year for the protection of Chinese Paddlefish and other species endangered by the Three Gorges Project. This could generate annual revenues greater than $\$1.00 \times 10^9$ (Xiao et al. 2009). These funds could be used to search for remaining fish, if any exist, and develop effective propagation for a long-term stocking program.

After two large specimens of the fish were found unexpectedly in 2002 and 2003, a series of measures were undertaken to save this species. These measures include the establishment of protected areas (Fan et al. 2006), fishing bans, a rescue network to save individuals caught as bycatch, and hydroacoustic and capture-based surveys (Zhang et al. 2009), as well as a cryopreservation study of sperm (Li et al. 2008) and gynogenesis studies (Zou et al. 2011a, 2011b) using North American Paddlefish as a surrogate. Additional planned measures include attempting to capture live specimens for use in intensive husbandry research (e.g., artificial gynogenesis, androgenesis, cloning, and surrogate broodstock technologies) to produce new individuals for stocking. Success of this work depends upon finding wild adult specimens.

Mississippi River Paddlefish

Status.—There is a single recognized species of paddlefish native to North America (Figure 2). Its historic range includes the Mississippi and Missouri River basins, the Lau-

Figure 2. Paddlefish of the Mississippi River. Top, adult entrained at Bonnet Carré diversion of the lower Mississippi River; bottom, preserved juvenile. Photo credit: J. J. Hoover.

rentian Great Lakes, and some smaller river basins terminating in the Gulf of Mexico (Jennings and Zigler 2009). The Paddlefish of the Mississippi River once reached sizes greater than 2 m TL and more than 70 kg (e.g., Nichols 1916). Because of its elongate, almost-flat rostrum and heterocercal tail (especially in juveniles), the Paddlefish is easily distinguishable from any other North American fish. Its large size and swift water habitat precluded extensive collections of museum specimens; therefore, comprehensive taxonomic studies do not exist. During the 19th century, however, Paddlefish were believed to represent two separate taxa: a smaller, longer-snouted toothed species with caudal lobes of unequal size and a larger, shorter-snouted toothless species with caudal lobes of equal size (LeSueur 1818; Rafinesque 1820; Kirtland 1843–1844, 1845–1847). These characters, however, are influenced by age of the fish. Genetic investigations showing a lack of differentiation among upper and middle basin populations (Carlson et al. 1982) concluded that the Paddlefish in the Mississippi River basin was monotypic. Subsequent investigations of genetics and morphology, however, hint at the existence of some distinct locally adapted populations of uncertain taxonomic status (Epifanio et al. 1996; Hoover et al. 2009a).

Paddlefish were abundant throughout the 1800s. However, commercial fisheries developed during the early 1900s were coincident with rapid declines of local populations (Stockard 1907; Wagner 1908; Alexander 1914). Subsequent alteration of rivers via channelization and impoundment (reservoirs, lock-and-dam structures), especially in the upper Mississippi River, is also likely a principal factor associated with rangewide declines (Carlson and Bonislawsky 1981). In addition, public perception of Paddlefish as a rough fish and as a competitor with game fish prompted some state agencies to control populations in the southern United States (e.g., Bryan 1942; Cottrell 1944). By the 1970s, populations had disappeared from the Great Lakes and eastern states, and Paddlefish were abundant only in parts of the Mississippi River drainage. Some of those populations, like those of the Mississippi Delta, have recovered during recent decades (Scholten and Bettoli 2005).

The Paddlefish is considered "vulnerable" by the International Union for Conservation of Nature (comparable to U.S. Fish and Wildlife Service/National Marine Fisheries Service terminology "threatened"), "secure" in the United States (i.e., no listing of imperilment by the U.S. Fish and Wildlife Service), "apparently secure" in the global scale by NatureServe, and "vulnerable" by the American Fisheries Society (comparable to the U.S. Fish and Wildlife Service's "special concern"; IUCN 2004; Jelks et al. 2008). Population status varies geographically. For the 26 states representing the historic range in the United States, populations in 2006 were increasing in 2, increasing or stable in 4, stable in 10, stable or decreasing in 2, decreasing in 1, extirpated in 4, and stable and/or unknown in 3 (Bettoli et al. 2009). Declining and imperiled populations are all at the periphery of the range, with four extirpations at the eastern edge (i.e., New York, Pennsylvania, Maryland, and North Carolina). In 2010, Paddlefish were designated by local governments as extirpated in two states and one province (Michigan, North Carolina, and Ontario), endangered in none, threatened in five states (Minnesota, Ohio, Texas, Virginia, and Wisconsin), and requiring some other level of recognition or protection in four states (North Dakota, Arkansas, Montana, and Alabama; NANFA 2010). Habitat degradation and overharvest continue to threaten viability of Paddlefish populations (Gerken and Paukert 2009; Quinn 2009). Because Paddlefish frequently swim close to the surface, they are prone to injury from boats, more so by small recreational vessels than commercial barges (Rosen and Hales 1980; Killgore et al. 2011). In addition, the establishment and proliferation of bigheaded carps threatens Paddlefish through depletion of planktonic food and competitive exclusion (Pegg et al. 2009).

Biology.—The Paddlefish is highly adapted for life in turbid, turbulent, fast-flowing rivers. Its rostrum is covered with a forward extension of their lateral line system. Its more than 40,000 ampullae of Lorenzini (Allis 1903; Kistler 1906; Jørgensen et al. 1972) make it an antennae for nonvisual (vibration and electrical) signals given off by the zooplankton on which the Paddlefish feeds (New and Bodznick 1985; Wilkens et al. 1997). Paddlefish swim continuously at moderate to fast speeds (1–4 body lengths/s) and are capable of moving great distances in short times (Burggren and Bemis 1992; Hoover et al. 2009b; Hoover et al. 2013; Patterson et al. 2013). These movements, along with their rapid growth and large size (Adams 1942; Reed et al. 1992), are energetically costly. Metabolic demands are offset, in part, by low-effort, cost-effective respiration (ram ventilation) and passive feeding on highly caloric zooplankton. The rostrum is believed to confer hydrodynamic advantages of increased lift with reduced drag (Hoover et al. 2009a, 2009b) at little additional risk or cost to the fish via its unique rhomboid-mesh endoskeleton of tiny cartilaginous stellate "bones," which are strong, resilient, and lightweight (Allison et al. 2013). The mouth is large, but unlike that of the Chinese Paddlefish, the upper jaw is firmly attached to the head with the lower jaw dropping open widely.

Life-history characteristics of the Paddlefish render it sensitive to habitat alteration and harvest (Jennings and Zigler 2009). Successful spawning takes place in swift water over firm substrates (typically gravel). Life spans are long (25–55 years), but age at maturity is high (4–14 years). Fecundity is high (egg masses comprise up to 25% of the body mass). Reproduction for a single female in the Mississippi River typically occurs at multiyear intervals but, based on data from other southern U.S. rivers, could occur annually. Because of these *K*-selected attributes, habitat modifications that reduce survival or harvest of reproductive adults may take several decades to fully recover due to lag time in compensatory recruitment.

Management.—Paddlefish are exploited primarily as a source of caviar, but also as a source of meat and protein (for fish meal). With the closure of sturgeon fisheries in most of the Mississippi River basin (to protect Pallid Sturgeon *Scaphirhynchus albus*), there has been increased demand by these industries for Paddlefish products. States formerly closed to Paddlefish harvest (e.g., Mississippi), have reopened commercial fisheries, but with greater control and oversight of harvest. In addition, there are recreational fisheries for Paddlefish throughout most of their range, but with small harvest limits (typically 1–2/d).

Populations are managed through a combination of harvest restrictions, stocking, and habitat restoration. The majority of states within the historic range (18/25) have either never had commercial fisheries for Paddlefish (7/18) or were closed sometime during the period 1925–1992 (11/18) (Scholten 2009). The remaining seven states allowing commercial harvest (Arkansas, Illinois, Indiana, Kentucky, Mississippi, Missouri, and Tennessee) use the following techniques: fishing seasons to reduce take of reproductive fish (three states), minimum size limits (four states), and gear restrictions (minimum mesh size and length of nets, and net attendance requirements). Stocking hatchery-reared fish to supplement natural recruitment has been extensively practiced in 12 states since 1994, 4 of which also support commercial fisheries (Grady and Elkington 2009). Numbers stocked typically range from 2,000 to 30,000/state, with more than 2 million released range-wide since 1983. Mean cost of producing a stockable fingerling is less than $5. Last, there are collaborative habitat restoration projects by the U.S. Army Corps of Engineers (USACE) and U.S. Fish and Wildlife Service (USFWS) to restore generic riverine fish habitat. This is being done primarily though the reconnection of isolated secondary channels (Boysen et al. 2012). Although not specific for Paddlefish, such habitats with restored flow may provide expanded spawning and rearing grounds for fishes.

Figure 3. Chinese Sturgeon of the Yangtze River. Top, captive adult in aquarium; bottom, breeding adult captured at Gezhouba spawning grounds. Photo credit: Q. Wei.

Sturgeons

Sturgeons (class Actinopterygii, subclass Chondrostei, order Acipenseriformes, family Acipenseridae) represent one of the most ancient groups of cartilaginous fishes (Gardiner 1984). Dated fossils indicate that sturgeons have existed since the Upper Cretaceous for more than 100 million years and have undergone little change in form since that time. All are cylindrical to pentagonal in cross section, with flat bellies. They have cuneiform or rounded rostra, substantially shorter than that of paddlefishes but similarly equipped with numerous ampullae for electroreception and barbels for chemoreception. Unlike those of paddlefishes, the barbels are conspicuous, long, and frequently ornate. Jaws are protrusible and unattached to the head. Prominent plates or scutes occur in five longitudinal rows along the body dorsally, laterally, and ventrolaterally. There are approximately 25 species of sturgeon in the world, of which 5 occur in China and 9 occur in the United States (Pikitch et al. 2005).

Yangtze River Sturgeons

Status.—There are two sturgeon species in the Yangtze River: the Chinese Sturgeon *Acipenser sinensis* (Figure 3) and the Dabry's Sturgeon (also known as the Yangtze Sturgeon) *A. dabryanus*, locally called *Sha Lazi* (Figure 4). These are similar in appearance, and until the 1960s, there was taxonomic confusion regarding the two species due, in part, to low numbers and small size of specimens available to ichthyologists (Nichols 1943; Wu 1963). Random amplified polymorphic DNA analysis showed that they are closely related

Figure 4. Dabry's Sturgeon of the Yangtze. Top, adult. Photo credit: Chinese Academy of Sciences and Shanghai Natural History Museum (1982). Bottom, juvenile. Photo credit: Q. Wei.

(Zhang et al. 1999), but differences in morphology, size, and migratory habits delineate the species (YARSG 1988). Chinese Sturgeon have 50–70 dorsal fin rays and 32–40 anal fin rays, uniform coloration laterally, a long rostrum, and less than 28 gill rakers. Dabry's Sturgeon have 40–49 dorsal fin rays and 27–30 anal fin rays, dark dorsolateral and white ventrolateral coloration, a short rostrum, and greater than 30 gill rakers. Recent descriptions of 25 juvenile Chinese Sturgeon (30–50 cm TL) and 14 adult Dabry's Sturgeon (also 30–50 cm TL) indicate substantial overlap in scute counts and fin ray counts but no overlap in gill raker counts: 14–28 for Chinese Sturgeon and 32–55 for Dabry's Sturgeon (Vescei and Peterson 2007). In addition, the body of juvenile Dabry's Sturgeon is very rough due to many bony plates on the skin (Fu 1988).

Both species were once widespread and abundant throughout the basin but have declined substantially in range and in numbers. Factors causing the declines include overfishing, dam construction, pollution, and river traffic (Wang et al. 2012a). Both Chinese Sturgeon and Dabry's Sturgeon were listed in appendix II in CITES (Convention on International Trade in Endangered Species of Wild Fauna and Flora) in 1997 and classified as critically endangered on the IUCN red list in 2010 (IUCN 2010). The Chinese government has listed both species as national first-level protected animals, the Chinese Sturgeon in 1988 and the Dabry's Sturgeon in 1989. Despite a long history of protection, the populations have not recovered.

Chinese Sturgeon reach 200–500 cm TL, weigh up to 600 kg, and are anadromous. Historically, they were recorded from the East China Sea, the Yellow Sea, the Taiwan Strait, and the large rivers draining into those seas, including the Yangtze, Pearl, Minjiang, Qiantang, and Yellow rivers (Bemis et al. 1997; Wei et al. 1997). The species was also recorded in southwestern Korea and western Kyushu of Japan (YARSG 1988; Wei et al. 1997). Currently, Chinese Sturgeon may be extirpated in all rivers except the Yangtze (Wei et al. 1997). In the Yangtze River, the species has declined substantially following the closure of the Gezhouba Dam in 1981, which blocked its spawning migration route (Yue and Chen 1998). Chinese Sturgeon in the Yangtze River spawn in the fall while Pearl River populations spawn in the spring. Due to this behavior, it was questioned whether they represented different species or were a subspecies (YARSG 1988). Interestingly, the type specimen in 1834 was likely from the Pearl River. Taxonomic distinctions among populations are uncertain, but at present, the Chinese Sturgeon in the Yangtze and Pearl rivers are believed to be the same species (Wei 2003).

The Dabry's Sturgeon is smaller (<130 cm TL and <25 kg) and potamodromous. Historically, it was found in headwater streams (e.g., the Jinsha River), to the middle reaches and large tributaries of the Yangtze River (e.g., the Minjiang, Tuojiang, Jialingjiang, Xiangjiang, and Hanjiang rivers), as well as the large lakes connecting to the Yangtze River. In all, its range included almost two-thirds of the total basin (YARSG 1988; Zhuang et al. 1997). The Dabry's Sturgeon was first described by Duméril in 1868 based on the type specimen collected from the Yangtze River by M. Dabry and now kept in the Museum National D'Histoire Naturelle, Paris (Zhang et al. 1993). Later, Günther (1873), Wu (1930), Nichols (1943), Wu (1963), and Fu (1988) studied and redescribed the biological characteristics of this species using additional specimens from the Yangtze River. Because it is endemic to the Yangtze and highly migratory, the Dabry's Sturgeon is assumed to consist of a single population.

Prior to 1981, commercial fisheries for Chinese Sturgeon were mainly in the upper and middle Yangtze River. During the 1970s, an estimated 10,000 Chinese Sturgeon migrated upstream to spawn (YARSG 1988). From 1972 to 1980, the average annual fishery was equivalent to about 517 individuals. In the first 2 years after the closure of the Gezhouba Dam in 1981, the annual fishery increased to 1,163 individuals because sturgeon gathered downstream of the dam in large numbers. Of that number, only 161 were captured upstream of the dam. In 1982, commercial capture of Chinese Sturgeon was banned and only scientific capture was authorized (Wei 2003). The breeding population was estimated at 2,176 (95% confidence interval = 946–4,169) based on results of mark-recapture studies in 1983 and 1984 (Ke and Wei 1992). From 1996 to 2001, Wei (2003) used mark-recapture methods to show that the breeding population fluctuated from 292 to 473 individuals, with an average of 363. In 2002, there were only 222 breeders captured before the spawn and 61 after the spawn (Qiao et al. 2006). In 2013 and 2014, no naturally spawning Chinese Sturgeon were observed, although tens of adult individuals were detected in the spawning reach by hydroacoustic methodology (Wei, unpublished data). A decline in juveniles was coincident with the decline in breeders. At the Xupu monitoring site near the estuary of Yangtze River, they declined from more than 10,000 individuals in the 1990s to 4 individuals in 2012 and 0 individuals in 2014 (Wei et al. 1997; Li et al. 2011; Main 2014).

Dabry's Sturgeon supported a commercial fishery as recently as the 1970s, when fishermen harvested 5,000 kg during the spring spawning migration in Sichuan Province (Zhuang et al. 1997). The decline of the Dabry's Sturgeon was initially caused by overharvest and subsequently exacerbated by the construction of the Gezhouba Dam in 1981 on the main stem of the Yangtze River. The dam, in the middle of the species' geographic range, blocked upstream migrations to spawning and foraging grounds and killed or injured downstream migrants as they passed through turbines at the dam (Yu et al. 1986; Zhuang 1994; Wei et al. 1997; Zhuang et al. 1997; Chang and Cao 1999). Since 1982, less than a hundred specimens have been captured in the Yangtze River annually, and there have been no captures downstream of the Gezhouba Dam since 1995 (Chen 2007; Zhang et al. 2011). Natural spawning and recruitment of Dabry's Sturgeon presumably ceased sometime around or after 2000.

Biology.—Gross morphology of the two sturgeons is consistent with that of other sturgeons around the world: fusiform body, flat breast and belly, slender, five rows of scutes, four barbels arranged in a single row, and a heterocercal tail. The snout is pointed and slightly upturned in juvenile Chinese Sturgeon but becomes broadly obtuse in older adults (YARSG 1988; Yue and Chen 1998). The ventral and lateral surfaces of the head are covered with electroreceptive ampullae of Lorenzini. Four barbels are arranged in one row on the underside of the snout. The dorsal fin is situated far rearward while the pectoral fin is well developed and located near the belly.

Chinese Sturgeon are anadromous. Between June and October, they historically migrated to the upper reaches of the Yangtze River and lower Jinsha River. After construction of the Gezhouba Dam, they migrate only as far as the tailwater of the dam, where they purportedly spawn. They migrate principally in the main channel and along the bottom of the river, but also rise to the water surface once about every 2–3 h (Watanabe et al. 2008; Wang et al. 2012b). After hatching, early life stages migrate downstream and reach the Yangtze Estuary after 6–8 months. Juveniles enter the sea and rear in the continental shelf waters and can travel as far as the eastern Sea of Korea (Wei, unpublished data). They forage in marine waters until they return to the Yangtze River. The prespawning staging habitat of Chinese Sturgeon is in the deep water downstream of spawning locations such as Yanzhiba Bar and Tiaoguan (Wang et al. 2012b). At least 16 historical spawning sites were known in the reach upstream of Fuling to the lower Jinsha River downstream of Xinshi, a 600-km reach (YARSG 1988; Wei et al. 1997). After the Gezhouba Dam was built in 1981, a single new spawning location was discovered below the dam in 1982. This location was apparently used by the entire remaining population. This site has been monitored annually and no spawning activities were detected in 2013 and 2014, suggesting that spawning failures may occur in some years, which will result in some year-classes becoming absent from the population. Suitable spawning locations must include rock or gravel substrates, variations in water depth and channel width, a bend in the river, and moderate distance with moderate slope (YARSG 1988; Zhang et al. 2007a, 2007b, 2009). In addition, a combination of the correct water temperature, water level, and current velocity is critical for spawning success. Females mature at 14–26 years old and males at 8–18 years old (YARSG 1988; Wei et al. 1997; Wang at al. 2012a). Females spawn at 4-year intervals while males spawn every 3 years (Wang et al. 2012a).

The Dabry's Sturgeon is potamodromous. It inhabits sandy shoals with silt substrates and low water velocities. When water levels rise in spring, fish move from the main channel into the tributaries to feed. Young individuals often stay in sandy shallows and occur frequently in stretches between Luzhou and Jiangjing, Sichuan Province, where the current is slow (Zhang et al. 1999). Although the Dabry's Sturgeon adapts to a variety of habitats, individuals prefer sublittoral areas 10–20 m from the riverbank, with a current velocity of about 1 m/s, water depth of 8–10 m, and substrates of sand and silt with abundant detritus and benthos for most of the year. Diets of juveniles and adults include fish and benthic invertebrates (Xiong 1988; Zeng 1990). Dabry's Sturgeon are often found in areas where wastewater from drainage pipes promote growth of benthos (Zhuang et al. 1997; Kynard et al. 2003). Dabry's Sturgeon are sedentary and do not move for substantial distances except during spawning runs, when they swim upstream individually (not as groups). After spawning, the fish gradually move to the sandy shoals where they resume feeding. Spawning occurs mostly during March or April in the 300-km-long reach between Yibin City in Sichuan Province and Jiangjin in Chongqing Municipality (Zhuang et al. 1997). A few records, however, suggest that some individuals spawn in the fall (Yang 1986; Zeng 1990). Spawning occurs on rubble, cobble and gravel substrates, in clear water with velocities of 1.2–1.5 m/s and depths of 5–15 m. Males mature in 4 to 7 years at sizes greater than 4.5 kg. Females mature in 6 to 8 years at sizes greater than 9 kg. Male Dabry's Sturgeon can spawn annually while most females cannot. Under aquaculture conditions, the reproductive intervals of females are 1–2 years. At water temperatures of 17–18°C, embryos hatch approximately 5 d after fertilization (Xie 1979; Xie et al. 1988).

Management.—Before construction of the Gezhouba Dam, commercial capture of Chinese Sturgeon occurred mainly in the upper and middle Yangtze River (YARSG 1988; Wei et al. 1997). Since then, methods have been initiated to protect Chinese Sturgeon, including a fishing ban on these species since 1982, seasonal localized bans on all fishing, and establishment of reserves. In addition, artificial propagation and release has been an effectively used sturgeon conservation method that has been carried out for more than 40 years.

In 1972, low numbers of Chinese Sturgeon larvae were released into the Sichuan Reach of the Yangtze River, representing the first release of cultivated individuals. Large-scale releases were implemented after 1983 (Wei 2003). In the early stage of those programs, only larvae were released because of the rather low survival rate of larger fry before 1997. In 1997, the survival rate of larger fry increased to more than 80%, and in 1998, more than 30,000 juvenile sturgeon, with a body length greater than 10 cm, were released. Afterwards, more than 100,000 larger sturgeon have been released annually. In the past 31 years, more than 6,800,000 juvenile Chinese Sturgeon have been released in the Yangtze River. To further long-term protection of this species, the Chinese government and researchers have formulated the Chinese Sturgeon Conservation Action Plan (2011–2030).

Controlled propagation and domestication of Dabry's Sturgeon began in the early 1970s (Xie 1979). Because commercial fishing for Dabry's Sturgeon in the Yangtze River was banned in the early 1980s, few individuals were available for aquaculture. Captive propagation, however, has been successful in recent years (Wei, unpublished data) and tens of thousands of F_2 generation larvae are produced each year. This suggests that commercial harvest of the species may become possible again sometime in the future. In addition to commercial fishing bans and aquaculture propagation, several other conservation strategies have been pursued for the recovery of the Dabry's Sturgeon. Sturgeon caught as bycatch are reported to officials and returned to the river immediately after capture. Also, a national nature reserve that covers 436.5 km of the mainstream upper Yangtze River (total area of 331.7 km^2) was established in 2000 (Fan et al. 2006). Since 2003, a generic fishing ban has been in effect in the upper Yangtze River basin from February through April. This measure effectively protects all potential spawning areas and habitats for the entire spawning season of the Dabry's Sturgeon (Zhuang et al. 1997; Zhang et al. 2011). Since 2007, restocking programs have been conducted with thousands of hatchery-produced fingerlings released into the upper and the middle Yangtze River each year. Studies on direct genetic parameters and indirect parameters showed that the continuous decline in wild populations of Dabry's Sturgeon has caused a loss of genetic diversity in the present populations. Thus, effective management is urgently needed to preserve genetic diversity for both the captive and wild populations (Wan et al. 2006).

Mississippi River Sturgeons

Status.—In the Mississippi River basin, there are three potamodromous extant native sturgeon species: Lake Sturgeon *Acipenser fulvescens* (Figure 5), Pallid Sturgeon (Figure 6), and Shovelnose Sturgeon *Scaphirhynchus platorynchus* (Figure 7). The Lake Sturgeon's historic range encompassed most of the states and provinces within the Great Lakes Basin, Hudson River basin, and the Mississippi River basin (Peterson et al. 2007). However, these populations have declined due to historic overharvest and habitat degradation (Noakes et al. 1999; Auer 2004; Wilson and McKinley 2004). Kyle Bales (Southeast Mis-

Figure 5. Lake Sturgeon of the Mississippi River. Photo credit: David Ostendorf.

Figure 6. Pallid Sturgeon of the Mississippi River. Photo credit: David Ostendorf.

Figure 7. Shovelnose Sturgeon of the Mississippi River. Photo credit: David Ostendorf.

souri State University, unpublished data) recently compiled a comprehensive status review of Lake Sturgeon throughout North America and reported that greater than 70% of the states and provinces within the Lake Sturgeon's historic range considered the population to be either threatened, endangered, or a species of special conservation concern. Furthermore, in an additional 20% of these states and provinces, Lake Sturgeon have not been captured recently, suggesting local extirpation (Bales, unpublished data).

In contrast to the Lake Sturgeon, the Pallid Sturgeon has a relatively restricted historical range that includes only the Missouri, Platte, Yellowstone, Mississippi and Atchafalaya rivers (Wilson and McKinley 2004). Similar to other sturgeons in the Mississippi River basin, Pallid Sturgeon populations have declined. These declines have been attributed to habitat degradation and overharvest (Pikitch et al. 2005). The Pallid Sturgeon was listed as endangered in 1990 under the Endangered Species Act (Dryer and Sandvol 1993).

The Shovelnose Sturgeon historically occurred throughout the majority of the Mississippi River, Ohio River, and Missouri River basins (Koch and Quist 2010). Their range extended as far north as Montana and southward to Louisiana and from Pennsylvania westward to New Mexico (Keenlyne 1997). Even though the Shovelnose Sturgeon has not experienced a substantial range reduction within the Mississippi River basin, population declines have been associated with multiple stressors, including habitat degradation and commercial harvest (Phelps et al. 2013). Furthermore, the Shovelnose Sturgeon has been listed as threatened in areas of sympatry with Pallid Sturgeon due to their similarity of appearance. However, other locations on the upper Mississippi River remain open to commercial harvest (Phelps et al. 2013).

The two genera of sturgeon are readily distinguished from one another based on rostrum, scutes, and tail characteristics (Vecsei and Peterson 2004). *Acipenser* sturgeons, including Lake Sturgeon, have rounded conical snouts, deep bodies with reduced scutes, and thick fleshy caudal peduncles, features well suited for fishes that inhabit sluggish habitats. In contrast, *Scaphirhynchus* sturgeons, including the Shovelnose and Pallid sturgeons, have flattened shovel-shaped snouts, slimmer bodies with well-defined and sharply keeled scutes, and long, narrow, fully armored caudal peduncles, features well suited for a fish inhabiting swift water. The body shape of *Scaphirhynchus* minimizes hydrodynamic lift and its rough surface maximizes friction with the river bottom, allowing the sturgeon to hold station in fast-flowing water and avoid being swept downstream (Webb 1994).

Biology.—Lake Sturgeon are long-lived (lifespan between 50 and 150 years), late maturing (>15 years), and periodic spawners (Peterson et al. 2007). Spawning begins when water temperatures reach 10–15°C. Spawning locations are largely unknown in the Mississippi River basin, but spawning is generally thought to occur on gravel bars with flowing, potentially shallower water (David Herzog, Missouri Department of Conservation, personal communication).

Pallid Sturgeon and Shovelnose Sturgeon are closely related to one another, and although Pallid Sturgeon can achieve much larger size, the two species can be difficult to differentiate morphologically. Differences in maximum size, barbel length, mouth size, and meristic (e.g., fin ray counts) exist. However, ontogenetic and latitudinal variation is substantial and can confound identifications even by experienced biologists (Murphy et al. 2007). An inability of commercial fishers to identify and release endangered Pallid Sturgeon can substantially damage their populations. It was this concern for the ability to distinguish between species that prompted the USFWS to list the Shovelnose Sturgeon as threatened (to prevent accidental commercial harvest) in the areas of the Mississippi River where their range overlaps with the endangered Pallid Sturgeon (Phelps et al. 2013).

The existence of morphologically intermediate forms suggests that hybridization may occur (and may be frequent) between Pallid and Shovelnose sturgeons. Genetic data both supports (Schrey and Heist 2007) and fails to support hybridization (Ray et al. 2007). Multivariate analyses of morphological data from field-collected specimens in the middle and lower Mississippi River indicated that the two species, although variable in appearance, are distinct (Murphy et al. 2007).

Long life spans (>20 years), late maturity (>7 years), and infrequent spawning (at least 2- and sometimes more than 3-year intervals) are life-history strategies common to the Pallid Sturgeon (Keenlyne and Jenkins 1993; Killgore et al. 2007a) and Shovelnose Sturgeon (Keenlyne 1997; Morrow et al. 1998; Koch and Quist 2010; Phelps et al. 2013). Despite these similarities, differences exist in their biology. Juvenile Pallid Sturgeon are less likely than Shovelnose Sturgeon to free-swim at high water velocities (Adams et al. 2003). Pallid Sturgeon adults are piscivorous while Shovelnose Sturgeon are not (Hoover et al. 2007). Pallid Sturgeon grow more rapidly than Shovelnose Sturgeon (Killgore et al. 2007b). Differences in locomotion and feeding behavior may be sufficient for ecological segregation of the two species and could, in part, account for differences in growth rates.

Management.—Because of the reported population declines, state or federal listing (e.g., threatened or endangered) has occurred for all three sturgeons. Because of this, much effort by state agencies, federal agencies, nongovernmental agencies, and universities has been devoted to protect or enhance these populations (Boreman 1997). The Lake Sturgeon is of conservation concern (endangered in Missouri), and the Pallid Sturgeon has been listed by the USFWS as an endangered species. While Shovelnose Sturgeon is the most abundant of the Mississippi River sturgeons, it is protected from harvest (classified as threatened) in a substantial portion of its range (Dryer and Sandvol 1993; Peterson et al. 2007; Phelps et al. 2013). Because of the reduced numbers within the Mississippi River basin, Lake Sturgeon are currently protected from commercial harvest. Nonetheless, incidental take may occur (Herzog, personal communication).

Many states within the Mississippi River basin have intensive Lake Sturgeon monitoring and stocking programs. For example, the Missouri Department of Conservation has been stocking Lake Sturgeon in the Mississippi River since the 1980s (MDC 2007). These stocking efforts have resulted in catches of Lake Sturgeon exceeding 25 kg and are reproductively viable (Bales, unpublished data). Pallid Sturgeon are monitored rangewide and intensively cultivated and stocked in some reaches. Initial stockings in 1994 and 1997 occurred in the lower Missouri River and middle Mississippi River; subsequent efforts have been concentrated in the upper Missouri River (Krentz 2005). Pallid Sturgeon are not part of a regular stocking program in the lower Mississippi River and have been stocked only once (USFWS 2000).

Regardless of location along the Mississippi River, federal no-take commercial or recreational harvest regulations exist for Pallid Sturgeon. Shovelnose Sturgeon are legally fished in some locations throughout the Mississippi River basin; however, commercial regulations are strict (length limits) and intended to protect spawning adults (Phelps et al. 2013). Limited stocking of Shovelnose Sturgeon has occurred previously but is not current practice in the Mississippi River basin.

Recovery and Conservation

Native sturgeons and paddlefishes in the Yangtze and Mississippi rivers face comparable challenges to their viability, including (1) impounded upper reaches (with concomitant loss of spawning and nursery areas), (2) pollution associated with urban centers, (3) river engineering (reducing off-channel habitat quantity), (4) overexploitation of fish populations (i.e., excessive take of reproductive adults), and (5) insufficient information on demography and taxonomic resolution of local populations. Despite these similarities, the status of the fishes in the two rivers is different. In the Yangtze River, sturgeons and paddlefish are all declining, critically imperiled, and, in the case of the Chinese Paddlefish, probably functionally extirpated or extinct. In the Mississippi River, sturgeons and paddlefish are all relatively abundant, at least in some reaches. This is certainly due to different historic and societal pressures on the rivers.

The Yangtze supports more than 30 major urban areas, many of which were established more than 2,000 years ago and the larger of which (Chonqing, Wuhan, Nanjing, and Shanghai) have collective populations of nearly 50 million people. The Mississippi River, in contrast, supports fewer than 15 major urban areas, none of which were established more than 300 years ago and the larger of which (Minneapolis–St. Paul, St. Louis, Memphis, and New Orleans) have collective populations of less than 10 million people. Exacerbating these pressures are cultural differences between China and the United States; artisanal fishing and consumption of locally caught fish are pervasive along the Yangtze, even in urban environments, but fishing and wild fish consumption in the Mississippi less so. Fishery resources of the Yangtze, then, have incurred more intensive damage over a longer period than those of the Mississippi, and much of the damage occurred at a time in history when the demography of large fish species was poorly understood by researchers and prior to development of effective aquaculture techniques for large riverine fishes.

Differences in fish population status are, in part, responsible for different utilization and management techniques emphasized in each of the two systems and the applicability of each in the corresponding river. In the Yangtze River, fishing pressure from artisanal fishermen has been implicated in population declines. However, it is unclear how much of the decline can be linked to fishing pressure, as opposed to other stressors. Intensive production of sturgeons at a large multispecies aquaculture center, Tianxia Company at Yidu (Hubei Province), has been sufficient to maintain (at low levels) populations of nonreproducing Chinese Sturgeon in the Yangtze River, an indication that similar stocking efforts with Pallid Sturgeon could be used to maintain nonreproductive populations in areas of the Mississippi Basin in which spawning or rearing habitat are limited (e.g., the upper Missouri River).

In the Mississippi River, Paddlefish have recovered principally through coordinated state-by-state efforts at harvest limits and assisted by information shared though a 28-state basinwide interagency organization (the Mississippi Interstate Cooperative Research Association). Also, in the Mississippi River, national resource agencies (e.g., U.S. Fish and Wildlife Service and U.S. Army Corps of Engineers) have worked with local (state) resource agencies to construct and evaluate river restoration projects (e.g., side-channel reconnection), which may have benefits for Paddlefish.

One management technique, already practiced in both rivers, offers great potential for species recovery and merits greater effort and agency resources: fish rescue. Chinese researchers rescue poached and accidentally captured paddlefish and sturgeons for use in aquaculture, captive propagation, and ultimately stocking. Researchers in the United States also rescue paddlefish and sturgeons, specifically those that have been stranded behind water diversions, for return to the river. Efforts have been shown effective for recovery of individual Paddlefish in the Mississippi River (e.g., Hoover et al. 2013), sturgeon populations in other systems (e.g., Thomas et al. 2013), and sturgeons in the Mississippi River (J. J. Hoover, unpublished data).

Both rivers would benefit by extensive collaboration through the development of a more robust exchange program between Chinese and U.S. scientists. The depth of knowledge that exists on both continents on the important megafishes of the Yangtze and Mississippi Rivers species has been inadequately transferred between continents, in either direction. We recommend that interested representatives from each country engage in a dialogue aimed at a mutual understanding and goal of species conservation within their native range, and share information regarding lessons learned and successful advances reached in megafauna species conservation.

Acknowledgments

Authorship, exclusive of the principle writers and organizers, was determined by alphabetical order. Author contributions were (1) concept, objectives, approach, editing, and summary sections: JJH, QEP; (2) Yangtze River Paddlefish section: QW, HZ; (3) Mississippi River Paddlefish section: JJH; (4) Yangtze River Sturgeon section: CW, LS, QW; (5) Mississippi River Sturgeon section: QEP; and (6) assistance with citations: KB. Life-history data, unless explicitly cited otherwise, are from FishBase (www.fishbase.org/). Conservation statuses, unless explicitly stated otherwise, are from the International Union for the Conservation of Nature Red List of Threatened Species (www.iucnredlist.org/). Funding for the principal authors was provided by the Missouri Department of Conservation Resource Science Division (QEP) and the U.S. Army Corps of Engineers Mississippi Valley Division (JJH). Permission to publish was granted by the Chief of Engineers.

References

Adams, L. A. 1942. Age determination and rate of growth in *Polyodon spathula* by means of growth rings of the otoliths and dentary bone. American Midland Naturalist 28:617–630.

Adams, S. R., G. L. Adams, and G. R. Parsons. 2003. Critical swimming speed and behavior of juvenile Shovelnose Sturgeon and Pallid Sturgeon. Transactions of the American Fisheries Society 132:392–397.

Alexander, M. L. 1914. The Paddlefish (*Polyodon spathula*) (commonly called "spoonbill cat"). Transactions of the American Fisheries Society 44:73–78.

Allis, E. A., Jr. 1903. On certain features of the lateral canals and cranial bones of *Polyodon folium*. Zoologische Jahrbücher 17:659–678.

Allison, P. G., J. F. Deang, A. J. Diaz, A. R. Poda, J. J. Hoover, M. F. Horstemeyer, and E. J. Perkins. 2013. Characterization of Paddlefish (*Polyodon spathula*) rostrum stellate bones. Bioinspired, Biomimetic, and Nanobiomaterials 3:1-17.

Auer, N. A. 2004. Conservation. Pages 252–276 *in* G. T. LeBreton, F. H. Beamish, and R. S McKinley, editors. Sturgeon and paddlefish of North America. Kluwer Academic Publishers, New York.

Bemis, W. E., and B. Kynard. 1997. Sturgeon rivers: an introduction to acipenseriform biogeography and life history. Environmental Biology of Fishes 48:167–183.

Bemis, W. E., E. K. Findeis, and L. Grande. 1997. An overview of Acipenseriformes. Environmental Biology of Fishes 48:25–71.

Bettoli, P. W., J. A. Kerns, and G. D. Scholten. 2009. Status of Paddlefish in the United States. Pages 23–38 *in* C. P. Paukert and G. D. Scholten, editors. Paddlefish management, propagation, and conservation in the 21st century. American Fisheries Society, Symposium 66, Bethesda, Maryland.

Boreman, J. 1997. Sensitivity of North American sturgeons and paddlefish to fish mortality. Environmental Biology of Fishes 48:399–405.

Boysen, K. A., K. J. Killgore, and J. J. Hoover. 2012. Ranking restoration alternatives for the lower Mississippi River: application of multicriteria decision analysis. U.S. Army Engineer Research and Development Center, EMRRP-ER-17, Vicksburg, Mississippi.

Bryan, P. 1942. Spoonbill fishing, new valley industry. Alabama Conservation 14:7, 15.

Burggren, W. W., and W. E. Bemis. 1992. Metabolism and ram ventilation in juvenile Paddlefish, *Polyodon spathula* (Chondrostei: Polyodontidae). Physiological Zoology 65:515–539.

Carlson, D. M., and P. S. Bonislawsky. 1981. The Paddlefish (*Polyodon spathula*) fisheries of the Midwestern United States. Fisheries 6:17–27.

Carlson, D. M., M. K. Kettler, S. E. Fisher, and G. S. Whitt. 1982. Low genetic variability in Paddlefish populations. Copeia 3:721–725.

Chang, J. B., and W. X. Cao. 1999. History and prospect of conservation on Chinese Sturgeon in the Yangtze River. Acta Hydrobiologica Sinica 236:713–720.

Chen, X. H. 2007. Biology and resources of acipenseriform fishes. China Ocean Press, Beijing.

Cope, E. D. 1883. A new chondrostean from the Eocene. American Naturalist 17:1152–1153.

Cope, E. D. 1885. Eocene paddlefish and Gonorhynchidae. American Naturalist 19:1090–1091.

Cope, E.D. 1886. On two new forms of polyodont and gonorhynchid fishes from the Eocene of the Rocky Mountains. Memoirs of the National Academy of Sciences 3(2):161–165.

Cottrell, T. 1944. Rough fish given hook treatment: Big Lake clubmen reduce the population of spoonbills, gars, and turtles. Alabama Conservationist 15(7):4.

Dryer, M. P., and A. J. Sandvol. 1993. Recovery plan for the Pallid Sturgeon (*Scaphirhynchus albus*). U.S. Fish and Wildlife Service, Denver.

Epifanio, J. M., J. B. Koppelman, M.A. Nedbal, and D. P. Phillip. 1996. Geographic variation of Paddlefish allozymes and mitochondrial DNA. Transactions of the American Fisheries Society 125:546–561.

Fan, X. G., Q. W. Wei, J. B. Chang, H. Rosenthal, J. X. He, D. Q. Chen, L. Shen, H. Du, and D. G. Yang. 2006. A review on conservation issues in the upper Yangtze River: a last chance for a big challenge: can Chinese Paddlefish (*Psephurus gladius*), Dabry's Sturgeon (*Acipenser dabryanus*) and other fish species still be saved? Journal of Applied Ichthyology 22:32–39.

Fu, C. R. 1988. Age and growth of Dabry's Sturgeon, Acipenser dabryanus. Pages 220–226 *in* The biology of sturgeons in Changjiang and their artificial propagation. Sichuan Scientific & Technical Publishing House, Chengdu, China. (In Chinese.)

Gao, Y., G. Cao, Y. Chen, and J. Chang. 2009. A review on conservation issue in Chinese Paddlefish, *Psephurus glasius*. Pages 2279–2284 *in* C. Zhang and H. Tang, editors. Advances in water resources and hydraulic engineering: proceedings of 16th IAHR-APD congress and 3rd symposium of IAHR-ISHS. Springer-Verlag, Berlin.

Gardiner, B. G. 1984. Sturgeons as living fossils. Pages 148–152 *in* N. Eldredge and S. M. Stanley, editors. Living fossils. Springer-Verlag, New York.

Gerken, J. E. and C. P. Paukert. 2009. Threats to paddlefish habitat: implications for conservation. Pages 173–183 *in* C. P Paukert and G. D. Scholten, editors. Paddlefish management, propagation, and conservation in the 21st century: building from 20 years of research and management. American Fisheries Society, Symposium 66, Bethesda, Maryland.

Grady, J. M., and B. S. Elkington. 2009. Establishing and maintaining Paddlefish populations by stocking. Pages 385–396 *in* C. P Paukert and G. D. Scholten, editors. Paddlefish management, propagation, and conservation in the 21st century: building from 20 years of research and management. American Fisheries Society, Symposium 66, Bethesda, Maryland.

Grande, L., and W. E. Bemis. 1991. Osteology and phylogenetic relationships of fossil and recent paddlefishes (Polyodontidae) with comments on the interrelationships of Acipenseriformes. Journal of Vertebrate Paleontology 11(supplement 1):1–121.

Grande, L., and W. E. Bemis. 1998. Interrelationships of Acipenseriformes with comments on "Chondrostei." Pages 85–115 *in* M. L. J. Stiassny, L. R. Parenti and G. D. Johnson, editors. Interrelationships of fishes. Academic Press, New York.

Günther, A.1873. Report on a collection of fishes from China. Annals and Magazine of Natural History 12:239–250.

Hoover, J. J., A. Turnage, and K. J. Killgore. 2009b. Swimming performance of juvenile Paddlefish: quantifying risk of entrainment. Pages 141–155 *in* C. P Paukert and G. D. Scholten, editors. Paddlefish management, propagation, and conservation in the 21st century: building from 20 years of research and management. American Fisheries Society, Symposium 66, Bethesda, Maryland.

Hoover, J. J., K. A. Boysen, C. E. Murphy, and S. G. George. 2009a. Morphological variation in juvenile paddlefish. Pages 157–191 *in* C. P Paukert and G. D. Scholten, editors. Paddlefish management, propagation, and conservation in the 21st century: building from 20 years of research and management. American Fisheries Society, Symposium 66, Bethesda, Maryland.

Hoover, J. J., S. G. George, and K. J. Killgore. 2013. A paddlefish entrained by the 2011 Mississippi River flood: rescue, recapture, and inferred swim speed. Southeastern Naturalist 12:N26–N30.

Hoover, J. J., S. G. George, and K. J. Killgore. 2007. Diet of Shovelnose Sturgeon and Pallid Sturgeon in the free-flowing Mississippi River. Journal of Applied Ichthyology 23:494–499.

IUCN (International Union for Conservation of Nature). 2004. *Polyodon spathula* (Duckbill Cat, Paddlefish, Spadefish, Spoonbill Cat). IUCN Red List of Threatened Species, version 2014.2. Available: www.iucnredlist.org/details/17938/0. (Oct 2015).

IUCN (International Union for Conservation of Nature). 2010. IUCN red list of threatened species, version 2010.3. Available: www.

iucnredlist.org/apps/redlist/details/231/0. (October 2010).

Jelks, H. L., S. J. Walsh, N. M. Burkhead, S. Contreras-Balderas, E. Díaz-Pardo, D. A. Hendrickson, J. Lyons, N. E. Mandrak, F. McCormick, J. S. Nelson, S. P. Platania, B. A. Porter, C. B. Renaud, J. J. Schmitter-Soto, E. B. Taylor, and M. L. Warren. Jr. 2008. Conservation status of imperiled North American freshwater and diadromous fishes. Fisheries 33:372–407.

Jennings, C. A., and S. J. Zigler. 2009. Biology and life history of the Paddlefish in North America: an update. Pages 1–22 *in* C. P Paukert and G. D. Scholten, editors. Paddlefish management, propagation, and conservation in the 21st century: building from 20 years of research and management. American Fisheries Society, Symposium 66, Bethesda, Maryland.

Jørgensen, J. M., Å. Flock, and J. Wersäll. 1972. The Lorenzian ampullae of *Polyodon spathula*. Zeitschrift fur Zellforschung und mikroskopische Anatomie 130:362–377.

Ke, F., and Q. Wei. 1992. Investigations on the structure of spawning population of Chinese Sturgeon (*Acipenser Sinensis*, Gray) and the estimate of its stock. Freshwater Fishes 4: 7–11. (In Chinese.)

Keenlyne, K. D. 1997. Life history and status of the Shovelnose Sturgeon, *Scaphirhynchus platorynchus*. Environmental Biology of Fishes 48:291–298.

Keenlyne, K. D., and L. G. Jenkins. 1993. Age at sexual maturity of the Pallid Sturgeon. Transactions of the American Fisheries Society 122:393–396.

Killgore, K. J., J. J. Hoover, S. G. George, B. R. Lewis, C. E. Murphy, and W. E. Lancaster. 2007a. Distribution, relative abundance, and movements of Pallid Sturgeon in the free-flowing Mississippi River. Journal of Applied Ichthyology 23:476–483.

Killgore, K. J., J. J. Hoover, J. P. Kirk, S. G. George, B. R. Lewis, and C. E. Murphy. 2007b. Age, growth, and mortality of Pallid Sturgeon in the free-flowing Mississippi River. Journal of Applied Ichthyology 23:452–456.

Killgore, K. J., L. E. Miranda, C. E. Murphy, D. M. Wolff, J. J. Hoover, T. M. Keevin, S. T. Maynord, and M. A. Cornish. 2011. Fish entrainment rates through towboat propellers in the upper Mississippi and Illinois rivers. Transactions of the American Fisheries Society 14:570–581.

Kirtland, J. P. 1843–1844. Descriptions of the fishes of the Ohio River and its tributaries. Boston Journal of Natural History 4:16–26, 231–308.

Kirtland, J. P. 1845–1847. Descriptions of the fishes of the Ohio River and its tributaries. Boston Journal of Natural History 5:21–32, 265–344.

Kistler, H. D. 1906. The primitive pores of *Polyodon spathula*. Journal of Comparative Neurology 16:294–298.

Koch, J. D., and M. C. Quist. 2010. Current status and trends in Shovelnose Sturgeon (*Scaphirhynchus platorynchus*) management and conservation. Journal of Applied Ichthyology 26:491–498.

Krentz S. 2005. Pallid Sturgeon spawning and stocking summary report, 1992–2004. U.S. Fish and Wildlife Service, Bismarck, North Dakota. Available: www.fws.gov/mountain-prairie/fisheries/missouriRiverFWCO/pallid%20spawning%20and%20stocking%20summary%200512.pdf. (Oct 2015).

Kynard, B., P. Zhuang, T. Zhang, and L. Z. Zhang. 2003. Ontogenetic behavior and migration of Dabry's Sturgeon, *Acipenser dabryanus*, from the Yangtze River, with notes on body color and development rate. Environmental Biology of Fishes 66:27–36.

LeSueur, C. A. 1818. Descriptions of several new species of North American fishes. Journal of the Academy of Natural Sciences of Philadelphia 1:222–235, 359–368.

Li, L., H. Zhang, Q. Wei, H. Du, and K. Hong. 2011. Occurrence time and amount variation of juvenile Chinese Sturgeon, *Acipenser sinensis* at Xupu, Changshu section of Yangtze River after closure of Three Gorges Dam. Journal of Fishery Sciences of China 18:611–618.

Li, P., Q. Wei, and L. Liu. 2008. DNA integrity of *Polyodon spathula* cryopreserved sperm. Journal of Applied Ichthyology 24:121–125.

Liu, C. H., A. G. Todd, X. M. Diao, and J. H. Liu. 1995. Biology of the Chinese Paddlefish *Psephurus gladius* (Martens). Pages 13–21 *in* A. D. Gershanovich and T. I. J. Smith, editors. Proceedings of the second international symposium on sturgeons. VNIRO Publishing, Moscow.

Liu, C. H., and Y. J. Zeng. 1988. Notes on the Chinese Paddlefish, *Psephurus gladius* (Martens). Copeia 1988:482–484.

Ma, J., Z. Deng, X. Deng, and M. Cai. 1996. Age determination and growth of Chinese Paddlefish *Psephurus gladius*. Acta Hydrobiologia Sinica 20:150–159.

Main, D. 2014. Rare Chinese Sturgeon not reproducing, close to extinction. Newsweek(September 15) [online]. Available: www.newsweek.

com/rare-chinese-sturgeon-not-reproducing-close-extinction-270756. (October 2014).

MDC (Missouri Department of Conservation). 2007. A plan for recovery of the Lake Sturgeon in Missouri. Missouri Department of Conservation, Jefferson City.

Morrow, J. V., Jr., J. P. Kirk, K. J. Killgore, and S. G. George. 1998. Age, growth, and mortality of Shovelnose Sturgeon in the lower Mississippi River. North American Journal of Fisheries Management 18:725–730.

Murphy, C. E., J. J. Hoover, S. G. George, and K. J. Killgore. 2007. Morphometric variation among river sturgeons (*Scaphirhynchus* spp.) of the middle and lower Mississippi River. Journal of Applied Ichthyology 23:313–323.

New, J. G., and D. Bodznick. 1985. Segregation of electroreceptive and mechanoreceptive lateral line afferents in the hindbrain of chondrostean fishes. Brain Research 336:89–98.

Nichols, J. T. 1916. A large *Polyodon* from Iowa. Copeia 34:65–65.

Nichols, J. T. 1943. The fresh-water fishes of China. American Museum of Natural History, Natural History of Central Asia IX, New York.

Noakes, D. L. G., F. W. H. Beamish, and A. Rossiter. 1999. Conservation implications of behavior and growth of the Lake Sturgeon, Acipenser fulvescens, in northern Ontario. Environmental Biology of Fishes 55:135–144.

NANFA (North American Natives Fishes Association). 2010. Listing status of North American freshwater fishes. Available: www.nanfa.org/bccconservation.shtml. (October 2015).

Patterson, J. T., S. D. Mims, and R. A. Wright. 2013. Effects of body mass and water temperature on routine metabolism of American Paddlefish *Polyodon spathula*. Journal of Fish Biology 82:1269–1280.

Pegg, M. A., J. H. Chick, and B. M. Pracheil. 2009. Potential effects of invasive species on Paddlefish. Pages 185–201 *in* C. P Paukert and G. D. Scholten, editors. Paddlefish management, propagation, and conservation in the 21st century: building from 20 years of research and management. American Fisheries Society, Symposium 66, Bethesda, Maryland.

Peterson, D. L., P. Vecsei, and C. A. Jennings. 2007. Ecology and biology of the Lake Sturgeon: a synthesis of current knowledge of a threatened North American Acipenseridae. Reviews of Fish Biology and Fisheries 17:59–76.

Phelps, Q. E., I. Vining, R. Dames, D. Herzog, S. Tripp, and M. Boone. 2013. A comparison of techniques to evaluate sturgeon mortality in the Mississippi River adjacent to Missouri and Illinois. North American Journal of Fisheries Management 33:754–761.

Pikitch, E. K., P. Doukakis, L. Lauck, P. Charkrabarty, and D. L. Erickson. 2005. Status, trends and management of sturgeon and paddlefish fisheries. Fish and Fisheries 6:233–265.

Ping, C. 1931. Preliminary notes on the fauna of Nanking. Contributions from the Biological Laboratory of the Science Society of China (Zoological Series) 7:173–201.

Qiao, Y., X. Tang, S. Brosse, J. Chang. 2006. Chinese Sturgeon (*Acipenser sinensis*) in the Yangtze River: a hydroaeoustic assessment of fish location and abundance on the last spawning ground. Journal of Applied Ichthyology 22:140–144.

Quinn, J.P. 2009. Harvest of Paddlefish in North America. Pages 203–221 *in* C. P Paukert and G. D. Scholten, editors. Paddlefish management, propagation, and conservation in the 21st century: building from 20 years of research and management. American Fisheries Society, Symposium 66, Bethesda, Maryland.

Rafinesque, C. S. 1820. Ichthyologica ohiensis, or natural history of the fishes inhabiting the River Ohio and its tributary streams, preceded by a physical description of the Ohio and its branches. Printed for the author by W.G. Hunt, Lexington, Kentucky.

Ray, J. M., C. B. Dillman, R. M. Wood, B. R. Kuhajda, and R. L. Mayden. 2007. Microsatellite variation among river sturgeons of the genus *Scaphirhynchus* (Actinopterygii: Acipenseridae): a preliminary assessment of hybridization. Journal of Applied Ichthyology 23 304–312.

Reed, B. C., W. E. Kelso, and D. A. Rutherford. 1992. Growth, fecundity, and mortality of Paddlefish in Louisiana. Transactions of the American Fisheries Society 121:378–384.

Rosen, R. A., and D. C. Hales. 1980. Occurrence of scarred paddlefish in the Missouri River, South Dakota-Nebraska. Progressive Fish-Culturist 42:82–85.

Scholten, G. R. 2009. Management of commercial Paddlefish fisheries in the United States. Pages 291-306 *in* C. P Paukert and G. D. Scholten, editors. Paddlefish management, propagation, and conservation in the 21st century: building from 20 years of research and management. American Fisheries Society, Symposium 66, Bethesda, Maryland.

Scholten, G. D., and P. W. Bettoli. 2005. Population characteristics and assessment of overfishing for an exploited paddlefish population in the lower Tennessee River. Transactions of the American Fisheries Society 134:1285–1298.

Schrey, A. W., and E. J. Heist. 2007. Stock structure of Pallid Sturgeon analyzed with microsatellite loci. Journal of Applied Ichthyology 23:297–303.

Sichuan Fisheries Research Institute. 1980. A preliminary report on a spawning ground for *Psephurus gladius*. Sichuan Fisheries 3:39–40. (In Chinese.)

Stockard, C. R. 1907. Observations on the natural history of *Polyodon spathula*. American Naturalist 41:753–766.

Tchang, T.-L. 1928. A review of the fishes of Nanking. Contributions from the Biological Laboratory of the Science Society of China 4.

Thomas, M. J., M. L. Peterson, N. Friedenberg, J. P. Van Eenennaam, J. R. Johnson, J. J. Hoover, and A. P. Klimley. 2013. Stranding of spawning run Green Sturgeon in the Sacramento River: post-rescue movements and potential population level effects. North American Journal of Fisheries Management 33:287–297.

USFWS (U.S. Fish and Wildlife Service). 2000. Biological opinion for the operation and maintenance of the 9-foot navigation channel on the upper Mississippi River system. U.S. Fish and Wildlife Service, Washington, D.C. Available: www.mvs.usace.army.mil/Portals/54/docs/recreation/rivers/compliance/Final%20BO.pdf. (Oct 2015).

Vescei, P., and D. L. Peterson. 2007. Morphology and ecology of Pacific sturgeons. Environmental Biology of Fishes 79:369–381.

Vecsei, P., and D. L. Peterson. 2004. Sturgeon ecomorphology: a descriptive approach. Pages 103–133 *in* G. T. LeBreton, F. H. Beamish, and R. S McKinley, editors. Sturgeon and paddlefish of North America. Kluwer Academic Publishers, Boston.

Wagner, G. 1908. Notes on the fish fauna of Lake Pepin. Transactions of the Wisconsin Academy of Sciences, Arts, and Letters 16:23–37.

Wan, Q. H., S. G. Fang, and Y. N. Li. 2006. The loss of genetic diversity in Dabry's Sturgeon (*Acipenser dabryanus*, Dumeril) as revealed by DNA fingerprinting. Aquatic Conservation: Marine and Freshwater Ecosystems 13:225–231.

Wang, C., B. Kynard, Q. Wei, H. Du, and H. Zhang. 2012a. Spatial distribution and habitat suitability indices for non-spawning and spawning adult Chinese Sturgeons below Gezhouba Dam, Yangtze River: effects of river alterations. Journal of Applied Ichthyology 29:31–40.

Wang, C. Y., Q. W. Wei, K. Kynard, H. Du, and H. Zhang. 2012b. Migrations and movements of adult Chinese Sturgeon *Acipenser sinensis* in the Yangtze River, China. Journal of Fish Biology 81:696–713.

Watanabe, Y., Q. W. Wei, D. G. Yang, X. H. Chen, H. Du, J. Yang, K. Sato, Y. Naito, and N. Miyazaki. 2008. Swimming behavior in relation to buoyancy in an open swimbladder fish, the Chinese Sturgeon. Journal of Zoology 275:381–390.

Webb, P. W. 1994. The biology of fish swimming. Pages 45–62 *in* L. Maddock, Q. Bone, and J. M. V. Rayner, editors. Mechanics and physiology of animal swimming. Cambridge University Press, Cambridge, UK.

Wei, Q. W. 2003. Reproductive behavioral ecology of Chinese Sturgeon (*Acipenser sinensis* Gray) with its stock assessment. Doctoral dissertation. Graduate School of Chinese Academy of Sciences, Beijing.

Wei, Q. W., F. E. Ke, M. J. Zhang, P. Zhuang, D. J. Luo, Y. Q. Zhou, and W. Yang. 1997. Biology, fisheries, and conservation of sturgeons and paddlefish in China. Environmental Biology of Fishes 48:241–255.

Wilkens, L. A., D. F. Russell, X. Pei, and G. Gurgens. 1997. The paddlefish rostrum functions as an electrosensory antenna in plankton feeding. Proceedings of the Royal Society of London B 264:1723–1729.

Williams, S. W. 1883. The middle kingdom, volume I. Charles Scribner's Sons, New York.

Wilson, J. A., and R. S. McKinley. 2004. Distribution, habitat and movements Pages 252–276 *in* G. T. LeBreton, F. H. Beamish, and, R. S McKinley, editors. Sturgeon and paddlefish of North America. Kluwer Academic Publishers, Boston.

Wu, H. W. 1930. On some fishes collected from the Yangtze valley. Sinensia 1:65–86.

Wu, X. C. 2005. The loss of genetic diversity in the Chinese Paddlefish (*Psephurus gladius* Martens) as revealed by DNA fingerprinting. Indian Academy of Sciences 84:323–327.

Wu, X. W. 1963. Economic animals of China: freshwater fishes. China Science Press, Beijing. (In Chinese.)

Xiao, J., M. Wang, G. Shi, Q. Yu, C. Mao, and D. Chen. 2009. Economic value evaluation of conserving the rare and endangered species affected by the Three Gorges Project. Biodiversity Science 17:257–265.

Xie, D. J. 1979. Artificial propagation and rearing of Dabry's Sturgeon, *Acipenser dabryanus* Sichuan Fisheries 1:2–8.

Xie, D. J., X. T. Ke, Y. P. Tian, and D. Y. Chen. 1988. Embryonic and larval development of Dabry's Sturgeon. Pages 238–240 *in* B. N. Shi, editor. The biology of sturgeons in Changjiang and their artificial propagation. Sichuan Scientific and Technical Publishing House, Chengdu, China.

Xiong, T. S. 1988. Feeding habit of Dabry's Sturgeon. Pages 228–229 *in* B. N. Shi, editor. The biology of sturgeons in Changjiang and their artificial propagation. Sichuan Scientific and Technical Publishing House, Chengdu, China.

Yang, G. R. 1986. Ichthyography of Hubei Province. Hubei Science & Technology Press, Wuhan, China.

YARSG (Yangtze Aquatic Resources Survey Group). 1988. The biology of the sturgeons and paddlefish in the Yangtze River and their artificial propagation. Sichuan Scientific and Technical Publishing House, Chengdu, China.

Yu, Z. T., Y. G. Xu, Z.L. Deng, C.S. Zhou, and X. Yang. 1986. Studies on the reproductive ecology of Chinese Sturgeon below Gezhouba Dam. Transactions of the Chinese Ichthyological Society 5:1–16.

Yue, P. and Y. Chen, 1998. China red data book of endangered animals: Pisces. Science Press, Beijing.

Zeng, Q. Z. 1990. Fisheries resources in Yangtze Valley. China Ocean Press, Beijing.

Zhang, H., Q. W. Wei, H. Du, and L. X. Li. 2011. Present status and risk for extinction of the Dabry's Sturgeon (*Acipenser dabryanus*) in the Yangtze River watershed: a concern for intensified rehabilitation needs. Journal of Applied Ichthyology 27:181–185.

Zhang, H., Q. W. Wei, H. Du, L. Shen, Y. H. Li, and Y. Zhao. 2009. Is there evidence that the Chinese Paddlefish (*Psephurus gladius*) still survives in the upper Yangtze River? Concerns inferred from hydroacoustic and capture surveys, 2006–2008. Journal of Applied Ichthyology 25:95–99.

Zhang, H., Q. W. Wei, D. G. Yang, H. Du, H. J. Zhang, and X. H. Chen. 2007a. An observation on water current profiles of spawning of *Acipenser sinensis* downward Gezhouba Dam. Journal of Fisheries Sciences of China 14:183–191.

Zhang, H., Q. W. Wei, D. G. Yang, H. Du, H. J. Zhang, and X. H. Chen. 2007b. Topography research on spawning grounds of *Acipenser sinensis* below Gezhouba Dam. Acta Ecologica Sinica 27:3945–3955.

Zhang, S. M., H. Deng, D. Q. Wang, and Q. J. Wu. 1999. Random amplified polymorphic DNA analysis of seven Acipenseridae species. Natural Science Progress 9:819–823.

Zhu, C., and N. Yu. 1987. A preliminary study on the morphology, growth, and feeding habits of young *Psephurus gladius* in the estuarine region of the Changjiang (Yangtze) River. Acta Hydrobiologica Sinica 11:289–298.

Zhuang, P. 1994. Effects of Three-Gorges Dam on the aquatic animals of the Yangtze River and strategies for their rescue and conservation. Freshwater Fish 24:26–29.

Zhuang, P., F. E. Ke, Q. W. Wei, X. F. He, and Y. J Cen. 1997: Biology and life history of Dabry's Sturgeon, *Acipenser dabryanus*, in the Yangtze River. Environmental Biology of Fishes 48:257–264.

Zou, Y. C., Q. W. Wei, and G. B. Pan. 2011a. Induction of meiotic gynogenesis in Paddlefish (*Polyodon spathula*) and its confirmation using microsatellite markers. Journal of Applied Ichthyology 27:496–500.

Zou, Y. C., Q. W. Wei, L.B. Zeng, and P. Li. 2011b. Production of gynogenetic diploid *Polyodon spathula* using fertilization with irradiated sperm followed by spontaneous diploidization and distant hybridization caused by heat shock. Journal of Applied Ichthyology 27:505–509.

American Fisheries Society Symposium 84:113–126, 2016

Bigheaded Carps of the Yangtze and Mississippi Rivers: Biology, Status, and Management

DUANE C. CHAPMAN
U.S. Geological Survey, Columbia Environmental Research Center
4200 New Haven Road, Columbia, Missouri 65201, USA

DAQING CHEN
Yangtze River Fisheries Research Institute, Chinese Academy of Fishery Sciences,
No. 8, 1st Wudayuan Road, East Lake Hi-Tech Development Zone,
Wuhan, Hubei 430223, P. R. China

JAN JEFFREY HOOVER
U.S. Army Engineer Research and Development Center
3909 Halls Ferry Road, Vicksburg, Mississippi 39180, USA

HAO DU
Yangtze River Fisheries Research Institute, Chinese Academy of Fishery Sciences,
No. 8, 1st Wudayuan Road, East Lake Hi-Tech Development Zone,
Wuhan, Hubei 430223, P. R. China

QUINTON E. PHELPS
Big Rivers and Wetlands Field Station, Missouri Department of Conservation
3815 East Jackson Boulevard, Jackson, Missouri 63755, USA

LI SHEN, CHENYOU WANG, QIWEI WEI, AND HUI ZHANG
Yangtze River Fisheries Research Institute, Chinese Academy of Fishery Sciences,
No. 8, 1st Wudayuan Road, East Lake Hi-Tech Development Zone,
Wuhan, Hubei 430223, P. R. China

Abstract.—Bighead Carp *Hypophthalmichthys nobilis* and Silver Carp *H. molitrix* are native in the Yangtze River and extremely important economically and culturally as food fishes; however, the two species have declined due to overfishing and anthropogenic modifications to hydrology and water quality. Bighead Carp and Silver Carp were imported to North America in the early 1970s, escaped confinement, and have now become undesirable and problematic invasive species. The two carps have become the most abundant fish species in many portions of their invaded range, which continues to expand. We compare the biology, status, and management of these species between their natal range in the Yangtze River and their invaded habitats of the Mississippi River basin.

Introduction

Several aquatic species have been introduced between China and the United States. Largemouth Bass *Micropterus salmoides*, Bluegill *Lepomis macrochirus*, Black Bullhead *Ameiurus melas*, Channel Catfish *Ictalurus punctatus*, and red swamp crayfish *Procambarus clarkii* are among the species introduced from the

* Corresponding author: dchapman@usgs.gov

Mississippi River basin to China. Bighead Carp *Hypophthalmichthys nobilis*, Silver Carp *H. molitrix*, Grass Carp *Ctenopharyngodon idella*, Black Carp *Mylopharyngodon piceus*, Northern Snakehead *Channa argus*, and the Oriental Weatherfish *Misgurnus anguillicaudatus* are species native to the Yangtze River that have been introduced to the Mississippi River basin. Of these fishes, the Bighead and Silver carps are possibly the most important fish species culturally and economically in China and are by far the most problematic in the Mississippi River basin. We here address the biology of Bighead and Silver carps and discuss their very different status and management in their native Yangtze River and in the United States, where they are considered invasive species.

Bigheaded carps (class Actinopterygii, subclass Teleostei, order Cypriniformes, family Cyprinidae, genus *Hypophthalmichthys*) are undocumented in the fossil record and, like many carps and minnows, are of comparatively recent origin. The family Cyprinidae originated in Asia (the earliest fossils appear to be mid-Eocene, or 50 million years old), but representatives of modern lineages appear later during the Oligocene (24–37 million years ago; Cavender 1991). *Ctenopharyngodon*, a genus closely related to bigheaded carps (Li et al. 2009), appears during the Pliocene (2–5 million years ago; Cavender 1991) and has a single extant species, Grass Carp *C. idella*. Bigheaded carps have a centrally placed dorsal fin and forked homocercal tail, and their skeleton is true bone.

The bigheaded carps are native to Asia and are well documented in the scientific literature (Kolar et al. 2007). The Bighead Carp and Silver Carp (Figure 1) are native to the Yangtze River basin and other large rivers of China. A third species of bigheaded carp, the Largescale Silver Carp *H. harmandi* inhabits a much smaller range, mostly in Vietnam. It is not present in either the Yangtze or Mississippi rivers and thus is not discussed further here. Bighead and Silver carps are among the most important aquaculture fishes worldwide and thus have escaped or been introduced to many locations around the globe (Kolar et al. 2007). Their native range extends from southern China to the Amur River system, which forms the northern border of China and the southern border of Russia (Xie and Chen 2001). The Silver Carp has a somewhat wider native range than Bighead Carp, ranging from the Red River in southern China into Siberia (from 21°N to 64°N latitude) (Xie and Chen 2001).

The genus of Bighead Carp has been variously described. *Hypophthalmichthys* is the genus accepted by the American Fisheries Society (Page et al. 2013) and has been used by most western scientists since phylogenetic analysis was conducted by Howes (1981) and molecular research by Li et al. (2009). However, the previous nomenclature, *Aristichthys nobilis*, continues to be used by many Asian scientists (as in Li et al. 2014).

Bighead and Silver carps are extremely important food fishes in China, and they have been cultured as food fish and introduced around the world for fishery enhancement (Kolar et al. 2007). Together with Grass Carp and Black Carp *Mylopharyngodon piceus*, they constitute a group known widely in China as the "four famous Chinese carps." In the Yangtze River basin, bigheaded carps have declined in abundance, and there are strong efforts to restore these species in the wild (Li 2006; Dudgeon 2011). In the Mississippi River basin, however, bigheaded carps have become extremely abundant and are among the most notorious nuisance fishes (Koel et al. 2000). Research and management of bigheaded carps in North America has been mostly focused on controlling their populations and preventing their spread into new areas (Conover et al. 2007; ACRCC 2013, 2014).

The Bighead Carp (Figure 1) has a large, scaleless head, a large mouth, and eyes located low on the head. It is deep-bodied, spindle-shaped, and moderately compressed, with a smooth keel between the anal and pelvic fins that does not extend anterior to the base of the pelvic fins. The premaxilla and protruding mandible form rigid bony lips. Bighead Carp often feed at the water surface, and the white mandible is often visible, assisting in

Figure 1. Bigheaded carps from the Mississippi River. Top, Bighead Carp. Bottom, Silver Carp. Note the longer head and larger pectoral fins of the Bighead Carp compared with those of the Silver Carp. Photo credit: U.S. Geological Survey.

identification and location of feeding fish (Kolar et al. 2007). Coloration of the body is dark gray above and cream-colored below, with dark gray to black irregular blotches on the back and sides. This color pattern develops when the fish is about 2 months old. Their scales are small and cycloid. The lateral line is complete and strongly convex ventrally and continues posteriorly along the middle of the caudal peduncle, with 99–118 scales. There are 24–29 scale rows above the lateral line and 15–24 scale rows below the lateral line. Vertebrae number from 37 to 44, and the dorsal and anal fins are without spines. The number of dorsal fin rays is 7–8, anal fin rays 12–13, pelvic fin rays 7–8, and pectoral fin rays 17–21. Pectoral fins extend posteriorly beyond the origin of the pelvic fins. Pharyngeal teeth are in a single row, with four on each arch. They have a spoon-like shape and a shallowly concave grinding surface (Li et al. 1990). The intestine is long (approximately five times the total length of the fish) and highly convoluted.

The Silver Carp (Figure 1) is deep-bodied and spindle-shaped but more laterally compressed than the Bighead Carp. The keel of the Silver Carp is longer than that of the Bighead Carp, extending from the throat to the vent. The keel is scaled anteriorly but is scaleless posteriorly. Adult coloration is typically gray-black dorsally, and the upper sides are olivaceous, grading to silver laterally and ventrally. The lower jaw has a small tubercle, and the upper jaw is slightly notched. The scales are small and cycloid, and lateral line scale counts are typically between 102 and 130. There are 27–38 scales above the lateral line, and scales below the lateral line number 14–19. Vertebrae number 36–39 (Li et al. 1990). Fins are dark and without true spines, as in the Bighead Carp. However, in individuals larger than fingerlings, the anterior ray of the pectoral is thickened, ossified, and finely serrated posteriorly. Dorsal and anal fins are without spines. Dorsal fin rays number between 7 and 9, anal fin rays 11–13, pelvic fin rays 8, and pectoral fin rays 16–19. Pharyngeal teeth are similar to those of the Bighead Carp—long, bluntly rounded, and slightly concave on the grinding surface (Li et al. 1990)—but those of Silver Carp have striations on the grinding surface that are visible with magnification. Gut length of Silver Carp is longer than that of the Bighead Carp, reflecting a greater algal component to its diet. However, gut length varies from less than 7 times to more than 13 times their body length, depending on season and principal foods (phytoplankton or zooplankton; Ke et al. 2008).

Biology

Bigheaded carps are fishes of large, turbid rivers, associated backwaters, and floodplain lakes (Yi et al. 1988a, 1988b). Adults select low-velocity or lentic areas when not engaged in reproductive activities (Kolar et al. 2007), but they spawn in turbulent, high-velocity areas (Yi et al. 1988a, 1988b). Embryos develop while drifting in the current, and the eggs are only slightly heavier than water, which allows them to be kept suspended in the current by turbulence until hatching. The larvae drift until gas bladder inflation, in which case they swim precociously and begin to move into off-channel habitats (Deters et al. 2013; George and Chapman 2013). In the Yangtze River, young bigheaded carps are mainly distributed in the lakes connected to the Yangtze River while older bigheaded carps are mainly distributed in the main stem of the Yangtze River (ITFRYR 1990). In the Mississippi River basin, young-of-year bigheaded carps are often found in very shallow wetlands while adult fish can be found in nearly every habitat available, using primarily low-velocity waters when not spawning (Kolar et al. 2007).

Because the eggs develop while drifting, a minimum river length of 100 km and a minimum velocity or turbulence to keep the eggs in suspension have been considered necessary for survival of the eggs and perhaps the larvae (Krykhtin and Gorbach 1981; Yi et al. 2006). However, recent models of egg drift (Garcia et al. 2013; Murphy and Jackson 2013) and recruitment of Grass Carp (which has similar reproductive requirements) in a 26-km reach of the Sandusky River, Ohio (Chapman et al. 2013) provide evidence that complete embryonic development and recruitment can occur in much shorter river lengths.

Males of bigheaded carps usually mature 1 year earlier than females; the age of the first sexual maturity in the Yangtze River was 3 years for males and 4 years for females of both species (Li 2001). In the lower Mississippi River, gravid females have been found that are less than 2 years old (J. J. Hoover, personal observation). In China, spawning of Bighead Carp is usually synchronous (most females releasing most of their eggs over a short period), but Silver Carp seem to be incremental spawners (spawning a portion of the annual production of the eggs in many separate spawning events over a longer period; ITFRYR 1990). In North America, incremental spawning has been observed in both species (Papoulias et al. 2006). Spawning pattern may be dependent on environmental conditions; in some years on the Missouri River, nearly all female fish of both species are spent by midsummer (D. C. Chapman, personal observation). In other years, such as the years studied by Papoulias et al. (2006), few substantial hydrograph peaks occurred during the optimal temperature period, possibly resulting in the observed protracted incremental spawning.

In the Yangtze River, Bighead Carp generally spawn between May and July (peaking in late May), whereas Silver Carp spawning is somewhat more protracted, with spawning occurring April to July (peaking from mid-May to June; ITFMCCYR 1982; Yi et al. 1988a). Temperature and hydrograph seem to be the important factors influencing bigheaded carp spawning (ITFMCCYR 1982; Yi et al. 1988a). In both China and North America, spawning seems to occur principally between 20°C and 28°C (Yi et al. 1988a; Schrank et al. 2001). In China, bigheaded carps spawn mostly at sharp increases in the hydrograph, and the spawning stops immediately after the hydrograph peaks (ITFMCCYR 1982; Yi et al. 1988a; Duan et al. 2009). In North American studies (Coulter et al. 2013; Deters et al. 2013), bigheaded carp spawning seems somewhat less tied to flow, and some spawning occurs during periods of stable flow or after peak flows.

Bigheaded carps have high fecundity. In the Yangtze River, fecundity of a Bighead Carp weighing 18.5 kg was 1.1 million eggs (Chang 1966). In another study on the Yangtze River, the average absolute fecundity for Bighead Carp was 1.3 million eggs, and the average relative fecundity was 74.0 eggs per gram weight (ITFMCCYR 1982). The gonadosomatic index (gonad weight/body weight × 100) for males in reproductive condition was between 0.38 and 0.41 and for females was between 4.07 and 17.7. Estimates of Silver Carp fecundity in the Yangtze River ranged from an average of 396,000 eggs per individual for fish between 65 and 70 cm total length (TL), to 1.65 million eggs per individual for fish between 85 and 90 cm TL. Bighead Carp eggs are faint yellow in color with a yolk diameter of about 1.6 mm. Water-hardened eggs are usually between 5.0 and 6.7 mm. Silver Carp eggs are steel gray or tawny with a yolk diameter between 1.3 and 1.5 mm. Most water hardened eggs are between 4.2 and 6.2 mm (ITFMCCYR 1982; Yi et al. 2006); however, the size range seems to include much smaller Silver Carp eggs in North America (George and Chapman 2013).

Spawning grounds of bigheaded carps in the Yangtze River are similar for both species and well delineated but differ between the upper reaches and middle reaches of the Yangtze River. In the upper reaches of the Yangtze River, spawning sites were commonly found in narrow river valleys and canyons, with a narrow and deep river channel with boulders and rocky islands. River segments where spawning occurred often had a sinuosity of 1.3–1.4. In the middle reaches of the Yangtze River, bigheaded carps commonly spawn in river bends with a deep channel and sandy shoals (Yi et al. 1988a).

In North America, bigheaded carps have been observed spawning at river confluences (DeGrandchamp et al. 2008), in rapid sections of the river thalweg, and in the turbulent water around hard points such as revetted banks and wing dikes (man-made rock structures designed to focus river flow; Chapman, personal observation). In the lower Missouri River, which is fast-flowing, channelized, and characterized by many wing dikes and revetments, spawning seems to be widespread and not confined to a few discrete sites (Deters et al. 2013), but in that study, egg production was correlated to river sinuosity, which varied from 1.1 to 2.1. Deters et al. (2013) also reported that despite very high numbers of eggs collected in the Missouri River, spawning in the sampled tributaries of the Missouri was insubstantial or nonexistent. This may reflect a preference, rather than a requirement, for the very large river. Coulter et al. (2013) described spawning of Silver Carp in upper portions of the Wabash River of Indiana, with discharges similar to some of the tributaries in Deters et al. (2013).

Bighead and Silver carps are pelagic and occur mostly in the upper water column (Kolar et al. 2007), but the presence of sand grains in the gut indicates that some benthic feeding occurs (Davis et al. 2006). Bighead Carp are languid, and they rarely jump from the water. In contrast, Silver Carp are easily startled and actively avoid nets by scattering and jumping (ITFMCCYR 1982). In North America, Silver Carp are notorious for their habit of jumping individually or en masse when a motorboat passes (sometimes into the boat and injuring passengers), and even jumping into manually powered boats. This would be a rare occurrence in China (Chapman and D. Chen, personal observations), but the reason for this difference in behavior is not understood.

Bigheaded carps are primarily pump filter feeders that feed on planktonic foods, but they also feed on detritus (Cremer and Smitherman 1980). The Bighead Carp has a double row of long, finely spaced, comb-like gill rakers on each gill arch. These, combined with a large epibranchial (or suprabranchial) organ, comprise a highly specialized feeding apparatus (Voropayev 1969). The Silver Carp has a similar arrangement, but the epibranchial organ is smaller and the gill rakers are extremely thin, with a length 200 times the width at the tip. Each row is united into a continuous band by a mucous membrane within which the distal portions of the gill rakers are distinguishable, but the proximal portions are fused into hardened structures (Yokote 1956). The smooth interior surface between the two rows of gill rakers has microscopic pores (Wilamovski 1972) that allow Silver Carp to feed on particles smaller than 10 μm (Cremer and Smitherman 1980). The outer surface of the fused gill rakers are composed of buttresses for the fine, inner filtering surface and have a sponge-like appearance (Kolar et al. 2007).

Larvae of both Bighead and Silver carps feed on rotifers and early instars of cladocerans (such as *Bosmina* and *Moina*) and on copepod nauplii and copepodites (Chang 1966). Ling (1967) found that 10–17-mm larvae consumed Cladocera, and larvae began to eat phytoplankton (mainly diatoms) at lengths between 18 and 23 mm, and at 24–30 mm they readily consumed zooplankton and phytoplankton. As the carp develop, their diets begin to differ, and the gill rakers that facilitate respective modes of feeding take adult form by the time the fish reach 25 cm (ITFMCCYR 1982).

Adult Bighead Carp feed mainly on zooplankton, including rotifers, cladocerans, copepods, and protozoa (Dong and Li 1994). When concentrations of zooplankton are low, Bighead Carp will switch to phytoplankton. Dong and Li (1994) described a large number of taste buds in the epithelium of the large epibranchial organ, which may aid Bighead Carp in locating high densities of zooplankton. Silver Carp adults also consume plankton and other particles that are harvested by filtration but can effectively filter and consume smaller particles than can Bighead Carp. The Silver Carp feeds mainly on phytoplankton and phytodetritus, and phytoplankton is normally the principal food over most of its life (ITFMCCYR 1982). Both in China and in North America, bigheaded carps feed in both winter and summer (ITFMCCYR 1982; Kolar et al. 2007), but in China, the feeding intensity was much lower in cold periods (ITFMCCYR 1982).

Bigheaded carps grow rapidly (Kolar et al. 2007). In natural rivers and lakes with abundant food organisms, Bighead Carp attain weights greater than 1.0 kg in less than a year and are capable of reaching 3.5–4.8 kg in 3 years. Maximum length may exceed 1.5 m TL, and maximum weight is around 50 kg. Silver Carp can also grow quickly. In the Yangtze River, Silver Carp can attain 0.5 kg in a year, 2 kg in 2 years, and 3.5–4.5 kg in 3 years (ITFMCCYR 1982). Their maximum weight is 30–40 kg.

Yangtze River Bigheaded Carps

Status

Bighead Carp and Silver Carp are the most important aquaculture and wild-caught food fishes in China. The earliest record of bigheaded carps in the Yangtze River can be traced to the Tang Dynasty (618–907 AD). By the time of the Song Dynasty (960–1279 AD), bigheaded carp aquaculture was widely distributed in the floodplains of almost all river systems in China. Because of geographical isolation, Bighead Carp and Silver Carp populations in Asia can each be divided into three strains (Amur River, Pearl River, and Yangtze River strains). The Yangtze River strains have the largest fishery yield and the highest genetic diversity (Li et al. 1990).

The Yangtze River is the evolutionary cradle of the bigheaded carps, and the middle reach of the Yangtze River has been the most important area in China in terms of harvest and of spawning and larvae production. In the 1960s, there were 36 spawning grounds in the main stem of the Yangtze River between Chongqing and Pengze, a distance of 1,695 km (Yi et al. 1988a). After completion of the Gezhouba Dam in the 1980s, six of those spawning areas were inundated by the reservoir. Annual larval production of Bighead Carp was 6.92×10^9 in the main stem of the Yangtze River prior to the construction of the Gezhouba Dam (Yi et al. 1988b) but decreased about 90% to 7.09×10^8 annually after construction (ITFMCCYR 1982). Over the same period, Silver Carp annual larval drift decreased about 98% from 2.99×10^{10} to 6.75×10^8.

Duan et al. (2009) provides a history of changes in bigheaded carp larval production before and after major dam construction on the Yangtze River. From 1997 to 2005, there were 10 spawning grounds for bigheaded carps from Yichang to the Chenglingji cross section below the dam (a distance of 400 km). Among these sites, Yichang, Zhijiang, Jiangkou, Hexue, Shishou, and Tiaoguan had the highest productivity. These sites were not inundated, but when compared to the 1980s (Yi et al. 1988a, 1988b), larval production decreased because of the dam-induced changes in temperature and hydrology. In 2002, the construction of the much larger Three Gorges Dam (TGD) caused great hydrological changes both upstream and downstream of the dam, resulting in a further decline in Bighead Carp larval production. Silver Carp were not as severely affected by this second dam. Prior to the construction of the TGD, Grass Carp and Black Carp had dominated the larval composition of the four famous Chinese carps. However, Silver Carp has dominated the larval composition since the construction of TGD. In 2007, Bighead Carp larval drift dropped below 1% of the four famous Chinese Carps at the Sanzhou Reach, and no Black Carp larvae were collected at all.

Management

The Yangtze River was once the major production area for bigheaded carps. Dam construction, floodplain disconnection, environmental pollution, and overfishing have all been implicated in the decline of the natural bigheaded carp population. To recover these populations, several measures have been carried out since 2000. At present, these efforts include the creation of a stocking program, establishment of nature reserves (no fishing zones), and altering dam operations to be more environmentally sensitive.

The enhancement of bigheaded carps through stocking was first launched in the early 1950s. Afterwards, additional stocking activities were increasingly carried out in the Yunnan, Sichuan, Guizhou, Chongqing, Hubei, Hunan, Jiangxi, Anhui, Jiangsu, and Shanghai portions of the Yangtze River basin due to the continued decline of natural bigheaded carps. From 2005 to 2007, more than 10 million of the four famous Chinese carps were released in the Yangtze River basin. The majority (61.3%) of these fish were 1.5–2.5 cm TL, with approximately 7% exceeding 10 cm TL. Only a few adult fish were released (Chen et al. 2012) during these stocking periods. Since 2009, the Yangtze River Fisheries Research Institute, Chinese Academy of Fishery Science has begun the release of large fish into the Yangtze River. This resulted in about 6,000 adults of the four famous Chinese carps being released between 2010 and 2013. The nature reserves of the four famous Chinese carps were first established in the early 1990s, and now the number of reserves has increased to 11.

To alleviate the effects of TGD hydrologic influence on the natural spawning of bigheaded carps in the middle Yangtze River, the dam has been operated in a more ecologically sensitive fashion since June 15, 2011. However, current ecological operation measures do not yet seem to be entirely sufficient for the recovery of fish populations, and efforts to improve the ecological operation of the TGD are still underway.

Mississippi River Bigheaded Carps

Status

Bighead Carp and Silver Carp are established in the Mississippi River basin. They were first imported into the United States in the early 1970s for experimental use in food fish production and for biological control of excess algae in sewage treatment facilities, aquaculture, and lakes (Kolar et al. 2007; Kelly et al. 2011). Attempts to use bigheaded carps for water quality control in North America outside of aquaculture mostly ceased by the early 1980s (see Lieberman 1996 for an exception), but aquaculturists found an application for Bighead Carp as a food fish in polyculture with Channel Catfish *Ictalurus punctatus*. They were widely used for this purpose (Engle and Brown 1999). Silver Carp were never substantially used for food fish production in North America (Kelly et al. 2011).

Silver Carp were first captured from open rivers within months of the original importation, but the first record of a Bighead Carp capture from the wild was not until 1981 (Kolar et al. 2007). Captures of both species remained rare until the early 1990s. Tucker et al. (1996) reported capture of 48 Bighead Carp between 1991 and 1994, including young-of-year fish that they concluded were evidence of wild reproduction. In some portions of the Mississippi River basin, exponential population growth of Bighead Carp began in the latter half of the 1990s. Since that time, both species have reached extraordinary abundance in the central part of the basin (Kolar et al. 2007) and continue to expand their range (USGS 2014). Presently, bigheaded carps are abundant (at least locally) from Louisiana to southern Iowa on the Mississippi River, as far upstream as Indiana on the Ohio River, and to the Nebraska–South Dakota border on the Missouri River. Individuals have been captured as far upstream on the Mississippi River as Pool 2 near Cottage Grove, Minnesota (MNDNR 2014) and on Mississippi River tributaries as far north as North Dakota (James River) and as far east as Ohio in the Ohio River (USGS 2014). Range expansion seems to have been hindered by dams on the upper Mississippi, Ohio, and Tennessee rivers because expansion into undammed reaches occurred more quickly.

Effects of Bigheaded Carp

The introduction or invasion of bigheaded carps (especially Silver Carp) causes substantial declines in abundance of crustacean zooplankton and a reduction in the size of phytoplankton, coincident with the proliferation of phytoplankton smaller than that which can be consumed by the carp (reviewed by Kolar et al. 2007). In the absence of large zooplankton, Silver Carp in the Mississippi basin feed on a wide variety of small organisms, including euglenoids, diatoms, green algae, cyanobacteria, and rotifers (Pongruktham et al. 2010). Euglenoids, cyanobacteria, and pinnate diatoms often survive passage through the gut of the fish.

The invasion of bigheaded carps has corresponded with a reduction in body condition of Bigmouth Buffalo *Ictiobus cyprinellus* and Gizzard Shad *Dorosoma cepedianum* (Irons et al. 2007), both of which have substantial diet overlap with bigheaded carps (S. J. Sampson, J. H. Chick, and M. A. Pegg, presentation at the 2004 Annual Meeting of the American Fisheries Society). There has been no study on the effect of bigheaded carps on river populations of the zooplanktivorous Paddlefish *Polyodon spathula* of North America. However, because Bighead Carp were shown to outcompete Paddlefish in ponds (Schrank et al. 2003) and because of the well-documented effect of bigheaded carps on crustacean zooplankton (thought to be the most important food of Paddlefish), it is likely that Paddlefish have been adversely affected (Pegg et al. 2009). This hypothesis is supported by observations of low-condition Paddlefish in water bodies that are intermittently connected to large rivers and that have high densities of bigheaded carp (Varble et al. 2007). In addition to competing with fishes that are planktivorous as adults, Silver Carp have been shown to have strong adverse effects on recruitment of fishes with pelagic early life history, especially percids, although the mechanism for this effect is unknown (Costa-Pierce 1992).

Management

In the mid-2000s, the U.S. Fish and Wildlife Service organized an Asian Carp Working Group consisting of stakeholders from federal and state agencies, industry, academia, and nongovernmental organizations. The working group completed a management plan for the control of Asian carps (Grass Carp, Black Carp, and the bigheaded carps; Conover et al. 2007), which was submitted to and approved by the Aquatic Nuisance Species Task Force, an intergovernmental organization dedicated to preventing and controlling aquatic nuisance species and implementing the Nonindigenous Aquatic Nuisance Prevention and Control Act of 1990. This plan recommended strategies to control the range expansion of Asian carps and to control existing populations. The plan was not funded in entirety, but some portions of the plan have been initiated by various federal and state agencies. In an effort to inhibit range expansion, both Bighead and Silver carps were listed as injurious species under the Lacey Act (1900), which prohibits interstate transport of live injurious species. Barriers to adult fish movement in large rivers, other than dams that also bar navigation, are difficult or impossible to achieve (Rasmussen 2011). However, sustained and prolonged swimming endurance of bigheaded carps less than 350 mm TL is less than 120 cm/s, raising the possibility that hydraulics might be used to prevent or displace young fish in some circumstances (Hoover 2012).

Electric barriers are now in use to prevent movement of bigheaded carps from Mississippi River tributaries into lakes in Iowa (IDNR 2013) and into Lake Michigan through the Chicago Area Waterway System (CAWS; Moy et al. 2011). In addition, the navigation lock on the Mississippi River at St. Anthony Falls in Minnesota was closed in June 2015 to impede the progression of bigheaded carps (Matos 2014). Underwater loudspeakers will be deployed at the Mississippi River Lock and Dam 8 (near the southern border of Minnesota) in an effort to deter bigheaded carp range expansion (Pennekamp 2014). In 2009, the Asian Carp Regional Coordinating Committee, chaired by the White House Council on Environmental Quality and the U.S. Environmental Protection Agency, was established to join federal and state efforts to prevent the establishment of bigheaded carps in the Laurentian Great Lakes. The Great Lakes are of special concern, in part because of the value of the fishery. The value of the recreational fishery on U.S. waters, alone, was estimated at almost US$2 $\times 10^9$, annually (USFWS and USCB 2014), and the U.S. commercial fishery produces $2.25 $\times 10^7$ in landings, annually (USCOE 2012). This committee specifically focused on the CAWS as the major entry point from the Mississippi River Basin to Lake Michigan. In 2014, alone, more than $6.1 $\times 10^7$ was dedicated to preventing the spread of bigheaded carps from the Mississippi River basin to the Great Lakes, aimed at helping fund monitoring, control, eradication, research, barriers construction, technology refinement, enforcement, and outreach efforts.

Harvest is the only method currently being practiced substantially as a control for existing populations of bigheaded carps. Commercial fishers are contracted to remove fish from the reaches of the Illinois River near the electric barrier in the CAWS in an effort to reduce the number of fish near the barrier (ACRCC 2014). Some state agencies encourage commercial harvest of bigheaded carps through incentives to processors (Anonymous 2010) or, in Kentucky, by a bigheaded carp commercial fishing tournament (KDFWR 2013). However, market development has detractors because of the potential for unwanted bycatch of native species, potential conflicts between commercial fishers and other users of the resource, and the lack of a clear exit strategy. A substantial industry based on the bigheaded carp fishery would result in the creation of jobs and infrastructure. However, once those jobs and infrastructure are in place, if technology is developed that allows the eradication of the carp, deployment of the technology would cause economic turmoil. The 2007 management plan recommends harvest only for the eradication of bigheaded carps or maximum possible reduction of their populations, and not management for long-term sustenance of commercial fisheries. In any case, the low value of landed bigheaded carp in the United States has limited the fishery. Markets have been slow to develop because the name "carp" is associated with poor-quality fish in North America and because the bone structure of bigheaded carps makes it difficult to deliver the boneless filet preferred by North American consumers (Varble and Secchi 2013)

A recent meta-analysis of carp population dynamics indicates that harvest that is sufficiently intensive (>70% of the population) and nonselective (for either species or for any size-class) could effectively reduce populations in the Mississippi basin (Tsehaye et al. 2013). Population models on which the analysis is based, however, are limited by inputs used to construct them, and for many subpopulations of carp, vital rates (e.g., longevity, survivorship, and fecundity) are unknown or uncertain.

Also problematic in such management scenarios is the potential for rapid genetic change in bigheaded carps in the Mississippi River. Such rapid evolution is well documented in invading species, and it can be enhanced by hybridization that occurs due to species invasions (Cox 2004). Hybridization and introgression between the Bighead and Silver carps and their fertile hybrids are uncommon in the Yangtze River but are extensive in the Mississippi River basin (>50% of individuals show evidence of hybridization and/or introgression), with aberrant and variable phenotypes occurring (Lamer et al. 2010, 2015). Such aberrancies and variation reduce the ability of managers to assess risk and forecast or manage populations with confidence because existing biological information on bigheaded carps from native or other introduced populations sometimes inadequately predicts the biology of bigheaded carps in North America (Coulter et al. 2013).

Given the current lack of tools other than barriers and harvest, alternative avenues for the potential control of bigheaded carps are being researched. These include, but are not limited to, the introduction of Trojan genes, which would limit the production of female carp (Teem et al. 2013); water cannons and other noisemaking repellants; directed toxicant formulations; feeding attractants; the development of Judas fish strategies (using telemetered fish to locate concentrations of carp, especially at the margins of their range where carp are not abundant) to enhance harvest; the development of new harvest gears; and assessment of the potential of native predators of juvenile carps as a control (ACRCC 2015).

China–U.S. Collaborations in Bigheaded Carp Research and Management

Information collected during decades of research on bigheaded carps in China has been extremely valuable in understanding the North American invasion. For example, the excellent bigheaded carp early life history information developed by Yi et al. (2006) provided the groundwork for biological research in the United States (Chapman and George 2011; Deters et al. 2013; George and Chapman 2013) and for North American carp risk assessments based on a combination of hydrologic modeling and biological data (Kocovsky et al. 2012; Garcia et al. 2013; Murphy and Jackson 2013). In addition, fishing methods for bigheaded carps are highly developed in China, and those techniques may have application for harvest and control in the United States. In late 2015, the Institute of Hydrobiology, Chinese Academy of Sciences, in Wuhan, China, together with regional fishing cooperatives (Wuhu Lake Fishery Group and Niushanhu Lake Fishery Group), hosted scientists from the U.S. Geological Survey and the Illinois Department of Natural Resources for transferring the technology of the Chinese unified method of fishing. This large-scale and complex fishing method incorporates many kinds of nets coupled with slowly (over weeks or months) driving fish to a collection point to harvest most of the target fishes in the fished water body. Habitats, regulatory concerns, and fishery goals differ between the United States and China; therefore, the efficacy of the unified method in the United States has not yet been determined. The first attempt at use of this method in the United States is scheduled for March 2016 in a large artificial backwater of the Illinois River, with hopes that this method or modifications of fishing methods used in China will allow for more efficient harvest of bigheaded carps.

References

ACRCC (Asian Carp Regional Coordinating Committee). 2013. Monitoring and response plan for Asian carp in the upper Illinois River and Chicago Area Waterway System, May 2013. Available: http://asiancarp.us/documents/mrp2013.pdf. (July 2014).

ACRCC (Asian Carp Regional Coordinating Committee). 2014. Asian Carp control strategy framework, June 2014. Available: http://asiancarp.us/documents/2014Framework.pdf. (December 2014).

ACRCC (Asian Carp Regional Coordinating Committee). 2015. Asian carp control strategy framework, June 2015. Available: www.asiancarp.us/documents/2015Framework.pdf. (October 2015).

Anonymous. 2010. Illinois company receives grant to export Asian carp to China. Circle of Blue (November 4). Available: www.circleofblue.org/waternews/2010/world/north-america/illinois-company-receives-grant-to-export-asian-carp-to-china/. (July 2014).

Cavender, T. M. 1991. The fossil record of the Cyprinidae. Pages 34–54 *in* I. J. Winfield and J. S. Nelson, editors. Cyprinid fishes: systematic, biology, and exploitation. Chapman and Hall, London.

Chang, Y. F. 1966. Culture of freshwater fish in China. *In* E. O. Gangstad, editor. 1980. Chinese fish culture, report 1. U.S. Army Waterways Experiment Station, Technical Report A-79, Washington, D.C.

Chapman, D. C., and A. E. George. 2011. Developmental rate and behavior of early life stages of Bighead and Silver carp. U.S. Geological Survey, Scientific Investigations Report 2011-5076, Reston, Virginia.

Chapman, D. C., J. J. Davis, J. A. Jenkins, P. M. Kocovsky, J. G. Miner, J. Farver, and P. R. Jackson. 2013. First evidence of Grass Carp recruitment in the Great Lakes basin. Journal of Great Lakes Research 39:547–554.

Chen, D. Q., S. J. Li, and K. Wang. 2012. Enhancement and conservation of inland fisheries resources in China. Environmental Biology of Fishes 93:531–545.

Conover, G., R. Simmonds, and M. Whalen, editors. 2007. Management and control plan for Bighead, Black, Grass, and Silver carps in the United States. Asian Carp Working Group, Aquatic Nuisance Species Task Force, Washington, D.C.

Costa-Pierce, B. A. 1992. Review of the spawning requirements and feeding ecology of Silver Carp (*Hypophthalmichthys molitrix*) and reevaluation of its use in fisheries and aquaculture. Reviews in Aquatic Sciences 6:257–273.

Coulter, A. A., D. Keller, J. J. Amberg, E. J. Bailey, and R. Goforth. 2013. Phenotypic plasticity in the spawning traits of bigheaded carp (*Hypophthalmichthys* spp.) in novel ecosystems. Freshwater Biology 58:1029–1037.

Cox, G. W. 2004. Alien species and evolution: the evolutionary ecology of exotic plants, animals, microbes, and interacting native species. Island Press, Washington, D.C.

Cremer, M. C., and R. O. Smitherman. 1980. Food habits and growth of Silver and Bighead carp in cages and ponds. Aquaculture 20:57–64.

Davis, B. M., P. L. Hudson, and D. C. Chapman. 2006. Zooplankton diet of Bighead and Silver carp in the lower Missouri River and one of its tributaries. Page 7 in Great Lakes in changing environment: 49th annual conference on Great Lakes research, International Association for Great Lakes Research, Ann Arbor, Michigan.

DeGrandchamp, K. L., J. E. Garvey, and R. E. Colombo. 2008. Movement and habitat selection by invasive Asian carps in a large river. Transactions of the American Fisheries Society 137:45–56.

Deters, J. E., D. C. Chapman, and B. McElroy. 2013. Location and timing of Asian carp spawning in the lower Missouri River. Environmental Biology of Fishes 96:617–629.

Dong, S. L., and D. S. Li. 1994. Comparative studies of the feeding selectivity of Silver Carp, *Hypophthalmichthys molitrix*, and Bighead Carp, *Aristichthys nobilis*. Journal of Fish Biology 44:621–626.

Duan, X. B., S. P. Liu, M. G. Huang, S. L. Qiu, Z. H. Li, K. Wang, and D. Q. Chen. 2009. Changes in abundance of larvae of the four domestic Chinese carps in the middle reach of the Yangtze River, China, before and after closing of the Three Gorges Dam. Environmental Biology of Fishes 86:13–22.

Dudgeon, D. 2011. Asian river fishes in the Anthropocene: threats and conservation challenges in an era of rapid environmental change. Journal of Fish Biology 79:1487–1524.

Engle, C. R., and D. Brown. 1999. Growth, yield,

dressout, and net returns of Bighead Carp *Hypophthalmichthys nobilis* stocked at three densities in fertilized earthen ponds. Journal of the World Aquaculture Society 30:371-379.

Garcia, T., P. R. Jackson, E. A. Murphy, A. J. Valocchia, and M. H. Garcia. 2013. Development of a fluvial egg drift simulator to evaluate the transport and dispersion of Asian carp eggs in rivers. Ecological Modeling 263:211-222.

George, A. E., and D. C. Chapman. 2013. Aspects of embryonic and larval development in Bighead Carp *Hypophthalmichthys nobilis* and Silver Carp *Hypophthalmichthys molitrix*. PLoS (Public Library of Science) One [online serial] 8(8):e73829. DOI: 10.1371/journal.pone.0073829.

Hoover, J. J. 2012. Swimming performance of Bighead Carp and Silver Carp: methodology, metrics, and management applications. U.S. Army Engineer Research and Development Center, Aquatic Nuisance Species Program, ERDC/TN-ANSRP-12-3. Vicksburg, Mississippi. Available: http://el.erdc.usace.army.mil/elpubs/pdf/ansrp12-3.pdf. (April 2016).

Howes, G. 1981. Anatomy and phylogeny of the Chinese major carps *Ctenopharyngodon* Steind., 1866 and *Hypophthalmichthys* Blkr., 1860. Bulletin of the British Museum (Natural History) Zoology Supplement 41:1-52.

IDNR (Iowa Department of Natural Resources). 2013. Electric fish barrier ready to protect Iowa Great Lakes. IDNR, Des Moines.

Irons, K. S., G. G. Sass, M. A. McClelland, and J. D. Stafford. 2007. Reduced condition factor of two native fish species coincident with invasion of non-native Asian carps in the Illinois River, U.S.A. Is this evidence for competition and reduced fitness? Journal of Fish Biology 71:258-273.

ITFRYR (Investigation Team of Fisheries Resources in the Yangtze River System). 1990. Fisheries resources in the Yangtze River system. Ocean Press, Beijing.

ITFMCCYR (Investigation Team of the Four Major Chinese Carp in the Yangtze River). 1982. A survey on spawning sites of the four major Chinese carps in the Yangtze River after damming at Gezhouba. Journal of Fisheries of China 6:287-304. (In Chinese with English abstract.)

Ke, Z., X. Ping, and L. Guo. 2008. Phenotypic plasticity in gut length in the planktivorous filter-feeding Silver Carp (*Hypophthalmichthys molitrix*). The Scientific World Journal 8:169-175.

Kelly, A. M., C. R. Engle, M. L. Armstrong, M. Freeze, and A. J. Mitchell. 2011. History of introductions and governmental involvement in promoting the use of Grass, Silver, and Bighead carps. Pages 163-174 *in* D. C. Chapman and M. H. Hoff, editors. Invasive Asian carps in North America. American Fisheries Society, Symposium 74, Bethesda, Maryland.

KDFWR (Kentucky Department of Fish and Wildlife Resources). 2013. Carp madness. KDFWR, Frankfort.

Kocovsky, P. M., D. C. Chapman, and J. E. McKenna. 2012. Thermal and hydrologic suitability of Lake Erie and its major tributaries for spawning of Asian carps. Journal of Great Lakes Research 38:159-166.

Koel, T. M., K. S. Irons, and E. Ratcliff. 2000. Asian carp invasion of the upper Mississippi River system. U.S. Geological Survey, Upper Midwest Environmental Sciences Center, Project Status Report 2000-05, La Crosse, Wisconsin.

Kolar, C. S., D. C. Chapman, W. R. Courtenay, C. M Housel, J. D. Williams, and D. P. Jennings. 2007. Bigheaded carps: a biological synopsis and environmental risk assessment. American Fisheries Society, Special Publication 33, Bethesda, Maryland.

Krykhtin, M. L., and E. I. Gorbach. 1981. Reproductive ecology of the Grass Carp, *Ctenopharyngodon idella*, and the Silver Carp, *Hypophthalmichthys molitrix*, in the Amur basin. Journal of Ichthyology 21:109-123.

Lacey Act. 1900. U.S. code, volume 16, sections 3371-3378.

Lamer, J. T., C. R. Dolan, J. L. Petersen, J. H. Chick, and J. M. Epifanio. 2010. Introgressive hybridization between Bighead Carp and Silver Carp in the Mississippi and Illinois rivers. North American Journal of Fisheries Management 30:1452-1461.

Lamer, J. T., B. C. Ruebush, Z. Arbieva, Z. H., M. A. McClelland, J. M. Epifanio, and G. G. Sass. 2015. Diagnostic SNPs reveal widespread introgressive hybridization between introduced Bighead and Silver carp in the Mississippi River basin. Molecular Ecology 24:3931-3943.

Li, S. F., L. Z. Wu, Q. Wang, Q. R. Chou, and Y. Chen. 1990. Comprehensive genetic study of Chinese carps. Shanghai Scientific and Technical Publishers, Shanghai, China.

Li, S. F. 2001. The impact of large reservoirs on fish biodiversity and fisheries in China. Pages 22-28 *in* S. S. De Silva, editor. 2000. Reservoir and culture-based fisheries: biology and management: proceedings of an international workshop, February, 15–18, Bangkok, Thailand. Australian Centre for International Agriculture Research, Canberra.

Li, S. F. 2006. Fish biodiversity and conservation of Yangtze River basin. Pages 1–8 *in* G. B. Cao and S. G. Kohl, editors. Proceedings of the second international symposium on ecology and fishery biodiversity in large rivers of northeast Asia and western North America. Heilongjiang Science and Technology Press, Harbin, China.

Li, S. F., J. W. Wu, Q. L. Yang, C. H. Wang, Q. Chen, D. C. Chapman, and G. Lu. 2009. A comparison of complete mitochondrial genomes of Silver Carp *Hypophthalmichthys molitrix* and Bighead Carp *Hypophthalmichthys nobilis*: implications for their taxonomic relationship and phylogeny. Journal of Fish Biology 74:1787–1803.

Li, W., J. Chen, P. Xie, J. He, X. Guo, X. Tuo, W. Zhang, and L. Wu. 2014. Rapid conversion and reversible conjugation of glutathione detoxification of microcystins and Bighead Carp (*Aristicthys nobilis*). Aquatic Toxicology 147:18–25.

Lieberman, D. M. 1996. Use of Silver Carp (*Hypophthalmichthys molitrix*) and Bighead Carp (*Aristichthys nobilis*) for algae control in a small pond: changes in water quality. Journal of Freshwater Ecology 11:391–397.

Ling, S. W. 1967. Feeds and feeding of warm-water fishes in ponds in Asia and the Far East. FAO (Food and Agriculture Organization of the United Nations) Fisheries Report 44:291–309.

Matos, A. 2014. St. Anthony lock closure is first step to slow down invasive carp. Star Tribune (May 27). Available: www.startribune.com/local/260794031.html. (July 2014).

MNDNR (Minnesota Department of Natural Resources). 2014. Bighead and Silver carp discovered in Pool 2 of the Mississippi River. MNDNR press release. Available: http://news.dnr.state.mn.us/2014/07/18/bighead-and-silver-carp-discovered-in-pool-2-of-mississippi-river/#more-14965. (July 2014).

Moy, P. B., I. Polls, and J. M. Dettmers. 2011. The Chicago Sanitary and Ship Canal aquatic nuisance species dispersal barrier. Pages 121–137 *in* D. C. Chapman and M. H. Hoff, editors. Invasive Asian carps in North America. American Fisheries Society, Symposium 74, Bethesda, Maryland.

Murphy, E. A., and P. R. Jackson. 2013. Hydraulic and water-quality data collection for the investigation of Great Lakes tributaries for Asian carp spawning and egg-transport suitability. U.S. Geological Survey, Scientific Investigations Report 2013-5106, Reston, Virginia.

Page, L. M., H. Espinosa-Pérez, L. T. Findley, C. R. Gilbert, R. N. Lea, N. E. Mandrak, R. L. Mayden, and J. S. Nelson. 2013. Common and scientific names of fishes from the United States, Canada, and Mexico, 7th edition. American Fisheries Society, Special Publication 34, Bethesda, Maryland.

Papoulias, D. M., D. C. Chapman, and D. E. Tillitt. 2006. Reproductive condition and occurrence of intersex in Bighead Carp and Silver Carp in the Missouri River. Hydrobiologia 571:355–360.

Pegg, M. A., J. H. Chick, and B. M. Pracheil. 2009. Potential effects of invasive species on Paddlefish. Pages 185–201 *in* C. P. Paukert and G. D. Scholten, editors. Paddlefish management, propagation, and conservation in the 21st century: building from 20 years of research and management. American Fisheries Society, Symposium 66, Bethesda, Maryland.

Pennekamp, T. 2014. Asian carp barrier. Guttenberg Press (April 28). Available: www.guttenbergpress.com/articles/2014/04/28/asian-carp-barrier. (July 2014).

Pongruktham, O., C. Ochs, and J. J. Hoover. 2010. Observations of Silver Carp (*Hypophthalmichthys molitrix*) planktivory in a floodplain lake of the lower Mississippi River basin. Journal of Freshwater Ecology 25:85–93.

Rasmussen, J. L. 2011. Regulations as a tool in Asian carp management. Pages 175–190 *in* D. C. Chapman and M. H. Hoff, editors. Invasive Asian carps in North America. American Fisheries Society, Symposium 74, Bethesda, Maryland.

Schrank, S. J., P. J. Braaten, and C. S. Guy. 2001. Spatiotemporal variation in density of larval Bighead Carp in the lower Missouri River. Transactions of the American Fisheries Society 130:809–814.

Schrank, S. J., C. S. Guy, and J. F. Fairchild. 2003. Competitive interactions between age-0 Bighead Carp and Paddlefish. Transactions of the American Fisheries Society 132:1222–1228.

Teem, J. L., J. B Gutierrez, and R. D. Parshad. 2013. A comparison of the Trojan Y Chromosome and daughterless carp eradication strategies. Biological Invasions 16:1217–1230.

Tsehaye, I., M. Catalano, G. Sass, D. Glover, and B. Roth. 2013. Prospects for fishery-induced collapse of invasive Asian carp in the Illinois River. Fisheries 38:445–454.

Tucker, J. K., F. A. Cronin, R. A. Hrabik, M. D. Petersen, and D. P. Herzog. 1996. The Bighead Carp (*Hypophthalmichthys nobilis*) in the Mississippi River. Journal of Freshwater Ecology 11:241–243.

USCOE (U.S. Army Corps of Engineers). 2012. Commercial fisheries baseline economic assessment: U.S. waters of the Great Lakes, upper Mississippi River, and Ohio River basins. Available: http://glmris.anl.gov/documents/docs/Commercial_Fisheries_Report.pdf. (March 2016).

USFWS (U.S. Fish and Wildlife Service) and USCB (U.S. Census Bureau). 2014. 2011 national survey of fishing, hunting, and wildlife-associated recreation. 2014 revision. Available: www.census.gov/prod/2012pubs/fhw11-nat.pdf. (March 2016).

USGS (U.S. Geological Survey). 2014. Nonindigenous aquatic species [online database]. Available: http://nas.er.usgs.gov. (July 2014).

Varble, K. A., J. J. Hoover, S. G. George, C. E. Murphy, and J. J. Killgore. 2007. Floodplain wetlands as nurseries for Silver Carp, *Hypophthalmichthys molitrix*: a conceptual model for use in managing local populations. U.S. Army Engineer Research and Development Center, Aquatic Nuisance Species Research Program, ANSRP-07-4, Vicksburg, Mississippi. Available: http://el.erdc.usace.army.mil/elpubs/pdf/ansrp07-4.pdf. (October 2015).

Varble, S., and S. Secchi. 2013. Human consumption as an invasive species management strategy. A preliminary assessment of the marketing potential of invasive Asia carp in the US. Appetite 65:58–67.

Voropayev, N. V. 1969. Morphological features and some fish cultural characteristics of *Aristichthys nobilis*, *Hypophthalmichthys molitrix* and of their hybrids. Trudy Vsesoyuznogo Nauchno Instituta-Issledouatel'skogo Rybnogo Khozgajstva I Okeanograffi 16:272–281.

Wilamovski, A. 1972. Structure of the gill apparatus and the suprabranchial organ of *Hypophthalmichthys molitrix* Val. Bamidgeh 24:87–99.

Xie, P., and Y. Y. Chen. 2001. Invasive carp in China's Plateau lakes. Science 224:999–1000.

Yi, B., Z. Yu, Z. Liang, S. Sujuan, Y. Xu, J. Chen, M. He, Y. Liu, Y, Hu, Z. Deng, S. Huang, J. Sun, R. Liu, and Y. Xiang. 1988a. The distribution, natural conditions, and breeding production of the spawning ground of four famous freshwater fishes on the main stream of the Yangtze River. Pages 1–46 *in* B. Yi, Z. Yu, and Z. Liang, editors. Gezhouba Water Control Project and four famous fishes in Yangtze River. Hubei Science and Technology Press, Wuhan, China.

Yi, B. L., Z. S. Liang, and Z. T. Yu. 1988b. The present situation of the spawning grounds of the four Chinese domestic fishes in Yangtze River after construction of the Gezhouba water control project. Pages 47–68 *in* B. Yi, Z. Yu, and Z. Liang, editors. Gezhouba Water Control Project and four famous fishes in Yangtze River. Hubei Science and Technology Press, Wuhan, China.

Yi, B., Z. Liang, Z. Yu, R. Lin, and M. He. 2006. A study of the early development of Grass Carp, Black Carp, Silver Carp, and Bighead Carp in the Yangtze River, China. Chapter 2 *in* D. C. Chapman, editor. Early development of four cyprinids native to the Yangtze River, China. U.S. Geological Survey, Data Series 239, Reston, Virginia.

Yokote, M. 1956. Morphological notes on the two Chinese carps, *Hypophthalmichthys molitrix* and *Aristichthys nobilis*. Bulletin of the Freshwater Fish Research Laboratory 6:61–70.

American Fisheries Society Symposium 84:127–137, 2016

Review of Population Status and Conservation Measures for Baiji and Yangtze Finless Porpoise

DING WANG*, KEXIONG WANG, YUJIANG HAO, AND JINSONG ZHENG
The Key Laboratory of Aquatic Biodiversity and Conservation
Institute of Hydrobiology of the Chinese Academy of Sciences
Wuhan 430072, P. R. China

Abstract.—Two species of cetaceans are endemic to the Yangtze River, the baiji or Yangtze River dolphin *Lipotes vexillifer* and the Yangtze finless porpoise *Neophocaena asiaeorientalis asiaeorientalis*. Over the past several decades, both species have suffered dramatic declines in abundance and concomitant reductions in the extent of their geographic ranges. In 2006, the baiji was presumed to be extinct when an extensive survey of the Yangtze River failed to find any evidence of its continued existence. Data from that same survey further indicated that the population size of the Yangtze finless porpoises had decreased from 2,700 in 1991 to approximately 1,800 individuals by 2006. Results of a more recent population survey of the Yangtze River conducted in 2012 revealed that this rapid decline has continued, with only approximately 1,000 Yangtze finless porpoises remaining across their former range. The population viability of both species is threatened primarily by overfishing and illegal fishing of their prey, heavy boating traffic, sand dredging, construction projects in or near the water, and pollution throughout the Yangtze River ecosystem. In an attempt to protect these species, several conservation strategies, including in situ conservation (i.e., on-site conservation), ex situ conservation (i.e., off-site conservation), and captive breeding programs have been implemented. In this review, we examine the effectiveness of these conservation strategies over the past three decades and synthesize these findings to make recommendations for the future protection of the Yangtze finless porpoise.

Introduction

The Yangtze River of central China is the largest river in Asia and the third largest river worldwide. The catchment area of the Yangtze River contains the most densely populated area in China, which includes approximately 40% of the nation's total population. Two cetaceans are endemic to the Yangtze River, the baiji or Yangtze River dolphin *Lipotes vexillifer* and the Yangtze finless porpoise (YFP) *Neophocaena asiaeorientalis asiaeorientalis*. Both species occur in the middle and lower reaches of the Yangtze River from Yichang to Shanghai, including the Poyang and Dongting Lakes, which both adjoin the river (Figure 1). The baiji once occurred in the Qiantang River (Figure 1) but has not been observed there since the 1950s (Zhou et al. 1977). As top predators in their food chains, the survival of both species depends heavily on habitat stability and resource availability. However, habitat quality in the Yangtze River, which is considered China's "golden channel," has greatly deteriorated due to extensive anthropogenic activities, including overfishing, illegal fishing, heavy boating traffic, sand dredging, water-related project constructions, and pollution. These detrimental activities have led to the likely extinction of the baiji (Turvey et al. 2007) and to rapid declines in the population

* Corresponding author: wangd@ihb.ac.cn

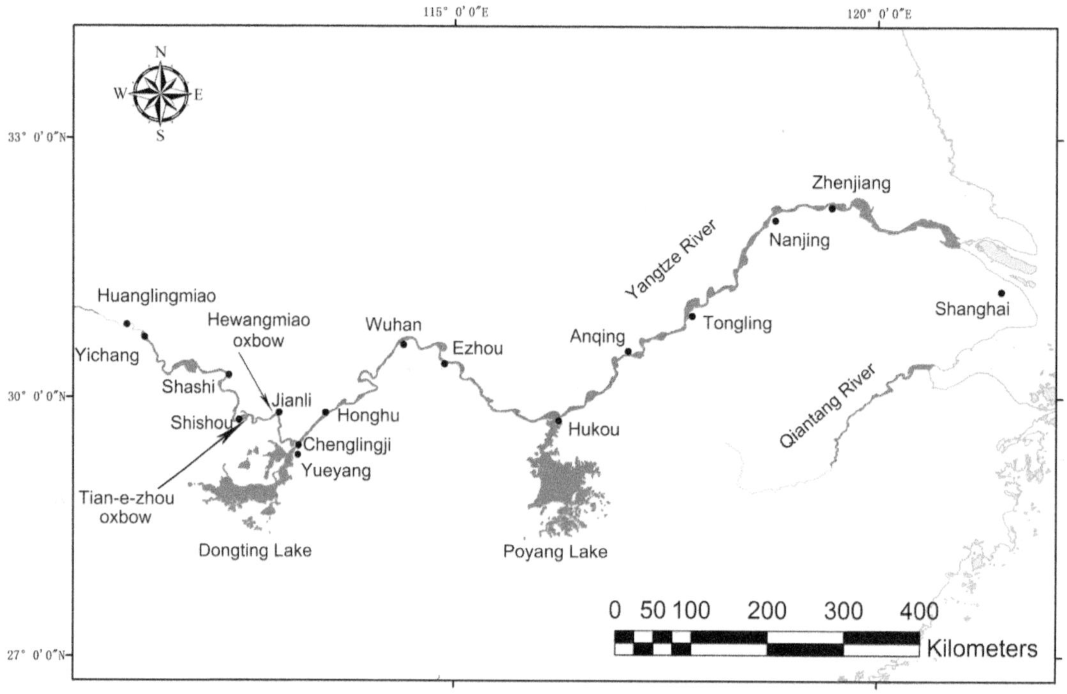

Figure 1. Historical distribution map of the baiji and the Yangtze finless porpoise (adapted from Wang 2009). The distribution of baiji is shown by the dark grey area along the Yangtze and Qiantang rivers and Dongting and Poyang lakes; the distribution of the Yangtze finless porpoise is shown by the dark grey area along the Yangtze River and Dongting and Poyang lakes.

abundance of the YFP. As a result, the YFP has been listed in the second order of protected animals in China since 1989 and was recently upgraded to critically endangered status in the *IUCN of Nature Red List of Threatened Species* (Wang et al. 2013).

Abundance and Distribution

The Baiji

The baiji is a member of the true river dolphins and is one of the world's rarest mammals (Dudgeon 2005). The baiji is the world's most threatened cetacean species, being listed as critically endangered by the International Union for the Conservation of Nature (Reeves et al. 2003). It is the sole representative of the family Lipotidae, which diverged from other cetaceans more than 20 million years ago (Nikaido et al. 2001). The first systematic modern surveys of baiji conducted during the late 1970s and early 1980s reported that there were around 300–400 individuals in existence (Zhou 1982; Lin et al. 1985; Chen and Hua 1987, 1989), with about 100 individuals residing in the Yangtze River reach between Hukou and Shanghai (Zhou and Li 1989; Figure 1). Subsequent surveys conducted in the following decades described the rapid demise of the species. By 1990, an estimated 200 were observed throughout the Yangtze River (Chen et al. 1993), with fewer than 100 individuals observed by 1995 (Liu et al. 1996). Turvey et al. (2007) reported that no baiji were observed anywhere in the Yangtze River in 2006. A subsequent survey conducted at the end of 2012 similarly failed to detect baiji anywhere in the Yangtze River, and the species is now presumed to be extinct (Mei et al. 2014).

The Yangtze Finless Porpoise

The first range-wide estimates of YFP abundance in the Yangtze River were obtained from numerous small-scale nonsystematic surveys conducted between 1984 and 1991. The population size estimated from these was approximately 2,700 YFPs, with 2,550 individuals residing in the Yangtze River main stem, which excluded the adjoining lakes (Zhang et al. 1993). In the late 1990s, surveys were conducted in different sections of the Yangtze River by various researchers using similar survey methods (Wang et al. 1998, 2000; Zhou et al. 1998a; Yang et al. 2000; Yu et al. 2001). From 1997 to 1999, annual synchronous surveys (i.e., surveys conducted in different reaches of the Yangtze River by several teams using the same survey methods) were conducted by the Ministry of Agriculture of China and the Institute of Hydrobiology (IHB) of the Chinese Academy of Sciences. These surveys indicated that the population of the YFP numbered approximately 2,000 (D. Wang, unpublished data; refer to Zhang et al. 2003 for details on survey design). Several years later in 2006, a systematic survey using a modified standard line-transect survey method was conducted throughout the geographic range of the YFP within the Yangtze River (but excluding Poyang and Dongting lakes). This extensive survey indicated that the YFP population within the entire Yangtze River main stem numbered between 1,000 and 1,200 individuals (Zhao et al. 2008). Zhao et al. (2008) estimated that an additional 600 individuals resided within Poyang and Dongting lakes. Results of these surveys indicated that the population of YFP in the Yangtze River had approximately halved in less than two decades (Zhao et al. 2008). A more recent survey has documented continued rapid decline in the YFP population. In late 2012, just 505 YFPs were estimated to reside in the Yangtze River main stem, with a total population size of approximately 1,040 animals (Mei et al. 2014).

Of the six extant species of porpoise from the family Phocoenidae, the YFP is the only population to inhabit freshwater (Gao and Zhou 1995). Thus, this YFP population has a far smaller potential habitat area compared to marine populations of finless porpoises, which occur in coastal waters from Japan to the Arabian Sea (Kasuya 1999). The present distribution of the YFPs is restricted to the Yangtze River main channel and its two largest adjoining lakes (Poyang and Dongting; Figure 1). Although previously the YFP was occasionally reported within some of the larger Yangtze River tributaries and their adjoining lakes (Zhang et al. 1993; Yang et al. 2000; Xiao and Zhang 2002), it appears absent in those areas today. In addition, most YFPs are concentrated in the middle and lower reaches of the Yangtze River, from Ezhou to Jiangyin in particular (Figure 1; Zhao et al. 2008, 2013). The porpoises residing in the upper reach of the Yangtze River between Yichang and Ezhou (Figure 1) appear to be at the highest risk of local extirpation, as population densities are lowest in this reach of the river (Zhao et al. 2008, 2013; Mei et al. 2012, 2014). The observed density in this reach decreased from 0.12 porpoises/km^2 in 2006 (Zhao et al. 2008) to 0.04 porpoises/km^2 in 2012 (Mei et al. 2014).

Another observation is that the distribution of YFPs in the upper section of the river appeared to be highly fragmented. For example, no porpoises were detected in the 150-km subsection between Yueyang and Shishou in 2006, and by 2012, two additional gaps in their distribution were identified between Yichang and Shashi and the section near Wuhan (see Figure 1; Zhao et al. 2008; Mei et al. 2014). Although the possibility of false negatives in determining the presence of YFP remains, the number of individuals in these reaches must be extremely low. Should local extinction occur in the upper reach of the river, the current range of this population on the Yangtze River would have decreased by approximately 700 km or 40% of its historical range. It should also be noted that the baiji disappeared first from the upper reaches of the Yangtze River (Zhou et al. 1977; Chen et al. 1997a; Zhang et al. 2003). Although limited photo identification studies suggest that baiji probably traveled hundreds of kilometers up and down the river (Zhou et al. 1998b), molecular results have indicated that YFPs exhibit different mitochondrial DNA haplotypes in different stretches of the river, which implies that these animals do not move far (Zheng et al. 2005). Even if all threats were eliminated and habitat conditions improved, there is little likelihood that YFPs from other areas would repopulate the upper reach of the Yangtze River (i.e., between Ezhou and Yichang; Figure 1). Unless the current trend in habitat deterioration is immediately reversed, there is a high likelihood that YFPs will soon disappear permanently from that area of the Yangtze River.

Past and Ongoing Conservation Initiatives

Chen and Hua (1989) initially proposed two measures for protecting the baiji: (1) creating in situ and ex situ reserves, and (2) initiating a captive breeding program. The need to implement these conservation measures was further promoted by scientists in subsequent workshops (Reeves et al. 2000; Wang et al. 2000) and has since been undertaken by the Chinese government and scientists. Below contains a summary of the outcomes and overall progress achieved by these conservation programs to date.

The first two national reserves aiming to protect the Yangtze River cetaceans were established in 1992. The Honghu Xin-Luo Baiji National Natural Reserve includes a 135-km section of the Yangtze River between Xintankou and Luoshan, which is located in Honghu City of Hubei Province. The Shishou Tian-e-Zhou Baiji Natural Reserve includes an 89-km section of the Yangtze River in Shishou and a connected 21-km section of the Tian-e-Zhou oxbow (Figure 1). In 1996, another five protecting stations were designated at Jianli, Chenglingji (a small town near Yueyang City), Hukou, Anqing, and Zhenjiang (Figure 1). Yueyang City established a local reserve in east Dongting Lake in 1996 that encompasses 66,700 ha (Figure 1). A provincial Yangtze freshwater cetacean natural reserve located in the Tongling section of Anhui Province was established in 2000 and was upgraded to a national reserve in 2006. It covers a 58-km section of the river within Tongling City (Figure 1). The Zhenjiang protection station was upgraded to a provincial reserve in 2003 and covers a 15-km section of a side-river channel at Zhenjiang (Figure 1). A provincial Poyang Lake YFP reserve was established in 2004 that encompasses an area of 8,600 ha (Figure 1). In 2007, the Anqing protectiom station was upgraded to a local city reserve and includes the entire 243 km of the Yangtze River that flows through Anqing (Figure 1).

As the habitat quality of the Yangtze River is unlikely to improve in the foreseeable future, ex situ reserves will be essential to ensure the continued survival of cetaceans. Preparations for the establishment of ex situ reserves were undertaken by IHB scientists as early as the mid-1980s, including site selection, investigation, and evaluation. Following a 4-year feasibility study, the Tian-e-Zhou oxbow, an old course of the Yangtze River (Figure 1), was considered to be most suitable based on the high-quality water and habitat, abundant biological productivity, and high fish production (Zhang et al. 1995). The first group of five YFP individuals (three females and two males) was introduced into the oxbow in 1990. Since then, several more groups of YFP that have been captured or rescued from some section of the Yangtze River have been translocated into this oxbow reserve (Wang et al. 2005, 2006a). One other smaller-scale seminatural reserve was established in Tongling of Anhui Province in 1994 (Figure 1). This reserve is located in a small channel (1.6 km long and 80–220 m wide) between two sandbars of the Yangtze River.

In addition to the creation of a network of protected YFP reserves, a captive breeding program was also initiated by the IHB. The aim of this program was to increase knowledge of YFP biology, particularly with respect to their reproductive biology, and to supplement ongoing conservation work with the wild population. Finally, in order to provide effective protection for the baiji and the YFP, a conservation administration network composed of government agencies and research institutions was organized by the Ministry of Agriculture of China, IHB, and World Wide Fund for Nature in 2008. This network is intended to promote and coordinate conservation efforts (including organizing surveys and training staff) and also will serve as a platform for information exchange and public education.

Threats

A number of anthropogenic factors are known or suspected to be responsible for the current population decline and range contraction of both Yangtze River cetaceans. These include the use of harmful fishing gears, overfishing, sand dredging, boat collisions, water pollution, and the construction of water infrastructure (Wang 2009). Turvey et al. (2007) concluded that entanglement in gears used in unregulated and unselective fishing (e.g., rolling hooks, electrofishing gear, and gill nets) was the primary factor responsible for the probable extinction of the baiji. These same factors also may explain much of the ongoing decline of the YFP, as has been speculated by several authors (Wang et al. 1998, 2000, 2005, 2006b; Wang 2009). Boat traffic, which has increased significantly in the Yangtze River and its adjacent lakes, also likely causes mortality of cetaceans from propeller strikes, while boat noise may interfere with their social communication and foraging efficiency (Wang et al. 1998, 2000, 2006b; Wang 2009). Widespread mining of the river and lake beds and riverbanks has destroyed important habitats of the YFP's main prey. These activities, which are mostly done illegally, also have adversely affected the primary productivity of the Yangtze River. This problem is especially significant in Poyang Lake, where around 450 YFPs reside (Xiao and Zhang 2000; Wang et al. 2006b; Zhao et al. 2008).

Riverine habitats are believed to be at greater risk from anthropogenic pollution than marine habitats. Cetaceans in rivers throughout the world generally occur in the most densely populated human environments (Reeves et al. 2000). Moreover, water development projects, especially dams, have had major impacts on these freshwater cetacean populations. For example, in the Yangtze River system, such structures create obvious impediments to the movement of YFPs between the main river channel and its adjoining lakes and tributaries (Liu et al. 2000; Smith and Reeves 2000). These structures also affect the dynamics and movement of prey species (Xie and Chen 1996). The Three Gorges Project (TGP) represents another example. Begun in 1997 and completed in 2009, the TGP is one of the world's largest hydropower dams, which has severely altered downstream hydrologic conditions throughout the Yangtze River (Wei et al. 2007; Tong et al. 2008). The TGP dam and its annual regulation are expected to moderate natural flood pulses in the Yangtze River, resulting in fewer, lower-magnitude flood events. Wei et al. (2007) also has suggested that the TGP will decrease fish spawning activities in the middle reaches of the Yangtze River. In addition, since filling began in 2004, the TGP reservoir has trapped a high proportion of the silt content carried by the river. Thus, the water discharged by the TGP reservoir will be less silt-saturated and more transparent, though it will have a greater capacity to scour the downstream riverbed and channel. Such modifications downstream of the TGP are expected to alter the spatial distribution of water resources in the complex river–lake ecosystem of the middle and lower Yangtze River (Wei et al. 2007). It is further predicted that these effects will adversely impact the habitats and forage of both the baiji and the YFP.

Although the relative importance of each of the above threats has not been thoroughly quantified, it is likely that all have cumulatively contributed to the decline of the YFP and to the likely extinction of the baiji. It is especially noteworthy that none of these negative impacts has ever been seriously addressed by government agencies, despite that most of these factors have escalated dramatically in recent decades. Consequently, we must reiterate that immediate action is urgently needed to minimize these threats to the Yangtze River's cetacean populations, with highest priority given to areas with greatest remaining abundances of the YFP.

Conservation Progress

Although protective regulations for the baiji, the YFP, and their habitats are in place, their effective enforcement remains an immense problem, particularly given the scale of the river and the density of the human population in this developing country. For example, despite bans on certain harmful fishing equipment, these gears continue to be used, even in established cetacean reserves (Turvey et al. 2007). Therefore, while the reserves may help to slow the process of extinction of both the baiji and the YFP, they alone cannot prevent the occurrence of harmful human activities. Therefore, the future success of in situ conservation is predicted to be highly limited without appropriate regulations and enforcement (Wang et al. 2006a; Turvey et al. 2007).

The YFP individuals moved into Tian-e-Zhou oxbow have lived in there freely without being affected by any factitious variables. For example, no artificial feeding has been done for these individuals as an adequate forage base appears to exist for YFP. This result confirms that these animals can both survive and successfully reproduce in this reserve. Approximately two or more calves have been born each year, with at least 31 calves born in the reserve through the end of 2007 (Wang 2009). More recent data indicates there are now more than 40 individuals in the reserve with 3–5 calves born each year. An estimated 80–100 individuals could be supported according to the fishery production in this reserve (Wang, unpublished data). In addition, five of the eight adult females observed in April 2008 were pregnant, which is an encouraging sign of what the reserve may be able to potentially accomplish (Wang 2009). These data further confirm that a viable YFP population capable of breeding and expanding has been established in this particular reserve. As such, the establishment of Shishou Tian-e-Zhou oxbow reserve represents the first successful example of ex situ preservation of a cetacean species (Wang et al. 2005) and demonstrated the oxbow's adequacy as an ex situ environment for cetaceans (Braulik et al. 2006). Using this success story as a template to guide future restoration efforts for YFPs would seem warranted.

Captive breeding of freshwater cetacean began by IHB scientists in the 1980s and was met with some success (one male baiji named QiQi survived in captivity for almost 23 years). The first two YFP individuals to enter the breeding program (a 1.5-year-old male and a 1.5-year-old female) were captured from the Yangtze River and introduced into the indoor pools at the IHB Dolphinarium in 1996. In later years, a 1.5-year-old female and an adult male were introduced into captivity from the Tian-e-Zhou oxbow in 1999 and 2004, respectively. With the addition of three young females from Poyang Lake in 2009 and 2011 (one in 2009 and two in 2011), there were seven animals (two males and five females) introduced into captivity. These animals have been the subject of extensive research on their rearing, behavior, acoustics, physiology, and reproductive biology (Chen et al. 1997a, 1997b, 2005; Akamatsu et al. 1998; Wei et al. 2004; Popov et al. 2005, 2006; Li et al. 2007, 2008; Wu et al. 2010a, 2010b). All of these individuals, with the exception of one female who died accidentally in 2007 and a male who died due to old age in 2014, remain in good health within the facility. In fact, the two individuals introduced in 1996 have survived in captivity for 18 years. This success story marks great progress for rearing the YFP in captivity, as no such achievement has been previously attained (Liu et al. 2002; Wang et al. 2005).

More successes of the captive breeding program came later. In 2005, the Dolphinarium witnessed the birth of the first YFP ever born in captivity—a male calf that is still alive and healthy today (Wang et al. 2005). In 2007, the same female gave birth to another male. Unfortunately, this female died accidently shortly after birth, and consequently, her second calf died 11 d later, despite efforts to hand-rear him. In 2008, the oldest female from the original 1996 introduction gave birth to a male calf, which died 5 d later because the mother did not lactate and was unable to nurse (Wang 2009). Nevertheless, the preliminary results of this captive breeding program have been encouraging and could become vital for the future survival of the YFP.

Future Prospects for Conservation

Because China continues on a path towards rapid economic development, it is likely that the quality of the Yangtze River ecosystem will continue to deteriorate due to anthropogenic influences. Current levels of anthropogenic impacts in the river have apparently already driven the baiji to likely extinction (Turvey et al. 2007). Hence, the key question that remains is how best to prevent the YFP from experiencing the same fate as the baiji.

In general, in situ protection should be a priority for YFP conservation. Despite the degradation of the habitats of the Yangtze River ecosystem, options to protect the YFP in its natural habitat should be actively pursued. Although a system of reserves does exist within the Yangtze River and its adjacent lakes, these reserves were originally designed to protect the baiji and do not adequately protect areas where the YFP is present and abundant. Spatial modifications to these existing reserves are needed to maximize the effectiveness on YFP populations (Zhao et al. 2013).

Most of the current reserves continue to face the ongoing and increasing impacts of anthropogenic activities, the majority of which are related to massive economic development that cannot be easily curtailed nor mitigated. For example, commercial river transportation throughout the Yangtze River valley cannot be halted (Chen et al. 1997a; Wang et al. 1998, 2000). Overfishing and illegal fishing remain widespread problems, even in the reserve areas, though these may be more feasible issues to address. However, under the current circumstances, we consider that the overall situation in the Yangtze River is not reversible in the foreseeable future. The strategy of species conservation has advanced to the stage of "saving the seeds" for any possible conservation program in the future. In this regard, captive breeding and ex situ conservation programs are both recognized as critically important measures that must be developed to save this unique subspecies. We recommend expanding the captive breeding population of the YFP to establish a sustainable group of individuals in captivity from which individuals could be exchanged between the captive and ex situ protected populations, and possibly even supplement the wild population in the future. A larger captive colony also would facilitate further research on the biology of the YFP, with special emphasis on the reproductive biology.

In addition to these two critical conservation measures, we further suggest establishing additional ex situ protected populations in habitats that are as close to natural habitats as possible (e.g., other Yangtze River oxbows). One scheme being considered is to connect the Tian-e-Zhou oxbow reserve with the nearby Heiwawu oxbow, which will increase the carrying capacity of the current Tian-e-Zhou reserve. Planning also is underway for the establishment of another ex situ reserve in an old course of the Yangtze River—the Hewangmiao oxbow near Jianli County (Figure 1). Further systematic surveys should be conducted to investigate other suitable sites. One large area worth investigating is the huge reservoir above the Three Gorges Dam, where the water flows are now much reduced compared to prereservoir conditions. Establishment of a viable YFP population in this area above Three Gorges Dam would allow individuals access to the upper reaches of the Yangtze River. To this end, an initial feasibility study has been completed by IHB in the Three Gorges Reservoir. This study concluded that the overall water quality was relatively good in the Three Gorges Reservoir, and the boat traffic intensity and underwater noise levels were far lower than those in the middle and lower Yangtze River. Furthermore, harmful fishing gears and illegal fishing were much less common when compared to downstream areas in the Yangtze River. Overall, this survey indicated that overall conditions in Three Gorges Reservoir were generally suitable for the YFP, and further studies were warranted. However, before establishing an ex situ reserve in the reservoir, further surveys should be conducted to investigate the likelihood of YFP mortality associated with entrainment or spillways.

Although commercial fishing in the Yangtze is not a large industry due to the massive growth of freshwater aquaculture (Wang 2009), conservation of Yangtze River fish stocks offers another avenue to aid the persistence of the YFP. Since 2003, commercial fishing has been prohibited in the middle and lower reaches of the Yangtze River for 3 months per year from April through June. This measure may have improved the status of many of the river's main fisheries, but it falls short of assuring sustainable fish stocks because fisherman increase their fishing effort during the 9-month legal fishing season. To truly protect the Yangtze River's fisheries resources, we suggest that commercial fishing be prohibited year-round in all reserves and, ideally, throughout the entire river. Furthermore, reconnection of the Yangtze River and its adjoining lakes also would greatly improve fisheries resources along the entire river (Wang 2009).

All of these proposed conservation measures have been suggested multiple times in countless workshops, published papers, and reports to the Chinese government. Most of the well-characterized threats posed by anthropogenic activities are still present, and at least some of them continue to grow throughout the Yangtze River valley. In an environment dominated by pressures for rapid economic development, the best compromise that can be pursued from the Chinese government is to establish a balance between development and conservation. But in a developing country like China, these goals are often at odds with economic development, which nearly always prevails. As a result, conservation-oriented recommendations and proposed programs often become nothing more than "conservation on paper" (e.g., Bearzi 2007). The success of any conservation program will be underpinned by the will of Chinese government agencies and the support of the Chinese public. In the end, we must eventually face the inevitable question of are we prepared to lose yet another large endemic mammal from the Yangtze River ecosystem? An even more pressing issue is whether we can afford the cost of losing not just one or two species of noteworthy mammals, but ultimately the loss of far greater biodiversity in the unique Yangtze River ecosystem. As conservation-minded scientists, we hope that lessons have been learned from the tragedy of the baiji and that decision and policy makers take immediate measures to ensure the long-term survival and growth of the YFP population.

Acknowledgments

This research was supported by grants from the National Natural Science Foundation of China (Nos. 31170501, 31070347) and the Knowledge Innovation Program of the Chinese Academy of Sciences (No. KSCX2-EW-Z-4).

References

Akamatsu, T., D. Wang, K. Nakamura, and K. Wang. 1998. Echolocation range of captive and free-ranging baiji (*Lipotes vexillifer*), finless porpoise (*Neophocaena phocaenoides*), and bottlenose dolphin (*Tursiops truncatus*). Journal of the Acoustical Society of America 104:2511–2516.

Bearzi, G. 2007. Marine conservation on paper. Conservation Biology 21:1–3.

Braulik, G. T., R. R. Reeves, D. Wang, S. Ellis, R.S. Wells, and D. Dudgeon. 2006. Report of the workshop on conservation of the baiji and Yangtze finless porpoise. Switzerland Baiji Foundation, Zurich.

Chen, D., Y. Hao, Q. Zhao, and D. Wang. 2005. Reproductive seasonality and maturity of male *Neophocaena phocaenoides asiaeorientalis* in captivity: a case study based on the hormone evidence. Marine Freshwater Behavior and Physiology 39:163–173.

Chen, P., and Y. Hua. 1987. Projected impacts of the Three Gorges Dam on the baiji, *Lipotes vexillifer*, and needs for conservation of the species. Pages 30–41 *in* W. F. Perrin, editor. A collection of articles on the impacts of the Three Gorges Dam project on aquatic ecosystem along the Changjiang and research on their countermeasures. Translated by C. H. Perrin. China Science Press, Beijing.

Chen, P., and Y. Hua. 1989. Distribution, population size and protection of Lipotes vexillifer. Pages 81–85 *in* W. F. Perrin, R. L. Brownell, Jr., K. Zhou, and J. Liu, editors. Biology and conservation of the river dolphins. Occasional Papers of the International Union for Conservation of Nature/SSC, No. 3, Gland, Switzerland.

Chen, P., R. Liu, D. Wang, and X. Zhang. 1997a. Biology, rearing and conservation of baiji. Science Press, Beijing.

Chen, D., Q. Zhao, and R. Liu. 1997b. Preliminary study on some hormones of *Neophocaena phocaenoides* in the Yangtze River. Acta Theriologica Sinica 17:43–47.

Chen, P., X. Zhang, Z. Wei, Q. Zhao, X. Wang, G. Zhang, and J. Yang. 1993. Appraisal of the influence upon baiji, *Lipotes vexillifer* by the Three-Gorge Project and conservation strategy. Acta Hydrobiologica Sinica 11:101–111.

Dudgeon, D. 2005. Last chance to see . . .: ex situ conservation and the fate of the baiji. Aquatic Conservation: Marine and Freshwater Ecosystems 15:105–108.

Gao, A., and K. Zhou. 1995. Geographical variation of external measurements and three subspecies of *Neophocaena Phocaenoides* in Chinese waters. Acta Theriologica Sinica 15:81–92.

Kasuya, T. 1999. Finless porpoise *Neophocaena phocaenoides* (G. Cuvier, 1829). Pages 411–442 *in* S. H. Ridgway and R. Harrison, editors. Handbook of marine mammals, volume 6: the second book of dolphins and the porpoises. Academic Press, San Diego, California.

Li, S. H., D. Wang, K. X. Wang, T. Akamatsu, Z. Q. Ma, and J. B. Han. 2007. Echolocation click sounds from wild inshore finless porpoise (*Neophocaena phocaenoides sunameri*) with comparisons to the sonar of riverine *N. p. asiaeorientalis*. Journal of the Acoustical Society of America 121:3938–3946.

Li, S. H., K. X. Wang, D. Wang, S. Y Dong, and T. Akamatsu. 2008. Simultaneous production of low- and high-frequency sounds by neonatal finless porpoises. Journal of the Acoustical Society of America 124:716–718.

Lin, K., P. Chen, and Y. Hua. 1985. Population size and conservation of *Lipotes vexillifer*. Acta Ecologica Sinica 5:77–85.

Liu, R., D. Wang, and K. Zhou. 2000. Effects of water development on river cetaceans in China. Pages 40–42 *in* R. R. Reeves, S. Brian, and K. Tasuya, editors. Biology and conservation of freshwater cetaceans in Asia. Occasional papers of the International Union for Conservation of Nature/SSC, No. 23, Gland, Switzerland.

Liu, R., K. Wang, and Q. Zhao. 2002. Rearing of cetaceans in captivity in China. Acta Theriologica Sinica 22:130–135.

Liu, R., X. Zhang, D. Wang, and J. Yang. 1996. Once again studies on the conservation of *Lipotes vexillifer* and *Neophocaena phocaenoides*. Resources and Environment in the Yangtze Valley 5:220–225.

Mei, Z. G., S. L. Huang, Y. J. Hao, S. T. Turvey, W. M. Gong, and D. Wang. 2012. Accelerating population decline of Yangtze finless porpoise (*Neophocaena phocaenoides asiaeorientalis*). Biological Conservation 153:192–200.

Mei, Z. G., X. Q. Zhang, S. L. Huang, X. J. Zhao, Y. J. Hao, L. Zhang, Z. Y. Qian, J. S. Zheng, K. Wang, and D. Wang. 2014. The Yangtze finless porpoise: on an accelerating path to extinction. Biological Conservation 172:117–123.

Nikaido, M., F. Matsuno, H. Hamilton, L. Robert, J. Brownell., Y. Cao, D. Wang, Z. Zhu, A. M. Shedlock, R. Ewan-Fordyce, M. Hasegawa, and N. Okada. 2001. Retroposon analysis of major cetacean lineages: the monophyly of toothed whales and the paraphyly of river dolphins. Proceedings of the National Academy of Sciences of the United States of America 98:7384–7389.

Popov, V., A. Supin, D. Wang., K. Wang, J. Xiao, and S. Li. 2005. Evoked-potential audiogram of the Yangtze finless porpoise *Neophocaena phocaenoides asiaeorientalis*. Journal of the Acoustical Society of America 117:2728–2731.

Popov, V., A. Supin, D. Wang, and K. Wang. 2006. Nonconstant quality of auditory filters in the porpoises, *Phocoena phocoena* and *Neophocaena phocaenoides* (Cetacea, Phocoenidae). Journal of the Acoustical Society of America 119:3173–3180.

Reeves, R. R., T. A. Jefferson, T. Kasuya, B. D. Smith, D. Wang, P. Wang, R. S. Wells, B. Würsig, and K. Zhou. 2000. Report of the workshop to develop a conservation action plan for the Yangtze River finless porpoise. Pages 67–80 *in* R. R. Reeves, B. D. Smith, and T. Kasuya, editors. Biology and conservation of freshwater cetaceans in Asia. Occasional papers of the International Union for Conservation of Nature/SSC, No. 23, Gland, Switzerland.

Reeves, R. R., B. D., Smith, E. A. Crespo, and G. Notarbartolo di Sciara. 2003. Dolphins, whales and porpoises: 2002–2010 conservation action plan for the world's cetaceans. International Union for Conservation of Nature/SSC, Gland, Switzerland.

Smith, B., and R. R. Reeves. 2000. Report of the workshop on the effects of water development on river cetaceans, 26–28 February 1997, Rajendrapur, Bangladesh. Pages 15–22 *in* R. R. Reeves, B. D. Smith, and T. Kasuya, editors. Biology and conservation of freshwater cetaceans in Asia. Occasional papers of the International Union for Conservation of Nature/SSC, No. 23, Gland, Switzerland.

Tong, H., Y. Zheng, and Q. Xu. 2008. Preliminary analysis of flow and sediment variation of the Yangtze River from Yichang to Wuhan. Yangtze River 39:37–41.

Turvey, S., R. L. Pitman., B. L. Taylor, J. Barlow, T. Akamatsu, L. A. Barrett, X. Zhao, R. R. Reeves, B. S. Stewart, K. Wang, Z. Wei, X. Zhang, L. T. Pusser, M. Richlen, J. R. Brandon, and D. Wang. 2007. First human-caused extinction of a cetacean species? Biology Letters 3:537–540.

Wang, D. 2009. Population status, threats and conservation of the Yangtze finless porpoise. Chinese Science Bulletin 54:3473–3484.

Wang, D., Y. Hao, K. Wang, Q. Zhao, D. Chen, Z. Wei, and X. Zhang. 2005. The first Yangtze finless porpoise successfully born in captivity. Environmental Science and Pollution Research 12:247–250.

Wang, D., R. Liu, X. Zhang, J. Yang, Z. Wei, Q. Zhao, and X. Wang. 2000. Status and conservation of the Yangtze finless porpoise. Pages 81–85 *in* R. R. Reeves, B. D. Smith, and T. Kasuya, editors. Biology and conservation of freshwater cetaceans in Asia. Occasional papers of the International Union for Conservation of Nature/SSC, No. 23, Gland, Switzerland.

Wang, D., S. T. Turvey, X. Zhao, and Z. Mei. 2013. *Neophocaena asiaeorientalis* ssp. *asiaeorientalis*. In IUCN red list of threatened species, version 2013.1. Available: www.iucnredlist.org/. (September 2013).

Wang, D., X. Zhang, and R. Liu. 1998. Conservation status and its future of baiji and Yangtze finless porpoise in China. Pages 218–226 *in* Z. L. Hua, B. Fu, and Y. Yang, editors. Ecology and environmental protection of large irrigation projects in Yangtze River in 21st Century. Environmental Science Press, Beijing.

Wang, D., X. Zhang, K. Wang, Z. Wei, B. Würsig, G.T. Braulik, and S. Ellis. 2006a. Conservation of the baiji: no simple solution. Conservation Biology 20:623–625.

Wang, K., D. Wang, X. Zhang, A. Pfluger, and L. Barrett. 2006b. Range-wide Yangtze freshwater dolphin expedition: the last chance to see baiji? Environmental Science and Pollution Research 13:418–424.

Wei, Q., D. Wang, and L. Wang. 2007. Aquatic biodiversity conservation. Pages 90–113 *in* G. S. Yang, L. D. Weng, and L. F. Li, editors. Yangtze conservation and development report. Changjiang Press, Wuhan, China.

Wei, Z., D. Wang, X. Zhang, K. Wang, D. Chen, Q. Zhao, X. Kuang, W. Gong, and X. Wang. 2004. Observation on some sexual behavior of the Yangtze finless porpoise (*Neophocaena phocaenoides asiaeorientalis*) in captivity. Acta Theriologica Sinica 24:98–102.

Wu, H., Y. Hao, X. Li, Q. Zhao, D. Chen, X. Kuang, Z. Kou, K. Feng, W. Gong, and D. Wang. 2010a. B-mode ultrasonographic evaluation of the testis in relation to serum testosterone concentration in male Yangtze finless porpoise (*Neophocaena phocaenoides asiaeorientalis*) during the breeding season. Theriogenology 73:383–391.

Wu, H., Y. Hao, X. Yu, Y. Xian, Q. Zhao, D. Chen, X. Kuang, Z. Kou, K. Feng, W. Gong, and D. Wang. 2010b. Variation in sexual behaviors in a group of captive male Yangtze finless porpoises (*Neophocaena phocaenoides asiaeorientalis*): motivated by physiological changes? Theriogenology 74:1467–1475.

Xiao, W., and X. Zhang. 2000. A preliminary study on the population size of Yangtze finless porpoise in Poyang Lake, Jiangxi. Chinese Biodiversity 8:106–111.

Xiao, W., and X. Zhang. 2002. Distribution and population size of Yangtze finless porpoise in Poyang Lake and its branches. Acta Theriologica Sinica 22:7–14.

Xie, P., and Y. Chen. 1996. Biodiversity problems in freshwater ecosystems in China: impact of human activities and loss of biodiversity. Pages 160–168 *in* J. MacKinnon and S. Wang, editors. Conserving China's biodiversity. China Environmental Science Press, Wuhan.

Yang, J., W. Xiao, X. Kuang, Z. Wei, and R. Liu. 2000. Studies on the distribution, population size and the activity of *Lipotes vexillifer* and *Neophocaena phocaenoides* in Dongting Lake and Poyang Lake. Resources and Environment in the Yangtze Basin 9:444–450.

Yu, D., M. Dong, J. Wang, and X. Zhang. 2001.

Population status of Yangtze finless porpoise in the Yangtze River section from Hukou to Nanjing. Acta Theriologica Sinica 21:174–179.

Zhang, X., R. Liu, Q. Zhao, G. Zhang, Z. Wei, X. Wang, and J. Yang. 1993. The population of finless porpoise in the middle and lower reaches of Yangtze River. Acta Theriologica Sinica 13:260–270.

Zhang, X., D. Wang, R. Liu, Z. Wei, Y. Hua, Y. Wang, Z. Chen, and L. Wang. 2003. The Yangtze River dolphin or baiji (*Lipotes vexillifer*): population status and conservation issues in the Yangtze River, China. Aquatic Conservation: Marine and Freshwater Ecosystems 13:51–64.

Zhang, X., Z. Wei, X. Wang, J. Yang, and P. Chen. 1995. Studies on the feasibility of establishment of a semi-natural reserve at Tian-e-Zhou (Swan) oxbow for baiji, *Lipotes vexillifer*. Acta Hydrobiologica Sinica 19:110–123.

Zhao, X., J. Barlow, B. L. Taylor, R. L. Pitman, K. Wang, Z. Wei, B. S. Stewart, S. T. Turvey, T. Akamatsu, R. R. Reeves, and D. Wang. 2008. Abundance and conservation status of the Yangtze finless porpoise in the Yangtze River, China. Biological Conservation 141:3006–3018.

Zhao, X., D. Wang, S. T. Turvey, B. Taylor, and T. Akamatsu. 2013. Distribution patterns of Yangtze finless porpoises in the Yangtze River: implications for reserve management. Animal Conservation 16:509–518.

Zheng, J., J. Xia, S. He, and D. Wang. 2005. Population genetic structure of the Yangtze finless porpoise (*Neophocaena phocaenoides asiaeorientalis*): implications for management and conservation. Biochemical Genetics 43:307–320.

Zhou, K. 1982. On the conservation of the baiji, *Lipotes vexillifer*. Journal of Nanjing Normal University (Natural Science Edition) 4:71–74.

Zhou, K., and Y. Li. 1989. Status and aspects of the ecology and behavior of the baiji, *Lipotes vexillifer*, in the lower Yangtze River. Pages 86–91 *in* W. F. Perrin, R. L. Brownell, Jr., K. Zhou, and J. Liu, editors. Biology and conservation of the river dolphins. Occasional Papers of the International Union for Conservation of Nature/SSC, No. 3, Gland, Switzerland.

Zhou, K., W. Qian, and Y. Li. 1977. Studies on the distribution of baiji, *Lipotes vexillifer* Miller. Acta Zoologica Sinica 23:72–79.

Zhou, K., J. Sun, A. Gao, and B. Würsig. 1998b. Baiji (*Lipotes vexillifer*) in the lower Yangtze River: movements, numbers threats and conservation needs. Aquatic Mammals 24:123–132.

Zhou, K., G. Yang, A. Gao, J. Sun, and X. Xu. 1998a. Population abundance and distribution characteristics of finless porpoise in the river section from Nanjing to Hukou of the Yangtze River. Journal of Nanjing Normal University (Natural Science) 21:91–98.

American Fisheries Society Symposium 84:139–167, 2016

Predicting Fish Invasions in the Yarlung Zangbo River of the Qinghai–Tibet Plateau, China

XIAOYUN SUI, XIAO ZHANG, YINTAO JIA, YIFENG CHEN*, AND DEKUI HE
Institute of Hydrobiology, Chinese Academy of Sciences
7 South Donghu Road, Wuhan, Hubei 430072, P. R. China

Abstract.—A total of 13 alien fish species were introduced to the middle reaches of the Yarlung Zangbo River with the development of aquaculture in recent years. Eight of these species now have established populations and five have not established. To explore the mechanism for alien fish species establishment in this river, we compiled 23 biological traits of those species and tested for differences in biological trait and ecological niche between the established and nonestablished fish species. The trait analysis results showed that successfully established fish species have smaller body length, smaller egg diameter, earlier sexual maturity, and wider diet breadth than the nonestablished species. Ecological niches between established and nonestablished fish species also differed significantly. Our results support the vacant niche hypothesis that the successfully established species could occupy the vacant niches in the introduced range. We also used a model to predict the invasion risk of 27 potentially introduced fish species in the middle reaches of the Yarlung Zangbo River. The model predicted that there was a higher risk of establishment for 15 of the 27 species.

Introduction

With many human activities, such as agriculture, aquaculture, and trade, more and more alien species are introduced intentionally and unintentionally across their natural dispersal barriers. Some of these introductions result in undesirable effects on global ecosystems (Mack et al. 2000; Christian 2001). Although most species do not establish in new ranges (Williamson and Fitter 1996), successfully established species have usually been difficult to eradicate from their introduced ranges (Sala et al. 2000; Strayer et al. 2006). Therefore, predicting the identity of potential invaders and their potential impacts and distribution have become important research topics (Kolar and Lodge 2001).

Biological traits are important indicators for the identification of a new invasive species (Radford and Cousens 2000; van Kleunen et al. 2010). For example, fish species with parental care and a large native range have an advantage during the establishment phase (Marchetti et al. 2004). Seed weight plays an important role in determining the invasiveness of herb seed plants in China (Chen et al. 2015). These studies suggest that biological traits could be applied in predicting the invasiveness of species. Invasion ecology has already witnessed considerable progress in theoretic and applied research (Ricciardi and Rasmussen 1998; Kolar and Lodge 2001). For instance, one study used biological traits to develop a quantitative model to predict establishment, spread, and potential nuisance factor of fishes in the Great Lakes (Kolar and Lodge 2002).

Ecological niche in the modern view is defined as where and under what conditions a species can survive and the interaction

* Corresponding author: chenyf@ihb.ac.cn

that occurs between species and the environment (Olden et al. 2006). The species can adapt their ecological niche through changing the morphology of organisms, physiological responses, and unique behavior (Odum 1983). The species' biological traits therefore can be used to effectively quantify the ecological niche within the community (Olden et al. 2006; Thuiller et al. 2012). In general, a successful invader will find its own niche within the new ecosystem. As the vacant niche hypothesis suggests, communities with more vacant niches are more likely to be invaded (Elton 2000).

We selected the Yarlung Zangbo River, located in the Qinghai–Tibet Plateau, as the study area. A sensitive and vulnerable environment and unique ichthyofauna make this place an ideal region for studying global change, ecological protection, and biological invasions (Liu and Chen 2000; Klein et al. 2004). A total of 13 alien fish species were introduced to the middle reaches of the Yarlung Zangbo River by 2011. We conducted analyses on 23 biological traits of these 13 alien fish species to test associations among establishment success, biological traits, and ecological niches. We also predicted the invasion risk of 27 potentially invasive fish species, not currently present within this study region, to provide insights into potential future invasions and provide relevant agencies with information to draft appropriate policies and prevention plans.

Materials and Methods

Study area

The Yarlung Zangbo River has unique ichthyofauna composed of three major groups: Schizothoracinae (Cypriniformes), Cobitidae (Cypriniformes), and Sisoridae (Siluriformes) (Wu and Wu 1992). Prior to 1976, the only alien fish in Tibet was Yadong Salmon *Salmo trutta fario*, which was introduced by British travelers from Europe and subsequently became established in the lower reaches of the Yarlung Zangbo River (Cai 1997). With the development of aquaculture and the change of human living habits, introductions of alien fish species increased. By 2011, 13 alien fish species were found in the basin. Our study area was the middle reach of the Yarlung Zangbo River (Figure 1), which is characterized by high habitat heterogeneity and wild distribution of alien fish. The middle reach of the river is relatively flat and has a variety of habitat types, such as wetlands and reservoirs, which provide plenty of habitat favorable to both indigenous and alien species. In previous surveys, eight species have established populations, in which they exhibited reproduction (i.e., larvae detected), successful recruitment into the juvenile stage, and self-sustaining populations. The other five species are currently not established (Table 1).

Phylogenetic relationships

Phylogenetic relationship among species is another important factor influencing the successful establishment of introduced species (Lee 2002; Alcaraz et al. 2005; van Kleunen et al. 2010). Species that are phylogenetically related (i.e., share common ancestry) are not independent, which violates the basic assumption of most traditional statistical approaches that data observations should be independent (Mims et al. 2010). Therefore, the correlation between phylogeny and biological traits should be tested prior to analyzing comparative data. We applied the empirical approach adopted by Grafen (1989) to assess the phylogenetic relationships among species: two species in the same genus were given a value of 1, two species in the same class were given a value of 0.5, two species in the same order were given a value of 0.33, and two species in different orders were given a value 0.25. We therefore developed a matrix of phylogenetic similarity used in the following analysis.

Biological traits

We compiled 23 biological traits of 13 kinds of alien fish (not native to the Yarlung Zangbo River) within the Yarlung Zangbo

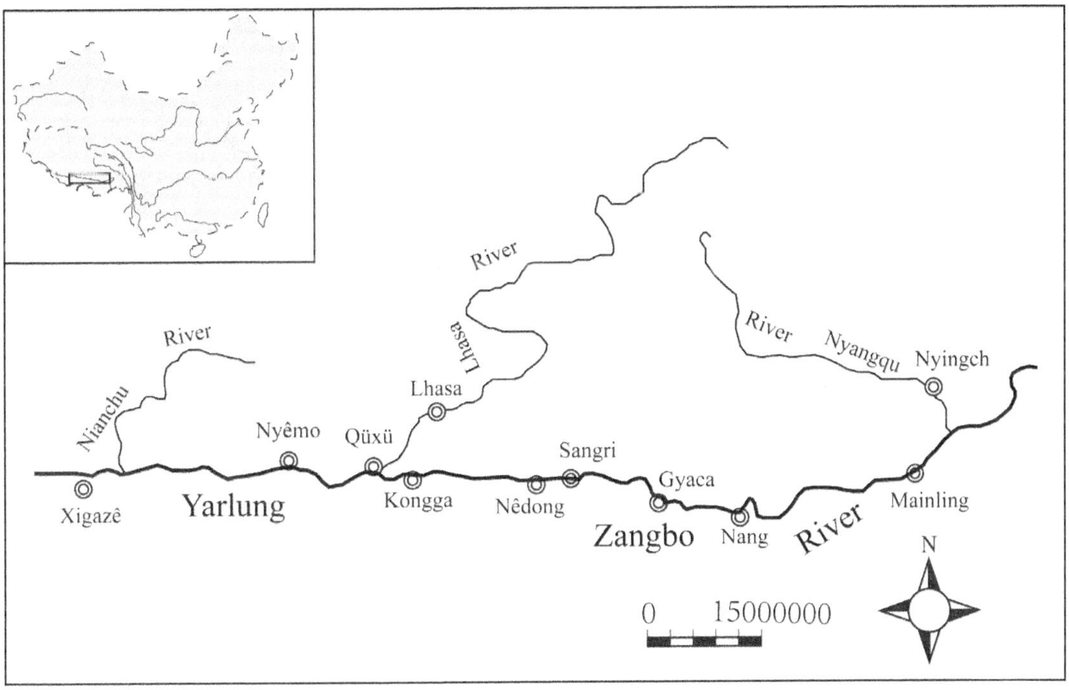

Figure 1. Location of study area of the Yarlung Zangbo River.

River basin, which were divided into four main categories: morphology, behavior, trophic traits, and life history traits (Table 2). Data used in the study were compiled from published scientific papers (literature and graduate theses), our unpublished data (unpublished fish ecological data in our laboratory), electronic databases (Fishbase and China Animal Scientific Database), and expertise from fish biologists (expert judgment regarding some traits that could not be determined from above data sources) to ensure data objectivity. The traits data included four types: continuous, normal, ordinal, and binary. For continuous type, the average or maximal data (i.e., the maximal body length) were used. Other types, according to adult performance, were assigned a single state (Olden et al. 2006). Detailed information about the fish traits are available in Appendices A and B.

Potentially introduced fish

All currently established alien fishes in the Yarlung Zangbo River are important Chinese aquaculture species. The rapid development of Chinese aquaculture elicits concern that more alien fishes will be introduced to the Yarlung Zangbo River. Due to the similar water temperature of the Yarlung Zangbo River and northeast rivers in China, it was expected that the fish inhabiting in northeastern rivers would be thermally preadapted to establish in the Yarlung Zangbo River, which suggested a high potential invasion risk. Therefore, 18 aquaculture and 9 indigenous fish species that were commonly found in northeastern China were selected to predict the possibility of establishment in the Yarlung Zangbo River basin.

Data analyses

Significant tests of phylogenetic impacts.—We followed Olden et al. (2006) and calculated a matrix of trait similarities for the 13 alien fish species within our study area according to 23 biological traits using Gower's dissimilarity coefficient. Gower's dissimilarity coefficient is a multivariate index to accommodate mixed data types (four data

Table 1. Failed and successful alien fishes in the middle reaches of the Yarlung Zangbo River.

Order	Family	Species	Common names	Establishment
Synbranchiformes	Synbranchidae	*Monopterus albus*	Asian Swamp Eel (also known as Swamp Eel)	Nonestablished
Cyprinodontiformes	Cyprinidintidae	*Oryzias latipes*	Japanese Rice Fish (also known as Japanese Medaka)	Established
Cypriniformes	Cyprinidae	*Abbottina rivularis*	Chinese False Gudgeon	Established
		Pseudorasbora parva	Stone Moroko	Established
		Carassius auratus	Goldfish	Established
		Cyprinus carpio	Common Carp	Established
		Hypophthalmichthys nobilis	Bighead Carp	Nonestablished
		Ctenopharyngodon idella	Grass Carp	Nonestablished
	Cobitidae	*Misgurnus anguillicaudatus*	Pond Loach (also known as Oriental Weatherfish)	Established
Perciformes	Channidae	*Channa argus*	Northern Snakehead	Nonestablished
	Cichlidae	*Oreochromis nilotica*	Nile Tilapia	Nonestablished
	Eleotridae	*Micropercops cinctus*		Established
Siluriformes	Siluridae	*Silurus asotus*	Amur Catfish	Established

Table 2. The 23 biological traits and the quantifying methods used in this research.

Groups	Traits	Units	Quantifying methods
Morphology	Maximum body length	mm	Maximum body length
	Shape factor		The ratio of body length to maximum body depth
	Swim factor		The ratio of minimum depth of the caudal peduncle to the maximum depth of the caudal fin, and smaller factors representing stronger swimming ability (following Webb 1984)
Behavior	Temperature preference		Cold (<18°C), cool (18–26°C), warm (>26°C)
	Substrate preference		Rubble, sand, silt/mud, or general
	Fluvial dependence		Reliance on flowing waters for completing life cycle
	Velocity preference		Slow, moderate, fast
	Vertical position		Benthic or not benthic
Trophic traits	Trophic guild		Herbivores, omnivores, carnivores
	Diet breadth		The number of all kinds of major diet items contains during the whole life cycle, including inorganic material, vegetative material, plankton, aquatic/terrestrial insects, oligochaetes/crustaceans/mollusks, fish/fish eggs (range 1–6)
Life history	Longevity	year	Maximum potential life span
	Age at maturity	year	Age at first maturity
	Length at maturity	mm	Body length in millimeters at first maturity
	Fecundity	unit	Total number of eggs per breeding season
	Relative fecundity	unit/g	Average fecundity per gram of body weight
	Fully yolked egg size	mm	Mean diameter of mature eggs (fully yolked)
	Spawning temperature	°C	Temperature at spawning
	Parental care		Representing the total energetic contribution to the offspring (following Winemiller 1989)
	Spawning substrate		Mineral substrate, vegetation, pelagic, or various
	Time to hatch	day	Mean time to egg hatch within the range of average postspawning water temperatures
	Body length at hatching	mm	Larvae length at hatching
	Body length at first age	mm	The average body length of the age 1
	Spawning type		Single spawner, fractional spawner, continuous spawner

types are included in this study: binary, ordinal, nominal, and continuous) and tolerate missing data (Legendre and Legendre 1998). Then, we correlated the matrix of trait dissimilarity with a matrix of phylogenetic dissimilarity (calculated as 1 minus the data of the matrix of phylogenetic similarity) using the Mantel test (Mantel 1967). The Mantel test can unfold both matrices into single column vectors to test their correlation. The Mantel test showed a nonsignificant correlation (Mantel's standardized R = 0.1182, P = 0.269) between the matrix of trait dissimilarity and phylogenetic dissimilarity, which suggested that the influence of phylogeny on life history traits was not significant. Hence, the influence was not considered in the subsequent statistical analysis.

Biological traits and ecological niche analyses.—We used both one-way analysis of variance (ANOVA) and the Mann-Whitney *U* nonparametric test to examine the differences of fish traits between established and nonestablished fish species.

We illustrated the ecological niches of both established and nonestablished alien fish species in a dimension-reduced space, which extracts from the 23 fish traits by performing principal coordinate analysis (PCoA) on the matrix of traits dissimilarity transformed by Gower's dissimilarity coefficient. The broken-stick rule was used to judge how many principal components were important to explaining the variance of the data (Legendre and Legendre 1998).

Predicting the successful establishment.—Following Kolar and Lodge (2001), at establishment stage a discriminant analysis was conducted on the fish traits data to develop an equation that could stepwise select the key biological trait attributes to predict fish species establishment. We then used this model to predict the influence of each trait on successful establishment.

The Mantel test and PCoA were performed using the Vegan and Labdsv packages and Gower's dissimilarity computation used FD package (Legendre and Legendre 1998) in R (R Core Team 2013). One-way ANOVA, Mann–Whitney *U*-test and discriminant analyses were conducted in SPSS 19.0 (IBM, Chicago).

Results

Trait differences between established and nonestablished fish species

The consistent results of ANOVA and Mann-Whitney *U* nonparametric test showed significant differences of biological traits between established and nonestablished fish species (Table 3). Established fishes tended to have the following characteristics: smaller body length, smaller egg diameter, earlier sexual maturity, and relatively wider diet breadth than nonestablished fishes.

Niche comparison between established and nonestablished fish

The results of PCoA showed a separation of niches from the established fish species to the nonestablished fish species (Figure 2). The first two principal components were considered important and explained 73.75% of the total trait variation. The ecological niches of established and nonestablished fish species were not different in the first coordinate axis, but there was a significant difference in the second coordinate axis. With the exception of the Nile Tilapia, the ecological niches of established fish species occupied the lower quadrant, especially concentrated in the lower left quadrant of the niches space. However, the nonestablished fish species took the top of the niches (Figure 3).

Prediction of establishment risk

Discriminant analysis revealed that a total of eight biological traits, including four of those identified in the ANOVA and Mann–Whitney *U*-test, discriminated between successfully established and nonestablished fishes with 100% accuracy (Table 4). Twenty-seven potentially introduced fishes were introduced to the model, which predicted that 15 species would have a high risk of establishment and the other 12 species a low risk (Table 5).

Table 3. Results of analysis of variance (ANOVA) and Mann-Whitney *U*-tests for biological traits of successfully established and nonestablished species.

Traits	Means		ANOVA	Mann-Whitney
	Established	Nonestablished	*P*-value	*P*-value
Maximum body length	415.08	1,031.72	0.025	0.03
Shape factor	4.75	7.96	0.352	0.724
Swim factor	0.46	0.56	0.347	0.524
Temperature preference	2.00	2.20	0.22	0.622
Substrate preference	3.75	3.60	0.606	0.724
Fluvial dependence	1.88	1.60	0.29	0.435
Velocity preference	1.25	1.40	0.606	0.724
Vertical position	1.50	1.60	0.751	0.833
Trophic guild	2.38	2.40	0.95	0.833
Diet breadth	4.50	3.40	0.082	0.093
Longevity	17.66	20.06	0.767	0.622
Age at maturity	1.06	2.80	0.004	0.019
Length at maturity	85.08	458.30	0.005	0.006
Fecundity	56,211.25	144,373.68	0.348	0.524
Fully yolked egg size	1.25	2.01	0.055	0.011
Spawning temperature	19.81	22.10	0.219	0.354
Parental care	1.50	2.40	0.206	0.284
Spawning substrate	1.88	2.00	0.83	0.833
Time to hatch	4.50	3.40	0.502	0.524
Body length at first age	83.88	176.01	0.053	0.093
Spawning type	1.75	1.60	0.695	0.833

Discussion

The invasion mechanism

Probability of establishment in the Yarlung Zangbo River was explained by two characters between established and nonestablished fish species: biological traits and ecological niches. Shorter body length, earlier maturity, and wider diet breadth, the characteristics of successful invaders in the Yarlung Zangbo River basin, are similar to the results of the study in Lake Fuxian in Yunnan–Guizhou Plateau (Li 2008) and in the Great Lakes (Kolar and Lodge 2002). In general, fish species with such characteristics could adapt well to disturbed environments and even thrive in small spatial scales (Winemiller and Rose 1992). The habitat heterogeneity of the middle reaches of the river was high. There were a large number of slow-flowing river branches, ponds, puddles, pools, and so forth where water level would fluctuate seasonally (Chen and Chen 2010). Large-bodied fishes did not always inhabit these water bodies (wild survey), which might provide many vacant niches for small-bodied fishes in the Yarlung Zangbo River. Japanese Rice Fish, Chinese False Gudgeon, Stone Moroko, Goldfish, Pond Loach, and *Micropercops cinctus* established because they were able to survive in the disturbed environment and occupy the vacant niches. Populations of Goldfish, Pond Loach, and *M. cinctus*, often became the dominant species in such habitats. Grass Carp, Black Carp *Mylopharyngodon piceus*, Northern Snakehead, and Nile Tilapia did not establish, even though these species had been introduced into Europe, Australia, and North America and had damaged the invaded ecosystems (Koehn 2004; Alcaraz et al. 2005; Zambrano et al. 2006; Herborg et al. 2007; DeVaney et al. 2009). There are two possible explanations why these four fish did not establish. First, they had ecological niches similar to indigenous species. Thus, they would have experienced more competition for resources. Second, the water temperature did not meet the thermal requirement at

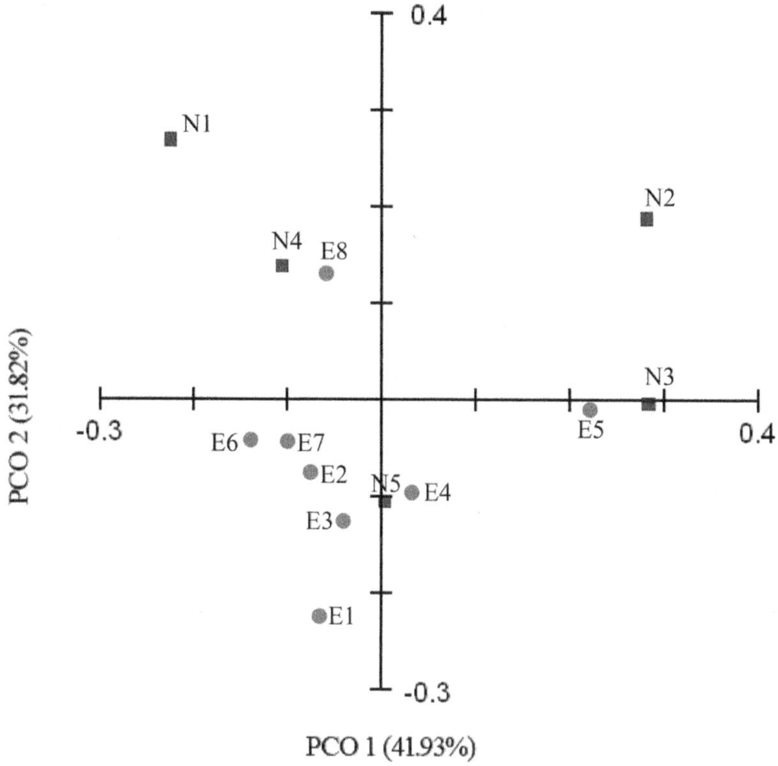

Figure 2. Two-dimensional ordination plots resulting from the principal coordinate analysis on the biological traits for the successfully established and the nonestablished fishes (circles and squares represent successfully established and nonestablished species, respectively). Legend: E1, Japanese Rice Fish; E2, Chinese False Gudgeon; E3, Stone Moroko; E4, Goldfish; E5, Common Carp; E6, Pond Loach; E7, *Micropercops cinctus*; E8, Amur Catfish; N1, Asian Swamp Eel; N2, Bighead Carp; N3, Grass Carp; N4, Northern Snakehead; N5, Nile Tilapia. PCO1 and PCO2, principal coordinates 1 and 2.

these fishes' critical life history stages, such as reproduction. It was reported that the Asian carps required a minimum of 18°C to complete spawning (Yi et al. 1988). However, the average monthly water temperatures (based on Lhasa hydrological station data [Xiong and Tang 1998]) were below 18°C, which might have resulted in their reproductive failure. Therefore, the key life history process could not be completed, which might have hindered the alien fish establishment.

Prevention of alien fishes

The study predicted a high probability of successful establishment for 15 fishes and a lower probability for 12 fishes. This is helpful for the protection and development of fish resources in the middle reaches of the Yarlung Zangbo River basin.

As a result of the diet preferences of people toward alien fishes such as Common Carp, Bighead Carp, and Grass Carp, there was a rapid increase in aquaculture of alien fishes within this region. These fishes may escape aquaculture facilities, which could damage the native ecosystem. We propose that the Black Carp and the Ussuri Catfish *Pelteobagrus ussuriensis* could be imported for aquaculture purposes because of their low risk of establishment and because they are important aquaculture species in China (Ou et al. 2014; Hu et al. 2015), but caution to limit escapement is needed because excessive escapement could result in undesirable environmental effects even if

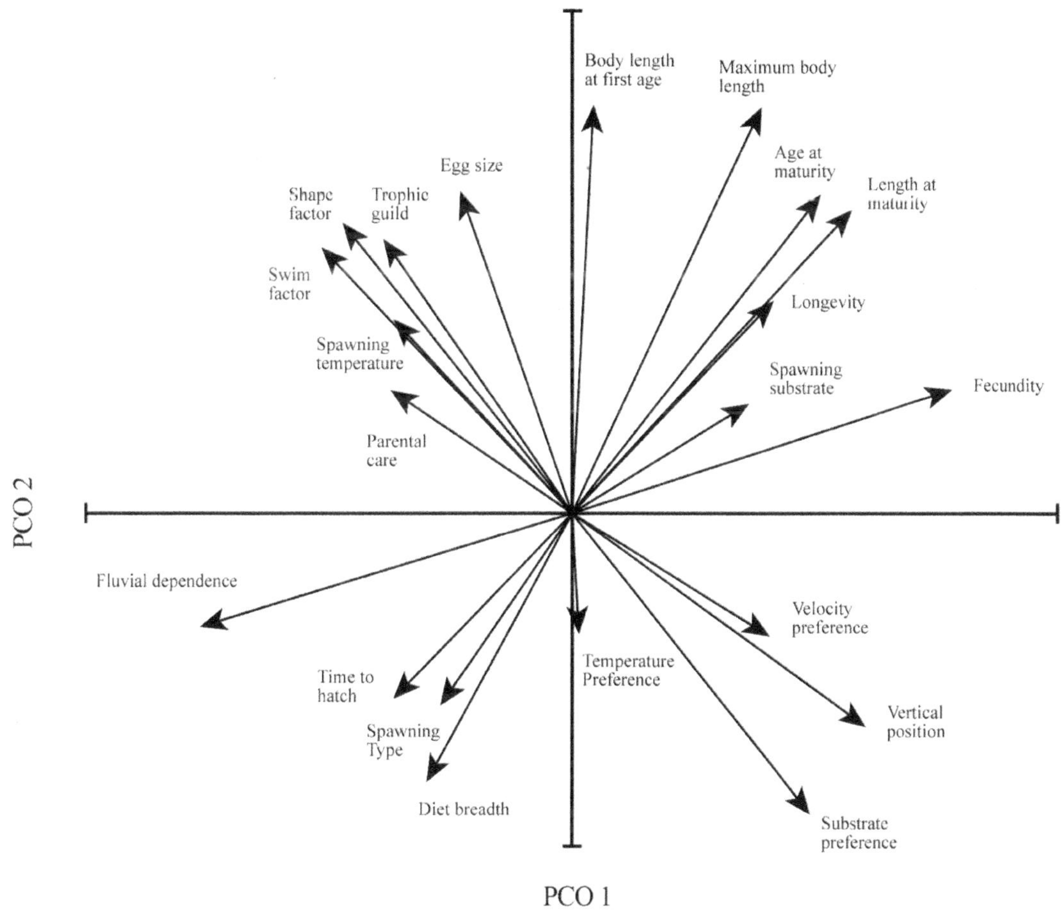

Figure 3. Eigenvector biplot of the traits on the first two principal components (PCO1 and PCO2).

Table 4. Variables contributing to discriminant function for establishment with the associated function coefficients.

Traits	Function coefficients
Maximum body length	0.016
Temperature preference	34.419
Fluvial dependence	40.596
Age at maturity	33.186
Length at maturity	-0.032
Fully yolked egg size	-16.532
Parental care	2.729
Spawning type	11.612
Constant	-202.863

the escaped fish do not reproduce. The model suggests that Rainbow Trout *Oncorhynchus mykiss* and Yellow Catfish *Tachysurus fulvidraco* should not be introduced because of their high risk of establishment, even though they are preferred aquaculture species in other regions in China (Tan et al. 2012; Yu et al. 2014). *Rhinogobius giurinus* and *R. cliffordpopei* (freshwater gobies, no English species names), *Neosalanx taihuensis* (an icefish, no English species name), and Mosquitofish *Gambusia affinis* should be particularly quarantined because they are *r*-selected species that may easily establish populations in the wild.

All the alien fishes in the middle reaches of the Yarlung Zangbo River have also invaded lakes and rivers in the Yunnan–Guizhou Plateau in the past and caused great losses

Table 5. Categories of the fishes potentially introduced to the middle reaches of Yarlung Zangbo River and their predicted risk. Species without common names listed have no accepted English common name.

High risk		Low risk	
Latin names	Common names	Latin names	Common names
Hemiculter leucisculus	Sharpbelly	*Anguilla japonica*	Japanese Eel
Culter alburnus		*Mylopharyngodon piceus*	Black Carp
Culter dabryi dabryi	Humpback	*Tinca tinca*	Tench
Claris lareza		*Megalobrama amblycephala*	Wuchang Bream
Tachysurus fulvidraco	Yellow Catfish	*Spinibarbus denticulatus*	
Tachysurus nitidus		*Labeo rohita*	Roho Labeo (also known as Rohu)
Oncorhynchus mykiss	Rainbow Trout	*Paramisgurnus dabryanus*	
Salvelinus malma	Dolly Varden	*Silurus meridionalis*	
Brachymystax lenok	Lenok	*Pseudobagrus usseriensis*	Ussuri Catfish
Hypomesus olidus	Pond Smelt	*Lota lota*	Burbot
Neosalanx taihuensis		*Perca fluviatilis*	European Perch (also known as Eurasian Perch)
Gambusia affinis	Mosquitofish	*Macropodus chinensis*	Paradisefish (also known as Roundtail Paradisefish)
Perccottus glenii	Rotan (also known as Chinese Sleeper)		
Rhinogobius giurinus			
Rhinogobius cliffordpopei			

to the local biodiversity (Yang 1996; Yan and Chen 2007). Barbless Carp *Cyprinus pellegrini*, a species native to Xingyun Lake, was outcompeted by the introduced Bighead Carp and hybridized with the introduced Common Carp, which together led to the catastrophic extinction of Barbless Carp (Chen et al. 1998). The author also claimed that the *R. giurinus* ate the eggs of indigenous fishes, decreasing their recruitment. In Fuxian Lake, also in the Yunnan–Guizhou Plateau, 26 alien fishes dominated, leading to a chain of ecological effects that permanently changed the original ichthyofauna (Xiong et al. 2008). We are concerned that the Yarlung Zangbo River may suffer a similar fate. This study may help relevant governmental agencies selectively introduce fish species according to the risk that they represent. We should learn from the lessons of the unplanned introductions of many alien fish species to the Yunnan–Guizhou Plateau lakes and protect the biodiversity of the aquatic ecosystem in the middle reaches of the Yarlung Zangbo River.

Potential limitation in this study

We should be cautious to interpret our results. Due to limited data availability, we could not select more traits that might provide more detail information about ecological characters of an alien species, such as tolerance, migration, and hydrograph requirements. Furthermore, because of diverse life history traits data source, the results would be sensitive to the original measurement errors (Gotelli and Pyron 1991). We would put more effort in the future to further conform our predicted results with more traits included and more reliable data.

Acknowledgments

This research was supported by the Natural Science Foundation of China (grant

41030208, 31472016) and National Basic Research Program of China (973 Program, grant 2009CB119200). We would like to thank Duane Chapman, Yushun Chen, and two anonymous reviewers for their constructive suggestions on improving the manuscript.

References

Alcaraz, C., A. Vila-Gispert, and E. García-Berthou. 2005. Profiling invasive fish species: the importance of phylogeny and human use. Diversity and Distributions 11:289–298.

Cai, B. 1997. Tibet's fish resources and their rational use. Chinese Fisheries Economy Research 4:38–40. (In Chinese.)

Chen, F., and Y. Chen. 2010. Investigation and protection strategies of fishes of Lhasa River. Acta Hydrobiologica Sinica 34:278–285. (In Chinese.)

Chen, L., S. Peng, and B. Yang. 2015. Predicting alien herb invasion with machine learning models: biogeographical and life-history traits both matter. Biological Invasions 17:2187–2198.

Chen, Y., J. Yang, and Z. Li. 1998. The diversity and present status of fishes in Yunnan Province. Chinese Biodiversity 6:272–277. (In Chinese.)

Christian, C. E. 2001. Consequences of a biological invasion reveal the importance of mutualism for plant communities. Nature (London) 413:635–639.

DeVaney, S. C., K. M. McNyset, J. B. Williams, A.T. Peterson, and E. O. Wiley. 2009. A tale of four "carp": invasion potential and ecological niche modeling. PLoS (Public Library of Science) One [online serial] 4(5):e5451.

Elton, C. S. 2000. The ecology of invasions by animals and plants. University of Chicago Press, Chicago.

Gotelli, N. J., and M. Pyron. 1991. Life history variation in North American freshwater minnows: effects of latitude and phylogeny. Oikos 1:30–40.

Grafen, A. 1989. The phylogenetic regression. Philosophical Transactions of the Royal Society of London B 326:119–157.

Herborg, L. M., N. E. Mandrak, B. C. Cudmore, and H. J. MacIsaac. 2007. Comparative distribution and invasion risk of snakehead (Channidae) and Asian carp (Cyprinidae) species in North America. Canadian Journal of Fisheries and Aquatic Sciences 64:1723–1735.

Hu, Y., Y. Huang, F. Feng, L. Zhong, Q. Ai, T. Xiao, and H. Wen. 2015. Effect of soybean meal replacement by cottonseed meal on growth, feed utilization and some blood physiological/biochemical indices of juvenile Black Carp, *Mylopharyngodon piceus*. Aquaculture Research 46:2490–2500.

Klein, J. A., J. Harte, and X. Zhao. 2004. Experimental warming causes large and rapid species loss, dampened by simulated grazing, on the Tibetan Plateau. Ecology Letters 7:1170–1179.

Koehn, J. D. 2004. Carp (*Cyprinus carpio*) as a powerful invader in Australian waterways. Freshwater Biology 49:882–894.

Kolar, C. S., and D. M. Lodge. 2001. Progress in invasion biology: predicting invaders. Trends in Ecology and Evolution 16:199–204.

Kolar, C. S., and D. M. Lodge. 2002. Ecological predictions and risk assessment for alien fishes in North America. Science 298:1233–1236.

Lee, C. E. 2002. Evolutionary genetics of invasive species. Trends in Ecology and Evolution 17:386–391.

Legendre, P., and L. Legendre. 1998. Numerical ecology, volume 1. Elsevier Press, Amsterdam.

Li, X. 2008. Population biology of an alien fish Pelteobagrus fulvidraco and invasive risk forecasting in Lake Fuxian, Yunnan. Institute of Hydrobiology, Chinese Academy of Sciences, Wuhan, China. (In Chinese.)

Liu, X., and B. Chen. 2000. Climatic warming in the Tibetan Plateau during recent decades. International Journal of Climatology 20:1729–1742.

Mack, R. N., D. Simberloff, W. M. Lonsdale, H. Evans, M. Clout, and F. A. Bazzaz. 2000. Biotic invasions: causes, epidemiology, global consequences, and control. Ecological Applications 10:689–710.

Mantel, N. 1967. The detection of disease clustering and a generalized regression approach. Cancer Research 27:209–220.

Marchetti, M. P., P. B. Moyle, and R. Levine. 2004. Invasive species profiling? Exploring the characteristics of non-native fishes across invasion stages in California. Freshwater Biology 49:646–661.

Mims, M. C., J. D. Olden, Z. R. Shattuck, and N. L. Poff. 2010. Life history trait diversity of native freshwater fishes in North America. Ecology of Freshwater Fish 19:390–400.

Odum, E. P. 1983. Basic ecology. Saunders College Pub, Rochester, New York.

Olden, J. D., N. L. Poff, and K. R. Bestgen. 2006. Life-history strategies predict fish invasions and extirpations in the Colorado River basin. Ecological Monographs 76:25–40.

Ou, T., X. Lei, L. He, F. Zhou, and Q. Zhang. 2014. Development of an Ussuri Catfish *Pseudobagrus ussuriensis* skin cell line displaying differential cytopathic effects to three aquatic animal viruses. Virus Research 189:56–62.

R Core Team. 2013. R: a language and environment for statistical computing. R Foundation for Statistical Computing, Vienna, Austria.

Radford, I. J., and R. D. Cousens. 2000. Invasiveness and comparative life-history traits of exotic and indigenous Senecio species in Australia. Oecologia 125:531–542.

Ricciardi, A., and J. B. Rasmussen. 1998. Predicting the identity and impact of future biological invaders: a priority for aquatic resource management. Canadian Journal of Fisheries and Aquatic Sciences 55:1759–1765.

Sala, O. E., F. S. Chapin, J. J. Armesto, E. Berlow, J. Bloomfield, R. Dirzo, E. Huber-Sanwald, L. F. Huenneke, R. B. Jackson, A. P. Kinzig, R. Leemans, D. M. Lodge, H. A. Mooney, M. Oesterheld, and N. L. Poff. 2000. Global biodiversity scenarios for the year 2100. Science 287:1770–1774.

Strayer, D. L., V. T. Eviner, J. M. Jeschke, and M. L. Pace. 2006. Understanding the long-term effects of species invasions. Trends in Ecology and Evolution 21:645–651.

Tan, X., P. Xie, Z. Luo, H. Lin, Y. Zhao, and W. Xi. 2012. Dietary manganese requirement of juvenile Yellow Catfish *Pelteobagrus fulvidraco*, and effects on whole body mineral composition and hepatic intermediary metabolism. Aquaculture 326:68–73.

Thuiller, W., N. Gassó, J. Pino, and M. Vilà. 2012. Ecological niche and species traits: key drivers of regional plant invader assemblages. Biological Invasions 14:1–18.

van Kleunen, M., E. Weber, and M. Fischer. 2010. A meta-analysis of trait differences between invasive and non-invasive plant species. Ecology Letters 13:235–245.

Webb, P. W. 1984. Form and function in fish swimming. Scientific American 251:72–82.

Williamson, M. and A. Fitter. 1996. The varying success of invaders. Ecology 77:1661–1666.

Winemiller, K. O. 1989. Patterns of variation in life history among South American fishes in seasonal environments. Oecologia 81:225–241.

Winemiller, K. O., and K. A. Rose. 1992. Patterns of life-history diversification in North American fishes: implications for population regulation. Canadian Journal of Fisheries and Aquatic Sciences 49:2196–2218.

Wu, Y., and C. Wu. 1992. The fishes of the Qinghai-Xizang Plateau. Sichuan Publishing House of Science and Technology Press, Chengdu, China. (In Chinese.)

Xiong, F., W. Li, and J. Pan. 2008. Present status of alien fishes and analysis of relative problems in Fuxian Lake in Yunnan Province. Acta Agriculture Jiangxi 20:92–94. (In Chinese.)

Xiong, Y., and Q. Tang. 1998. The hydrology of river in cold area of the Tibetan Plateau. Page 118 *in* Y. Xiong, and Q. Tang, editors. The hydrology of river in China. Science Press, Beijng. (In Chinese.)

Yan, Y., and Y. Chen. 2007. Plasticity in reproductive tactics of *Ctenogobius giurinus* in Lake Fuxian. Acta Hydrobiologica Sinica 31:414–418. (In Chinese.)

Yang, J. 1996. Exotic fish and native fish in Yunnan: the manner and extent of the impact and related research questions. Protecting China's biodiversity (II). China Environmental Science Press, Beijing. (In Chinese.)

Yi, B., Z. Yu, Z. Liang, S. Shen, Y. Xu, J. Chen, M. He, Y. Liu, Y. Hu, Z. Deng, S. Huang, J. Sun, R. Liu, and Y. Xiang. 1988. The distribution, natural conditions and breeding production of the spawning grounds of four domestic freshwater fishes on the mainstream on the Yangtze River. Pages 1–46 *in* B. Yi, Z. Yu, and Z. Liang, editors. Project and four domestic fishes in Yangtze River. Hubei Science and Technology Press, Wuhan, China. (In Chinese.)

Yu, Z. , M. Deng, Y. Geng, Y. Zhou, K. Wang, D. Chen, Z. Zhong, X. Peng, and X. Huang. 2014. An outbreak of infectious haematopoietic necrosis virus (IHNV) infection in cultured Rainbow Trout (*Oncorhynchus mykiss*) in southwest China. Aquaculture Research, DOI: 10.1111/are.12680.

Zambrano, L., E. Martínez-Meyer, N. Menezes, and A. T. Peterson. 2006. Invasive potential of Common Carp (*Cyprinus carpio*) and Nile Tilapia (*Oreochromis niloticus*) in American freshwater systems. Canadian Journal of Fisheries and Aquatic Sciences 63:1903–1910.

Appendix A: Trait Values for the Alien Fishes in the Yarlung Zangbo River Examined in Our Study

Table A.1. Trait values for the alien fishes in the middle reaches of the Yarlung Zangbo River.

Order	Family	Species	Morphology: Maximum body length	Morphology: Shape factor	Morphology: Swim factor	References
Cypriniformes	Cyprinidae	Chinese False Gudgeon *Abbottina rivularis*	189	5.2	0.31	10, 17, 29, 86, 87
		Stone Moroko *Pseudorasbora parva*	110	3.7	0.3	13, 17, 41
		Common Carp *Cyprinus carpio*	972	3.1	0.42	10, 17, 65, 78
		Goldfish *Carassius auratus*	337.6	2.8	0.43	6, 10, 78, 82, 88
		Bighead Carp *Hypophthalmichthys nobilis*	1,493.6	3.2	0.32	5, 10, 16, 25, 37
		Grass Carp *Ctenoparyngodon idellus*	1,023.9	3.7	0.42	10, 48
	Cobitidae	Pond Loach (also known as Oriental Weatherfish) *Misgurnus anguillicaudatus*	300	7.2	0.64	11, 16, 57
Siluriformes	Siluridae	Amur Catfish *Silurus asotus*	1,300	6.1	0.55	11, 17, 40, 46
Cyprinodontiformes	Cyprinidintidae	Japanese Rice Fish (also known as Japanese Medaka) *Oryzias latipes*	32	4.4	0.45	11, 17, 22, 83
Sybranchiformes	Synbranchidae	Swamp Eel *Monopterus albus*	1,041.1	24.7	1	4, 11, 32, 72, 73, 77
Perciformes	Channidae	Northern Snakehead *Channa argus*	1,000	5.47	0.58	11, 16, 17
	Cichlidae	Nile Tilapia *Oreochromis nilotica*	600	2.75	0.48	11, 17
	Eleotridae	*Micropercops cinctus*	80	5.5	0.55	8, 11, 16, 17

Species	Behavior: Temperature preference	Behavior: Substrate preference	Behavior: Fluvial dependence	Behavior: Velocity preference	Behavior: Vertical position	References
Chinese False Gudgeon	cool	general	NO	moderate	benthic	10, 12, 87
Stone Moroko	cool	general	NO	slow	not benthic	10, 13, 41, 42
Common Carp	cool	general	YES	moderate	not benthic	9, 21, 33, 58, 68
Goldfish	cool	general	NO	slow	not benthic	10, 14, 33
Bighead Carp	cool	general	YES	slow	not benthic	10, 15, 20, 45, 49, 58
Grass Carp	cool	general	YES	moderate	not benthic	10, 58, 81
Pond Loach	cool	mud	NO	slow	benthic	11, 64, 76

Table A.1. Continued.

Species	Behavior: Temperature preference	Substrate preference	Fluvial dependence	Velocity preference	Vertical position	References
Amur Catfish	cool	mud	NO	slow	benthic	11, 18, 46
Japanese Rice Fish	cool	general	NO	slow	not benthic	11, 22, 58, 83
Swamp Eel	cool	mud	NO	slow	benthic	11, 31, 69, 70, 73
Northern Snakehead	cool	mud	NO	slow	benthic	11, 34, 51, 77
Nile Tilapia	warm	general	NO	moderate	not benthic	27, 80
Micropercops cinctus	cool	general	NO	slow	benthic	8, 11, 81

Species	Trophic traits: Trophic guild	Diet breadth	Life history: Longevity	Age at maturity	Length at maturity	References
Chinese False Gudgeon	carnivores	4	9.3	1	60.5	10, 29, 86
Stone Moroko	omnivores	5	5	1	25.24	10, 29, 41, 42, 74
Common Carp	omnivores	5	38	2	250	7, 9, 19, 21, 62, 75, 79
Goldfish	omnivores	5	41	1	70	6, 10, 29, 62
Bighead Carp	carnivores	1	32.6	4	900	5, 24, 28, 39, 43, 47, 85
Grass Carp	herbivores	4	21	4	593	10, 29, 81
Pond Loach	omnivores	5	17	1	87	26, 29, 60, 76
Amur Catfish	carnivores	4	26	1	139	11, 18, 29, 46, 61
Japanese Rice Fish	omnivores	5	4	0.5	20	22, 58, 83
Swamp Eel	carnivores	4	26.1	3	300	11, 29, 30, 31, 66, 71
Northern Snakehead	carnivores	4	11.6	2	370	11, 29, 51, 56, 67, 77
Nile Tilapia	omnivores	4	9	1	128.5	11, 27, 35, 59, 80
Micropercops cinctus	carnivores	3	1	1	28.9	8, 11, 81

Species	Life history: Fecundity	Egg size	Spawning temperature	Parental care	Spawning substrate	References
Chinese False Gudgeon	1,744	1.1	21	3	mineral	12, 29, 87
Stone Moroko	2,156	1.6	24	3	mineral	10, 29, 41, 42

Table A.1. Continued.

	Life history					
Species	Fecundity	Egg size	Spawning temperature	Parental care	Spawning substrate	References
Common Carp	388,686	1.3	18	1	vegetation	9, 21, 29, 75
Goldfish	45,800	1.5	17	1	vegetation	6, 23, 29, 58
Bighead Carp	400,000	1.46	22	1	pelagic	58, 63, 64
Grass Carp	300,000	1.6	18	1	pelagic	36, 58, 64
Pond Loach	277	1.3	18	1	vegetation	26, 29, 58, 60, 64
Amur Catfish	7,608	1.56	19	1	vegetation	29, 46, 61
Japanese Rice Fish	3,000	0.9	22.5	0	mineral	22, 58, 83
Swamp Eel	718.4	3.7	28.5	3	mineral	2, 31, 52, 66, 69, 71, 84, 89
Northern Snakehead	19,814	1.65	22	4	vegetation	11, 29, 38, 51, 77
Nile Tilapia	1,336	1.65	20	3	mineral	1, 27, 35, 53
Micropercops cinctus	419	0.76	19	2	various	81

	Life history				
Species	Time to hatch	Body length at hatching	Body length at first age	Spawning type	References
Chinese False Gudgeon	6	4	41.75	signal	3, 12, 87
Stone Moroko	3	4.6	28.66	fractional	3, 41, 42, 58, 74
Common Carp	2	5.3	170.80	fractional	3, 9, 44, 58
Goldfish	5	4.9	79.04	fractional	3, 6, 44, 58
Bighead Carp	2	7	242.81	signal	3, 44, 58
Grass Carp	2	6.3	31.57	signal	3, 36, 58
Pond Loach	2	3.7	81	fractional	3, 54, 58, 60
Amur Catfish	2	4.98	201.1	signal	3, 18, 58, 61
Japanese Rice Fish	11	4.6	30	continuous	3, 22, 58, 83
Swamp Eel	7	13	233.93	fractional	2, 3, 31, 50, 66, 69, 71
Northern Snakehead	2	4.1	212.3	fractional	3, 11, 51, 55, 77
Nile Tilapia	4	4.97	159.45	fractional	27, 35, 44, 53
Micropercops cinctus	5		38.65	signal	81

References

[1] Babiker, M., and H. Ibrahim. 2006. Studies on the biology of reproduction in the cichlid *Tilapia nilotica* (L.): gonadal maturation and fecundity. Journal of Fish Biology 14:437–448.

[2] Cao, K., M. Shu, and H. Dong. 1988. The study on the individual fecundity and first batch of eggs of *Monopterus albus*. Journal of Fishery Sciences 3:1–6. (In Chinese.)

[3] Cao, W. 2007. Fish resources of early life history stages in Yangtze River. China Water and Power Press, Beijing. (In Chinese.)

[4] Chen, H. 1998. Age determination and growth of *Monopterus albus*. Journal of Fisheries of China 22:296–302. (In Chinese.)

[5] Chen, M., Y. He. Z. Chen, and J. Gao. 1984. On the growth rate of some commercial fishes in the Qiantang River. Acta Ecologica Sinica 4:181–187. (In Chinese.)

[6] Chen, P. 2005. An ecological study of *Carassius auratus* in Lake Liang-tze. Acta Hydrobiologica Sinica 4:411–419. (In Chinese.)

[7] Chen, Y. 1965. The age and growth of *Cyprinus carpio* L. in Daihai Lake, Inner Mongolia. Oceanologia et Limnologa Sinica 7:253–264. (In Chinese.)

[8] Chen, Y. 1998. Fish in Hengduan Mountains. Science Press, Beijing. (In Chinese.)

[9] Chiang, I., and P. Chen. 1960. On the biology of the Common Carp *Cyprinus carpio* of Liang-tse Lake. Acta Hydrobiologica Sinica 1:43–56. (In Chinese.)

[10] Chu, X., and Y. Chen. 1989. The fishes of Yunnan, China: part I. Science Press, Beijing. (In Chinese.)

[11] Chu, X., and Y. Chen. 1990. The fishes of Yunnan, China: part II. Science Press, Beijing. (In Chinese.)

[12] Deng, Q. 1990. An observation of the propagation habit of the *Abbottina rivularia*. Journal of Sichuan Teachers College 3:200–203. (In Chinese.)

[13] Dong, C. 2005. Preliminary observation on the biology of *Pseudorasbora parva*. Fisheries Science and Technology Information 1:50–53. (In Chinese.)

[14] Duan, Z., and Y. Sun. 1994. The growth and stock assessment of *Carassius auratus auratus* L. in Wanghu Lake. Journal of Lake Science 6:257–266. (In Chinese.)

[15] Feng, J., D. Chang, J. Hui, R. Geng, R. Li, Z. Li, Z. Qing, and A. Hu. 2006. Growth and resource assessment of Bighead Carp in Nanwan Reservoir. Henan Fisheries 1:30–33. (In Chinese.)

[16] Institute of Hydrobiology of Hubei Province. 1976. Fishes of Yangtze River. Sciences Press, Beijing. (In Chinese.)

[17] Froese, R., and D. Pauly, editors. 2010. FishBase. www.fishbase.org.

[18] Fu, Z., and X. Ma. 1999. Feeding and breeding technology of juvenile catfish. Scientific Fish Farming 5:22–23. (In Chinese.)

[19] Gong, S., and X. Hu. 1996. The study on growth of *Cyprinus carpio* in Wang Lake. Journal of Hydroecology 2:22–25. (In Chinese.)

[20] Hu, H., X. Zhang, X, Cao, X. Bing, and X. Geng. 2010. The preliminary study on the growth of carp fish in Lihu Lake. Chinese Agricultural Science Bulletin 26:337–339. (In Chinese.)

[21] Huang, S., Q. Wu, B. Yi, Y. Lu, and M. Ren. 1959. Breeding, feeding habits and growth of carp in the Heilongjiang River basin. Acta Hydrobiologica Sinica 2:210–214. (In Chinese.)

[22] Huang, Y. 1988. Biological characteristics and feeding management techniques of *Oryzias latipes*. Chinese Journal of Zoology 6:10–14. (In Chinese.)

[23] Huang, Y., X. Zhou, and P. Xu. 1980. Beeding and protection of Crucian Carp in Baiyangdian. Chinese Journal of Zoology 1:46–49. (In Chinese.)

[24] Jiang, H., K. Chen, F. Zou, Z. Wen, and Z. Dai. 2004. Study on age and growth of Bighead Carp in Zhongzhou fishing ground, Yueyang. Current Fisheries 29:35–37. (In Chinese.)

[25] Jiang, Z., K. Qin, S. Shu, L. Wu, D. Zhao, and B. Zou. 1994. The age, growth and resourse of Silver Carp (*Hypophthalmichchys molitrix*) and Bighead Carp (*Aristichthgs nobilis*) in Biliuhe Reservoir. Journal of Dalian Fisheries University 9:8–14. (In Chinese.)

[26] Lei, F., and B. Wang. 1990. Studies on reproduction and growth of loach. Acta Hydrobiologica Sinica 14:60–67. (In Chinese.)

[27] Li, J., and S. Li. 2001. Introduction and research advances of *Oreochromis niloticus* in Chinese mainland. Journal of Fisheries of China 25:90–95 (In Chinese.)

[28] Li, J., and W. Shi. 1999. The growth of Silver Carp and Bighead Carp and their resource utilization. Journal of Dalian Fisheries University 14:57–62. (In Chinese.)

[29] Li, M., and Z. Yang. 1991. The age, growth and reproduction of fishes in Yuqiao Reservoir. Acta Ecologica Sinica 11:269–273. (In Chinese.)

[30] Li, M. 2002. Progress in biological research of *Monopterus albus*. Fisheries Science and Technology of Guangxi 3:32–38. (In Chinese.)

[31] Li, Z. 2007. Studies on the growth, reproduction and genetic diversity of *Monopterus albus* (Zuiew) in Poyang Lake. Master's thesis. Nanchang University, Nanchang, China. (In Chinese.)

[32] Li, Z., Y. Ou, X. Wu, Y. Dai, and H. Huo. 2008. The growth characteristics of *Monopterus albus* in Poyang Lake. Chinese Journal of Zoology 43:113–120. (In Chinese.)

[33] Liang, Z., and S. Yang. 1993. Studies on the growth characteristics of Common Carp (*Cyprinus carpio*) and Crucian Carp (*Carassius auratus*) in Biliuhe Reservoir and its utilization in fishery. Journal of Dalian Fisheries College 2:33–42. (In Chinese.)

[34] Liu, J.,Y. Cui, and J. Liu. 2002. The optimum temperatures for Mandarin Fish (*Siniperca chuatsi*) and the Northern Snakehead (*Hanna argus*). Acta Hydrobiologica Sinica 26:433–437. (In Chinese.)

[35] Liu, Y., G. Liu, S. Chen, and C. Liu. 1983. Study on the gonodal development of *Tilapia nilotica*. Acta Hydrobiologica Sinica 8:17–32. (In Chinese.)

[36] Liu, Y., S. Chen, Y. Wang, and H. Gu. 2005. Study on the type of spawning of the Grass Carp, *Ctenoparyngodon idellus*. Acta Hydrobiologica Sinica 6:247–260. (In Chinese.)

[37] Ma, C., X. Dai, W. Zheng, and J. Tao. 2011. Study on the growth of Silver Carp and Bighead Carp in Dianshan Lake. Transactions of Oceanology and Limnology 3:69–75. (In Chinese.)

[38] Ma, T., and C. Xie. 2000. The study on individual fecundity of *Ophiocephalus argus* in Liangzi Lake. Reservoir Fisheries 20:1–3. (In Chinese.)

[39] Mu, H., J.Yao, S. Ma, Y. Wang, and Z. Liang. 2012. Study on the age structure and growth of Silver Carp and Bighead Carp in the Hongfeng Lake of Guiyang City. Guangdong Agricultural Sciences 39:122–125. (In Chinese.)

[40] Peng, J., Z. Cao, and S. Fu. 2007. Phenotypic variation in Chinese Catfish (*Silurusasotus linnaeus*). Journal of Chongqing Normal University (Natural Science Edition) 24: 69–71 (In Chinese.)

[41] Peng, Z. 2008. Preliminary study on biological characteristics of *Pseudorasbora parva*. Inland Fisheries 32:33-33. (In Chinese.)

[42] Qin, Y., L. Li, and R. Huang. 2005. Biological characteristics and cultivation technology of *Pseudorasbora parva*. Jiangsu Agricultural Sciences 3:114–116. (In Chinese.)

[43] Ruan, J. 1986. Studies on several aspects of growth of Silver Carp and Bighead Carp in Lake Donghu, Wuhan. Acta Hydrobiologica Sinica 10:252–263. (In Chinese.)

[44] Shen, Y. 2008. Fish aquaculture. China Agriculture Press, Beijing. (In Chinese.)

[45] Shen, Y., J. Yang, B. Qi, S. Zhang, X. Wang, and H. Liu. 2001. Growth and methods of increase fish yield of Silver Carp and Bighead Carp in Mengjiaduan Reservoir. Journal of Hydroecology 21:26–28. (In Chinese.)

[46] Shi, B. 1980. Biological characters introduction of the catfish in Jialing River. Journal of Southwest China Normal University (Natural Science Edition) 2:53–59. (In Chinese.)

[47] Sun, J., J. Shen, W. Song, L. Hu, and C. Wan. 2007. The growth of Silver Carp and Bighead Carp in Yunlonghu Reservoir. Journal of Hydroecology 26:56–58. (In Chinese.)

[48] Tian, J., and D. Wang. 1989. Study on the characters and the growth of the wild Grass Carp in Yangtze River. Journal of Henan Normal University (Natural Science) 2:60–64. (In Chinese.)

[49] Wang, C., and J. Li. 2008. A study on reasonable utilization of resources of Silver Carp and Bighead Carp. Modern Fisheries Information 23:13–15. (In Chinese.)

[50] Wang, D. 1992. Special aquatic culture. Higher Education Press, Beijing. (In Chinese.)

[51] Wang, G. 2000. The biological characteristics and breeding technology of *Ophiocephalus argus*. Freshwater Fisheries 30:10–11. (In Chinese.)

[52] Wang, G., L. Lou, and S, Yang. 2004. A preliminary study of fecundity and parent selection of *Monopterus albus* in Yuanshui River downstream. Current Fisheries 29:35–36. (In Chinese.)

[53] Wang, L., and Q. Chou. 1981. Observation on the embryonic and larval development of *Tilapia nilotica*. Acta Zoologica Sinica 27:327–336. (In Chinese.)

[54] Wang, M. 2000. The study of development of the digestive organs, feeding and growth of juvenile *Misgurnus anguillicaudatus*. Journal of Hydroecology 20:10–11. (In Chinese.)

[55] Wang, M., and K. Cao. 1994. Study on the early development of the digestive organs and daily diet of juvenile *Ophiocephalus argus*. Journal of Hydroecology 6:9–11. (In Chinese.)

[56] Wang, M., and K. Cao. 1997. Study on feeding habits and feeding rate of *Ophiocephalus argus*. Journal of Hydroecology 5:12–15. (In Chinese.)
[57] Wang, M., W. Wang, and J. Yan. 2001. Comparative study on age and growth of *Misgurnus anguillicaudatus* and *Paramisgurnus dabryanus*. Journal of Hydroecology 21:7–9. (In Chinese.)
[58] Wang, S., G. Li, and Y. Cao. 2001. The fauna of Hebei: fish. Hebei Science and Technology Press, Shijiazhuang, China. (In Chinese.)
[59] Wang, T. 1981. Feeding observation of Nile Tilapia. Freshwater Fisheries 2:47–48. (In Chinese.)
[60] Wang, W., H. Ding, Z. Mao, X. Chang, B. Qing and T. Xu. 2003. On biological characteristics and technology of artificial reproduction of loach in Qinling Mountain. Journal of Shaanxi Normal University (Natural Science Edition) 31:153–157. (In Chinese.)
[61] Wei, G., and L. Huang. 1997. Studies on the reproductive biology of *Silurus asotus*. Journal of Fisheries of China 21:225–232. (In Chinese.)
[62] Wen, Z., and M. Zeng. 1993. Studies on age and growth of *Cyprinus carpio* and *Carassius auratus* in Diaocha Lake. Freshwater Fisheries 23:20–22. (In Chinese.)
[63] Wu, H., and J., Tan. 2000. The study on mature egg size correlation of *Hypophthalmichthys molitrix* and *Aristichthys nobilis*. Current Fisheries 25:7–8. (In Chinese.)
[64] Wu, X., G. Yang, P. Yue, and H. Huang. 1963. The economic fauna of China: freshwater fishes. Science Press, Beijing. (In Chinese.)
[65] Xi, Y., and K. Cao. 1995. On the age and growth of the carp (*Cyprinus carpio* L.) in Caohe Reservoir compared with that of the carp in Danjiangkou Reservoir. Journal of Zhejiang College of Fisheries 14:74–84. (In Chinese.)
[66] Xiao, Y. 1993. Studies on individual growth of *Monopterus albus*. Inland Fisheries 6:7–9. (In Chinese.)
[67] Xie, C., Z. Xia, B. Zhu, and H. Jin. 1997. The structure of catch and growth characteristics of Chinese Snakehead Fish (*Channa argus*) in Baoan Lake. Journal of Huazhong Agricultural University 16:367–373. (In Chinese.)
[68] Xiong, F., S. Liu, X. Duan, and D. Chen. 2009. Age and growth of *Cyprinus carpio* L. in Poyang Lake. Journal of Hydroecology 2:66–70. (In Chinese.)
[69] Yang, D., F. Chen, B. Liu, and D. Li. 1994. On the type of spawning and fecundity of the eel (*Monopterus albus*). Journal of Hubei Agricultural College 14:40–44. (In Chinese.)
[70] Yang, D., F. Chen, D. Li, and B. Liu. 1993. A preliminary study on the growth characteristics of eel (*Monopterus allous*). Journal of Hubei Agricultural College 13:194–199. (In Chinese.)
[71] Yang, D., F. Chen, D. Li, and B. Liu. 1999. The embryonic development and fingerling culture of eel (*Monopterus albus*). Journal of Hubei Agricultural College 19:149–153. (In Chinese.)
[72] Yang, M. 1993. Studies on age and growth of *Monopterus albus*. Freshwater Fisheries 23:43–45. (In Chinese.)
[73] Yang, M. 1997. Studies on hyoid and the growth of *Monopterus albus*. Chinese Journal of Zoology 32:12–14. (In Chinese.)
[74] Yang, Z., and M. Li. 1989. Studies on biology of *Pseudorasbora parva* in Tianjin area. Chinese Journal of Zoology 24:11–14. (In Chinese.)
[75] Ye, F., P. Luo, and X. Ye. 1992. The biology of carp, *Cyprinus carpio* in Dongjiang River. Journal of Zhanjiang Fisheries College 12:22–28. (In Chinese.)
[76] Yin, J., Z. Tian, and Z. Zhao. 2000. A preliminary research on feeding habits of *Misgurnus anguillicaudatus*. Reservoir Fisheries 20:15–16. (In Chinese.)
[77] Yu, H. 2007. Studies on biology and artificial propagation of *Ophiocephalus argus* in Poyang Lake. Master's thesis. Nanchang University, Nanchang, China. (In Chinese.)
[78] Yue, P. 2000. Fauna Sinica (Osteichthyes): Cypriniformes, volume III. Science Press, Beijing. (In Chinese.)
[79] Zhang, J., Q. Fan, and W. Wang. 1998. Study on the population dynamics of Common Carp in Laojianghe Lake, a breed resources pool of four main kinds of Chinese carps. Journal of Huazhong Agricultural University 17:395–400. (In Chinese.)
[80] Zhang, P., X. Lin, J. Sun., G. Chen, Y. He, C. Xu, and Y. Wang. 2012. Annuli and age-related factors of the alien species Nile Tilapia (*Oreochromis niloticus*) in Dongjiang River, Guangdong Province. Journal of Fishery Sciences of China 19:416–422. (In Chinese.)
[81] Zhang, T. 2005. Life-history strategies, trophic patterns and community structure in the fishes of Lake Biandantang. Doctoral dissertation, Graduate School of the Chinese Academy of Sciences, Wuhan, China. (In Chinese.)

[82] Zhang, X., D. Zhang, Z. Chen, J. Guo, Z. Wang, and J. Cai. 1997. Studies on age and growth of Crucian Carp from Yellow River in Ningxia. Journal of Ningxia Agricultural College 18:62–66. (In Chinese.)

[83] Zhao, H. 2006. A new species of Cyprinodontiformes in Ningxia—*Oryzias latipes*. Freshwater Fisheries 36:10–12. (In Chinese.)

[84] Zhao, Y., and X. Ke. 1992. Observation on the reproductive biology of *Monopterus albus*. Sichuan Journal of Zoology 11:7–8. (In Chinese.)

[85] Zheng, S., B. Peng, and J. Xin. 2002. The study on biological characters of dry Silver Carp and Bighead Carp in Tara Reservoir. Inner Mongolia Agricultural Science and Technology 2:18–20. (In Chinese.)

[86] Zhou, C., Q. Deng, L. Ren, and J. Hu. 1998. A biological study on *Abbottina rivularis*. Journal of Sichuan Teachers College (Natural Science Edition) 19:307–311. (In Chinese.)

[87] Zhou, C., Q. Zhu, P. Cheng, Y. Xiong, and D. Tan. 2008. Biology of the Siamese algae-eater *Abbotina rivularis* in Panzhihua Reach, Jinshajiang River. Acta Zoologica Sinica 54:218–224. (In Chinese.)

[88] Zhu, Q., Z. Wu, and H. Liu. 2011. Age and growth characteristics of Crucian Carp in Poyang Lake. Jiangxi Fishery Sciences and Technology 4:25–29 (In Chinese.)

[89] Zu, G., Z. Wei, X. Wang, X. Huang, Y. Zhang, and Z. Deng. 2005. Preliminary research on the biological characteristic of Mud Eel's reproduction of Huaihe Valley in Anhui Province. Journal of Anhui Agricultural University 32:478–482. (In Chinese.)

Appendix B: Trait Values for the Potential Introduced Species to the Yarlung Zangbo River Examined in Our Study

Table B.1. Trait values for 27 fishes that have a potential risk to introduce into the middle reaches of the Yarlung Zangbo River.

Order	Family	Species	Morphology: Maximum body length	Morphology: Shape factor	Morphology: Swim factor	References
Anguilliformes	Anguillidae	Japanese Eel *Anguilla japonica*	1,011	19	0.8	6, 72, 74
Cypriniformes	Cyprinidae	*Spinibarbus denticulatus*	690	3.4	0.43	6
		Black Carp *Mylopharyngodon piceus*	1,720	3.7	0.42	6, 7, 72, 76
		Tench *Tinca tinca*	700	3.2	0.54	1, 7, 72
		Roho Labeo (also known as Rohu) *Labeo rohita*	2,000	3	0.45	7, 36
		Sharpbelly *Hemiculter leucisculus*	283.84	4.02	0.32	6, 25, 26, 59, 72
		Wuchang Bream *Megalobrama amblycephala*	550	2.2	0.45	41, 54, 72
		Culter alburnus	1,404.9	3.9	0.35	6, 14, 59, 72, 82
		Humpback *C. dabryi*	473	4	0.38	72
	Cobitidae	*Paramisgurnus dabryanus*	294.4	6.43	0.86	60, 72
Siluriformes	Bagridae	Yellow Catfish *Tachysurus fulvidraco*	300	3.6	0.41	7, 72
		Tachysurus nitidus	246	4.7	0.33	6, 72
		Ussuri Catfish *Pseudobagrus ussuriensis*	1,000	5.8	0.35	6, 7
	Siluridae	*Silurus meridionalis*	1,309.5	5.6	0.5	6, 72
		Tawada *Clarias lazera*	1,700	6.73	0.66	7
Salmoniformes	Osmeridae	Pond Smelt *Hypomesus olidus*	200	5.8	0.37	7, 72
	Salangidae	*Neosalanx taihuensis*	82			7, 53, 72
	Salmonoidea	Rainbow Trout *Oncorhynchus mykiss*	1,200	3.94	0.44	7, 40, 72
		Dolly Varden *Salvelinus malma malma*	1,270	3.5	0.4	7, 72
		Lenok *Brachymystax lenok*	750	3.95	0.3	7, 72
Gadiformes	Gadidae	Burbot *Lota lota*	1,520	5.8	0.36	7, 72
Cyprinodontiformes	Poeciliidae	Mosquitofish *Gambusia affinis*	60	5.42	0.48	4, 42
Perciformes	Percidae	European Perch (also known as Eurasian Perch) *Perca fluviatilis*	600	3.6	0.5	7, 72

Table B.1. Continued.

Order	Family	Species	Morphology: Maximum body length	Morphology: Shape factor	Morphology: Swim factor	References
Perciformes	Eleotridae	Ratan (also known as Chinese Sleeper) *Perccottus glenii*	250	3.8	0.57	4, 7, 11
	Gobiidae	*Rhinogobius giurinus*	80	5.2	0.68	7, 72
		Rhinogobius cliffordpopei	80	5.2	0.57	7, 72
	Belontiidae	Paradisefish (also known as Roundtail Paradisefish) *Macropodus chinensis*	67	2.7	0.73	7, 72

Species	Behavior: Temperature preference	Behavior: Substrate preference	Behavior: Fluvial dependence	Behavior: Velocity preference	Behavior: Vertical position	References
Japanese Eel	cool		YES	moderate	not benthic	6, 7, 74
Spinibarbus denticulatus	warm	genera	NO	fast	not benthic	7, 6, 18, 20
Black Carp	cool	genera	YES	moderate	not benthic	6, 7, 75
Tench	cool	mud	NO	slow	benthic	7, 31
Roho Labeo	warm	genera	NO	moderate	not benthic	6, 7, 36, 51
Sharpbelly	cool	genera	NO	moderate	not benthic	6, 7, 25, 72
Wuchang Bream	cool	mud	NO	slow	not benthic	6, 7, 23, 35, 72
Culter alburnus	cool	mud	YES	slow	not benthic	6, 7, 72, 82
Humpback	cool	sand	NO	slow	not benthic	7, 61, 72
Paramisgurnus dabryanus	cool	mud	NO	slow	benthic	7, 72, 87
Yellow Catfish	cool	mud	NO	moderate	benthic	6, 7, 37, 38, 72
Tachysurus nitidus	cool	genera	NO	moderate	not benthic	6, 7, 72, 86
Ussuri Catfish	cool	sand	NO	slow	benthic	7, 52
Silurus meridionalis	cool	genera	YES	moderate	benthic	6, 7, 48, 71, 72
Tawada	cool	mud	NO	slow	benthic	7, 34
Pond Smelt	cold	genera	NO	slow	not benthic	7, 49, 72
Neosalanx taihuensis	cold	genera	NO	slow	not benthic	1, 7, 9, 11, 13, 72, 80
Rainbow Trout	cold	rubble	YES	fast	not benthic	7, 16, 40, 72
Dolly Varden	cold	rubble	YES	moderate	not benthic	7, 19, 72

Table B.1. Continued.

	Behavior					
Species	Temperature preference	Substrate preference	Fluvial dependence	Velocity preference	Vertical position	References
Lenok	cold	sand	YES	moderate	not benthic	7, 72
Burbot	cold	sand	YES	moderate	not benthic	7, 63, 72
Mosquitofish	cool	genera	NO	slow	not benthic	4, 7, 24, 42
European Perch	cold	sand	NO	moderate	not benthic	7, 45, 72
Ratan	cold	genera	NO	slow	benthic	7, 72
Rhinogobius giurinus	cool	sand	NO	slow	benthic	6, 7, 44, 57, 72
Rhinogobius cliffordpopei	cool	sand	NO	slow	benthic	7, 57, 72
Paradisefish	cool	mud	NO	slow	not benthic	7, 72

	Trophic traits		Life history			
Species	Trophic guild	Diet breadth	Longevity	Age at maturity	Length at maturity	References
Japanese Eel	carnivores	3	16	5	700	6, 17, 72
Spinibarbus denticulatus	herbivores	4	12	5	39.63	6, 18, 55, 56, 20
Black Carp	carnivores	3	32.1	6	895	6, 72
Tench	omnivores	5	20	3	192	1, 6, 72
Roho Labeo	herbivores	3	10	2	434	6, 36, 51
Sharpbelly	omnivores	4	15.5	1	115	6, 25, 50, 72, 75
Wuchang Bream	herbivores	2	6	2	350	6, 41, 54, 72
Culter alburnus	carnivores	4		2	384.2	6, 10, 15, 39, 46, 72, 82
Humpback	carnivores	4	7	1	120	6, 61, 72, 79
Paramisgurnus dabryanus	omnivores	5	6	2	104	6, 58, 60, 72, 87
Yellow Catfish	carnivores	3		1	102.9	6, 37, 38, 64, 72
Tachysurus nitidus	carnivores	3	5	1	90	6, 72, 86
Ussuri Catfish	carnivores	4	11	3	200	6, 43, 52
Silurus meridionalis	carnivores	2	18	4	665	6, 7, 48, 70, 71,72
Tawada	carnivores	6	8	1	400	6, 29, 34
Pond Smelt	carnivores	3	10	1	55	6, 22, 72, 73
Neosalanx taihuensis	carnivores	1	1	1	60.47	6, 9, 33, 72, 80, 81, 84, 85

Table B.1. Continued.

	Trophic traits		Life history			
Species	Trophic guild	Diet breadth	Longevity	Age at maturity	Length at maturity	References
Rainbow Trout	carnivores	4	11	3	300	6, 13, 40, 72
Dolly Varden	carnivores	3	12	3	164	6, 19, 72
Lenok	carnivores	3	20	4	182	6, 72
Burbot	carnivores	4	20	4	300	6, 72
Mosquitofish	carnivores	3	1	0.3	16	4, 6, 42
European Perch	carnivores	3	22	3	144	6, 72
Ratan	carnivores	4	7	2	55	6, 21, 72
Rhinogobius giurinus	omnivores	5	4	1	25	6, 44, 72
Rhinogobius cliffordpopei	omnivores	5	3	1	22	6, 28, 72
Paradisefish	omnivores	4	4	1	32.5	6, 47, 72

	Life history					
Species	Fecundity	Relative fecundity	Egg size	Spawning temperature	Parental care	References
Japanese Eel	1,000,000		1.04	23	1	7, 17, 66
Spinibarbus denticulatus	26,000	11	2.998	26	0	2, 20, 56, 68, 83
Black Carp	1,000,000	52.6	1.6	18	1	2, 7, 72
Tench	271,155	262.7	0.9	22	1	7, 31, 32
Roho Labero			1.03	26	0	7, 36, 51, 67
Sharpbelly	30,293	657.3	0.9	25	0	2, 25, 50
Wuchang Bream	242,673	168	1.1	20	1	2, 23, 54, 72
Culter alburnus	300,000	300	0.9	24	1	2, 10, 14, 82
Humpback	19,907	219	1.4	19	1	2, 72
Paramisgurnus dabryanus	25,686	772.5	0.8	21	1	2, 72, 78, 87
Yellow Catfish	4,472	64.98	1	20	2	2, 37, 38, 64, 72
Tachysurus nitidus	1,536.5	147.38	2.2	23.5	2	62, 72
Ussuri Catfish	4,487		2	20	2	43, 44, 52
Silurus meridionalis	114,929.13	12.16	2.039	17	1	2, 48, 69, 70, 71, 72
Tawada	147,937		1.8	24	0	29, 30, 34

Table B.1. Continued.

	Life history					
Species	Fecundity	Relative fecundity	Egg size	Spawning temperature	Parental care	References
Pond Smelt	4,330	1,240.9	0.97	12	0	7, 22, 72, 73
Neosalanx taihuensis	1,413	1,220	0.63	16	0	9, 11, 72, 81, 84
Rainbow Trout	9,939	3	5.24	6.3	1	7, 16, 40, 72
Dolly Varden	348.5		4.7	7	1	19, 72,
Lenok	9,043.9	2.52	3.79	8	1	7, 72
Burbot	765,657	577	0.73	0.15	1	7, 8, 63, 72
Mosquitofish	150	313.8	1.4	20	3	4, 7, 42
European Perch	24,043	80	1.25	7	1	7, 45
Ratan	1,000	153.8	2.55	18	2	7, 21, 72
Rhinogobius giurinus	5,250	3,358	1.43	24	3	2, 44, 72, 77
Rhinogobius cliffordpopei	770	1,647.17	0.64	24.9	2	28, 72
Paradisefish	9,500		0.8	22.25	2	2, 47, 72

	Life history					
Species	Spawning substrate	Time to hatch	Body length at hatching	Body length at first age	Spawning type	References
Japanese Eel		1.5	3.5	350.3	single	7
Spinibarbus denticulatus	mineral	2	8.93	20.255	fractional	2, 83
Black Carp	vegetation	1.5	6.9	74	single	2, 7, 65
Tench	vegetation	2	3.86	156.8	fractional	1, 7, 31
Roho Labero	various	1			fractional	7, 51
Sharpbelly	mineral	1	2.3	96.9	fractional	27, 50
Wuchang Bream	vegetation	2	3.8	180	fractional	2, 3, 23, 41, 54, 65, 72
Culter alburnus	mineral	1	4.3	171.5	single	2, 5, 12
Humpback	vegetation	1.5	4	129	fractional	2, 72
Paramisgurnus dabryanus	vegetation	1.5	3.6	75	single	2, 72, 87
Yellow Catfish	mineral	2.5	5.2	112.9	single	2, 38, 72
Tachysurus nitidus	mineral	2		95.5	fractional	72
Ussuri Catfish	mineral	3	5.3	110	single	52

Table B.1. Continued.

Species	Life history					References
	Spawning substrate	Time to hatch	Body length at hatching	Body length at first age	Spawning type	
Silurus meridionalis	mineral	2	6.146	338	single	2, 48, 69, 70, 72
Tawada	various	1	3.1	350	fractional	29, 30, 34
Pond Smelt	mineral	12	4.25	74	single	7, 72, 73
Neosalanx taihuensis	mineral	4		62.9	single	9, 72, 84
Rainbow Trout	mineral	52	16	280	single	7, 72
Dolly Varden	mineral	73.5	18.8	83	single	7, 19, 72
Lenok	mineral	26		280	single	72
Burbot	mineral	42	2.9	154	single	7, 63, 72
Mosquitofish	various	20	8	40	continuous	4, 7, 42
European Perch	vegetation	11	4.5	67.5	single	7
Ratan	various	6	5.5		single	72
Rhinogobius giurinus	mineral	4.5	2.51	43.95	single	2, 72
Rhinogobius cliffordpopei	mineral	2.5	3.61	40	single	28, 57, 72
Paradisefish	pelagic	2	2.235	50	continuous	2, 7, 47

References

[1] Altindağ, A., S. Yİğİt, S. Ahiska, and Ş. Zkurt. 1998. The growth features of Tench (*Tinca tinca* L., 1758) in the Kesikköprü Dam Lake. Turkish Journal of Zoology 22:311-318.

[2] Cao, W., J. Chang, Y. Qiao, and Z. Duan. 2007. Fish resources of early life history stages in Yangtze River. China Water Power Press, Beijing. (In Chinese.)

[3] Chen, D., S. Liu, X. Duan, and F. Xiong. 2002. A preliminary study of the fisheries biology of main commercial fishes in the middle and upper reaches of the Yangtze River. Acta Hydrobiologica Sinica 26:618-622. (In Chinese.)

[4] Chen, G. 2010. Interspecific relationship between the invasive species *Gambusia affinis* and the native endangered species *Tanichthys albonubes*. Doctoral dissertation. Graduate School of the Chinese Academy of Sciences, Beijing. (In Chinese.)

[5] Chen, P. 1959. An ecological study of *Erythroculter dabryi* in Lake Liang-Tze. Acta Hydrobiologica Sinica 4:403-410. (In Chinese.)

[6] Chen, Y. 1998. Fish in Hengduan Mountains. Science Press, Beijing. (In Chinese.)

[7] Froese, R., and D. Pauly, editors. 2010. FishBase. www.fishbase.org.

[8] Gao, X., G. Xiao, L. Chen, and L. Zhang. 2012. Study on embryonic development of *Lota lota*. Freshwater Fisheries 41:85-90. (In Chinese.)

[9] Gong, S., H. Xu, X. Zhang, C. Liu, X. He, and J. Liu. 1999. Study on the reproductive biology of *Neosalanx taihuensis* in Fushui Reservoir. Journal of Huazhong Agricultural University 18:173-175. (In Chinese.)

[10] Gong, S., Y. Chen, and C. Chen. 1990. Study on growth and population control of *Erythroculter ilishaeformis* in Bacha Reservoir. Reservoir Fisheries 1:23-26. (In Chinese.)

[11] Gong, S., Z. Luo, J. Liu, X. He, and X. Zhang. 1999. Study on the individual fecundity of *Neosalanx taihuensis* in Fushui Reservoir. Reservoir Fisheries 19:36-38. (In Chinese.)

[12] Gu, Z., J. Zhu, Y. Jia, Y. Pan, X. Huang, G. Xu, and Y. Yang.2008. Research on embryonic and post-embryonic development of *Erythroculter ilishaeformis* Bleeker of Taihu Lake. Journal of Fishery Sciences of China 15:204-214. (In Chinese.)

[13] Guo, Y., L. Cai, and Y. Ma. 2007. Preliminary study on the growth and developing of Rainbow Trout (*Oncorhynchus mykiss*) in the Lake of Sailimu. Chinese Journal of Fisheries 20:1-6. (In Chinese.)

[14] Han, Y., X. Wang, and H. Yin. 2005. Growth pattern of Topmouth Culter (*Erythroculter ilishaeformis*) in Xingkai Lake. Journal of Dalian Fisheries University 20:218-221. (In Chinese.)

[15] Hu, Q., and S. Gong. 2002. Study on age and growth of *Erythroculter ilishaeformis* in Wu Lake. Reservoir Fisheries 20:46-47. (In Chinese.)

[16] Huang J, Z. Fan, Z. Jia, Y. Li, and G. Xu. 2005. The observation method of sinkable and large diameter egg and the embryonic development of *Oncorhyncus mykiss*. Journal of Economic Animal 9:235-238. (In Chinese.)

[17] Huang, D., and C. Zheng. 1997. Exploring on the life cycle and artificial breeding of eel (*Anguilla*). Progress in Biotechnology 17:3-12. (In Chinese.)

[18] Huang, L. 2009. Comparison on the main economic characters with three breeds of fingerlings of *Spinibarbus oshima*. Journal of Fujian Fisheries 2:44-47. (In Chinese.)

[19] Huang, Q., D. Zhang, J. Zhou, and C. Liu.1999. Embryonic development time and absolute fecundity of Dolly Varden charr (*S. malma*) in Yalu River. Journal of Jilin Agricultural University 21:68-69. (In Chinese.)

[20] Jiang, L., W. Liang, G. Huang, P. Feng, and Q. Deng. 2006. Study on the artificial propagation of *Spinibarbus denticulatus*. Southwest China Journal of Agricultural Sciences 19:1172-1175. (In Chinese.)

[21] Jin, G., Y. Liu, and P. Yang. 2010. Preliminary study on the artificial reproduction of *Perccottus glehni*. Modern Agricultural Sciences and Technology 15:351-359. (In Chinese.)

[22] Li, B., and Y. Xie. 1992. Effects of temperature, salinity and alkalinity on embryo development of pond smelt. Chinese Journal of Ecology 11:18-21. (In Chinese.)

[23] Li, C., Y. Li, C. Peng, Y. Peng, and P. Yuan. 1992. Biological studies of *Megalobrama amblycephalayih* at the Jiangkou Reservoir and the development utilization of this fish resource. Jiangxi Science 2:108-114. (In Chinese.)

[24] Li, J., X. Li, X. Tan, Y. Li, M. He, J. Luo, J. Lin, and S. Su. 2009. Species diversity of fish community

of Provincial Xijiang River Rare Fishes Natural Reserve in Zhaoqing City, Guangdong Province. Journal of Lake Sciences 21:556–562. (In Chinese.)

[25] Li, Q., J. Zhao, L. Zhong, and Z. Lan. 2009. The study of growth of *Hemiculter leucisculus* in the Beijiang River, Guangdong. Journal of Guangzhou University (Natural Science Edition) 8:38–41. (In Chinese.)

[26] Li, Q., W. Li, and J. Zhao. 2009. Histological studies on the gonadial development of *Hemiculter leucisculus*. Journal of Guangzhou University (Natural Science Edition) 8:63–67. (In Chinese.)

[27] Li, Q., Z. Lan, J. Zhao, L. Chen, and L. Zhong. 2011. Study on life-history pattern of *Hemiculter leucisculus* in the Beijiang River. Sichuan Journal of Zoology 30:348–352. (In Chinese.)

[28] Li, Y., and X. Zhang. 1990. Biological study on the *Ctenogobius cliffordpopei* in mountainous areas of Ji County. Acta Scientiarum Naturalium University Nankaiensis 4:68–74. (In Chinese.)

[29] Lin, G., and J. Xiong. 1995. Studies on the differentiation and development in the first sex-cycle of the ovary of *Clarias lazera*. Zoological Research 16:365–372. (In Chinese.)

[30] Lin, G., and L. Hua. 1999. Study on the early embryonic development of *Clarias lazera*. Journal of Nanchang University (Natural Science Edition) 23:333 338. (In Chinese.)

[31] Ling, Q., and H. Zhang. 2003. The study on the embryonic development and feeding of the yolk-sac larvae of *Tinca tinca*. Journal of Fisheries of China 27:43–48. (In Chinese.)

[32] Ling, Q., S. Li, X. Liang, J. Yin, and X. Liu. 2006. Observations of annual histological changes in ovary of *Tinca tinca*. Journal of Shanghai Fisheries University 15:36–40. (In Chinese.)

[33] Liu, E., C. Bao, L. Wu, and P. Cao. 2006. Preliminary analysis on the diet composition of *Neosalanx taihuensis* and inter-specific relationship with other fish. Reservoir Fisheries 26:41–44. (In Chinese.)

[34] Liu, G., and X. Zhang. 2006. Effective techniques of breeding *Clarias leather*. Hebei Fisheries 2:33–44. (In Chinese.)

[35] Liu, J., and B. He. 1992. Cultivation of freshwater fishes in China. Science Press, Beijing. (In Chinese.)

[36] Liu, J., Z. Luo, G. Xie, L. Lin, and H. Zhong. 1982. Artificial reproduction and embryonic development of *Labeo rohita*. Freshwater Fisheries 3:6–9. (In Chinese.)

[37] Liu, S. 1997. A study on the biology of *Pseudobagrus fulvidraco* in Poyang Lake. Chinese Journal of Zoology 32:10–16. (In Chinese.)

[38] Liu, W. 2004. Studies on reproduction physiology and some biological characteristics of *Pelteobagrus fulvidraco*. Master's thesis. Hunan Normal University, Changsha, China. (In Chinese.)

[39] Lu, F. 1995. Growth of *Erythroculter mongolicus* and *Erythroculter ilishaeformis* in Laojiang River. Reservoir Fisheries 75:29–33. (In Chinese.)

[40] Ma, Z., and G. Zhu. 2009. Biology and breeding technology of *Oncorhynchus mykiss*. Heilongjiang Aquatic Products 4:23–28. (In Chinese.)

[41] Ouyang, M., and D. Chen. 2001. Study on biology of *Megalobrama amblycephala* in Poyang Lake. Acta Agriculturae Jiangxi 13:47–50. (In Chinese.)

[42] Pan, J., B. Su., and W. Zheng. 1980. Biological characteristics of *Gambusia affinis* and the prospects for its use for mosquito control. Journal of South China Normal University (Natural Science Edition) 1:117–137. (In Chinese.)

[43] Pang, H., J. Deng, X. Yu, and H. Wang. 2005. Reproductive biological characters of *Leiocassis ussuriensis* and its artificial breeding technology. Heilongjiang Aquatic Products 110:1–2. (In Chinese.)

[44] Qi, W., Y. Guo, X. Li, and Y. Hao. 2008. Reproductive character, embryonic and larval development of *Ctenogobius giurinus*. Chinese Journal of Zoology 43:13–24. (In Chinese.)

[45] Qiao, D., Q. Ling, J. Yin, Y. Li, X. Cai, and Z. He. 2006. Development of embryo and yolk-sac stage larva of perch (*Perca fluviatilis*). Journal of Biology 23:34–38. (In Chinese.)

[46] Shao, L., and G. Jia. 1990. Age and growth of the whitefish *Erythroculter ilishaeformis*. Journal of Zhejiang College of Fisheries 9:95–102. (In Chinese.)

[47] Shao, W., and L. Yi. 1991. Reproduction and growth of *Macropod chinensis* and its predacity on mosquito larvae. Oceanologia et Limnologia Sinica 22:181–186. (In Chinese.)

[48] Shi, B. 1982. Biological study of *Silurus soldatovi meridionalis* Chen in Jialing River. Sichuan Journal of Zoology 1:45–52. (In Chinese.)

[49] Sui, X., Z. Lu, Y. Shen, H. Li, and Y. Cao. 2007. Study on the embryonic development of *Hypomesus olidus* in Tagang Reservoir. Reservoir Fisheries 27:79–81. (In Chinese.)

[50] Sun, Z. 1987. Biological study on *Hemiculter leucisculus*. Journal of Jilin Agricultural University 9:66–69. (In Chinese.)

[51] Tan, H. 1983. Observation of main features, habitats and diet of *Labeo rohita*. Fisheries Science and Technology of Guangxi 2:16–17. (In Chinese.)

[52] Tao, G., Z. Ma, Q. Zhang, C. Zhang, and Q. Wan. 2012. Preliminary study on the artificial reproduction technique of *Pseudobagrus ussuriensis*. Fisheries Science and Technology Information 39:174–178. (In Chinese.)

[53] Tao, J., X. Dai, S. Tian, and C. Ma. 2011. Community diversity and growth characteristics of wild fishes in Dianshan Lake. Hunan Agricultural Sciences 4:137–141. (In Chinese.)

[54] Tsao, W. 1960. A biological study of *Magalobrama amblycephala* and *M. terminalis* of Liang-Tse Lake. Acta Hydrobiologica Sinica 1:57–82. (In Chinese.)

[55] Wang, C., Z. Yi, X. Lin, and X. Chen. 2007. Analysis of food habits of *Spinibarbus denticulatus* in its life history. Chinese Journal of Zoology 42:101–107. (In Chinese.)

[56] Wang, G., S. Pang, J. Xie, and D. Yu. 2006. The growth characteristics of *Spinibarbus denticulatus denticulatus* in Pearl River system. Chinese Journal of Zoology 41:66–71. (In Chinese.)

[57] Wang, H., and Y. Guo. 2009. Embryonic development of *Rhinogobius cliffordpopei*. Sichuan Journal of Zoology 28:184–188. (In Chinese.)

[58] Wang, M., W. Wang, and J. Yan. 2001. Comparisons of age and growth between *Misgurnus anguillicaudatus* and *Paramisgurnus dabryanus* Sauvage. Reservoir Fisheries 21:7–9. (In Chinese.)

[59] Wang, T., H. Wang, G. Sun, D. Huang, and J. Shen. 2012. Length-weight and length-length relationships for some Yangtze River fishes in Tian-e-zhou Oxbow, China. Journal of Applied Ichthyology 28:660–662.

[60] Wang, Y., and D. Li. 2005. Preliminary study on the ecological characters of *Paramisgurnus dabryanus* Sauvage in Nansi Lake. Reservoir Fisheries 25:9–10. (In Chinese.)

[61] Wang, Y., B. Xiong, X. Ma, M. Xu, L. Zhang, X. Wang, and Q. Hu. 2007. Fecundity of *Culter dabryi* in Daoguanhe Reservoir, Wuhan, China. Oceanologia et Limnologia Sinica 38:180–186. (In Chinese.)

[62] Wei, G., L. Huang, Y. Chen, L. Lu, and G. Jiang. 2002. Observation on the embryonic development of *Pseudobagrus nitidus*. Journal of Southwest China Normal University (Natural Science Edition) 27:567–571.

[63] Xiao, G., Z. Zhao, X. Gao, Y. Jiang, Y. Tian, L. Chen, X. Han, and L. Zhang. 2011. The morphological characteristics of most embryonic development in freshwater cod *Lota lota*. Chinese Journal of Fisheries 24:34–40. (In Chinese.)

[64] Xiao, T., H. Zhang, X. Wang, K. Xiao, and Z. Dai. 2003. Biological characteristics of *Pelteobagrus fulvidracoin* in Dongting Lake. Chinese Journal of Zoology 38:83–88. (In Chinese.)

[65] Xie, C., and W. Cheng. 1995. Growth of three fishes in Wang Lake. Reservoir Fisheries 4:13–15. (In Chinese.)

[66] Xie, G., and B. Qi. 1995. Relationship between embryonic development of the eel (*Anguilla japonica*) and water temperature and salinity. Journal of Fishery Sciences of China 2:1–7. (In Chinese.)

[67] Xie, G., J. Liu, L. Lin, and H. Zhong. 2005. Study on the relationship between embryo development of *Labco rohita* and water temperature. Freshwater Fisheries 2:13–16. (In Chinese.)

[68] Xie, G., K. Chen, Y. Hu, G. Deng, and C. Li. 2003. Relationship between embryonic development of *Spinibarbus denticulatus* and water temperature and salinity. Journal of Dalian Fisheries University 18:95–98. (In Chinese.)

[69] Xie, X. 1986. On the embryonic development of *Silurus soldatovi meridionalis*. Journal of Southwest China Normal University (Natural Science) 3:72–78. (In Chinese.)

[70] Xie, X. 1989. On the development of larva of *Silurus soldatovi meridionalis* Chen. Acta Hydrobiologica Sinica 13:124–133. (In Chinese.)

[71] Xie, X., X. He, and T. Long. 1996. Reproductive biology of *Silurus meridionalis*: time, environmental conditions and behaviour of spawning. Acta Hydrobiologica Sinica 20:17–24. (In Chinese.)

[72] Xie, Y. 2007. The freshwater fishes of northeastern China. Liaoning Science & Technology Press, Shengyang, China. (In Chinese.)

[73] Xie, Y., and X. Pu. 1984. The biological aspects of Pond Smelt (*Hypomesus olidus* (pallas)) in the Shuifeng Reservoir. Acta Hydrobiologica Sinica 8:457–468. (In Chinese.)

[74] Xie, Z., H. Guo, W. Tang, K. Wei, L. Shen, J. Wu, and W. Chen. 2010. Age structure and growth characteristics at seaward migrating periods of *Anguilla japonica* in the Yangtze River estuary. Journal of Fisheries of China 34:245–254. (In Chinese.)

[75] Xie, Z., X. Wu, L. Zhuang, and D. Li. 1986. Investigations of the biology of *Hemiculter leucisculus* in Fenhe Reservoir. Periodical of Ocean University of China 16:54–69. (In Chinese.)

[76] Xu, Z., H. Shi, and J. Ye. 1983. Total length and weight relationship of one-year-old Black Carp (*Mylopharyngodon piceus*). Freshwater Fisheries 6:18–19. (In Chinese.)

[77] Yan, Y., and Y. Chen. 2007. Plasticity in reproductive tactics of *Ctenogobius giurinus* in Lake Fuxian. Acta Hydrobiologica Sinica 31:414–418. (In Chinese.)

[78] Yang, F. 2009. Histological studies on the ovarian development of *Paramisgurnus dabryanus*. Journal of Hydroecology 2:78–81. (In Chinese.)

[79] Yang, J., X. Zhang, and L. Sun. 2009. A study on the growth characters of *Culter dabryi* in Poyang Lake, Jiangxi. Acta Agriculturae Universitatis Jiangxiensis 31:336–340. (In Chinese.)

[80] Yang, Q., E. Liu, C. Bao, C. Wu, S. Wang, H. Li, and X. Song. 2009. Analysis of the component of the food for *Neosalanx tangkahkeii taihuensis* in Chaohu Lake. Journal of Anhui Agricultural Sciences 36:15922–15924. (In Chinese.)

[81] Yang, Z., Z. Li, J. Liu, T. Zhang, S. Ye, and H. Zhang. 2012. A comparative study on reproductive characteristics of different spawning stocks of the icefish (*Neosalanx taihuensis*) in the Danjiangkou Reservoir. Freshwater Fisheries 42:58–62. (In Chinese.)

[82] Yao, W., J. Hu, and X. Wu. 1987. The propagation of whitefish (*Erythroculter ilishaeformis*) in Chaohu Lake. Journal of Fisheries of China 12:101–109. (In Chinese.)

[83] Yi, Z., X. Chen, C. Wang, and X. Lin. 2004. Embryonic development in *Spinibarbus denticulatus denticulatus*. Journal of Fishery Sciences of China 11:65–69. (In Chinese.)

[84] Yin, G., K. Cao, Z. Yu, and S. Gong. 1997. The reproduction, feeding habits and growth of salangid fishes *Neosalanx taihuensis* and *N. oligodontis* in Wanghu Lake. Journal of Lake Sciences 9:63–70. (In Chinese.)

[85] You, Y., Q. You, Z. Guo, and X. Zhang. 1999. Comparison of feeding habit and evaluation of feeding biological capacity at different latitudes of *Neosalanx taihuensis*. Journal of Southwest University for Nationalities (Natural Science Edition) 25:176–180. (In Chinese.)

[86] Yuan, G., H. Ru, and X. Liu. 2011. Feeding habits of *Pelteobagrus nitidus* in Lake Dongting. Acta Hydrobiologica Sinica 35:270–275. (In Chinese.)

[87] Zhu, S., Q. Wan, F. Li, X. Gan, H. Zhang, X. Duan and B. Cao. 2007. Fecundity measurement and artificial propagation of *Paramisgurnus dabryanus*. Modern Agricultural Science and Technology 13:171–172. (In Chinese.)

American Fisheries Society Symposium 84:169–177, 2016

Current Fish Resources in the Upper Reach of the Yangtze River and Conservation Strategies

Xin Gao and Huanzhang Liu*
Institute of Hydrobiology, Chinese Academy of Sciences
No. 7 South Donghu Road, Wuhan, Hubei, 430072, P. R. China

Abstract.—In this paper, we report fish resources in the upper reach of the Yangtze River from 2011 to 2012. The five sampling locations were Yibin, Hejiang, Mudong, Wanzhou, and Zigui. Yibin and Hejiang are the most upstream locations; Mudong is at the upper end of the Three Gorges Reservoir while Wanzhou and Zigui are in the inundated areas of the Three Gorges Reservoir. During our investigations, we collected 56, 94, 93, 67, and 64 fish species, respectively, from these five locations. Among them, 13, 21, 17, 4, and 4 are endemic to the upper Yangtze River. Catch per unit effort calculated with pooled data were 1,221.6–6,433.9 g/(d · boat). Because Yibin and Hejiang are still in the riverine state, endemic species remain. These areas are also in the national protected areas range. We suggest that commercial fishing should be banned in these areas to protect endemic fish species. However, for Wanzhou and Zigui, because they are in the inundated reservoir area, dominance has shifted from riverine to lacustrine species. Fisheries strategies should also be changed to target on different species, employ different fishing gears, and apply different management strategies.

Introduction

The Yangtze River is the third longest river in the world (about 6,300 km), and it ranks third in discharge ($9.513 \times 10^{9}11 m^3$/year) and 12th in watershed area (1.8×10^6 km^2). It flows eastwards from the glaciers on the Qinghai–Tibet Plateau, crossing 11 provinces and cities, and eventually enters the East China Sea at the city of Shanghai. The fall of the Yangtze River between the headwater and estuary is approximately 5,000 m.

The Yangtze River is rich in aquatic biodiversity and supports a highly diverse fish fauna, with about 400 fish species and subspecies, among which 147 are endemic. Eight fish species have legally been classified as Chinese national protected animals; Chinese Sturgeon *Acipenser sinensis*, Chinese Paddlefish *Psephurus gladius*, and Yangtze Sturgeon *A. dabryanus* have been listed as critically endangered species by the International Union for Conservation of Nature and Natural Resources Red List of Threatened Species (www.iucnredlist.org). The upper reaches of the Yangtze River, from the headwaters to Yichang City in Hubei Province, wind through the hills and exhibit complex geological environments, characterized by habitat heterogeneity and rich biodiversity. Distributed in the upper Yangtze River are 286 fish species and subspecies, among which 124 species are endemic (He et al. 2011) and account for 84.3% of total number of endemic fish species in the Yangtze River.

More than 50,000 dams have been constructed in the Yangtze River watershed since 1950 (Xu and Milliman 2009). These dams induce substantial adverse consequences for the riverine ecosystem and biodiversity, despite providing services such as

* Corresponding author: hzliu@ihb.ac.cn

power generation, flood control, increasing navigation capability, irrigation, and water resources supplies (New and Xie 2008). The Three Gorges Dam (TGD) at Yichang City in the middle reach of the Yangtze River is one of the largest hydroelectric projects in the world. Construction of the TGD began in 1993 and was completed in 2009. The TGD has an elevation of 185 m above sea level (ASL) at the top. It was filled to 145 m ASL from 2003 and further filled to 175 m ASL in 2009, inundating a reach of approximately 600 km between Chongqing City and Yichang City, forming the well-known Three Gorges Reservoir (TGR). According to the reservoir operation plan, the TGR is to maintain a low water elevation of 145 m ASL in May to prevent flooding and to store water up to 175 m ASL for navigation and power generation during October and December.

The TGD significantly influenced the fish resources of the Yangtze River (Wu et al. 2004; Li et al. 2013). The impoundment of the TGR could decrease the population of the endemic fishes in the upper Yangtze River by inundating the habitat and spawning area of those fishes. Fish assemblages could be shifted from lotic to lentic fish in the area of TGR. The four major Chinese carps (Black Carp *Mylopharyngodon piceus*, Grass Carp *Ctenopharyngodon idella*, Silver Carp *Hypophthalmichthys molitrix*, and Bighead Carp *H. nobilis* [also known as *Aristichthys nobilis*]) are economically important fish species. Yi et al. (1988) found 36 major spawning grounds for the four major Chinese carps in the Yangtze River, of which eight were above TGD and could have been inundated by the impoundment.

For conservation interest of fish resources in the upper Yangtze River, understanding the current fish resources is the primary measure and the first priority. However, data of the current fish resources are not yet thoroughly and statistically analyzed and presented, which hampers the development of effective conservation and management. In this study, we present the data collected in 2011 and 2012. Our aim was to describe the fish resource status in the upper Yangtze River and provide conservation suggestions.

Methods

Study Area

In view of modern geography, the source of the Yangtze River, named as the Tuotuo River, flows out from snow-capped Geladandong Mountain in the Tanggula Range, southwestern Qinhai Province. After passing Dangqu, it is known as the Tongtian River. From the Batang joint in Yushu County to Yibin City, Sichuan Province, it is called the Jinsha River. After the confluence of the Jinsha and Min rivers in Yibin City, it is called the Yangtze River. Traditionally, "upper Yangtze River" is used to describe the river between the cities of Yibin and Yichang. In this study, we sampled fish in the upper Yangtze River from 2011 to 2012. The five sampling localities were Yibin, Hejiang, Mudong, Wanzhou, and Zigui. Among them, Yibin and Hejiang are the most upstream, Mudong is at the upper end of the TGR, and Wanzhou and Zigui are inundated by the TGR.

Fish Sampling

We conducted fish samples at three types of habitat in the main channel of the upper Yangtze River: riverine, transitional, and lacustrine, based on different flow characters. For these five sites, the Zigui and Wanzhou reaches, 5 m and 286 km upstream of the TGD, respectively, have been inundated in the lacustrine pool since 2003.The Mudong reach, 578 km upstream of the TGD, is located in the transitional zone. The Hejiang and Yibin reaches, 760 and 962 km from the TGD, respectively, are still in a riverine state at 175 m ASL (Figure 1). Fish were collected from fishing boats in the five sites from May to June 2011 and from September to October 2012. Sampling took place for about 15–20 d at each site.

We employed 5–10 fishing boats at each site. Fish were collected with the local fishing boats using 1.5-cm-mesh nylon trap nets or 3–6-cm-mesh drift gill nets (100–200 m long × 1.0–2.3 m high) from approximately 0700 to 0800 hours every field day. Normally, in the lacustrine zone of the TGR, trap nets are the

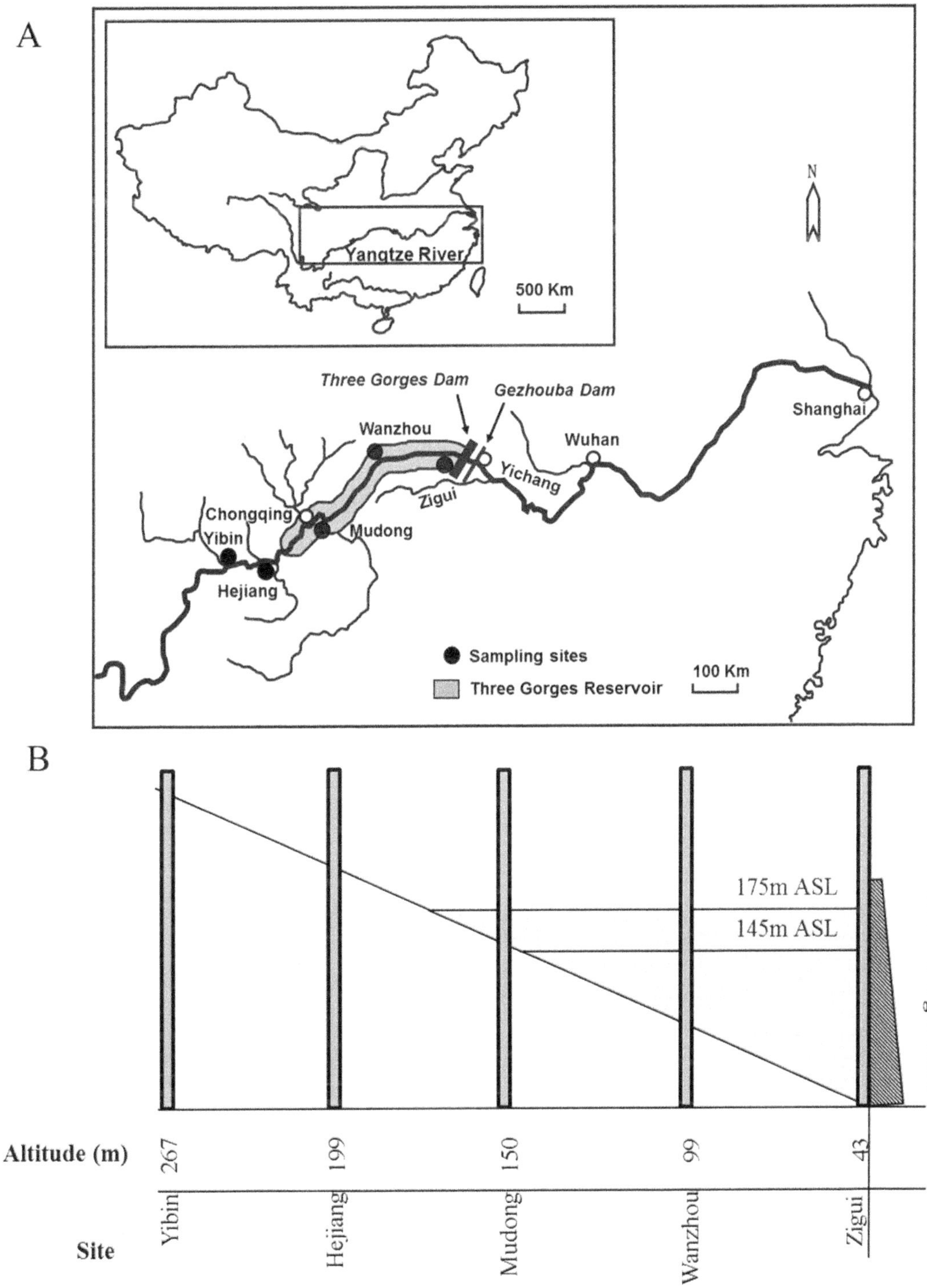

Figure 1. (A) Map shows the sampling sites in the upper Yangtze River; (B) altitudes of the sampling sites and the flood control limited water level is 145 m above sea level (ASL) and the normal pool level is 175 m ASL.

main fishing gears and one fishing boat harvested the nets once daily. Fishermen set the trap nets in scattered locations in Wanzhou and Zigui reaches at 0600 hours and then collected all the fishes at 0600 hours the next day. In the riverine and transitional zones, like the Mudong, Hejiang, and Yibin reaches, fishermen preferred using drift gill nets in the range of about a 20–30-km reach, and the fishing generally lasted about 12 hours each day. In this study, catch per unit effort was expressed as g/(d · boat). Fish were identified, counted, and weighed to the nearest gram. The top five species in biomass were considered the dominant species each sampling site.

Data Analysis

The data collected from the five sampling sites were used to estimate the index of relative importance (IRI%) to identify proportion of the endemic fishes to total catches.

$$\mathrm{IRI}\% = \left(\frac{(W_i\% + N_i\%) \times F_i\%}{\sum_{i=1}^{n} (W_i\% + N_i\%) \times F_i\%} \right) \times 100, \tag{1}$$

where $W_i\%$ and $N_i\%$ are percentage weight and number of ith species of total catches and $F_i\%$ is percentage frequencies of occurrence of ith species in the total number of samples (Jutagate et al. 2005).

The data collected from the five sampling sites were analyzed to determine the spatial distribution of the current fish resources in the upper Yangtze River. The cluster analysis and nonmetric multidimensional scaling ordination analysis, based on the Bray-Curtis similarity matrix (Clarke 1993), were used to estimate the similarity of the fish assemblages among these five sampling sites. The data obtained in the same year were pooled in the analyses. Relative biomass data (percentage of total biomass over all species) for each species during each site was used to calculate the Bray-Curtis similarity coefficient. We performed cluster and nondimensional scaling analyses with the PRIMER 5 software package (Clarke and Warwick 1994).

Results

During the investigation, 88,360 individual fish, weighing 4,148 kg, were collected in the upper Yangtze River. They belonged to 132 species from 80 genera and 24 families. Of the 132 species, 30 species were endemic to the upper reaches of the Yangtze River. In terms of the number of species collected, Cyprinidae was the dominant family (53.8%), followed by Cobitidae (12.1%) and Bagridae (7.6%).

We collected 56, 94, 93, 67, and 64 fish species, respectively, from the Yibin, Hejiang, Mudong, Wanzhou, and Zigui reaches (Figure 2). From the results, Longfin Grugeon *Rhinogobio ventralis*, Elongate Loach *Leptobotia elongata*, Largemouth Bronze Gudgeon *Coreius guichenoti*, Darkbarbel Catfish *Pelteobagrus vachelli*, and Bronze Gudgeon *Coreius heterodon* were the dominant fishes in biomass in Yibin reach. Longsnout Catfish *Leiocassis longirostris*, Largemouth Bronze Gudgeon, Darkbarbel Catfish, Silver Carp, and Common Carp *Cyprinus carpio* were the dominant fishes in biomass in Hejiang reach. Bronze Gudgeon, Darkbarbel Catfish, Largemouth Bronze Gudgeon, Roundbody Gudgeon *Rhinogobio cylindricus*, and Common Carp dominated in the biomass of the catches in the Mudong reach. Longnose Gudgeon *Saurogobio dabryi*, Southern Catfish *Silurus meridionalis*, Silver Carp, Common Carp, and *Hemiculter bleekeri* accounted for 55.0% of the biomass of the catches in abundance in Wanzhou reach. Bighead Carp, Silver Carp, Common Carp, Goldfish *Carassius auratus*, and *H. bleekeri* made up 65.9% of the biomass in Zigui reach (Table 1). Catch per unit effort calculated with pooled data were 1,221.6 g/(d · boat) in the Yibin reach., 6,433.9 g/(d · boat) in the Heijiang reach, 3,113.2 g/(d · boat) in the Mudong reach, 2,631.3 g/(d · boat) in the Wanzhou reach (four endemic species collected), and 6,196.4 g/(d · boat) in the Zigui reach (four endemic species collected; Figure 2). In the Yibin reach, 13 species endemic to the upper Yangtze River were collected, with an index of relative importance (IRI) of the endemic fish of 35.98%. In the Heijiang reach, 21 endemic species were collected (IRI = 20.24%). In the Mudong reach, 17 endemic species

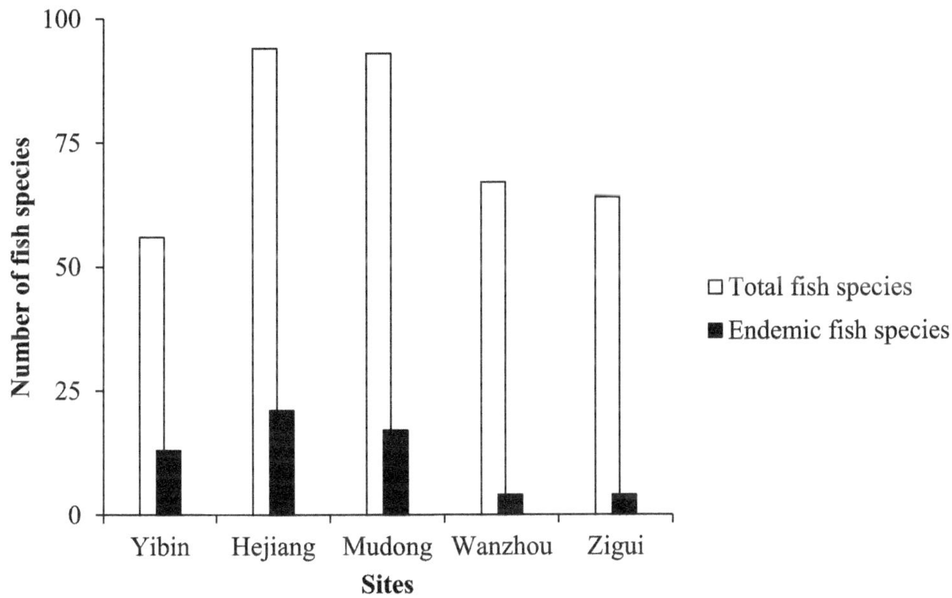

Figure 2. Number of total fish species and endemic fish species collected in the upper reaches of the Yangtze River.

were collected (IRI = 24.59%). In the more lacustrine reaches, the number of endemic species collected was much lower; with only four endemic species each in the Wanzhou (IRI = 0.73%) and Zigui (IRI = 0.01%) reaches (Figures 3 and 4).

The results of the cluster analysis revealed that the fish assemblages fell into three groups. Group 1 included the Yibin and Hejiang reaches. Group 2 consisted only of the Mudong reach. Group 3 was composed of the Wanzhou and Zigui reaches. The average similarity between groups 1 and 2 was 39.1%, and it was 26.7% between groups 2 and 3 (Figure 5A). The nonmetric multidimensional scaling ordination analysis resulted in the evident differences in the species compositions among these five sampling sites. The ordinations were strong with a low stress value of zero (Figure 5B).

Discussion

Current Fish Resources in the Upper Yangtze River

Our investigation showed significant differences in fish species composition between the riverine reaches and the inundated area of the TGR in 2011 and 2012. This indicates that impoundment of the TGR has broken the river continuum of the upper Yangtze River. In the riverine reaches, like the Yibin and Hejiang reaches, the dominant fishes were within the lotic group, such as Longfin Grugeon, Elongate Loach, Roundbody Gudgeon, Largemouth Bronze Gudgeon, and Bronze Gudgeon. With 13 endemic species found in the Yibin reach and 21 in the Hejiang reach, their abundance accounted for a large proportion of the total catches.

At the upper end of the TGR in the Mudong reach, despite the varied environment due to TGR regulation, some investigations suggested that the composition of the fish assemblages tended to have certain persistence and the dominant group mainly consisted of riverine species (Yang et al. 2012). Our results also showed that many lotic and endemic fish species still remain in the Mudong reach. However, the Wanzhou and Zigui reaches have been in lacustrine since 2003, leading to the shift in dominance from riverine to lacustrine species. In these inundated areas of the TGR, our results indicated that

Table 1. Weight contribution (%) of the dominant fishes in the catch in the five sampling sites of the upper Yangtze River.

Species	Yibin	Hejiang	Mudong	Wanzhou	Zigui	Endemic	Habitat
Longsnout Catfish		5.1					Lotic
Longfin Gudgeon	5.4					*	Lotic
Elongate Loach	11.0					*	Lotic
Roundbody Gudgeon			24.1			*	Lotic
Largemouth Bronze Gudgeon	20.9	9.5	8.9			*	Lotic
Bighead Carp					17.0		General
Darkbarbel Catfish	18.1	26.2	4.1				Lentic
Bronze Gudgeon	10.4		24.3				Lotic
Longnose Gudgeon				8.4			General
Southern Catfish				7.9			Lentic
Silver Carp		15.7		11.8	11.0		General
Common Carp		5.6	7.8	10.9	23.5		General
Goldfish					8.0		General
Hemiculter bleekeri				16.0	6.5		Lentic

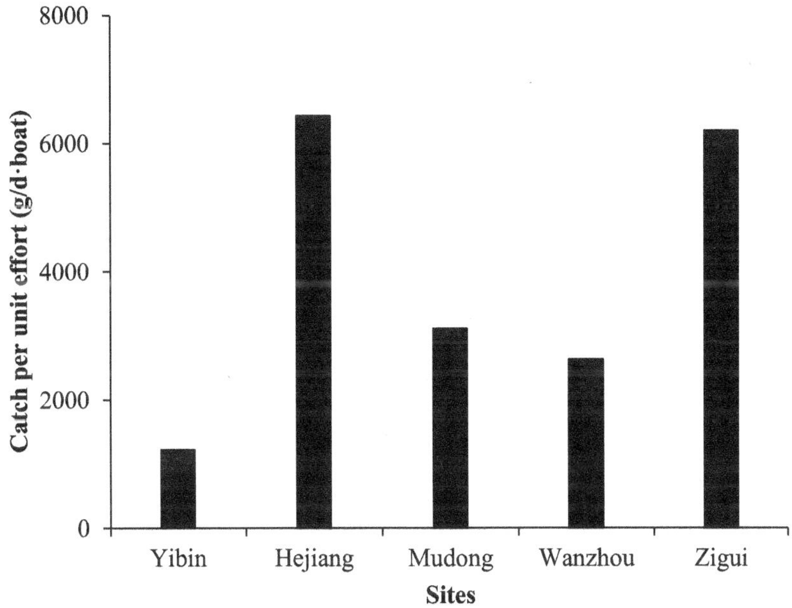

Figure 3. Catch per unit effort in the upper reaches of the Yangtze River.

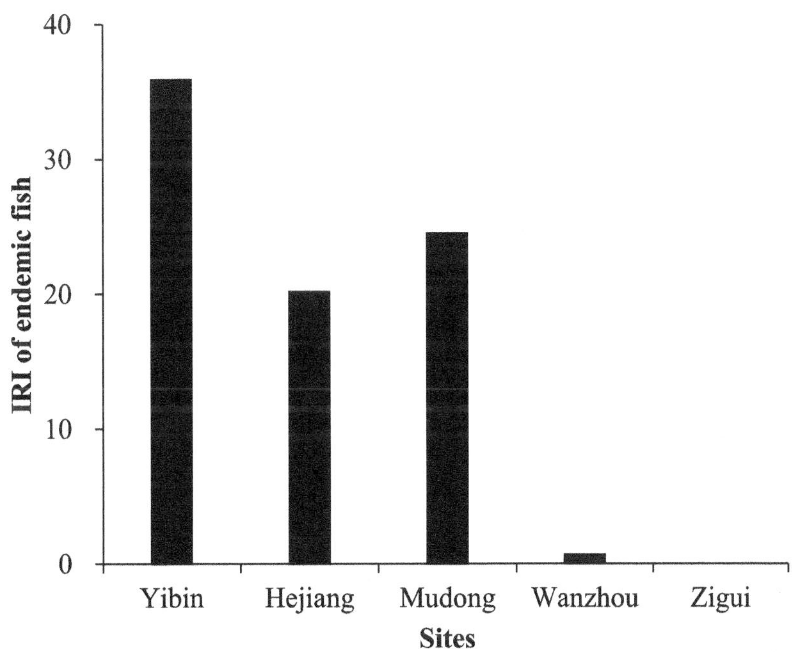

Figure 4. Index of relative importance (IRI) of the endemic fish in the upper reaches of the Yangtze River.

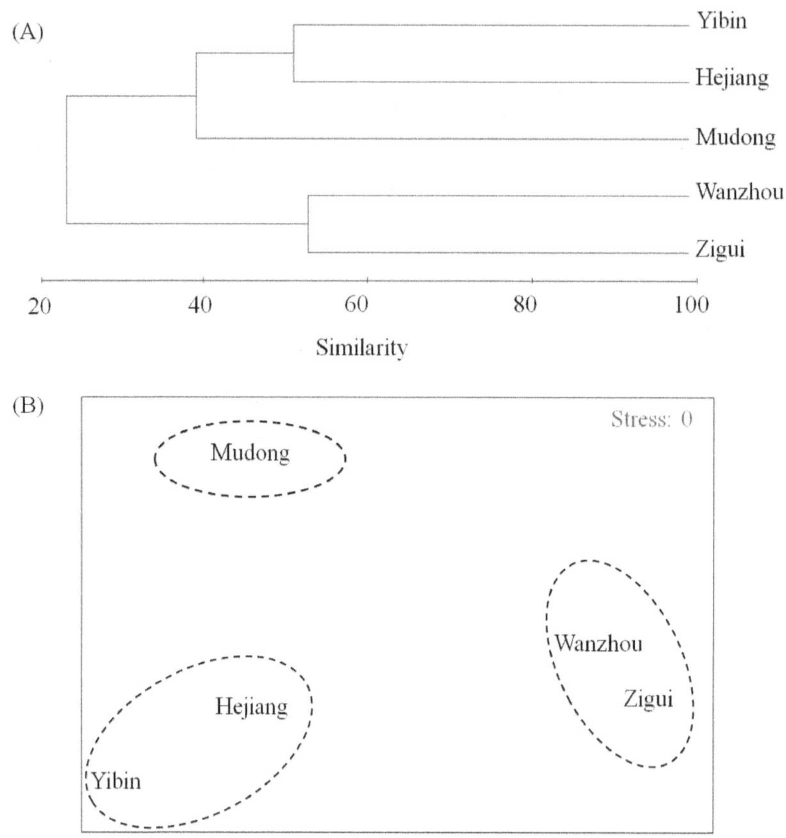

Figure 5. Spatial distribution of the current fish resources in the upper Yangtze River by using the methods of **(A)** cluster analysis and **(B)** nonmetric multidimensional scaling.

the lentic and generalist fishes, such as Bighead Carp, Silver Carp, Common Carp, and *Hemiculter bleekeri* were dominant in 2011 and 2012. In accordance with the previous study (Gao et al. 2010), few endemic fish inhabited in the lacustrine zone of the TGR because they could not adapt to the slow flow and moved upstream to the riverine reaches.

Conservation Measures

To mitigate the impact of the construction of the TGD, the Chinese Ministry of Agriculture approved the establishment of a national natural protected area for rare and endemic fishes of the upper Yangtze River, which extends from Yibin to Chongqing. For conservation interest, we suggest banning commercial fishing in this area. Although the transitional zone of the TGR was not included in the protected area, we suggest banning commercial fishing there as well because many endemic fishes inhabit the Mudong reach, such as Largemouth Bronze Gudgeon and Roundbody Gudgeon. Wanzhou and Zigui reaches in the inundated reservoir provide a habitat for lacustrine species. The fisheries strategies should be changed to fishing for these lacustrine species, employing different fishing gears and management strategies.

References

Clarke, K. R. 1993. Non-parametric multivariate analyses of changes in community structure. Australian Journal of Ecology 18:117–143.

Clarke, K. R., and R. M. Warwick. 1994. Similarity-based testing for community pattern: the two-way layout with no replication. Marine Biology 118:167–176.

Gao X., Y. Zeng, J. Wang, H. Liu.2010. Immediate impacts of the second impoundment on fish communities in the Three Gorges Reservoir. Environmental Biology of Fishes 87:163-173.

He, Y., J. Wang, S. Lek, W. Cao, and S. Lek-Ang. 2011.Structure of endemic fish assemblage in the upper Yangtze River basin. River Research and Applications 27:59-75.

Jutagate, T., C. Krudpan, P. Ngamsnae, T. Lamkom, and K. Payooha. 2005. Changes in the fish catches during a trial opening of sluice gates on a run-of-the-river reservoir in Thailand. Fisheries Management and Ecology 12:57-62.

Li, K., C. Zhu, L. Wu, and L. Huang. 2013. Problems caused by the Three Gorges Dam construction in the Yangtze River basin: a review. Environmental Reviews 21(3):127-135.

New, T., and Z. Xie. 2008. Impacts of large dams on riparian vegetation: applying global experience to the case of China's Three Gorges Dam. Biodiversity and Conservation 17:3149-3163.

Wu, J., J. Huang, X. Han, X. Gao, F. He, M. Jiang, Z. Jiang, R. B. Primack, Z. Shen. 2004.The Three Gorges Dam: an ecological perspective. Frontiers in Ecology and the Environment 2(5):241-248.

Xu, K., and J. D. Milliman. 2009. Seasonal variations of sediment discharge from the Yangtze River before and after impoundment of the Three Gorges Dam. Geomorphology 104(3):276-283.

Yang, S., X. Gao, M. Li, B. Ma, H. Liu. 2012.Interannual variations of the fish assemblage in the transitional zone of the Three Gorges Reservoir: persistence and stability. Environmental Biology of Fishes 93:295-304.

Yi B., Z. Yu, Z. Liang, S. Shen, Y. Xu, J. Chen. 1988. The distribution, natural conditions and breeding production of the spawning groups of four famous freshwater fishes on the main stream on the Yangtze River. Pages 1-46 *in* B. Yi, Z. Yu, and Z. Liang, editors. Gezhouba water control project and four famous fishes in the Yangtze River. Hubei Science and Technology Press, Wuhan, China.

American Fisheries Society Symposium 84:179–191, 2016

Changes in Fisheries Resources in the Hanjiang River and Danjiangkou Reservoir, China

JING YUAN

State Key Laboratory of Freshwater Ecology and Biotechnology, Institute of Hydrobiology Chinese Academy of Sciences, 7 South Donghu Road, Wuhan 430072, P. R. China
and
National Research Center for Freshwater Fishery Engineering 7 Donghu Road, Wuhan 430072, P. R. China

YUGUO XIA

State Key Laboratory of Freshwater Ecology and Biotechnology, Institute of Hydrobiology Chinese Academy of Sciences, 7 South Donghu Road, Wuhan 430072, P. R. China
and
University of Chinese Academy of Sciences 19A Yuquan Road, Shijingshan Distict, Beijing 100049, P. R. China

ZHONGJIE LI

State Key Laboratory of Freshwater Ecology and Biotechnology, Institute of Hydrobiology Chinese Academy of Sciences, 7 South Donghu Road, Wuhan 430072, P. R. China
and
National Research Center for Freshwater Fishery Engineering 7 Donghu Road, Wuhan 430072, P. R. China

ZHAN YIN

State Key Laboratory of Freshwater Ecology and Biotechnology, Institute of Hydrobiology Chinese Academy of Sciences, 7 South Donghu Road, Wuhan 430072, P. R. China

JIASHOU LIU*

State Key Laboratory of Freshwater Ecology and Biotechnology, Institute of Hydrobiology Chinese Academy of Sciences, 7 South Donghu Road, Wuhan 430072, P. R. China
and
National Research Center for Freshwater Fishery Engineering 7 Donghu RoadWuhan 430072, P. R. China

Abstract.—The Hanjiang River is the largest tributary of the Yangtze River and contains Danjiangkou Dam, which forms Danjiangkou Reservoir in the middle and upper reaches of the river. During the past 50 years, fisheries resources have changed significantly in the middle and lower Hanjiang River and in Danjiangkou Reservoir. Spawning grounds for major carps and other commercially important fishes have disappeared. Downstream of Danjiangkou Dam, total egg quantity spawned by major carps and other commercially im-

* Corresponding author: jsliu@ihb.ac.cn

portant fishes have decreased while similar measures from small-bodied fishes have increased. Important commercial fishes have experienced delayed spawning times and decreased growth. Overall catches of commercial fishes have decreased downstream of the dam, though increased upstream. Some exotic fishes captured in the Danjiangkou Reservoir were likely escapees from cage-culture fish farms in the reservoir or from land-based fish farms around the reservoir. Changes in fisheries resources were presumed related to reservoir management strategies, which produced a narrower range of year-round water temperatures and caused decreases in seasonal water flow variation downstream, overfishing, and eutrophication in the reservoir. To implement a national water diversion policy in central China, the height of the Danjiangkou Dam was increased 15 m in 2012, which significantly increased the impoundment area of the Danjiangkou Reservoir at the end of 2014. Further changes in fisheries resources can be expected in the future. At the present time, management and conservation strategies for fisheries resources need to be developed to ensure future fisheries sustainability for both the Hanjiang River and the Danjiangkou Reservoir.

Introduction

The Yangtze River is the longest river in China and contains numerous large tributaries. The Hanjiang River originates in the Qingling Mountains and flows through the Shaanxi and Hubei provinces. The Hanjiang River is the largest tributary of the Yangtze River, having a total length of 1,557 km and a watershed area of 159,000 km^2 (Du et al. 2005). Historically, the middle and lower reaches of the Hanjiang River contained fish faunas similar to the middle and lower reaches of the Yangtze River (HPIH 1976; Yu et al. 1981). In 1959, the middle reach of the Hanjiang River was dammed at Danjiangkou, just downstream of the confluence with the Danjiang River. When the hydroelectric project was completed in 1973, the dam height was 162 m above sea level and the impounded area was called Danjiangkou Reservoir. This reservoir, with an original surface area of 846 km^2 and storage volume of 1.745×10^{10} m^3, is now the second largest reservoir in China following the Three Gorges Reservoir, which was impounded at a height of 175 m in 2008. Since construction of Danjiangkou Reservoir, fisheries resources in the middle and lower reaches of the Hanjiang River have changed significantly. These fisheries changes have included the species composition of fish catches, species abundances, and the availability of suitable spawning habitats.

In order to relieve water shortages in the municipal cities of Beijing and Tianjin in northern China, the Chinese government developed the South–North Water Diversion Project in 2002, the largest water-diversion project in the world. This ambitious project consists of three diversion routes: eastern, middle, and western. The diversion route from Danjiangkou Reservoir to northern China is called the middle route. During 2005–2009, the dam's height was increased 15 m from 162 to 177 m above sea level in order to implement the South–North Water Diversion Project, which was finished in October 2014. After the Danjiangkou Dam's height was increased, water in the Danjiangkou Reservoir could flow naturally to Beijing and Tianjin through the Henan and Hebei provinces. This project also increased the impounded area of Danjiangkou Reservoir from 846 to nearly 1,000 km^2. It is expected that further effects on fish fauna in the Hanjiang River are likely, given that 9.5×10^9 m^3 of water is diverted north from Danjiangkou Reservoir annually, equaling 23% of the total inflow of 4.11×10^{10} m^3 upstream of Danjiangkou Dam (Shen and Liu 1998; Gao and Gao 2010). Fish faunal changes are believed more likely downstream of Danjiangkou Dam due to the expected decreases in water quantity associated with the water-diversion project.

Fish Community Composition Changes in the Hanjiang River

A total of 84 fish species were recorded in the middle and lower reaches of the Hanjiang River from the 1970s to the present (Table 1). However, fish community composition, as reflected by patterns of species presence and absence over time, had become slightly less similar through the three sampling periods (1976–1978, 2003–2004, and 2012–2013). Jaccard's index, which describes the similarity of two communities (Jaccard 1912; Taylor 2004), was 95% when comparing 1976–1978 and 2003–2004 but decreased to 90% when comparing 1976–1978 and 2012–2013 (Table 2). Although 90% still reflects relatively high similarity, Jaccard's values are based only on presence–absence data and do not reflect shifts in fish community structure. For instance, two large-sized piscivorous fishes frequently collected historically, *Ochetobius elongates* and Long Spiky-head Carp *Luciobrama macrocephalus*, were not collected in the two more recent investigations in 2003–2004 and 2012–2013 (Table 1). Similarly, the Chinese Longsnout Catfish *Leiocassis longirostris* was not collected in the most recent investigation (2012–2013) after having been collected in 1976–1978 and 2003–2004 (Table 1). Conversely, several exotic fishes, including Channel Catfish *Ictalurus punctatus*, Largemouth Bass *Micropterus salmoides*, and Paddlefish *Polyodon spathula*, have been recently found in Danjiangkou Reservoir and are occasionally collected in the middle and lower reaches of the Hanjiang River. All of these exotic fishes have undoubtedly escaped from cage-culture farms in the region of Danjiangkou Reservoir or from commercial aquaculture facilities along the Hanjiang River.

Temporal patterns in fish species abundances have indicated several notable changes in fish community composition that have occurred during the past few decades. In 1976–1977, the most abundant fishes in the catch in the Hanjiang River were Grass Carp *Ctenopharyngodon idella*, Bronze Gudgeon *Coreius heterodon*, White Amur Bream *Parabramis pekinensis*, Common Carp *Cyprinus carpio*, Mongolian Redfin *Culter mongolicus*, Smallscale Yellowfin *Xenocypris microlepis*, *Rhinogobio typus*, and *R. cylindricus*, respectively (Yu et al. 1981; Yu 1982). However, nearly 30 years later in 2003–2004, the most abundant fishes in catch ranked as Common Carp, Goldfish *Carassius auratus*, White Amur Bream, Yellow Catfish *Pelteobagrus fulvidraco*, Grass Carp, and Barbel Chub *Squaliobarbus curriculus*, which comprised 86% of the total catch (Li et al. 2005). Spearman's rank correlation coefficient (Nash et al. 2013) correlating species abundances from 2003–2004 with those from 1976–1978 was only 0.31 ($P = 0.24$). This finding indicated some structural differences in fish communities through time, probably being driven by the more abundant species. In addition, some species consistently exhibited the trends of decreasing ages and smaller sizes between 1976–1978 and 2003–2004. For example, the average ages of Grass Carp, Common Carp, White Amur Bream, Barbel Chub, Goldfish, and Yellow Catfish collected during 1976–1978 (Yu et al. 1981) were noticeably younger than those collected during 2003–2004 (Li et al. 2005) (Table 3).

Changes in Spawning Fish Composition, Spawning Habitat, and Egg Quantity in the Hanjiang River

Fishes that spawn drifting eggs are the most important group of commercial fishes in both the Yangtze and Hanjiang rivers (Li et al. 2010; Wan et al. 2011). There were 25 fish species that spawn drifting eggs collected in the middle and lower reaches of the Hanjiang River during the 1976–1978 surveys (Zhou et al. 1980). These species belonged to the families Cyprinidae, Homalopteridae, Cobitidae, and Serranidae, of which 17 species were commercially harvested (Zhou et al. 1980). However, by 2004, only 16 species from two families were recorded that spawn drifting eggs in the same area of the Hanjiang

Table 1. Changes in fish species composition in the middle and lower Hanjiang River across three time periods: 1976–1978 (Zhou et al. 1980; Yu et al. 1981), 2003–2004 (Li et al. 2005), and 2012–2013 (present study).

No.	Order	Family	Species	1976–1978	2003–2004	2012–2013
1	Acipenseriformes	Polydontidae	Paddlefish *Polyodon spathula*	–	–	+
2	Clupeiforms	Engraulidae	Shortjaw Tapertail Anchovy *Coilia brachygnathus*	+	+	+
3	Salmoniforms	Salangidae	Clearhead Icefish *Protosalanx hyalocranius*	+	+	+
4			Icefish *Neosalanx taihuensis*	–	+	+
5	Anguilliforms	Anguilidae	Japonese Eel *Anguilla japonica*	+	+	+
6		Cobitidae	Chinese Spined Loach *Botia superciliaris*	–	+	–
7			Barred Loach *Parabotia fasciata*	+	+	+
8			*Leptobotia taeniaps*	+	+	+
9			Pond Loach *Misgurnus anguillicaubatus*	+	+	+
10	Cypriniformes	Homalopteridae	*Lepturichthys fimbriata*	+	+	+
11		Cyprinidae	Chinese False Gudgeon *Abbottina rivularis*	+	+	+
12			*Acheilognathus macropterus*	+	+	+
13			Chinese Bleak *Aphyocypris chinensis*	+	+	+
14			Goldfish *Carassius auratus*	+	+	+
15			Bronze Gudgeon *Coreius heterodon*	+	+	+
16			Grass Carp *Ctenopharyngodon idella*	+	+	+
17			Topmouth Culter *Culter erythropterus*	+	+	+
18			Mongolian Redfin *C. mongolicus*	+	+	+
19			*Culter ilishaeformis*	+	+	+
20			*Culter dabryi*	+	+	+
21			*Culter oxycephaloides*	+	+	+
22			Common Carp *Cyprinus carpio*	+	+	+
23			*Distoechodon tunirostris*	+	+	+
24			*Gobiobotia filifer*	+	+	+
25			Yellowcheek *Elopichthys bambusa*	+	+	+
26			Barbel Steed *Hemibarbus labeo*	+	+	+
27			Spotted Steed *H. maculatus*	+	+	+
28			*Hemiculter bleekeri*	+	+	+
29			Sharpbelly *H. leucisculus*	+	+	+
30			Bighead Carp *Hypophthalmichthys nobi1is*	+	+	+
31			Silver Carp *H. molitrix*	+	+	+
32			Long Spiky-head Carp *Luciobrama macrocephalus*	+	–	–
33			Black Bream *Megalobrama skolkevii*	+	+	+
34			Bluntnose Black Bream *M. amblycephala*	+	+	+

Table 1. Continued.

No.	Order	Family	Species	1976–1978	2003–2004	2012–2013
35	Cypriniformes	Cyprinidae	Black Carp *Mylopharyngodon piceus*	+	+	+
36			*Ochetobius elongatus*	+	–	–
37			Chinese Hooksnout Carp *Opsariichthys bidens*	+	+	+
38			White Amur Bream *Parabramis pekinensis*	+	+	+
39			*Pseudobrama simoni*	+	+	+
40			*Pseudogobio vaillanti*	+	+	+
41			*Pseudolaubuca sinensis*	+	+	+
42			*Pseudolaubuca engrauliss*	+	+	+
43			Stone Moroko *Pseudorasbora parva*	+	+	+
44			*Rhinogobio typus*	+	+	+
45			*Rhinogobio cylindricus*	+	+	+
46			Rose Bitterling *Rhodeus ocellatus*	+	+	+
47			Chinese Fat Minnow *Sarcocheilichthys sinensis*	+	+	+
48			*Sarcocheilichthys nigripinnis*	+	+	+
49			Longnose Gudgeon *Saurogobio dabryi*	+	+	+
50			*Saurogobio gracilicaudatus*	+	+	+
51			Largescale Shoveljaw Fish *Scaphesthes macrolepis*	+	+	+
52			*Sinibrama wui*	+	+	+
53			*Squalidus argentatus*	+	+	+
54			Barbel Chub *Squaliobarbus curriculus*	+	+	+
55			*Toxabramis swinhonis*	+	+	+
56			*Xenocypris davidi*	+	+	+
57			Smallscale Yellowfin *X. microlepis*	+	+	+
58			Freshwater Yellowfin *X. argentea*	+	+	+
59			Freshwater Minnow *Zacco platypus*	+	+	+
60	Silurifomes	Siluridae	Amur Catfish *Silurus asotus*	+	+	+
61			Southern Catfish *S. meridionalis*	–	–	+
62		Bagridae	*Leiocassis crassilabris*	+	+	+
63			Chinese Longsnout Catfish *L. longirostris*	+	+	–
64			Largefin Longbarbel Catfish *Mystus macropterus*	+	+	+
65			Yellow Catfish *Pelteobagrus fulvidraco*	+	+	+
66			Darkbarbel Catfish *P. vachelli*	+	+	+
67			Shining Catfish *P. nitidus*	+	+	+
68			*Pseudobagrus truncatus*	+	+	+
69			Roundtail Bullhead *Pseudobagrus tenuis*	+	+	+

Table 1. Continued.

No,	Order	Family	Species	1976–1978	2003–2004	2012–2013
70	Silurifomes	Amblycitidae	Liobargrus marginatus	+	+	+
71		Ictaluridae	Channel Catfish *Ictalurus punctatus*	–	–	+
72		Sisoridae	*Glyptothorax sinensis*	+	+	+
73	Beloniformes	Hemirhamphidae	*Hemirhamphus kurumeus*	+	+	+
74	Synbranchiformes	Synbranchidae	Asian Swamp Eel *Monopterus albus*	+	+	+
75	Perciformes	Serranidae	Big-eye Mandarin Fish *Siniperca knerii*	+	+	+
76			Mandarin Fish *S. chuatsi*	+	+	+
77			Golden Mandarin Fish *S. scherzeri*	+	+	+
78		Centrarchidae	Largemouth Bass *Micropterus salmoides*	+	–	+
79		Eleotridae	*Hypseleotris swinhonis*	+	+	+
80			Dark Sleeper *Odontobutis obscura*	+	+	+
81		Gobiidae	*Ctenogobius giurinus*	+	+	+
82		Belontiidae	Roundtail Paradise Fish *Macropodus ocellatus*	+	+	+
83		Channidae	Northern Snakehead *Channa argus*	+	+	+
84		Mastacembelidae	Lesser Spiny Eel *Mastacembelus aculeatus*	+	+	+

Table 2. Jaccard's index computed for faunal similarity of the middle and lower Hanjiang River during the 1976–1978, 2003–2004, and 2012–2013 time periods.

Year	1976–1978	2003–2004	2012–2013
1976–1978	–	95%	90%
2003–2004	–	–	93%
2012–2013	–	–	–

River (Li et al. 2006). Among these 16 species, only eight species were classified as commercially harvested (Li et al. 2006). Similarly, in 1976, there were nine well-known spawning grounds for fishes that spawned drifting eggs in the middle and lower reaches of the Hanjiang River (Zhou et al. 1980). In 2004, the spawning grounds of the major Chinese carps, namely Grass Carp, Black Carp, Silver Carp, and Bighead Carp, had disappeared in five of these nine known spawning grounds. Over the same time period, the number of spawning grounds for small-bodied fishes increased substantially (Li et al. 2006). The fry percentages of *Rhodeus sinensis*, *Pseudolaubuca engraulis*, and Stone Moroko, three small-bodied fishes, reached 26%, 19%, and 19% of the total fry number, respectively, and spawning grounds of these small-bodied fishes were distributed along the entire river in the middle reaches (Li et al. 2006).

The total quantity of drifting eggs in the middle reaches of the Hanjiang River increased 1.5-fold between 1976 and 2004 (Zhou et al. 1980; Li et al. 2006). However, the total egg quantity of the four major carps and other economically important fishes decreased from 3.14×10^9 in 1976 to 0.46×10^9 in 2004 while the total egg quantity of small-bodied fishes increased from 1.56×10^9 to 1.632×10^{10} during the same time period (Zhou et al. 1980; Li et al. 2006).

Changes in Fisheries Resources and Commercial Harvest Production in Danjiangkou Reservoir

Prior to dam construction, 34 fish species were found in the Hanjiang River reach that would later become Danjiangkou Reservoir (Bolotskiy et al. 1959). Following construction of the dam, a total of 70 fish species have been recorded, including 67 native species (Yuan and Huang 1989) and three exotic species (i.e., Channel Catfish, Largemouth Bass, and Paddlefish). The fishes collected represented five orders, 13 families, and 54 genera. Among these species, 44 species were from the family Cyprinidae (comprising 63% of the total species), nine species from Bagridae, three species each from Cobitidae and Serranidae, two species each from Siluridae and Salangidae, and one species each from Homalopteridae, Synbranchidae, Engraulidae, Eleotridae, Gobiidae, Channidae, and Mastacembelidae.

During the past 30 years, commercial fish harvest production in Danjiangkou Reservoir has gradually increased from about 20 kg/ha to about 75 kg/ha (Figure 1). The major fishes harvested were *Culter* spp., Silver Carp, Bighead Carp, Common Carp, Goldfish, and Icefish, which is a small, transpar-

Table 3. Comparison of average ages of fishes in catches collected from the middle and lower Hanjiang River during the 1976–1978 (Yu et al. 1981) and 2003–2004 (Li et al. 2005) time periods.

	Average ages					
Sampling time (reference)	Grass Carp	Common Carp	White Amur Bream	Barbel Chub	Goldfish	Yellow Catfish
1976–1978	3.7	3.3	3.6	3.6	2.3	2.4
2003–2004	2.9	2.1	2.7	3.2	1.2	1.4

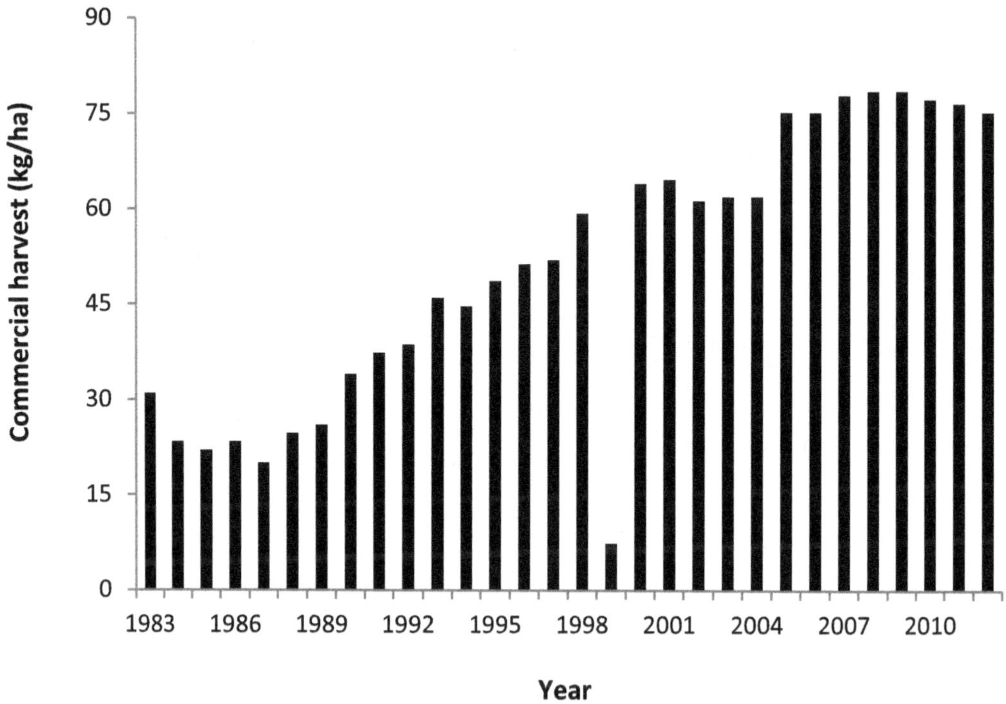

Figure 1. Commercial fisheries production as reflected by harvests in Danjiangkou Reservoir, 1983–2012.

ent species introduced to the reservoir in the late 1990s that has become commercially important (Yang et al. 2012). Cage-fish culture is an increasing industry within the Danjiangkou Reservoir (Figure 2). Fish production associated with cage-culture farms has increased from less than 200 metric tons in 1983 to more than 30,000 metric tons by 2011 (Figure 3). The main cage-cultured species included *Culter ilishaeformis*, Yellowcheek, Channel Catfish, and Largemouth Bass. In these cage-culture systems, small, low-valued fishes, such as Silver Carp, or frozen fishes are used as feeds for *C. ilishaeformis* and Yellowcheek. By contrast, Channel Catfish and Largemouth Bass are commonly fed formulated feeds. Fish production from these cage-culture systems typically ranges from 50 to 80 kg/m^3 (Hua Zhang, Fisheries Bureau of Danjiankou City, Hubei Province, P. R. China, unpublished data). This production is in the average level for cage culture in China (Liu and Huang 1998; Wang et al. 2014), but the value from the cage culture is much higher than those in other places since the price of *C. ilishaeformis* and Yellowcheek is about 3–5 times that of major carp species (Zhang, personal communication).

Possible Reasons for Fisheries Resources Changes in the Hanjiang River

Large dams impact river ecosystems and services worldwide (Dynesius and Nilsson 1994). Fisheries resources are impacted most by flow and temperature alterations, water chemistry changes, and migration blockage produced by dam construction (McAllister et al. 2001; Nilsson et al. 2005; Xie et al. 2007). A decade ago, Fu et al. (2003) warned that large water-diversion projects, such as the South–North Water Diversion Project, were likely to affect a broad array of native species throughout the Yangtze River basin. Following dam construction, the total egg quantity spawned by the four major commercial carp species (Grass Carp, Black Carp, Silver Carp,

Figure 2. Cage-fish culture facility in Danjiangkou Reservoir, China.

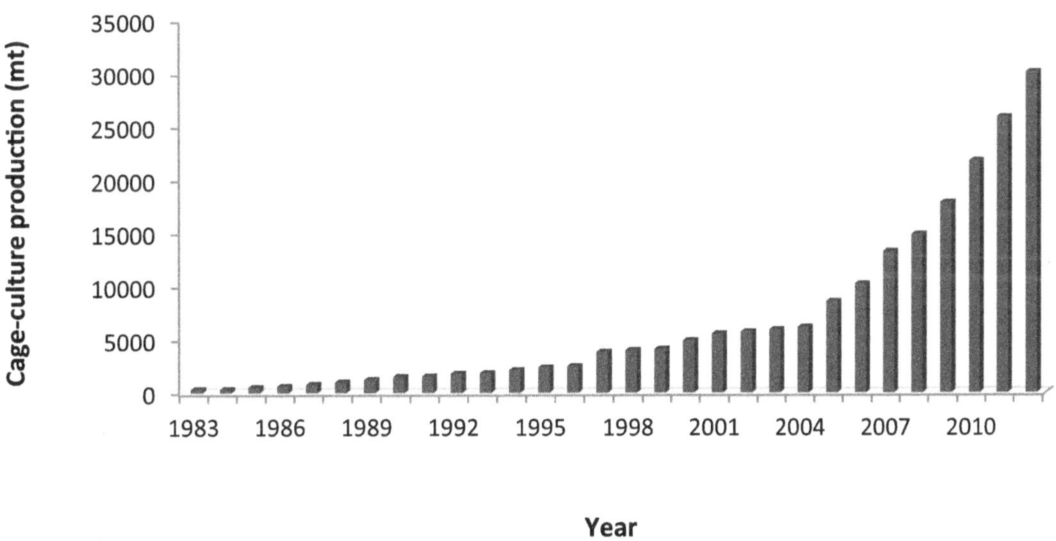

Figure 3. Cage-culture fisheries production in Danjiangkou Reservoir, 1983–2012.

and Bighead Carp) in the middle and lower Hanjiang River decreased significantly, which was consistent with the predictions of Fu et al. (2003). Spawning success of these carps is being negatively impacted, likely due to the narrower range of year-round water temperatures and water levels downstream of Danjiangkou Dam. Lack of spawning success for carps also may be related to the decreased range of flow fluctuations during the spawning season downstream of the dam, as has been observed at the Three Gorges Dam (Xie et al. 2007).

To implement the South–North Water Division Project, it was necessary to raise the height of Danjiangkou Dam 15 m during 2005–2009. Completed in late 2014, this project diverts 9.5×10^9 m^3 of water north from Danjiangkou Reservoir every year, which is equivalent to nearly one-fourth of the total flow of the Hanjiang and Danjiang rivers above the dam. With further decreases in water flow of the middle and lower reaches of the Hanjiang River, resources for the four major carps are predicted to continue decreasing. The habitat losses for these economically important carps must be fully considered and appropriate conservation strategies must be adopted. Given that sexual maturation of these carp species occurs in 5–7 years (Liu and He 1992), we suggest that establishment of conservation areas aimed at restoring major carp spawning habitats be prioritized for the reaches downstream of Danjiangkou Dam.

There are other factors that may have affected fisheries in the Hanjiang River. During the past few decades, modification of powered fishing boats and advanced fishing facilities have translated into significantly increased fishing effort and fish harvests. Li et al. (2006) estimated that there were at least 174 motor-powered fishing boats working from six locations along the Hanjiang River. Although there is a closed fishing season in the Hanjiang River from early April through late June, and government regulations prohibit harmful fishing techniques such as bombing, toxic fishing, and electric fishing, fishing effort and minimum harvest sizes are not strongly regulated. Thus, overfishing is another possible reason for decreases in abundances and sizes of many commercially important fishes. We strongly suggest implementation and enforcement of fisheries regulations that restrict total harvest, effort, and minimum harvest size for each boat and species along the entire Hanjiang River. These regulations combined with the closed fishing season will aid the restoration and protection of these valuable fisheries resources.

Another possibility for the decline of the four carp species is fragmentation of their habitats throughout the Hanjiang River. In the Hanjiang River basin, there are many natural lakes that were historically connected to the river. The four major carps regularly used these lakes for overwintering and growth, followed each year by migrations back to the main river for spawning during late spring and early summer (Liu and He 1992). However, widespread construction of sluice gates in the 1960–1970s has largely disconnected these lakes from the Hanjiang River main channel. Although obvious that disconnection blocks fish movements between lakes and the river, it is theorized that broodstock recruitment also has been affected, which is suspected to be further inhibiting the fisheries resources (Zhou et al. 1987; Chang and Cao 1999). We suggest the opening of the sluice gates during the spawning season of the major carps to allow broodstock spawning migrations between the lakes and Hanjiang River main channel. Opening of these gates also will facilitate movement of carp fry and juveniles from main river channel back to nursery habitats contained in these lakes (Zhou et al. 1987).

Eutrophication in the middle and lower reaches of the Hanjiang River is another potential threat to fisheries resources. During the past few decades, the frequency of reported algal blooms has increased from zero prior to the 1980s to once or twice during the 1990s. However, several dozen algal blooms have occurred since 2000 (Xie et al. 2006; Zhan Yin, Institute of Hydrobiology, Chinese Academy of Sciences, unpublished data). Eutrophication can increase fish production within limits, with a corresponding increase in system primary productivity (Liu and He 1992). Eutrophication may partly explain the significant increases of many small-bodied fishes such as *Rhodeus sinensis*, *Pseudolaubuca engraulis*, and Stone Moroku throughout the Hanjiang River. With the initiation of the South–North Water Diversion Project, further water pollution and subsequent eutrophication could affect future fisheries conservation efforts. The effects are predicted more for the middle and lower reaches of the Hanjiang River in light of the expected decreases in water flows.

Although exotic species are important to Chinese aquaculture (Liu and Li 2010), they may create ecological disasters in some natural waters (Liu 2014). Three exotic species common to Chinese aquaculture (Channel Catfish, Largemouth Bass, and Paddlefish) are being more routinely collected in Danjiangkou Reservoir and the Hanjiang River. Because the Hanjiang River provides an easy corridor to the Yangtze River, the potential ecosystem-level effects of these three exotic species on the Yangtze River should not be neglected. Furthermore, each species represents a different trophic guild, with Channel Catfish being an omnivore, Largemouth Bass a piscivore, and Paddlefish a planktivore. Stricter regulations and farm inspection programs should be developed that minimize the escape potential of these exotic species into the wild.

In conclusion, fisheries resources have changed significantly in the middle and lower Hanjiang River during the past 50 years. In particular, spawning grounds for the four major economically important carps and several other large-bodied fishes have significantly decreased. While total egg quantity spawned by the major carps and other commercially important fishes has decreased, production by small-bodied, economically less important fishes has increased downstream of Danjiangkou Dam. Important commercial fishes have experienced delayed spawning times and decreased growth. Overall catches of commercial fishes have decreased downstream of the Danjiangkou Dam, though they have increased upstream of the dam in Danjiangkou Reservoir. Changes in Hanjiang River fisheries resources also have presumably been affected by several factors, including Danjiangkou Dam, sluice gates separating floodplain lakes and the main river, overharvest, and eutrophication. With the decreased inflows in the middle and lower Hanjiang River due to the implementation of the South–North Water Diversion Project finished in late 2014, further changes in fisheries resources are expected in the future. Management and conservation strategies for fisheries resources need to be developed to ensure future fisheries sustainability both for the Hanjiang River and the Danjiangkou Reservoir.

Acknowledgments

This study was funded by the National Special Project for Water Pollution Control (No. 2012ZX07205002-02-04) and the National Science and Technology Supporting Program (No. 2012BAD25B08). We are grateful to Michael Eggleton and two anonymous reviewers for their valuable comments and suggestions.

References

Bolotskiy, E. B., X. Wu, G. Bai, M. Ge, Q. Wang, S. Wang, and S. Chen. 1959. Investigation of aquatic organisms in the Danjiangkou Reservoir area and the opinion on the fisheries utilization. Acta Hydrobiologia Sinica 1959:33–56. (In Chinese.)

Chang, J., and W. Cao. 1999. Fishery significance of the river-communicating lakes and strategies for the management of fish resources. Resources and Environment in the Yangtze Basin 8:153–157. (In Chinese with English abstract.)

Dynesius, M., and C. Nilsson. 1994. Fragmentation and flow regulation of river systems in the northern third of the world. Science 266:753–762.

Du, Y., X. Wang, and S. Cai. 2005. Effect and countermeasure of the Middle Route Project of South to North Water Transfer on ecology and environment in the middle and lower reaches of Hanjiang River. Science, Technology and Society 20:477–482. (In Chinese)

Fu, C., J. Wu, J. Chen, Q. Wu, and G. Lei. 2003. Freshwater fish biodiversity in the Yangtze River basin of China: patterns, threats and conservation. Biodiversity and Conservation 2:1649–1685.

Gao, Y., and J. Gao. 2010. Comprehensive assessment of eco-environment impact of the South-to-North Water Transfer Middle Route Project on the middle-lower Hanjiang River basin. Progress in Geography 29:59–64. (In Chinese with English abstract.)

HPIH (Hubei Provincial Institute of Hydrobiology). 1976. Fishes of the Yangtze River. Science Press, Beijing. (In Chinese.)

Jaccard, P. 1912. The distribution of flora in the alpine zone. New Phytologist 11:37–50.

Li, M., W. Jiang, X. Gao, Z. Duan, and H. Liu. 2010. Status of early life history stages at Wuxue cross-section of the Yangtze River. Acta Hydrobiologia Sinica 34:1211–1217. (In Chinese.)

Li, X., D. Huang, W. Xie, X. Chang, H. Yang, Y. Zhang, and J. He. 2005. Status of fish resources in the middle reaches of the Hanjiang River. Journal of Lake Science 17:366–372. (In Chinese with English abstract.)

Li X., D. Huang, W. Xie, S. Xie, X. Chang, H. Yang, Y. Zhang, and J. He. 2006. Current status of spawning grounds of fish with pelagic eggs in the middle reaches of Hanjiang River. Journal of Dalian Fisheries University 21:105–111. (In Chinese with English abstract.)

Liu, F. 2014. Exotic fishes in the Yangtze River basin and their damages. China Nature 2014:38–40.

Liu, J., and B. He. 1992. Cultivation of freshwater fishes of China. Science Press, Beijing. (In Chinese.)

Liu, J., and Y. Huang. 1998. Fisheries and fish culture practices in Fuqiaohe Reservoir, China. International Review on Hydrobiology 83:569–576.

Liu, J., and Z. Li. 2010. The role of exotics in Chinese aquaculture. Pages 173–185 *in* S. S. De Silva and B. F. Davy, editors. Success stories in Asian aquaculture. Springer, Dordrecht, The Netherlands.

McAllister, D., J. Craig, N. Davidson, D. Murray, and M. Seddon. 2001. Biodiversity impacts of large dams. International Union for Conservation of Nature and United Nations Environmental Programme, Background Paper 1, Gland, Switzerland.

Nash, K. L., N. A. J. Graham, and D. R. Bellwood. 2013. Fish foraging patterns, vulnerability to fishing, and implications for the management of ecosystem function across scales. Ecological Applications 23:1632–1644.

Nilsson, C., C. A. Reidy, M. Dynesius, and C. Revenga. 2005. Fragmentation and flow regulation of the world's large river systems. Science 308:405–408.

Shen, D., and C. Liu. 1998. Effects of different scales of MR-SNWTP on the down stream of the Danjiangkou Reservoir. Acta Geographica Sinica 53:341–348. (In Chinese with English abstract.)

Taylor, E. B. 2004. An analysis of homogenization and differentiation of Canadian freshwater fish faunas with an emphasis on British Columbia. Canadian Journal of Fisheries and Aquatic Sciences 61:68–79.

Wan, L., Y. Cai, H. Tang, Z. Yang, X. Zhang, H. Zheng, and Y. Qiao. 2011. Preliminary study on the larval resources of fishes spawning drifting eggs in the middle and lower reaches of the Hanjiang River. Journal of Hydroecology 32:53–57. (In Chinese with English abstract.)

Wang, Q., L. Cheng, J. Liu, Z. Li, and S. S . De Silva. 2014. Freshwater aquaculture in PR China: trends and prospects. Reviews in Aquaculture 5:1–20.

Xie, M., X. Wang, G. Guan, and M. Hu. 2006. Studies on the reason of water blooming and possible strategies in the middle and lower reaches of the Hanjiang River. Yangtze River 37:43–45. (In Chinese.)

Xie, S., Z. Li, J. Liu, S. Xie, H. Wang, and B. R. Murphy. 2007. Fisheries of the Yangtze River show immediate impacts of the Three Gorges Dam. Fisheries 32:343–344.

Yang, Z., W. Li, Z. Li, J. Liu, T. Zhang, S. Ye, and H. Zhang. 2012. A comparative study on reproductive characteristics of different spawning stocks of the Icefish (*Neosalanx taihuensis*) in the Danjiangkou Reservoir. Freshwater Fisheries 42:58–62. (In Chinese with English abstract.)

Yu, Z. 1982. Fish resource investigation in the middle and lower reaches of the Hanjiang River and impact assessment of hydropower project on the fish resources. Reservoir Fisheries 1982:19–26. (In Chinese.)

Yu, Z., Z. Deng, and Y. Xu. 1981. Fish resources in the Hanjiang River after the construction of the Danjiangkou Hydroproject. Collections of Ichthyological Papers I:77–96, Beijing Science Press, Beijing.

Yuan, F., and D. Huang. 1989. Fish resources and composition analysis in the Danjiangkou Reservoir. Reservoir Fisheries 1989:35–41.

Zhou C., C. Liang, and H. Huang. 1980. Ecological features of the spawning of certain fishes in the Hanjiang River after the construction of dams. Acta Hydrobiologia 7:175–188. (In Chinese with English abstract.)

Zhou, J., N. Wang, S. Zhang, B. Yi, and X. Nie. 1987. Studies on the aquatic organisms and fish enhancement in Lake Qingling. Oceanologia et Limnologia Sinica 18:442–449. (In Chinese with English abstract.)

Part 3 Anthropogenic Stressors, Floodplains, and River Restorations

American Fisheries Society Symposium 84:195–233, 2016

Managing the Mississippi River in a Nonstationary World: Past Practices and Future Challenges

JONATHAN W. F. REMO*
Department of Geography and Environmental Resources, Southern Illinois University 1000 Fanner Drive, Carbondale, Illinois 62901, USA

Abstract.—Substantial investment in the engineering of the Mississippi River and its tributaries over the past ~200 years by the U.S. federal government has resulted in its transformation into one of the most regulated and intensively managed river systems in the world. The regulation and management of the Mississippi River system for a select set of ecosystem services has provided substantial economic benefits to the United States. These economic benefits include the transportation of goods, the reduction of flood risk to floodplain communities and agricultural lands, securing or increasing water supply, and reliable energy production. However, these benefits have come with substantial externalities such as altered hydrology, altered river hydraulics, decreased sediment load, changes in the pattern of delta sedimentation, increased nutrient loads, reduced habitat diversity, decreased ecological diversity, and degraded water quality. In this chapter, the natural, management, and engineering histories of the Mississippi River are reviewed. The purpose of this review is to describe the evolution of river management for harnessing the current services the river provides and the environmental and social externalities resulting from the management actions. It is hoped this holistic review of the Mississippi River will assist in leading toward a more complete and integrated vision for the management of this coupled natural–human river system.

Introduction

Large rivers, like the Mississippi and the Yangtze, are important engines for both the global economy and environment. These rivers are economically important because they provide society with vital services such as transportation, power generation, and water supply. These services are the foundation for industrial and agricultural growth needed for economic development and financial stability. Large-river ecosystems are generally the most biologically diverse on Earth and are primary components of the global biogeochemical cycle. They assimilate and supply the chemical and sediment loads of their large basins to their deltas and the continental shelf where they support productive and diverse marine ecosystems. These ecosystems are an important source of food and other resources globally (Stanley and Warne 1997; Day et al. 2007).

The importance of large rivers to human society is evident from the distribution of both modern and ancient population centers, which are located along the banks and within the deltas of large rivers. Engineering works have been constructed along large rivers to enhance certain ecosystem services, including transportation, energy production, water supply, and the mitigation of flooding. The engineering of large rivers for the facilitation of these services has had significant economic benefits to society. However, these economic benefits have come with externalities that have in many cases counterbalanced the benefits gained (Schmidt et al. 1998). In

* Corresponding author: diamict@siu.edu

recent decades, the tradeoffs have become apparent between the set of ecosystem services for which large rivers have been managed and the resulting ecological, economic, and social externalities. Consequently, expensive and politically contentious measures are being undertaken to counterbalance the externalities, including rehabilitation efforts for protection of costal lands, improvement of water quality, and preservation of endangered species (Bernhardt et al. 2005; Lake et al. 2007; O'Donnell and Galat 2007).

The intent of this book chapter is to provide a review of the Mississippi River's natural, management, and engineering histories. Special attention will be paid to the river engineering and other anthropogenic effects on the Mississippi River and its delta. This chapter will specifically review (1) management and engineering history of the Mississippi River; (2) effects of the management and associated river engineering on the hydrology, sediment transport, geomorphology, and ecology of the river; (3) some specific anthropogenic effects on the Mississippi River from beyond the river corridor; and (4) current and future river management challenges facing society.

The Physical Geography of the Mississippi River Basin

The Mississippi River basin (MRB) covers 41% of the conterminous United States and totals 3,224,600 km^2. The Mississippi River extends 3,770 km from its headwaters at Lake Itasca, Minnesota to the Head of Passes, Louisiana where it empties into the Gulf of Mexico. The Mississippi River, in combination with its tributary the Missouri River, is one of the longest river systems in the world. Other major tributaries to the Mississippi River include the Ohio, Illinois, Arkansas, and Red rivers. In this paper, the designation "upper," "middle," and "lower" are used in reference of the Mississippi River upstream of St. Louis, Missouri, from St. Louis to the confluence of the Ohio River, and the Ohio River to the Gulf of Mexico, respectively (Figure 1).

The Mississippi River, in combination with its distributary the Atchafalaya River, discharges annually 580 km^3 of water into the Gulf of Mexico. In the MRB, the principal sources of water and sediment are decoupled. The Ohio River, which makes up about 17% of the MRB area, contributes approximately half of the annual discharge because it drains the more humid portion of the MRB. In contrast the Missouri River, which makes up 43% of the MRB area, contributes only 12% of the annual discharge (Meade 1995; Meade and Moody 2010). The primary sources of sediment in the MRB are the Mississippi River tributaries, chiefly the Missouri River which drains the Great Plains. The Great Plains tributaries carry proportionally larger sediment loads. This is because the Great Plain's climate is largely semiarid, which results in lower-vegetation-density land cover but still receives enough precipitation to mobilize a substantial amount of sediment (Langbein and Schumm 1958).

The majority of the MRB and its associated topography developed during the Cenozoic (<65 million years). During this period, the MRB has developed into one of the largest drainage basins (~3,224,000 km^2) on Earth. The current course and physical character of the Mississippi River and its major northern tributaries (the Missouri, Ohio, and Illinois rivers) were significantly influenced by the most recent Quaternary glaciations (e.g., Illinoian, 130,000–190,000 years ago, and Wisconsinan, 11,000–85,000 years ago). These glaciations covered significant portions of these subbasins (Figure 1). The MRB's major vegetation biomes (forests in the eastern and mountainous regions; prairies in western regions) were established ~10,000 years before present (Knox 2007).

The preengineered river channel morphology and flow regime of a particular Mississippi River segment (>100 km in length) largely dictates its engineering history. The varying channel morphologies of the Mississippi River are described here to provide context for why a particular suite of river-engineering tools were employed along a given segment. The upper Mississippi River's lon-

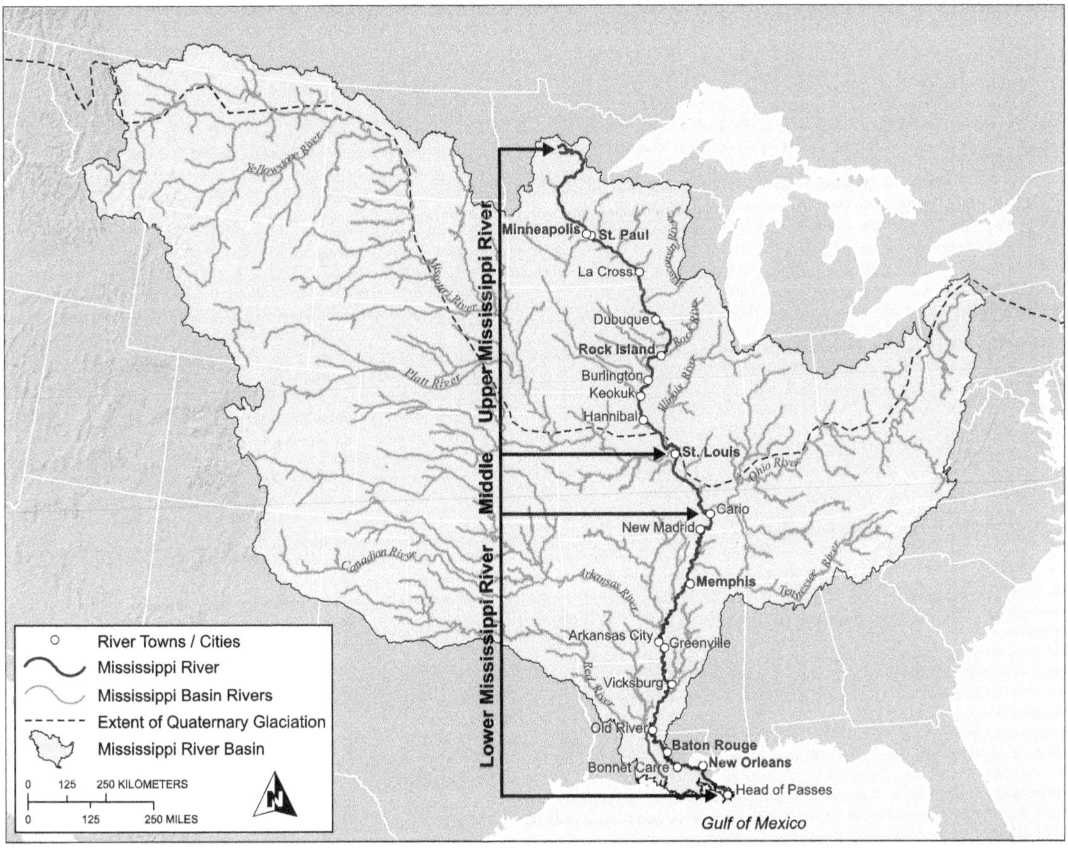

Figure 1. Mississippi River basin with its major subdivision, major tributaries, and significant locations labeled. Dashed line shows the approximate location of the maximum extent of Quaternary glaciation.

gitudinal profile does not have the common concave-upward morphology that is typical of most large alluvial rivers. The Mississippi River from its headwater source in the Lake Itasca area to its junction with the Missouri River is convex and comprised of a series of distinct segments that reflect the complex influence of the Quaternary glaciations in the region. The majority of its convexity resulted from a shortening of the river caused by diversion from its former course across northwestern Illinois to the present Illinois River and then southwestward to where the Illinois River joins the present day Mississippi River. The convexities in the Mississippi River's profile occur in the rock gorges between Dubuque and Keokuk, Iowa. These rock gorges are the result of Quaternary diversions and subsequent incision of the upper Mississippi River (UMR) across bedrock uplands. As a consequence, the modern river flows across resistant bedrock in these segments. Other local convexities in the longitudinal profile of UMR are the result of alluvial fans that have formed where tributaries have deposited large amount of Holocene sediments in the river valley (Knox 2007). Such fans can be found at the confluences of the Chippewa, Wisconsin, and Illinois rivers (Figures 1 and 2).

The UMR can be divided into segments of similar geomorphology. Here, the UMR is generalized into three distinct segments. The first segment consists of the Mississippi River's headwaters at Lake Itasca to approximately the confluence of the Minnesota River south of Minneapolis, Minnesota. Through this segment, the UMR has a relatively low

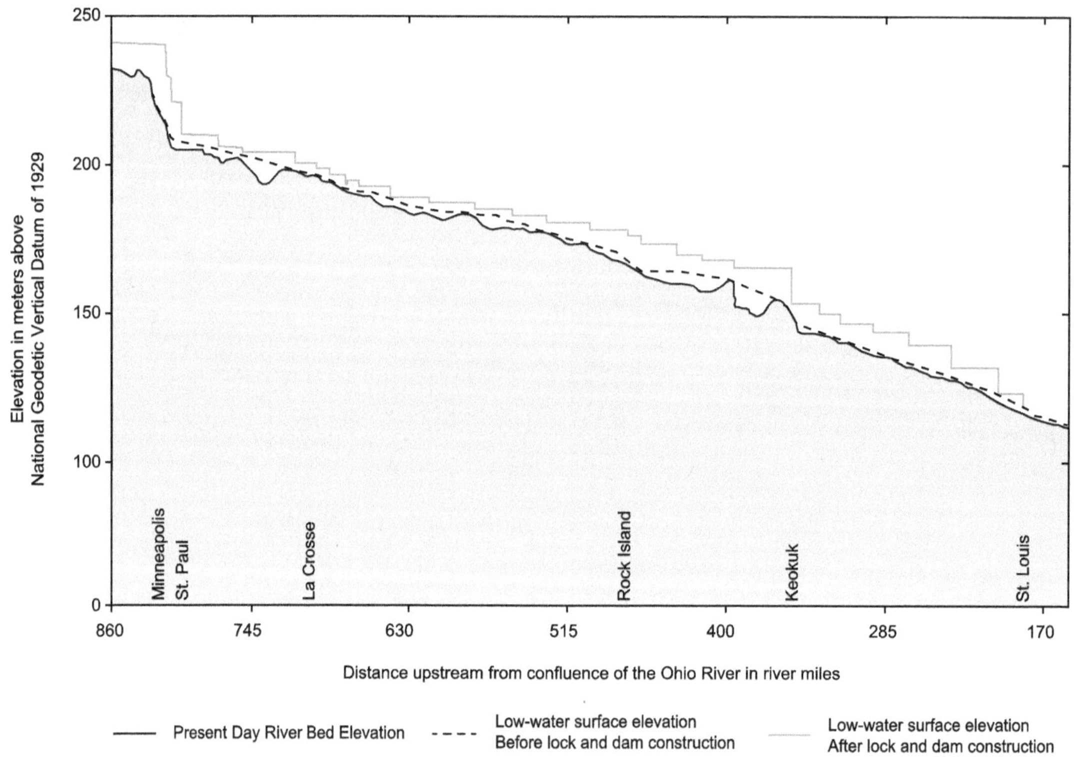

Figure 2. Upper Mississippi River bed profile and low-water surface-elevation profiles before and after construction of navigation locks and dams (after Alexander et al. 2012).

gradient and meandering channel form that winds through late Wisconsinan-age glacial-morain topography. The second segment extends from the Minnesota River mouth to the Rock Island Gorge in northwestern Illinois. This segment has an island-braided channel form. Large quantities of sandy bed load are delivered from the tributaries along this segment, accounting for the numerous islands and historically wide shallow channel morphology that existed prior to river regulation. The third segment of the UMR is relatively geomorphically complex and extends from the Rock Island Rapids to the mouth of the Missouri River. This complex segment contains both straight and meandering subsegments. The straight subsegments (>10 km to <100 km in length) are the result of the late Quaternary diversion and incision of the river across two major bedrock uplands. The meandering subsegments are located between the straight bedrock-lined subsegments where the newer river segments rejoin the ancient Mississippi River valley. Within these meandering subsegments, large suspended sediment loads are brought in from tributaries that drain landscapes covered by thick, easily erodible loess deposits. The large suspended sediment load contributes appreciably to numerous islands along theses subsegments as deposition of sediment during high flows is enhanced by vegetation on islands and bars (Knox 2007).

Like the UMR, the middle Mississippi River does not have the typical concave-upward stream profile. This segment of the Mississippi is relatively straight and prior to river engineering its channels had an island-braided morphology. The steep upper portion of this segment is the result of an alluvial fan that is found at the mouth of the Missouri River (Knox 2007). The Thebes Gap and Grand Tower Gorges located 70 and 130 km upstream of the Ohio River confluence, respectively, are the result of late Quaternary channel avulsion and subsequent river incision through low passes in bedrock ridges (Figure 3; Knox 2007; Rittenour et al. 2007).

The longitudinal profile of the lower Mississippi River (LMR) between the confluence with the Ohio to ~40 km north of Vicksburg shows relatively little convexity or concavity, except very locally (Figure 4; Schumm et al. 1994). The profile through this segment is generally straight and relatively steep, which reflects the energy gradient needed to transport the large volume of sand bed load. Downstream of Vicksburg, a modest concavity in the LMR profile is present as the river grades into the Gulf of Mexico. Despite the LMRs relatively straight down-valley profile, there was considerable variability in the river's planform prior to river regulation. The variability of the LMR's planform is attributed to local geological influences, including outcrops of Tertiary age bedrock, Pleistocene gravels, clay plugs within old meander cutoffs, and at least two zones of major surface deformation caused by tectonic uplift. In particular, these uplifts have contributed to abrupt differences in river width and sinuosity between local subsegments whose up-

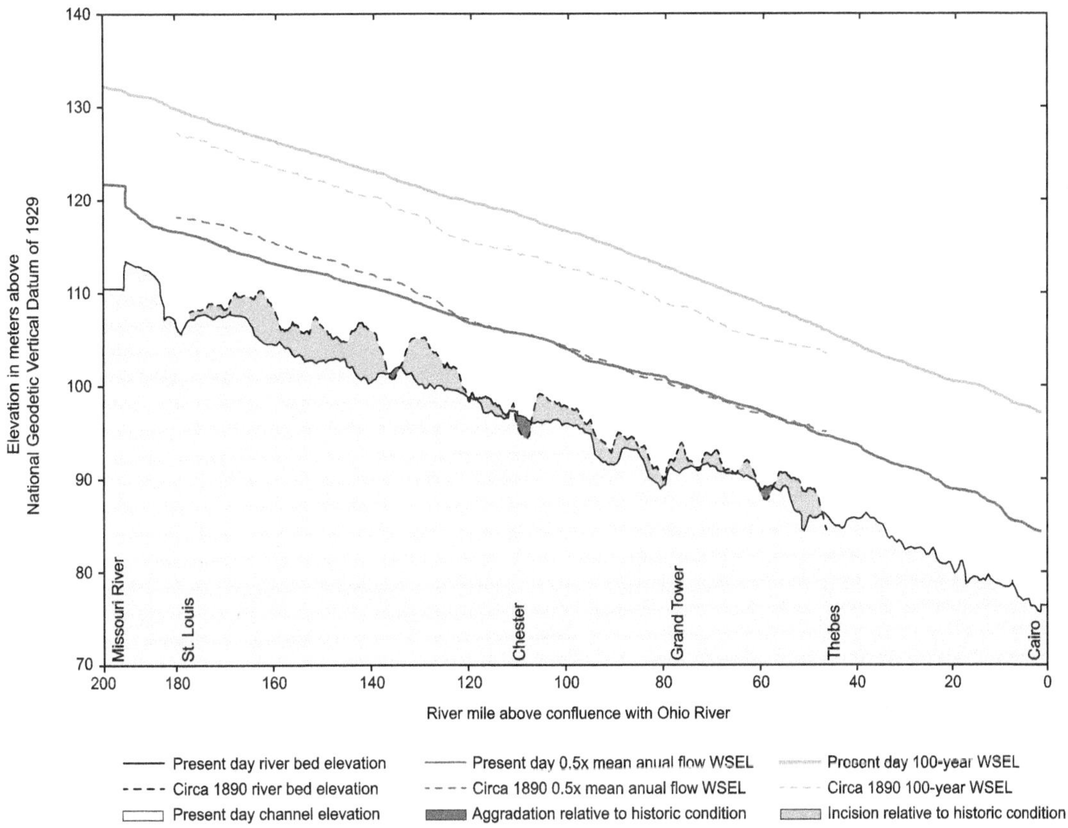

Figure 3. Middle Mississippi River bed profiles and water surface elevation (WSEL) for modern 100-year flood (~25,800 m^3/s) and low flow (0.5 × mean annual flow ~5,400 m^3/s) for river conditions circa 1890 and near modern conditions (circa 2000). Note riverbed incision between historic and modern reference conditions. Low-flow WSELs either decrease or remained the same between the historic and modern conditions. Decreases in low flows are attributed to channel bed incision. WSELs for the 100-year flood increase along the middle Mississippi River. Increases in WSELs for the 100-year flood are attributed to decreased carrying capacity of the river channel and its floodplain. Riverbed elevations and WSELs from Remo et al. (2009).

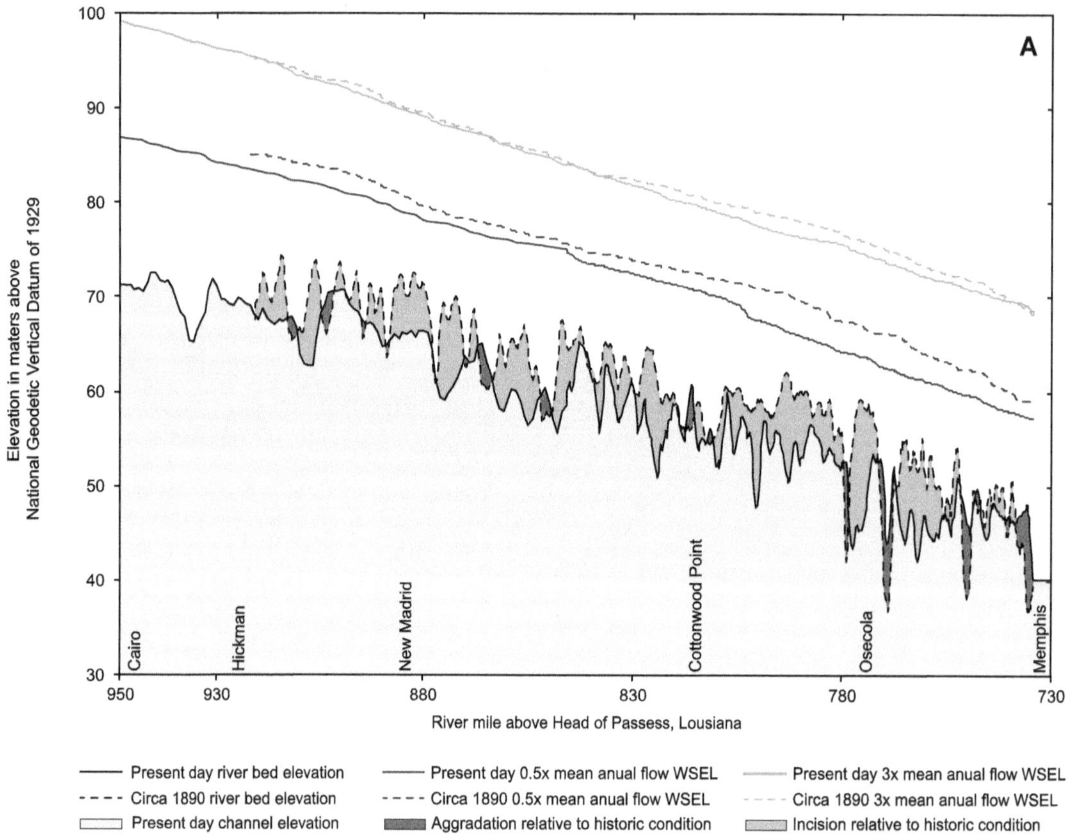

Figure 4. **(A)** Lower Mississippi riverbed profile and water surface elevation (WSEL) between Cario, Illinois and Memphis, Tennessee for modern 3.0 × mean annual flow (moderate flood ≈ 41,000 m^3/s) and 0.5 × mean annual flow (low flow ≈ 13,700 m^3/s) for circa 1890 and near-modern (mid-1990s) river conditions. Note riverbed incision and corresponding lowering of WSELs between historic and near modern reference conditions. Riverbed elevations and WSELs from Remo et al. (2009). **(B)** Lower Mississippi Riverbed profile and water surface elevations (WSEL) between Arkansas City, Arkansas and Natchez, Mississippi for modern 3.0 × mean annual flow (moderate flood ≈ 48,900 m^3/s) and 0.5 × mean annual flow (low flow ≈ 16,300 m^3/s) for circa 1970 and near-modern (mid-1990s) conditions. Note riverbed incision between Arkansas City and Vicksburg and riverbed aggradation between Vicksburg and Natchez. Also note corresponding increases and decreases in WSELs between areas with incision and aggradation. Riverbed elevations and WSELs from Remo et al. (2009).

stream and downstream boundaries were usually associated with discernable changes in floodplain slope prior to river engineering. The segments of the LMR most affected by the tectonic activity are at New Madrid, Missouri and between Greenville and Vicksburg, Mississippi. Along the LMR, steeper floodplain gradients were associated with greater bank erosion rates because of the corresponding adjustments in channel sinuosity. In contrast, the straight segments of the LMR were generally associated with low-valley slopes and channel adjustments tended to be accommodated by changes in channel width (Figure 4; Schumm et al. 1994; Knox 2007). Prior to river regulation, the most geomorphically active segment along the LMR was between the Arkansas and Red rivers. The relatively higher rate of geomorphic activity along this segment was attributed to historically large

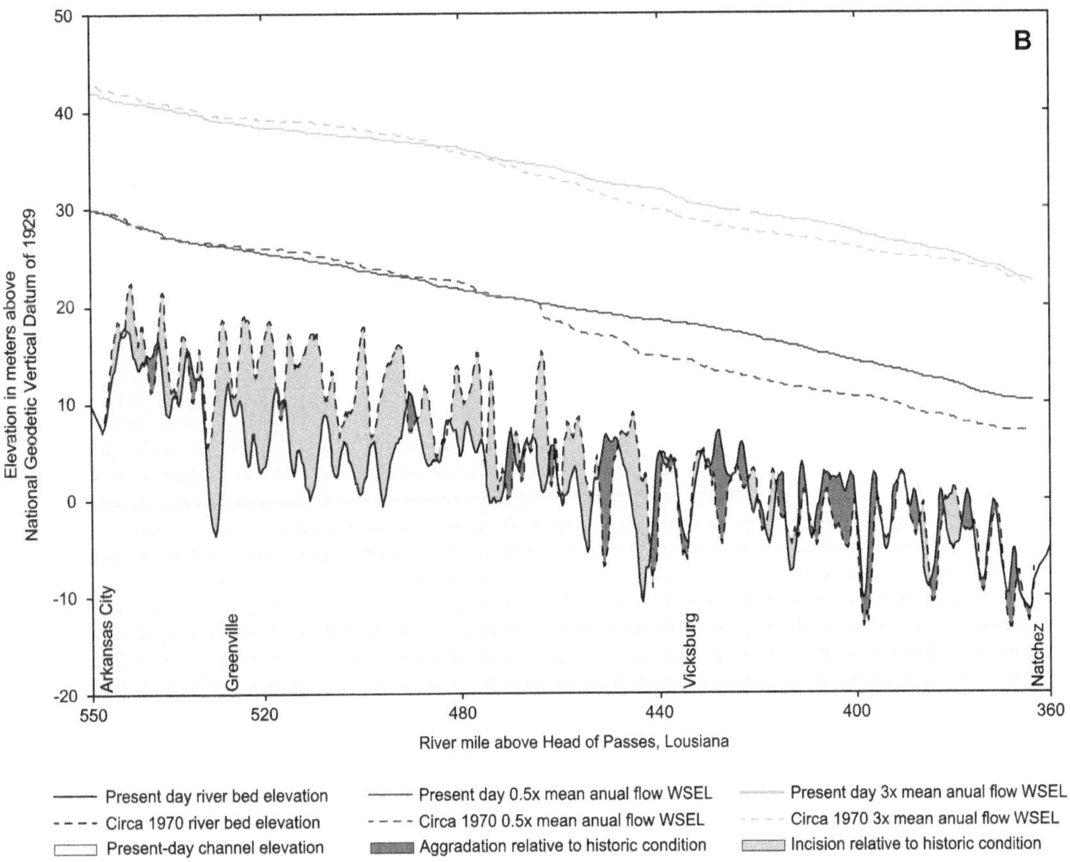

Figure 4. Continued.

input of medium textured sand from the Arkansas River (Schumm et al. 1994).

Management and Engineering History of the Mississippi River

The Mississippi River has been actively managed for navigation and the mitigation of flood risk for more than 200 years. Navigation and flood-risk management activities during the 19th century through today on the Mississippi have been largely carried out by the U.S. Army Corps of Engineers (USACE). The management of the Mississippi for these twin objectives has led to an intensively engineered system in which the hydrology, hydraulics, and riverine ecosystems have been fundamentally altered. Relatively recent (circa 1970) U.S. environmental laws such as the Endanger Species Act (ESA), the Clean Water Act (CWA), and the National Environmental Policy Act (NEPA) coupled with the environmental degradation attributable to navigation and structural flood-mitigation projects have resulted in a third management objective, rehabilitation of the Mississippi River's ecosystem. Other authorized uses in which the Mississippi River and its tributaries are managed include hydroelectric power generation, water supply, irrigation, and recreation (USACE 1994).

Navigation

The Mississippi River and its tributaries have been major transportation routes since prehistory. The first legal document decreeing the Mississippi River and its navigable tributaries as government-protected transportation routes was the Northwest Ordinance of 1787. The Northwest Ordinance states, "[T]he navigable waters leading into the Mississippi and St. Lawrence, and the carrying places between the same, shall be common highways and forever free." This ordinance is attributed with establishing commercial navigation as the preeminent user of the Mississippi River (Rasmussen 1994).

The use of federal funds to make improvements to the Mississippi River and its navigable tributaries did not begin until after the 1824 Gibbons v. Ogden U.S. Supreme Court decision. The Gibbons versus Ogden decision established congressional supremacy in navigation affairs pursuant to the Commerce Clause in the U.S. Constitution (Niebling et al. 2014). Soon after the Gibbons versus Odgen decision, the U.S. Congress tasked the Army Engineer Corps (forerunner to the USACE) with improving navigation through two laws: (1) the General Survey Act, which directed the Army Engineer Corps to survey potential navigation routes; and (2) the First River and Harbors Act, which appropriated US$75,000 for the improvement of navigation on the Mississippi and Ohio rivers by removing navigation hazards such as snags (large submerged woody debris) and sand bars (Dobney 1977).

During the last quarter of the 19th century through the mid-20th century, the U.S. Congress has authorized the USACE to establish and maintain a navigation channel with specific minimum channel widths and depths. Through this period, navigation channel depths and widths were increased to improve the cost efficiency of navigation on the Mississippi River and its navigable tributaries (Table 1; Anfinson 2003).

Early attempts during the late 19th and early 20th century to establish uniform navigation channel (minimum 1.8 m deep and 76.2 m wide) along most segments of the UMR employed an engineering approach commonly referred to as "river training." River training uses structures such as pile dikes, wing dikes, and revetments to train the river into a single-thread, self-scouring channel, which maintains the desired navigation channel dimensions with minimal maintenance dredging (Figure 5). Pile and wing dikes were constructed perpendicular or sometimes oblique from the riverbank to trap sediment in order to constrict the river's flow into a single channel. Willow mats, rock, and articulated concrete revetment were used to armor the riverbanks to prevent bank erosion and channel migration (USACE 2006a). In bedrock-floored segments of the UMR, river training was not sufficient to meet the navigational channel requirements. Navigation locks and dams were built at the

Table 1. Authorization history of navigation channel depths and widths along the Mississippi River. Current channel width and depth authorizations are in bold. LMR = lower Mississippi River, UMR = upper Mississippi River, and MMR = middle Mississippi River.

River segment	Authorization	Navigation channel depth (m)	Navigation channel width (m)
LMR	1896 River and Harbors Act	2.7	76.2
UMR and MMR	1878 Rivers and Harbors Act	1.4	76.2
UMR and MMR	1907 Rivers and Harbor Act	1.8	76.2
LMR	1928 River and Harbors Act	2.7	91.4
UMR and MMR	**1930 River and Harbors Act**	**2.7**	**91.4**
Cairo to Baton Rouge	**1944 River and Harbors Act**	**3.7**	**91.4**
Baton Rouge to New Orleans	**1945 River and Harbors Act**	**12.2**	**152.4**
New Orleans to Head of Passes	**1945 River and Harbors Act**	**12.2**	**304.8**

Figure 5. Examples of river-training structures applied on the Mississippi River: **(A)** pile dike (modified from Alexander et al. 2012), **(B)** revetment (modified from Moore 1972), **(C)** wing dike, **(D)** L-dike, **(E)** bendway weirs (modified from USACE 2012), **(F)** chevron dikes (modified from USACE 2011), **(G)** detached dikes, and **(H)** "W" dike and round points. Arrows indicate general flow direction.

Rock River Rapids near Rock Island, Illinois and Des Moines Rapids near Keokuk, Iowa to achieve the navigational channel requirements through these bedrock floored subsegments (Figures 1 and 2; Anfinson 2003).

As the construction to establish the first navigation channel neared completion in the 1920s, it was understood that this channel design would be inadequate to facilitate reliable and cost-effective shipping on the UMR. A proposal was made to increase the minimum navigation channel depth and width to 2.7 and 91.4 m, respectively (Anfinson 2003). Due to the seasonal low-water levels and geomorphic channel complexity along the majority of the UMR, a new system of river engineering was required to facilitate the enlarged navigation channel. This new system was comprised of a continuous series of 29 locks and dams that extended the 2.7-m-deep and 91.4-m-wide navigation channel from St. Louis, Missouri to Minneapolis, Minnesota (Rasmussen 1994).

Minimum channel depth and width requirements for the navigation channel through the middle Mississippi River (MMR) have been met using river training. The MMR has had the greatest density of river-training structures constructed (~1.4 km per kilometer of river) and the widest variety of structures employed on the Mississippi River (Huthoff et al. 2013). The river-training structures used along the MMR have included pile dikes, wing dikes, L-head dikes, revetments, bendway weirs, chevron dikes, detached dikes, root-less dikes, round points, and "W" dikes (Figure 5). Pile and wing dikes were constructed and used in the same manner as along the UMR. L-head dikes are river-training structures with a section extending downstream from the channel and generally parallel to the channel line. Like wing dikes, the purpose of L-head dikes is to focus flow into a single channel (Parchure 2005). Starting in the 1980s the St. Louis District of the USACE began developing new types of river-training structures to further reduce maintenance dredging of the navigation channel. The first of these new structures were bendway weirs. Bendway weirs are underwater rock structures used on the outside of river bends to maintain the width and depth requirements of the navigation channel through bends. Since the early 2000s, chevron dikes have been constructed along the MMR to maintain or further deepen navigation channel in order minimize dredging (Gordon 2004). Chevrons where also one of the first dike designs with dual purpose of navigation channel maintenance and physical-habitat enhancement (Davinroy et al. 2011). Additional types of dual-purpose structures include "W" dikes, detached dikes, root-less dikes, and round points (Figure 5).

Flows through the MMR are supplemented by the Missouri River Mainstem Reservoir System between March and late November. The MMR's highly engineered river channel coupled with the additional flow from the Missouri River allows for year round navigation on the MMR (USACE 2004). However, during extreme low-flow periods, significant dredging or bedrock excavation can be required to maintain navigation channel depths along the MMR.

Large-scale river training to support navigation on the LMR largely occurred after Mississippi River artificial cutoff program. Between 1932 and 1942, this program constructed 14 artificial cutoffs between Memphis, Tennessee and Baton Rouge, Louisiana. The purpose of the artificial cutoff program was to reduce flood levels and shorten travel distance along this sinuous segment of the LMR. These cutoffs shortened the LMR by 243 km (Moore 1972; Winkley 1977). The effects of the artificial cutoff program resulted in an increased valley slope and stream power along this segment causing significant channel instability (Biedenharn and Watson 1997). Channel-bed incision related to LMR cutoff program extends from Caruthersville, Missouri to at least Baton Rouge, Louisiana (Jemberie et al. 2008; Remo et al. 2009). As a result of the channel instability, significant dredging has been needed to maintain the LMR's navigation channel. In order to reduce the dredging and increase the stability of the navigation channel, more than 900 dike structures and nearly continuous revetment along both banks of the LMR were constructed between the mid-1930s and early 2000s. Bendway weirs have also been constructed in several LMR bends during the 1990s through the early 2000s to improve the navigation channel in these areas (Remo et al. 2009). Like the MMR, additional wing dike, bendway weirs, chevrons, and round points have been either constructed or proposed for the dual purpose of navigation channel maintenance and riverine habitat enhancement along the LMR (USACE 2013).

Flood-Risk Management

The earliest organized attempt to mitigate Mississippi River flooding occurred in 1724 when the French colonial governor Pierre Dugué de Boisbriant compelled riverfront property owners in New Orleans and adjacent parishes to construct a 3-ft-high levee to protect the city. Three years later, in 1727, Dugué de Boisbriant declared New Orleans to be flood proof after the completion of the levee. Despite de Boisbriant's declaration, New Orleans was flooded less than a decade later in 1735 (Clay 1986; McPhee 1989). Throughout the rest of the 18th century and into the mid-19th century, several small levee systems were constructed along the Mississippi River by levee boards, private property owners, and/or local governments. Due chiefly to a lack of adequate funding and engineering, these locally constructed levees were often inadequate to protect floodplain lands from inundation (Camillo and Pearcy 2004).

With increasing floodplain populations and corresponding escalation in flood losses in the Mississippi River Valley during the mid to late 19th century, the U.S. Congress took the first steps toward the federalization of flood-risk management by passing the Swamp Lands Acts of 1849 and 1850. These acts transferred federally owned wetlands along the Mississippi River to the states with the stipulation that funds received from the sale of these federal lands were to be used for building levees and associated flood mitigation infrastructure (Camillo and Pearcy 2004).

In addition to the Swamp Land Acts, U.S. Congress also reaffirmed its commitment to federal management of the Mississippi River by appropriating $50,000 for surveys geared towards the preparation of plans for improving navigation and flood-risk management along the LMR. This appropriation was split between the civilian and military engineering communities, which led to the development of competing plans. A prominent civil engineer, Charles Ellet, Jr., educated at the École Polytechnique in Paris, France, headed the civilian effort, and Lieutenant Colonel Stephen H. Long and Captain Andrew Humphreys led the military planning. Ellet's report, completed in 1852, proposed various river engineering techniques to both accommodate and control the river. The key elements of Ellet's plan were (1) improvement of the existing levees, (2) prevention of cutoffs, (3) the construction of headwater reservoirs along the UMR and other large tributaries, (4) enlargement of natural outlets through Bayou Plaquemine and the Atchafalaya River, and (5) the creation of an artificial outlet from the LMR to Lake Borgne. After than more than a decade of research and monitoring of the hydrology and hydraulics of the LMR, the military engineers released their plan in a report commonly referred to as the Delta Survey (i.e., Humphreys and Abbott 1861). In the Delta Survey (DS), military engineers dismissed several of Ellet's suggestions for LMR valley flood mitigation as too expensive or too dangerous to be attempted. In particular, the DS dismissed the ideas of artificial outlets and headwater reservoirs. Instead, the DS recommended an approach that almost exclusively relied on levees and a few natural outlets to prevent inundation of the LMR floodplain (Camillo and Pearcy 2004). The DS marked the beginning of the levees-only policy for U.S. flood-risk management along the LMR and elsewhere in the United States (Barry 1997; Klein and Zellmer 2007).

After the DSR's publication, construction of new levees and the enhancement of existing levees increased along the LMR. In 1879, the U.S. Congress established the Mississippi River Commission (MRC) to maintain a navigable waterway and assist with levee construction. The MRC was specifically designated to set levee design criteria (e.g., levee heights, widths, and construction material) for the Mississippi River levee systems. It also provided engineering assistance to local agencies in the construction and maintenance of the levee systems (Galloway 2000). Levee heights through this period were generally based off high-water profiles from previous floods. Local levee commissioners and engineers recognized, but ignored, that the construction of new or higher levees often increased flood heights (Davis 1982).

Inconsistent funding of levee construction in addition to setbacks from several levee-damaging floods between the 1880s and the mid-1910s led to slow progress on the LMR levee system. However from the late 1910s through the mid-1920s, levee construction along the LMR was hastened by unprecedented funding by the U.S. Congress through the 1917 and 1923 Flood Control Acts (FCAs) and the development of mechanized earthmoving equipment. By the fall of 1926, the LMR levee systems were nearly complete. In its annual report released on October 8, 1926, the MRC concluded that adequate flood protection had been achieved along the LMR (Camillo and Pearcy 2004).

The inadequacy of levees-only policy for flood mitigation along the LMR was quickly realized a few months later during the winter and early spring of 1927. Between January and April of 1927, three flood waves moved through the LMR causing several levees to be overtopped and breached or, in some cases, fail. The Great Mississippi Flood of 1927 resulted in the inundation of more than 67,400 km^2 of floodplain, 500 deaths, displacement of approximately 600,000 people, and up to \$1.0 × 10^9 in property damage (1927 dollars; Barry 1997; USACE 1998; Galloway 2000). This flood moved the U.S. Congress to assign the USACE and MRC full responsibility for flood-risk management along the LMR valley. They also passed the FCA of 1928, which began the funding for comprehensive flood-risk management project along the LMR. This project eventually became known as the Mississippi River and Tributaries Project (MR&T; Galloway 2000).

The MR&T is a multifaceted program for flood-risk management and navigation along the LMR. This program is comprised of four main components: (1) cutoffs to speed up flows in order to reduce flood levels; (2) enhanced levees (e.g., raised and strengthened) to serve as the principal protection from inundation; (3) four floodways to relieve flooding pressure on the LMR's levee systems downstream of Cairo, Illinois and in vicinity of New Orleans, Louisiana (Figure 6); and (4) construction of reservoirs to control flows into the LMR from its tributaries (USACE 1998).

The FCA of 1928 is not only important as a turning point away from the levees-only policy, but also in the way that levee systems (i.e., levees, flood walls, and associated pump stations) on the LMR are funded. It is estimated that the local levee boards on the LMR had spent approximately \$300 million (1927 dollars) on their levee systems prior the 1927 flood. The U.S. Congress determined that local levee boards had spent enough money on flood mitigation and absolved them from any obligation in the construction of the new levee systems. In addition, the U.S. Congress also released the levee boards from the responsibility for the majority of future repairs to the MR&T levee system. The 1928 FCA only required the levee boards to provide satisfactory assurance that they would maintain and conduct minor repairs on the completed levee systems and provide the federal government with right of way for levees at no cost if they were located along the main stem of the LMR (Niebling et al. 2014).

Following major floods throughout the United States in 1936, the U.S. Congress authorized and funded a major structural flood-risk reduction program that included the construction of hundreds of new dams and levees. Within the MRB alone, more than 250 dams were constructed for flood-risk management and other uses by the federal

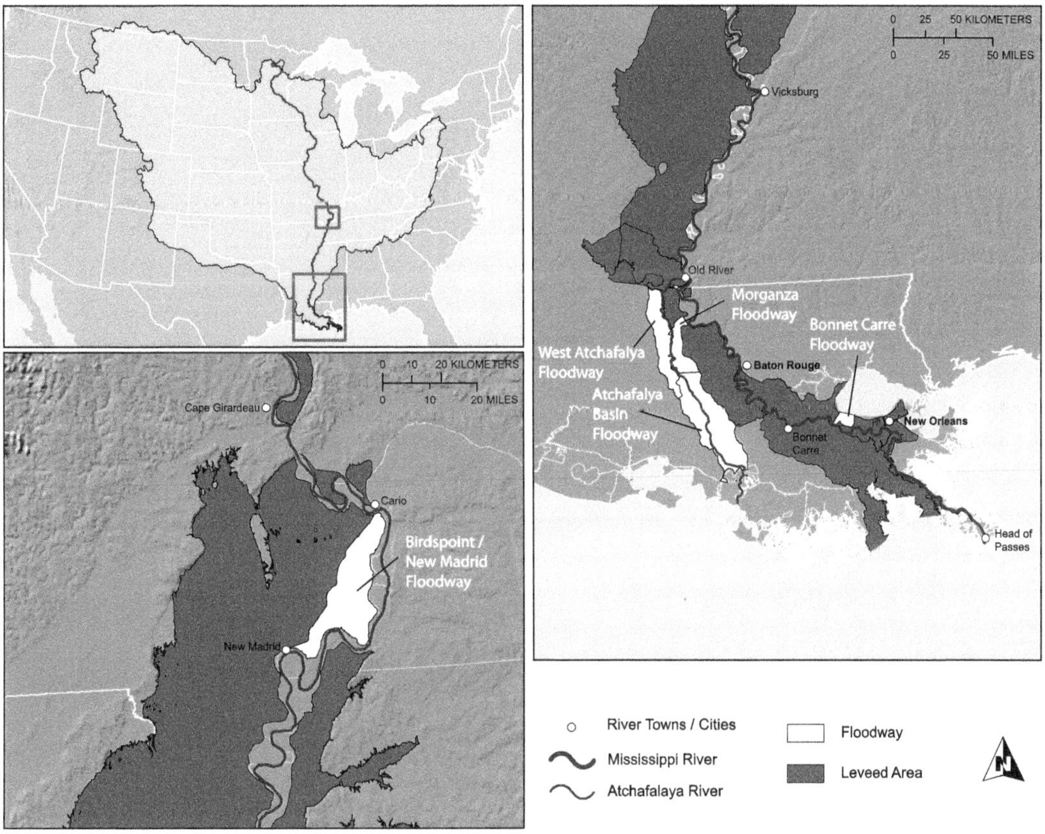

Figure 6. Location of the Mississippi River and tributaries floodways on the lower Mississippi River.

government. These dams were constructed not only to provide mitigation immediately downstream of the dam, but also to reduce flows reaching the LMR valley (Holmes 1979; Galloway 2000). Levee-protection levels outside LMR valley were tied to land-use conditions. Unlike the MR&T project, funding for these levee projects were a mix of both federal and local funding sources. The mix in funding sources coupled with a lack of centralized planning resulted in an assortment of levees with varying protection levels (Galloway 2000). Unlike the MR&T levee systems, levee districts (boards) or jurisdictions were responsible for the full operation and maintenance costs for these federally constructed levees. Federal assistance is available presently for flood fighting and major repairs to the levees that meet USACE standards. However, the levee district or local jurisdictions must fulfill a significant cost match (≥25% of the total cost of the project) in order to receive the federal assistance (USACE 1995).

Today, along the main stem of the Mississippi River, there are approximately 90 levee systems comprised of more than 4,600 km of levees and floodwalls (SAST 1995; USACE 2014b). These flood-mitigation structures protect ~93,800 km^2 of floodplain from inundation. Of the 93,800 km^2 of levee protected floodplain, 90,700 km^2 (97%) are located within the LMR valley (Figure 7; USACE 2008, 2014b). Along the UMR and MMR, levee protection levels (heights) range from less than 10-year flood for nonfederal levees that protect some agricultural lands up to the 500-year protection level for urban and industrial areas near St. Louis, Missouri

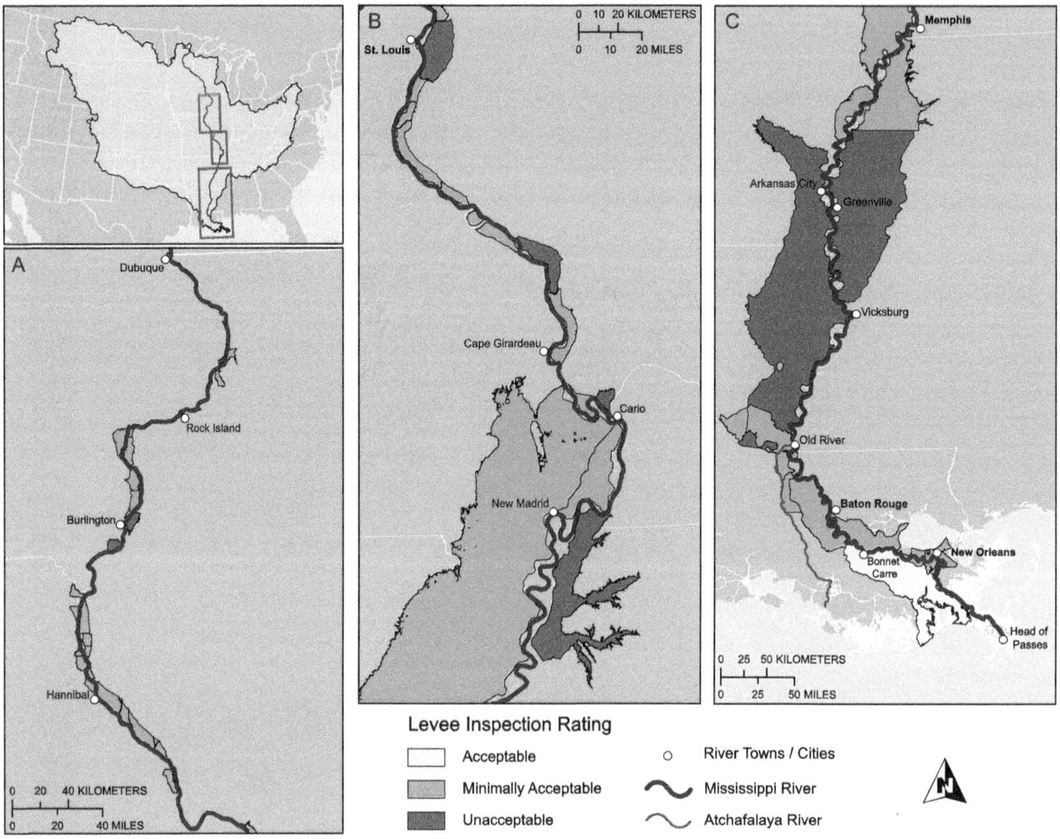

Figure 7. Location of the Mississippi River levee protected areas and the levee systems' inspection rating. (Levee protected areas and levee system ratings from USACE 2014a.)

(USACE 2004). Flood protection levels of the MR&T levees are based on a 60-year-old project design flood (PDF). The PDF is a hypothetical flood constructed from multiple historic flood-generating-rainfall events within the MRB. The PDF at confluence of the MMR and Ohio River at Cairo, Illinois is 70,800 m^3/s. It then increases up to 81,800 m^3/s below the confluence of the Arkansas River and LMR near Arkansas City, Arkansas. Below Arkansas City, the PDF decreases to 77,000 m^3/s by Vicksburg, Mississippi. The decrease of 4,800 m^3/s between Arkansas City and Vicksburg is attributed floodwater storage within the floodplain and tributaries (i.e., backwater storage) along this river segment. Water-diversion structures then divert up to 17,600, 17,000 and 7,100 m^3/s at Old River, Morganza, and Bonnet Carre, respectively, to reduce the flow of the LMR to within the limited floodwater carrying capacity (35,400 m^3/s) of the floodway between New Orleans and Head of Passes, Louisiana (Figure 1). Estimation of the exceedance probability for the MR&T PDF has not been undertaken by the USACE (USACE 1998).

Environmental Management

Federal environmental management of the Mississippi River has largely been born out of the suite of environmental laws passed by the U.S. Congress circa 1970. These laws include the National Environmental Policy Act of 1969 (NEPA), the Clean Water Act of 1972 (CWA), and the Endangered Species Act of 1973 (ESA). The environmental law that has had the most impact on the management of the Mississippi River is NEPA. This law requires all federal agencies to consider the values of environmental conservation for all substantial actions and prescribed procedural processes to ensure that the environmental values are respected. All federal agencies are required to systematically assess the environmental impacts of their proposed actions and consider alternative ways of accomplishing their missions in ways that are less damaging to the environment. On the Mississippi River and its navigable tributaries, the USACE must comply with NEPA regulations for all new river-engineering projects, as well as any significant change in the operation and management of existing levees, river training, or flow-control structures. The next most influential environmental law affecting the management of the Mississippi River is the ESA. The purpose of the ESA is to protect and recover imperiled species and the ecosystems upon which they depend. The USACE is required to ensure that its projects and operation procedures comply with this law through consultation with the U.S. Fish and Wildlife Service. The CWA, administered by the U.S. Environmental Protection Agency, comes into play in the management of the Mississippi River when potential pollutants are discharged into the river (such as dredge materials) or river management activities that affect water quality.

While the federal environmental management of the Mississippi River and its navigable tributaries largely began in the 1970s, state or regional environmental management of UMR and MMR has been ongoing since at least the formation of the Upper Mississippi River Conservation Committee (UMRCC) in 1943. The UMRCC was formed by the states within the UMR basin (Iowa, Illinois, Minnesota, and Wisconsin) to promote the preservation and sustainable utilization of the natural and recreation resource of the river and to formulate policies, plans, and programs for conducting cooperative studies (Garvey et al. 2010). However, it was not until the U.S. Congress designated the UMR, MMR, and Illinois River as a "nationally significant ecosystem" and "nationally significant waterway" for commercial navigation in the Water Resource Development Act (WRDA) of 1986 that environmental management of these waterways became a priority on par with navigation and flood-risk management.

The 1986 WRDA launched the Environmental Management Program (EMP) on the UMR and MMR. However, the permanence of the EMP was not established until the 1999 WRDA, which has allowed for nearly continuous collection of long-term monitoring data at six intensely sampled study subsegments that span the UMR, MMR, and Illinois River. The data collect in this monitoring effort is intended to evaluated trends in hydrologic, geomorphic, and water and sediment quality parameters along with floodplain habitats and fish populations. The USACE, which is responsible for the EMP, distributes funds to the U.S. Geological Survey for administration of the Long-Term Resource Monitoring Program (LTRMP), in cooperation with natural resource agencies from the five states located along these river segments. The LTRMP collects data using a stratified random sampling design in six intensively sampled study segments and assesses trends in the hydrology, geomorphology, floodplain habitats, water and sediment quality, vegetation, and fish populations (Sparks 2010).

The USACE directly manages the Habitat Rehabilitation and Enhancement Program (HREP) component of EMP, in consultation with the states, U.S. Fish and Wildlife Service and other federal partners. The HREP started out by making improvements to existing federal and state fish and wildlife management areas. These improvements included limited floodplain reconnection using water control structures (i.e., pumps and gates) to reconnect a formerly levee-protected floodplain, bank stabilization, construction of stone barriers to stabilize islands from wind-induced erosion, and dredging of aggrading side channels. More recent projects have employed an adaptive management cycle: implementing a new approach, monitoring the outcomes, and using the results to improve the approach (Sparks 2010). An example of such a project was the water-level drawdowns at multiple navigation dams along the UMR to expose mudflats, to dry and compress flocculent sediments, and to stimulate the growth of moist soil plants (Moore et al. 2010). However, the overwhelming majority of habitat-rehabilitation or habitat-enhancement projects have not been monitored to measure project performance or project success (O'Donnell and Galat 2007; Remo et al. 2013). Annual funding for LTMRP is ~$5–7 million and for HREP is $10–14 million (Sparks 2010).

Unlike the UMR and MMR, active environmental management on the LMR is a relatively recent undertaking. The environmental policy on the LMR for the MR&T project has been to "avoid and minimize" undesirable environmental effects (USACE 1998, 2013). Under the avoid and minimize policy, cumulative effects of river engineering projects on endangered species or the broader ecological health of the LMR were not addressed (USACE 2013). Starting in 2001, meetings to discuss LMR river-engineering projects and related environmental conservation issues were initiated between the Mississippi River Valley Division (MRVD) of the USACE, the Southeast Region of the USFWS, and the Lower Mississippi River Conservation Committee (LMRCC). The LMRCC is comprised of 12 state natural resource and environmental quality agencies from Missouri, Kentucky, Tennessee, Arkansas, Mississippi, and Louisiana. In 2005, the USACE with its LMRCC partners began to assess the ecological health and potential for habitat rehabilitation along the LMR (USACE 2013), a process that began on the UMR and MMR nearly 20 years prior. Finally, in 2013, the MRVD of the USACE published the first LMR conservation plan for the endangered interior least tern *Sternula antillarum*, the Pallid Sturgeon *Scaphirhynchus albus*, and the fat pocketbook *Potamilus capax*, in part for compliance with the ESA. The slow uptake of environmental management on the LMR can be attributed to LMR not being a part of the "nationally significant ecosystem" designation in the 1986 WRDA, which led to federal funding and kick-started habitat-rehabilitation projects and long-term environmental monitoring on the UMR and MMR.

Effects of River Engineering on the Mississippi River System and Its Delta

In this section, I discuss the cumulative effects of river engineering on the hydrology, hydraulics, sediment transport, geomorphology, river-delta dynamics, and riverine biota of the Mississippi River. Even though this summary of river engineering effects is divided into physical and biological components, in reality these components operate as a system through a complex series of interactions and feedback loops. Several of the natural interactions and feedback loops that have developed over the geologic history of the Mississippi River have been altered by river engineering. Many of these interactions and feedback loops are still not well understood by scientists. The summaries of river-engineering effects in this section area a simplification of the complexity of the system intended to distill the effects of river engineering into the mostly completely known and documented impacts.

Hydrology and Hydraulics

River engineering has significantly changed the hydraulics of many Mississippi River segments. On the UMR, the system of locks and dams has raised the water-surface elevations (WSELs) of low flows to maintain the 2.7-m-deep navigation channel. The elevated WSELs created by these structures has resulted in an overconnection of side channels and backwaters throughout the majority of the UMR and has prevented seasonal drying of wetlands (USGS 1999; Koel 2004; Johnson and Hagerty 2008). In addition, there has been a loss of sediment storage near the locks and dams, which has increased the number of days that low-floodplain surfaces are inundated (Grugbaugh and Anderson 1989). For flood flows along the UMR, Jemberie et al. (2008) found that navigation dam emplacement had little to no discernable impact upon flood WSELs. However, between Rock Island, Illinois and St. Louis, Missouri, where large levees were constructed during the 20th century, some hydrologic monitoring stations have shown significant increases in flood heights (up to 1.7 m; Jemberie et al. 2008).

Along the MMR, river-training structures have concentrated flow into a narrower and deeper channel increasing the river's stream power. The increased stream power, in combination with reduced sediment supplies, has resulted in channel-bed incision, which has progressively lowered WSELs for lower-magnitude discharges along the majority of the MMR (Jemberie et al. 2008). This channel-bed incision has impaired the connection of the MMR's main channel with its few remaining side channels (Johnson and Hagerty 2008). However, despite this incision, the direct and indirect physical effects of river-training structures, in combination with levees, have reduced flood conveyance along MMR resulting in increases in flood WSELs up to ~4.0 m (Figure 3; Belt 1975; Stevens et al. 1975; Pinter et al. 2001, 2009, 2010; Criss and Shock 2001; Remo and Pinter 2007; Jemberie; et al. 2008; Remo et al. 2009).

On the LMR, the MR&T cutoff program lowered WSELs for all magnitudes of discharge by up to ~5.0 m between New Madrid, Missouri and Vicksburg, Mississippi. The lowering of the WSELs is attributed to the channel-bed incision, which resulted in an increase in stream power due to the shorting of the river's length. However, up to 40% of the stage reductions from the cutoff program have been reclaimed between 80 km (50 river miles) upstream of Vicksburg to Natchez, Mississippi (Figure 4A, 4B; Harmar et al. 2005; Jemberie et al. 2008; Remo et al. 2009).

Prior to construction of the Mississippi River levee systems, adjacent low-lying portions of the floodplains were inundated semiannually, and the entire valley bottom was inundated during large flood events (Yin and Nelson 1996; USACE 1998). The levee-system construction has resulted in the isolation of the Mississippi River from the majority of its floodplain between Rock Island, Illinois to just upstream of the Head of Passes, Louisiana (USACE 1998; Theiling and Nestler 2010). On the heavily leveed portions of the UMR and the MMR, between 50% and 75% of the natural floodplain area is protected to greater than 50-year flood-protection level (SAST 1995; Remo et al. 2009). On the LMR, the floodplain area open to inundation has been reduced by ~90% (Figure 7; Kesel 2003; Remo et al. 2009). However, despite the large percentage of floodplain protected from inundation by LMR's levee systems, more than 7,200 km^2 of floodplain between the mainline levees (i.e., batture areas) remain connected to the river between Cairo, Illinois and New Orleans, Louisiana (~1,400 km). This is more than twice the floodplain area that remains connected along the entire UMR and MMR combined (3,200 km^2 of floodplain along ~1,350 km of river between Minneapolis, Minnesota and Cairo, Illinois). The LMR batture lands contain a diverse array of terrestrial and aquatic habitats important for many riverine organisms (Baker et al. 1991) and provide important ecosystem services such as nutrient sequestration (Schramm et al. 2009), recreational opportunities, and floodwater storage (USACE 1998; DeHann et al. 2012).

Sediment Transport

Before the wide-spread establishment of European-style agriculture and major flow modification to the rivers within the MRB (e.g., circa 1800), the Mississippi River is estimated to have transported 400 million metric tons, on average, of sediment to coastal Louisiana annually (Meade and Parker 1985; Kesel et al. 1992). Between 1987 and 2006, annual sediment delivery to coastal Louisiana averaged 170 million metric tons, a 60% decrease from 1800 (Rebich and Demcheck 2007; Meade and Moody 2010). The majority of the decrease in sediment transport along the MMR and LMR has been attributed to the construction of large dams on the Missouri and Arkansas rivers (Meade and Parker 1985; Keown et al. 1986).

River-training structures have also been attributed with reducing sediment transport along the MMR and LMR. River-training structures along these segments have caused a significant amount of land accretion. In addition, the emplacement of revetments to prevent bank erosion has likely further reduced sediment supplies. Revetments along the LMR have been attributed with a 90% reduction in bank caving, which was an important source of the total sediment load of the LMR prior to the large-scale river training undertaken to mitigate the channel degradation caused by the MR&T cutoff program (Kesel et al. 1992; Kesel 2003; Meade and Moody 2010). Meade and Moody (2010) have suggested that the major reduction (~50 million metric tons) in suspended sediment discharge observed along the LMR during the mid to late 1960s may have been related to these channel stabilization efforts.

Since about 1970, sediment transport along the LMR has continued to gradually decline by ~1 million metric tons per year. Meade and Moody (2010) attributed this continual loss in their sediment budget to improvements in agricultural conservation practices and dredging. However, this explanation for the deficit in their sediment budget has a few weaknesses. In the case of agricultural conservation practices, most of the response of the landscape to soil conservation generally occurs as adjustments within the individual tributary and not as a decrease in sediment load to the trunk river (Trimble 1983, 1999). As for the dredging, dredging along the MMR and LMR are primarily focused on sandy deposits within the relatively higher-velocity navigation channel (Winkley 1977; Winkley and Brooks 1980; Pinter et al. 2004) and not fine-grained materials, which comprise the Mississippi River's suspended sediment load.

An alternative to Meade and Moody's (2010) land conservation and dredging explanation to account for the deficit in their sediment budget is the hypothesis that the batture lands are a substantial sediment sink for suspended sediment. Since 1970, suspended discharge has decreased by ~40 million metric tons along the LMR (Meade and Moody 2010). While this is a sustainable amount of sediment, 40 million metric tons (or 0.064 km^3 assuming a density of 1.6 metric ton per cubed meter for silty clay) of sediment distributed evenly over 7,700 km^2 of batture lands along the MMR and LMR equates to less than 1 cm of deposition over a 40-year period or a deposition rate of 0.25 mm/year. Such a low sedimentation rate would likely go relatively unnoticed within the largely wooded batture lands. There is also antidotal evidence of substantial localized in filling of floodplain water bodies and wing-dike fields within the batture lands during the past 40 years (Baker et al. 1991; USACE 2013) lending credence to this hypothesis. While more work needs to be undertaken to test the batture land sediment-sink hypothesis, it is a plausible alternative to Meade and Moody's (2010) land conservation and dredging explanation for the closure of their sediment budget.

Geomorphology

The morphology of the Mississippi River has been affected by river engineering in three main ways: (1) channel simplification, (2) channel incision, and (3) disconnection of its floodplain from the river channel. Before major-river engineering, the Mississippi River along the majority of its length was more complex and dynamic. The river would undergo substantial channel adjustments annually, which resulted in a physically diverse channel and floodplain structure. This diverse physical structure created significant diversity in riverine and floodplain habitat heterogeneity (Yin and Nelson 1996; Moody et al. 2003; De Jager and Rohweder 2012; Remo et al. 2013). Channel simplification was accomplished through a combination of river-channel straightening and the closing of side channels to create a relatively uniform single-thread channel (Winkley 1994; Yin and Nelson 1996).

During the historic period, the UMR's and MMR's channel position and river length have been relatively constant compared to the LMR. However, island area has decreased along both the UMR and MMR. Within the pooled segments of the UMR, wind-driven waves have eroded away many islands. The sediment from these islands has filled in many of the floodplain lakes and impounded areas immediately upstream of the locks and dams (USGS 1999). Island area along the MMR has decreased from ~960 km^2 in 1821 to less than ~170 km^2 by 1970 (Stevens et al. 1975). This reduction in island area is the result of the use of closing dikes (i.e., dike structures that restrict flow through side channels) and other river-training structures that force the MMR into a single-thread channel (Remo and Pinter 2007). The LMR channel cutoff program initially shortened the river channel by ~330 km. However, natural readjustment of the LMR during and after the cutoff program resulted in a net channel shorting of only 235 km. Additional adjustment of the LMR to the river shortening has been largely ameliorated through the use of bank revetment emplaced during the 1960s and the early 1970s (Winkley 1977; Biedenharn et al. 2000). The current LMR channel is both shorter and, on average, wider than before the channel cutoff program (Kesel 2003).

Historic changes in riverbed elevation along the Mississippi River largely reflect their engineering history. Along the UMR, the riverbed has been locally scoured and aggraded. Scour has been generally associated with segments below the navigation dams and areas of repeated channel dredging (Alexander et al. 2012). Areas of local channel aggradation along the UMR can be found near tributary mouths, areas where flow is divided between the main and side channels and in long, straight segments (Pinter et al. 2004). Specific gauge analysis and retro modeling along the MMR and LMR generally suggests a net pattern of channel incision since the late 19th or early 20th centuries (Jemberie et al. 2008; Remo et al. 2009). The likely causes of channel incision along the MMR are the combined effects of increases in stream power due to river training (Jemberie et al. 2008) and a decrease in sediment supply from the Missouri River (Meade and Moody 2010). Exceptions to the pattern of incision along the MMR are local bed-rock-floored segments of the channel between Chester and Thebes, Illinois (Remo et al. 2009). Channel incision along the LMR at the subsegment (>10 km and <100 km) and reach scales (<10 km) is attributed to the decrease in sediment supply and an increase in stream power resulting from the MR&T channel cutoff program and local river training. Subsegments of the LMR that have aggraded or continue to aggrade are attributed to deposition of sediment mobilized as a result of the cutoff program (Figure 4A, 4B; Harmar et al. 2005; Jemberie et al. 2008).

Mississippi River Delta

Six delta complexes comprise the Mississippi River delta and associated coastal wetlands (Figure 8; Blum and Roberts 2012). Each of these deltas is the result of 6,000–8,000 years of cyclical deposition and erosion processes (Roberts 1997; Coleman et al. 1998). The deposition cycle is the river-dominated phase of delta development where the delta grows in area and expands out onto the floor of the Gulf of Mexico. The erosional cycle is a marine-dominated phase where the delta complex is abandoned by the river and delta then subsides into the gulf. Wave action reworks and erodes the delta resulting in a reduction in its area. There are two delta complexes currently active on the Mississippi River Delta Plain (Figure 8). The first of these, the Plaquenmines-Balize delta complex, is transitioning from a river-dominated to a marine-dominated phase. Since about 1500 AD, the Atchafalaya distributary channel has been capturing discharge and sediment from the Mississippi River, resulting in the creation of the newest delta complex on the Mississippi River delta plain, the Atchafalaya-Wax Lake delta (Mossa 1996).

The recent appearance (early 20th century) and subsequent growth of Wax Lake Lobe of the Atchafalaya-Wax Lake delta complex at the mouth of the Atchafalaya River have been attributed to river engineering. In the mid-20th century, a diversion canal from the Mississippi River to the Atchafalaya River was constructed. This canal hastened the capture of the Mississippi River's flow by its Atchafalaya distributary, compelling

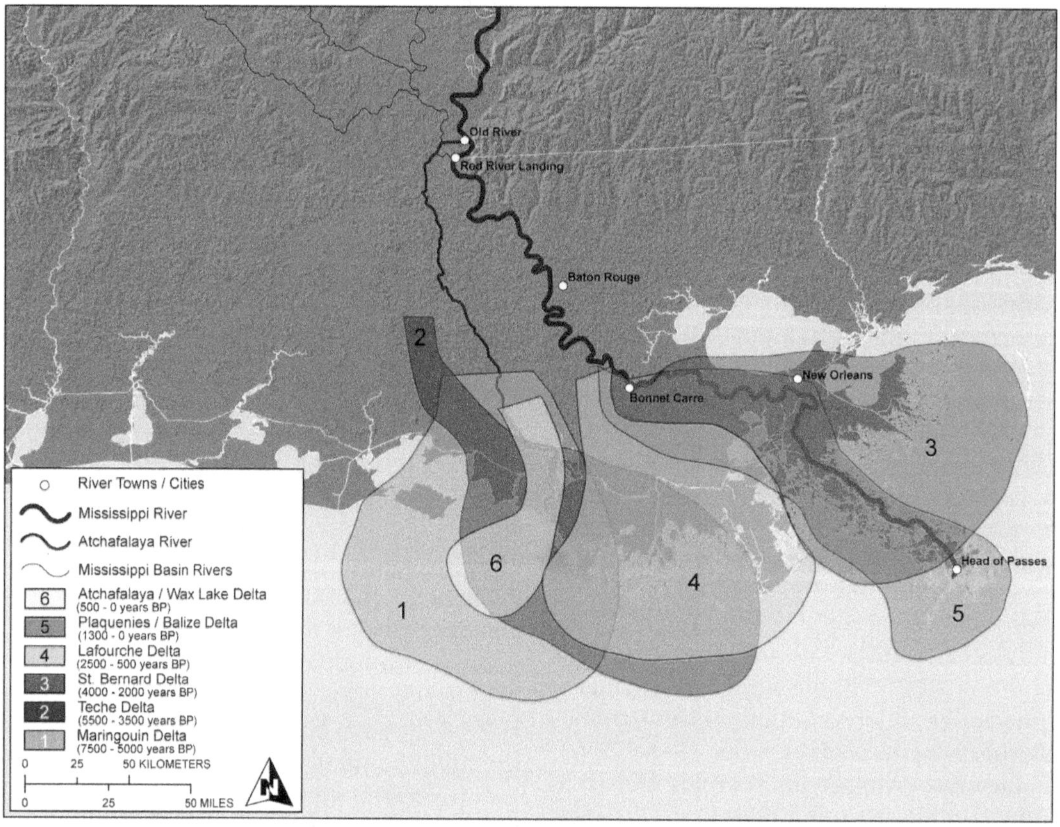

Figure 8. Approximate extent and depositional periods of the Mississippi River delta complexes (after Blum and Roberts 2009).

the USACE to build what is now a system of controls structures located near Old River, Louisiana (Figures 1 and 6; McPhee 1989). The Old River Control Structure currently regulates the flow of water and sediment to the Atchafalaya River and Atchafalaya-Wax Lake delta complex. The Old River Control Structure currently passes no more than 30% of the total discharge of the Mississippi River to the Atchafalaya River (Mossa 1996). However, this amount of flow has allowed for continued development of the Atchafalaya-Wax Lake delta complex (Couvillion et al. 2011).

Approximately 4,900 km^2 of coastal lands have been lost since the early 20th century in Louisiana (Day et al. 2007). Although deltas are prograding in some areas, there is a net loss because of land subsidence and sea level rise. The majority of these lands were wetlands whose losses are mainly attributed to their conversion to open water. However, a lesser amount of wetland loss can be ascribed to the draining of wetlands (Britsch and Dunbar 1993). Land loss rates in Louisiana have accelerated from 17.4 km^2/year in the early 20th century to as much as 100 km^2/year by the 1970s (Britsch and Dunbar 1993; Boesch et al. 1994). Since the 1970s, land loss rates have decreased to 43 km^3/year, with periodically higher rates associated with recent hurricanes (Couvillion et al. 2011).

Wetland and coastal land losses in Louisiana are the result of both natural and human-induced processes. The stability of the delta surface and associated wetlands is an intricate balance between relative depositional rates that drive vertical and horizontal growth of the delta plain and the processes of submergence and erosion, which result in aerial land loss. Renewal of wetland surface with sediment and sediment stabilization from vegetation growth are the ways in which the elevation of a delta's surface remains resilient against subsidence, sea level rise, and storm surge (Paola et al. 2011). Land loss in the Plaquenmines-Balize delta complex has been attributed to the construction alteration of sediment delivery to and the construction of manmade canals within the delta. Levees along its distributary channels are attributed with the elimination of overbank flooding onto the delta plain, which prevents the replenishment of the delta top with sediment. The sediments carried by these channels are deposited into the deeper waters of the Gulf of Mexico (Paola et al. 2011). In addition, reductions in sediment delivery to the Plaquemines-Balize delta complex are ascribed chiefly to the construction of reservoirs on the main-stem Missouri, the diversion of sediment through the Old River Control Structure, and other river engineering activities that have reduced bank erosion or created sediment sinks within the river corridor (Bass and Turner 1997; Meade and Moody 2010). Manmade canals, constructed largely by the petroleum industry, have resulted in saltwater intrusion, which kills wetland plants increasing wetland loss, and accelerated subsidence from petroleum fluid extraction (Mallman and Zoback 2007; Paola et al. 2011).

In 1978, the state of Louisiana began to regulate development activities that contributed to coastal land loss. Then, in 1990, the U.S. Congress enacted the Coastal Wetlands Planning and Protection Act, which has allotted between $30 million and $80 million in funding annually for coastal land loss mitigation. Since that time, more than 150 coastal restoration or protection projects have been undertaken in Louisiana. These projects have been attributed with creating more than 450 km^2 of wetlands and protecting another 2,200 km^2 of coastal lands (Alexander et al. 2012).

After the ruinous hurricane season of 2005, the state of Louisiana and the USACE have undertaken a comprehensive coastal protection and restoration planning effort. This planning effort involves multiple lines of defense, including coastal restoration, structural flood mitigation (i.e., levees and floodwalls), and nonstructural flood mitigation (i.e., voluntary buyouts and raising of existing structures; USACE 2009). Even if the coastal restoration strategies are successful, delta land loss will continue because sediment deliveries to the delta are likely to be insufficient to keep up with the combined forces of land subsidence and sea level rise unless there is a significant increase in the sediment supply from the MRB (up to ~180 million metric tons per year in 2100; Blum and Roberts 2009, 2012).

Effects of River Engineering on Fish and Other Riverine Biota

The Mississippi River supports a rich fish community. The richness of this fish community is attributed to the Mississippi's north–south geographic orientation and its large discharge gradient. Temperatures and discharge both increase from north to south along the Mississippi River. The Mississippi's headwaters begin in a cool temperate climate. As the Mississippi River flows southward, the climate slowly changes to subtropical as the river nears its outlet in the Gulf of Mexico (Schramm 2004). The Mississippi River contains 195 species of freshwater fish, representing nearly a third of the freshwaters species in North America (Fremling et al. 1989). The majority of these fishes require a diverse range of habitats to complete their life cycle (Wiener et al. 1998), yet the quality and quantity of these habitats have decreased along much of the Mississippi River (Koel 2004; Johnson and Hagerty 2008; DuBowy 2013).

When the navigation dams were first constructed on the UMR, there was an increase in habitat diversity resulting from the initial inundation of terrestrial areas. Since this initial increase in habitat diversity, substantial amounts of this diversity have been lost. Losses in habitat diversity are attributed to wind-fetch erosion of islands within the navigation pools, altered river hydraulics, and infilling of backwaters and floodplain water bodies with sediment (Johnson and Hagerty 2008; De Jager and Rohweder 2012). Altered river hydraulics has been identified as a potential cause for the decline of some fish species that require higher flow velocity habitats (Wiener et al. 1998). In addition, reproduction of migratory fish, such as the Paddlefish *Polyodon spathula*, may be adversely effected by navigation dams that restrict access to suitable spawning habitat (Wilcox et al. 2004; Garvey et al. 2010).

Along the UMR and MMR, fish species diversity has been correlated with physical habitat diversity (Koel 2004; Chick et al. 2005). Koel (2004) found that northern segments of the UMR, which were more hydrologically connected to their floodplains, and secondary channels had higher fish species richness and habitat diversity then the southern leveed segments of the UMR and the MMR. Along the MMR, habitat loss has been identified as one of the potential factors for the decline of the endangered Pallid Sturgeon Scaphirhynchus albus (USFWS 2000; Hurley et al. 2004). To improve habitat conditions in the most impacted segments of the UMR and MMR, Koel (2004) suggested that side-channel and floodplain reconnection and habitat rehabilitation could help reestablish the functional habitat heterogeneity across these highly altered riverine corridors.

The conclusion drawn from fish community studies (i.e., Koel 2004; Chick et al. 2005) and broader ecological studies (i.e., USGS 1999; Johnson and Hagerty 2008) by environmental managers was that the modern, highly engineered Mississippi River corridor has resulted in shifts in the natural habitat distribution within the river and this is likely to have had detrimental effects on the fish and other communities. Based on these conclusions, environmental mangers and river engineers have focused on the rehabilitation of physical habitat and physical habitat diversity for the preservation of threated and endangered species (Theiling 1995; USFWS 2000). Along the more degraded (i.e., substantially diminished channel form complexity) portions of the UMR and MMR, the USACE is using the same or some modified version of river training structures once used to simplify the channel for the facilitation of navigation to increase habitat and its heterogeneity. Very little effort has been undertaken to explore other options to increase physical habitat and physical habitat diversity such as side-channel or floodplain reconnection along these highly degraded river reaches. In addition, very little, if any, research has been undertaken to further explore whether or not habitat heterogeneity is the primary factor limiting the recovery of endangered or threatened aquatic organisms (Remo et al. 2013).

Along the LMR, it has been asserted that fish that inhabit higher-velocity habitats have not declined as severely as they have on the impounded UMR (Pflieger 1975; Baker et al. 1991). However, studies to test this assertion have only recently begun or are still in planning stages (USACE 2013). At present, the Pallid Sturgeon has been the only LMR fish species identified by the USWFS as endangered. As in the MMR, the endangered status of the Pallid Sturgeon is attributed, in part, to losses in habitat and habitat heterogeneity within the highly engineered LMR channel (USACE 2013). The use of river training structures for the enhancement of physical-aquatic habitat and its diversity are currently being explored for the LMR (DuBowy 2013; USACE 2013).

Levees have eliminated or substantially reduced the frequency of inundation of large floodplain areas (~94,000 km^2; Figure 7). The exclusion or greatly reduced frequency of inundation on levee-protected floodplains has eliminated many diverse, seasonally flooded habitats from the riverine ecosystem (Johnson and Hagerty 2008; De Jager and Rohweder 2012). In addition, the isolation of the river from its floodplain substantially alters the flood pulse, which is thought to be important to the health of many riverine ecosystems. Inundation during flood pulses allows for the vital exchange of nutrients between the river, its floodplain, and its floodplain water bodies (Junk et al. 1989; Galat et al. 1998). While floodplains are important components of the Mississippi River ecosystem, the amount of floodplain required to maintain crucial ecosystem services such as nutrient cycling, biotic diversity, and water supply is largely unknown (Johnson and Hagerty 2008).

Despite the large area of floodplain protected from inundation by Mississippi River floodwaters, more than 7,300 km^2 of floodplain remains connected along the LMR. These floodplain areas provide important services to society, such as nutrient sequestration (Schramm et al. 2009), flood-water storage, recreation opportunities, and habitat for commercial fishes (DeHaan et al. 2012; USACE 2013). For example, Schramm et al. (2009) has estimated that the LMR floodplain soils and associated riverine biota are able to sequester 542 kg nitrogen/ha, suggesting that the batture land along the LMR performs a very important role in nutrient dynamics. In addition to nutrient sequestration, the LMR batture lands are also important for the storage of floodwaters and have been attributed, in part, to the success of the MR&T Project's prevention of more than $612 billion in flood damages (DeHaan et al. 2012).

Effects to the Mississippi River from beyond the Riverine Corridor

This section presents the anthropomorphic and natural forces that are provoking change in the Mississippi River and its ecosystem from beyond the riverine corridor. These forces include point and nonpoint source pollution and climate change.

Water Quality

Water quality in the Mississippi River has largely transcended traditional river corridor management. Contamination of the Mississippi River with metals such as mercury and lead closely followed settlement of the MRB by European immigrants during the 19th century. Metal contamination in the Mississippi River peaked in the 1960s and has decreased since then along most river segments. DDT, its metabolites, and PCBs biomagnify to concentrations that adversely affect wildlife in upper tropic levels. U.S. environmental regulations, beginning with the 1972 CWA, have reduced discharges of waste from point sources, improved wastewater treatment, and reduced or banned the production of certain contaminants, including DDT, PCBs, and lead. Responses to water regulatory measures have reduced ecosystem contamination and biotic exposure, leading to the recovery of affected wildlife. However, the pace of recovery has been slow. Affected populations, such as the double-crested cormorant *Phalacrocorax auritus*, the American white pelican *Pelecanus erythrorhynchos*, and the American bald eagle *Haliaeetus leucocephalus*, have required decades to recover from reproductive system impairment related to chronic exposure to toxic substances. Full riverine ecosystem recovery is expected to take several more decades, if not centuries (Wiener and Sandheinrich 2010).

Past successes in managing point-source pollution impacts on the Mississippi River's water quality have not extended to nonpoint-source pollution. The current regulatory framework provided under the CWA does not adequately address nonpoint-source pollution (Wiener and Sandheinrich 2010). The gap in the CWA regulatory framework is demonstrated by the large hypoxic zone in the northern Gulf of Mexico, which is largely caused by the export of nitrogen and phosphorus from agricultural watersheds within the MRB (NRC 2008). Industrial agriculture within the MRB, in particular the UMR subbasin, uses substantial amounts of nitrogen-based (nitrate, nitrite, and ammonium) fertilizers (NRC 2008). In many locations, agricultural practices such as conventional tillage and tile drainage enhance the loss of nutrients from agricultural fields (McIsaac et al. 2002; DuBowy 2013). Approximately 90% of the nitrogen load reaching the Gulf of Mexico from the Mississippi River is nonpoint source, of which 54% of the total nitrogen load is attributable to fertilizer and mineralized soil nitrogen (Goolsby et al. 1999; NRC 2008). The remaining 10% of the nitrogen load reaching the Gulf of Mexico is comprised from municipal and industrial point sources. Corn and soybean cultivation is the largest contributor of nitrogen (52%) to the

Gulf of Mexico, followed by atmospheric deposition (16%; Alexander et al. 2008). A recent study by Sprague et al. (2011) showed that little progress has been made in reducing nitrate levels in the Mississippi River. Flow-normalized nitrate concentrations have increased by 9–76% between 1980 and 2008 in five of eight long-term monitoring sites within the MRB. Among the three sites in which concentrations decreased, decreases were very modest (1% to 3%; Sprague et al. 2011). Phosphorus loads are also relatively high within the Mississippi River. The contribution of phosphorus to the Gulf of Mexico is equally divided among the UMR, MMR, LMR, Ohio River, and Missouri River subbasins. Phosphorus in the northern Gulf of Mexico originates primarily from animal manure on pasture and range lands (37%), followed by agricultural lands that grow corn and soybean (25%; Alexander et al. 2008).

The discharge of nitrogen- and phosphorous-laden loaded waters from the Mississippi River into the Gulf of Mexico have increased algal growth, resulting in elevated turbidity and depleted dissolved oxygen levels in the lower 10–50% of the water column (Rabalais et al. 2002). The area of the northern Gulf of Mexico affected by hypoxic conditions is generally referred to as the "dead zone" because few marine organisms can survive in water with low oxygen concentrations (dissolved oxygen <2.0 mg/L; Rabalais and Turner 2001). The 5-year (2008–2013) average size of the hypoxic zone is 14,700 km^2. However, the hypoxic zone varies considerably from year to year. For example, in 2011 and 2012, hypoxic zones encompassed approximately 20,000 and 7,600 km^2, respectively (Mississippi River Gulf of Mexico Watershed Nutrient Task Force 2014). The size of the hypoxic zone depends on both natural and anthropogenic factors. These factors include inflow of poorly oxygenated waters from the deep gulf, river loading of organic carbon, channelization, coastal land loss, nitrogen flux from the atmosphere or deep gulf waters, and climate-induced alterations in discharge from the MRB (Rabalais and Turner 2006). The large difference in the hypoxic zone's size between 2011 and 2012 were attributed, at least in part, to difference in the amount of water discharged from the Mississippi River. In 2011, a very large flood and above-average flow conditions throughout much of the year contributed to nutrient loading into the northern gulf, which in turn created a larger than normal hypoxic zone. The smaller than normal hypoxic zone in 2012 was attributed to summer drought conditions and associated below average flows from the Mississippi River, resulting in greatly reduced nutrient inputs into the northern Gulf of Mexico (Mississippi River Gulf of Mexico Watershed Nutrient Task Force 2014).

Variations in primary production in the surface waters of the northern Gulf of Mexico are largely driven by the delivery of nitrogen from the Mississippi and Atchafalaya rivers (Rabalais et al. 2002). However, the annual cycle of nitrogen and phosphorus loads substantially affects these patterns of primary production over the continental shelf, while light is the limiting factor affecting primary production in the delta regions of the Mississippi and Atchafalaya rivers. The light limitation in the delta regions is attributed to the presence of suspended terrigenous sediments and dissolved organic matter (Lohrenz et al. 1999; Fennel et al. 2011; Quigg et al. 2011). Between the light- and nitrogen-limited regions of the northern Gulf, there is a substantial area (3,400 km^2 on average) where phosphorous limits primary production. This phosphorus limitation occurs during the annual peak in primary production between May to July. The phosphorous limitation on primary production in this intermediate area between light- and nitrogen-limited regions of the northern Gulf is largely driven by the timing and magnitude of inputs from the Mississippi and Atchafalaya rivers (Laurent et al. 2012; Laurent and Fennel 2014).

Climate Change

Paleohydrologic evidence, historical data, and regional climate modeling indicate precipitation, runoff, and flood frequencies throughout the MRB have changed in the past and are likely to change in the future. Geological evidence suggests that modest increases in temperature within the UMR basin have had large effects on river discharges in the Holocene recent geologic past (<10,000 years before present; Knox 1999, 2003, 2007). During the 20th century, there has been a 50% increase in the frequency of days with heavy precipitation (>101.6 mm) in the UMR basin. Summer and winter precipitation across UMR basin has been above average during the past 30 years, resulting in the wettest period in the past century (Kunkel et al. 2008).

Based on past changes in climate, specifically modest increases in temperature (~1.5°C), Knox (2009) concluded the 21st century will likely be characterized by extraordinary variability in river discharges within the UMR basin. These river discharges will likely contain periods with unusually high frequency of large floods separated by periods of smaller floods and occasional very low discharges associated with short-term droughts (Knox 2009). Climate models indicate that precipitation in the UMR basin will increase between 2009 and 2099 during the winter and spring months, with summer rainfall remaining the same or slightly less (Wuebbles and Hayhoe 2004; Wuebbles et al. 2009; Chien et al. 2013). The frequency and magnitude of floods are expected to increase because of increased storm intensity during the summer. This intensification in storm strength may increase precipitation by 21% between the 1990s and 2050, resulting in up-to-50% increase in stream flows in the UMR basin because this more frequent precipitation would likely fall on near-saturated ground, increasing runoff (Jha et al. 2004).

Along the north–south oriented Mississippi River Valley, southern species are expected to expand their ranges northward as the climate warms, while northern species ranges are expected to contract against the headwaters in Minnesota. In addition, changes in the timing, duration, and frequency of both high and low flows are expected to occur with climate change. These changes will certainly favor some species over others (Sparks 2010). For example, Tucker et al. (2008) have already noted a shift in dominance in turtle species along the lower Illinois River. They attributed the change in dominance in the turtle population along the lower Illinois River to a 10-year trend of earlier warming in the spring, which extends the reproductive season of the red-eared slider Trachemys scripta elegans. Changes in the timing of the seasonal flood pulse may also favor fish species that are early- , middle-, or late-season spawners that utilize the floodplain (Hoover and Killgore 1998).

River Management Challenges

Engineering and management of the Mississippi for navigation and flood-risk reduction have created many challenges. The greatest of these challenges are "wicked problems," where unintended consequences of river engineering, such as altered hydrology and sediment transport, have combined with other, even larger-scale human and natural coupled forces such as climate-change-driven sea level rise. In this section, acute river management challenges, including river infrastructure, climate change, riverine ecosystem rehabilitation, and delta-land loss, are discussed.

Navigation

Navigation has been the primary focus of river management on the Mississippi for nearly 200 years. It is asserted that a reliable navigation channel on the Mississippi River is essential for the U.S. economy because it is the heart of the U.S. inland system of navigation, which carries more than $70 billion in goods and commodities each year (Camillo and Pearcy 2004; DeHaan et al. 2012). Historically, funding for inland navigation engineering and infrastructure was born by the U.S. tax payer. In the 1986 WRDA, an inland waterway system Inland Waterways Trust Fund (IWTF) was established to help distribute the costs of the navigation infrastructure maintenance between the U.S. taxpayer and the user, the navigation industry. The IWTF has been supported by a tax on commercial transportation fuel of $0.20/gal since 1994. The annual contributions from the IWTF and matching tax payer funds are approximately $160 million, annually (NRC 2013a).

Despite the $160 million in annual funding from the U.S. Government and the IWTF, the principle navigation management challenge is the maintenance of the Mississippi River's aging navigation infrastructure (Niebling et al. 2014). The most acute of the navigation maintenance issues is associated with the aging locks and dams along the UMR. These locks and dams were constructed during the early and mid-20th century. Most of these structures (17 out of 26) have been undermaintained because of a lack of adequate funding (ITMS Capital Investment Strategy Team 2010). For the nine locks and dams that have been rehabilitated, their life expectancy has been extended into the middle of the 21st century. Currently, there is a maintenance backlog of nearly $3 billion for the locks and dams along the UMR (Dorothy 2013). Without adequate investment, lock and dam performance will likely continue to decrease, resulting in more delays and/or unexpected closures of portions of the Mississippi River and other portions of the inland waterway (NRC 2013a).

A recent National Resource Council study (NRC 2013a) outlined six options for USACE's water resource infrastructure challenges. These include (1) business as usual, where degraded navigation system service and the consequences of gradual or sudden failure of infrastructure components are expected; (2) increased federal funding for operations, maintenance, and rehabilitation; (3) divestment or decommission parts of the USACE's infrastructure to reduce operation and maintenance obligations; (4) increased revenue from USACE project beneficiaries, (5) expanded opportunities in operating and maintaining portions of existing infrastructure; and (6) some combination of the five proceeding options. The National Resource Council study (NRC 2013a) concluded that the business-as-usual option is the most likely outcome in the short to medium term. If federal leadership and additional funds are not forthcoming, at least two new river management challenges will arise. The first will be to determine which portions of the Mississippi River's navigation system to maintain and which parts should be decommissioned for cost savings. The second will be how to adapt the current navigation-focused management structure to address new challenges that will inevitably arise from new resource demands that will be placed on the Mississippi River and its ecosystem.

As discussed in the previous section, climate change is expected to result in more variability in the hydrologic extremes throughout much of the MRB (i.e., large floods and droughts; Wuebbles et al. 2009; Mishra and Singh 2010) and changes in mean discharge (Chien et al. 2013). These changes could exacerbate the navigation infrastructure maintenance challenges. For example, low river levels during droughts have periodically impacted navigation by temporarily closing portions of the river to navigation for emergency dredging. Increased dredging frequency would require more funding. An increase in the frequency or magnitude of large floods could result in morphological changes, such as channel avulsions, to the river channel, requiring major repairs to the navigation channel or possibly new expensive control structures. For example, during the Great Midwestern Flood of 1993 near Miller City, Illinois (upstream of Cairo, Illinois) a channel avulsion (meander cutoff) nearly occurred during the height of the flood. If the USACE had not intervened to stop this meander cutoff, the Mississippi River would have been shortened by 13 mi. This new channel would have been too steep for safe commercial navigation. Closing this reach of the Mississippi River would have prevented through navigation from the LMR to the MMR, Missouri River, UMR, and Illinois River. The construction of a new lock and dam would likely have been required for reestablishing safe navigation along this reach of the Mississippi River (Chrzastowski et al. 1994).

While increase in frequency of droughts and floods would likely cause more frequent, but localized, effects on navigation, a potentially larger and possibly system-wide challenge for river engineers would be changes to the channel-maintaining discharges (i.e., bank-full discharge or ~2-year flow event). Bank-full discharge often serves as a consistent morphological index that can be related to channel-forming flows that maintain the channel geometry as it exists under a particular climatic regime. Small variations in channel-maintaining discharge can cause substantial changes in sediment transport and consequently in channel morphology of alluvial rivers with fine-grained bed material (Ashmore and Church 2001). The Mississippi River and many of its navigable tributaries are chiefly fine-grained alluvial rivers that have the potential to rapidly respond (years to decades) to climate-driven changes in their flow regime. There is ample geomorphic evidence that certain segments or subsegments of the Mississippi River have undergone significant changes in channel form in response to climate drive changes in their flow regime during the Holocene (<10,000 years; Knox 2007). Changes in this discharge condition would alter sediment transport and channel morphology system-wide along the Mississippi River and potentially require adjustment or enhancements to river-training and navigation structures to maintain the navigation channel. The new, adjusted, or enhanced river-training and navigation structures would require substantial additional investment beyond the current multibillion dollar maintenance back log for the Mississippi River navigation system.

Flood-Risk Management

The USACE has successfully used the flood-mitigation tools of the MR&T project to protect the LMR Valley from inundation for more than 80 years. The USACE estimates that more than $14 billion (unadjusted costs) have been spent on the MR&T project since 1928. This massive project is not slated for full completion until at least 2030 because the funding of the remaining portions of the project have not been fully appropriated by the U.S. Congress (MRC 2011). As of December 2012, the USACE estimates a 44–1 benefit–cost ratio with ~$\$6.12 \times 10^{11}$ flood losses avoided for the flood mitigation portion of the MR&T project (DeHaan et al. 2012). While this benefit appears impressive, this economic assessment is not a full-cost accounting of the MR&T project. The benefit–cost ratio and loss-avoidance estimate does not account for the maintenance or operation of the system, flood-fighting costs, or other environmental externalities. The maintenance or operation of the system and flood-fighting costs alone are often quite high. For example, expenditures related to the 2011 flood along the LMR alone exceeded $\$1.6 \times 10^9$. Sixty million dollars were spent on the flood fight and $1,600 million were required for the postflood repairs to the MR&T system. These expenditures were in addition to the MR&T's annual $153 million operations and maintenance budget (DeHaan et al. 2012).

Despite the success of the MR&T project for flood-risk reduction on the LMR, replication of such robust system of levees and floodways along the UMR and MMR has proven to be economically and consequently politically infeasible. For example, a recent study by the USACE found that raising all the existing federally constructed levees from their current protection levels to the 500-year protection level along the UMR and MMR would have a benefit of five cents for every dollar spent (USACE 2008). The majority of the levees and floodwalls along the UMR and MMR today were constructed or enhanced to USACE standards with federal dollars between the 1950s and early 1980s. Urban areas along the UMR and MMR have federally constructed levees that protect to between the 100-year and 500-year flood levels. Agricultural areas with federally constructed levees are generally protected to ~50-year flood level (SAST 1995). However, unlike the MR&T levee systems, once the federally constructed UMR and MMR levee systems were complete, the full responsibility of their maintenance was turned over to the local jurisdictions (generally a levee and drainage district), which are also in charge of collecting taxes for the maintenance of these systems (U.S. Floodplain Management Review Committee 1994; USACE 1995).

Despite the nearly $\$7.00 \times 10^{11}$ of flood damages prevented by the federally constructed levees and floodwalls along the UMR and MMR ($\$8.30 \times 10^{10}$; USACE 2008) and the MR&T project on the LMR ($612 billion; DeHaan et al. 2012), as in many areas across the United States levee maintenance along the Mississippi River has been chronically underfunded (NCLS 2009). Evidence of this underfunding comes from recent levee inspections, which have found structural and/or operating issues with the majority of the Mississippi River levee systems. These inspections have led to an "unacceptable" rating for 15 (~17%) out of approximately 90 levees systems along Mississippi River. The overwhelming majority of levee systems along the Mississippi, 68 out of 90 levee systems (~75%), were rated as "minimally acceptable" (Figure 7; USACE 2014b). Unacceptable ratings are generally interpreted to mean that a levee system has one or more structural or operational deficiency that could result in it not performing to design specification during a large flood. A minimally acceptable rating is generally interpreted as a levee system having at least one minor deficiency that will not seriously impair the functioning of the item during the next flood event. However, the overall reliability of a minimally acceptable levee system will be lowered because of the minor deficiency (USACE 2006b).

In the case of the UMR and MMR levee systems, the underfunding of levee maintenance suggests that either an insufficient tax base or insufficient taxes are being collected to support proper maintenance of many levee systems. Without the large initial federal investment, most levees along the UMR and MMR would likely have either not have been built or would have been smaller. The smaller levee systems would likely be less costly to maintain and would not have attracted significant floodplain development because of their lower flood-protection levels (Remo et al. 2012). The rationale for constructing the larger agricultural levees along the UMR and MMR was that people and businesses would move into the new, more robustly protected areas. This would increase the tax base, which would support the maintenance of these levees (O'Neill 2006). However, for many UMR and MMR levee systems, the anticipated level of floodplain development appears not to have occurred.

The reasons for lack of the anticipated development within the agricultural levee districts along the UMR and MMR are twofold. First, most agricultural levees do not meet the 100-year regulatory flood elevation. As such, these levee-protected areas are mapped within the National Flood Insurance Program's 100-year regulatory floodplain. In jurisdictions that participate in the National Flood Insurance Program, there are restrictions on new construction in addition to the flood insurance requirements for any new or existing residential buildings purchased using loans from a federally insured financial institution. Restrictions on new construction and the requirement of flood insurance generally dissuade floodplain development (NRC 2013b). Second, continued industrialization and mechanization of agriculture has resulted in the need for fewer farmers and farm laborers (Calvin and Martin 2010).The need for fewer farmers and farm laborers has resulted in fewer levee district residences and consequently less tax revenue to support levee maintenance.

Climate change may also substantially affect flood-risk management through changes to the frequency and magnitude of floods. Like most water management throughout the developed world, flood-risk management on the Mississippi has been based on the assumption of stationarity. Stationarity assumes that hydrologic variables, such as annual stream flow or peak annual discharge, have a time-invariant probability density function whose properties can be determined from the instrumental (gaging) record (Milly et al. 2008). Milly et al. (2008) have asserted that stationarity is dead because of substantial anthropogenic change to the Earth's climate.

Decisions concerning existing and future long-lived investments to flood-mitigation infrastructure must take climate change into account. However, accounting for climate change is complicated for at least two reasons. First, climate change will likely require new river infrastructure to be able to cope with a larger range of discharge conditions. Infrastructure design will be more difficult and construction more expensive. Second, uncertainty in future climate prevents use of the output from a single or even an ensemble of climate models as an input for infrastructure design (Hallegatte 2009). While a successor to water-resources risk and planning methods that do not presume stationarity is sorely needed, it is highly likely that the desired climate information and robust nonstationary probabilistic models will not be available soon (Milly et al. 2008; Hallegatte 2009). Instead of making flood-mitigation decisions based solely on the climate conditions projected by models, future infrastructure should be made more robust to climate-driven changes in river discharges. This aim implies that river managers, planners, and engineers should enhance their practices and decision-making frameworks by adapting more robust uncertainty-management methods. If a broader range of uncertainty is taken into account in long-term planning, decision making, and design, many river infrastructure projects will likely be better adapted in the future and climate change effects would be more manageable.

As the UMR and MMR levee systems continue to age and new challenges arise (e.g., climate change driven changes in flood discharges), maintaining current flood-protection levels is likely unsustainable without substantial financial support from state and federal governments for the majority of levee districts. The longer levee maintenance remains underfunded, the larger the challenges will be to maintain current levee flood-protection levels. If the underfunding of levee maintenance continues, risk to public safety and infrastructure will only continue to increase (NCLS 2009).

Environmental Management

Habitat rehabilitation and enhancement projects have been on going along the UMR and MMR since the implementation of HERP in the mid-1990s. More than $200 million have been spent on rehabilitation and enhancement projects along these river segments (O'Donnell and Galat 2007; Sparks 2010). Despite the large amount of money spent on habitat rehabilitation and enhancement protects over the past 20 years, the LTMRP decadal status and trend reports (USGS 1999; Johnson and Hagerty 2008) have not shown statistically significant trends in the riverine ecosystem health indicators. The lack of ecosystem-level evidence for improved ecological health along the UMR and MMR is problematic for the justification of past, present, and future habitat projects. This is because the USACE's implementation guidance for Section 2039 of the 2007 WRDA explicitly indicates that rehabilitation projects include plans for monitoring the success of habitat projects (DuBowy 2013). Many of the individual habitat projects implemented along the UMR and MMR have lacked both predefined project objectives (i.e., goals, targets, and metrics) and the systematic data collection required from which to assess an individual project's performance (O'Donnell and Galat 2007; Remo et al. 2013). Without appropriate monitoring efforts, there can be no adequate evaluation of the success of these habitat projects.

On the LMR, the problem of evaluating project success appears especially acute (DuBowy 2013). However, far fewer projects have been under taken along this segment of the Mississippi River compared to the UMR and MMR. The absence of predefined habitat enhancement goals and little to no project monitoring has resulted in a lack of scientific accountability for many LMR habitat rehabilitation projects (DuBowy 2013).

The absence of project goals and adequate project monitoring precludes the use of effective adaptive environmental management of Mississippi River. Future habitat enhancement projects must include specific goals and rigorous monitoring, which is cost effective, using the best scientific protocols (DuBowy 2013; Remo et al. 2013). Without meaningful scientific assessment of projects' ecological benefits, funding of habitat rehabilitation or enhancement projects may be at risk of being discontinued.

Delta Land Loss

Efforts by federal and state agencies to conserve and restore the Mississippi River delta began more than 30 years ago. In the past decade, these efforts have accelerated due to the growing awareness of the implications of global sea-level rise and the importance of coastal wetlands in mitigating storm-surge impacts on coastal communities. Despite these efforts, land losses are expected to continue because upstream sediment supplies are insufficient to keep up with the combined forces of delta-plain subsidence and sea-level rise (Blum and Roberts 2009, 2012).

Over the next 50 years, estimates of delta land loss range from 1,350 km^2 (Morton and Bernier 2010) to 1,900 km^2 (State of Louisiana 2012). Recent work has begun to tackle the fundamental research needed to plan the portfolios of diversions likely needed for maximized regeneration of the Mississippi River delta (e.g., State of Louisiana 2012; Kenney et al. 2013). Modeling suggests that between 600 km^2 (Kim et al. 2009) and 980 km^2 (Kenney et al. 2013) of the delta can be constructed to help offset the projected future land losses. However, most of these plans do not address adverse effects on navigation (Allison et al. 2013) or flood-risk management. Due to current U.S. river management policy, the effect of diversions on navigation and flood-risk management will likely need to be minimized to the point where there are no adverse effects to these management objectives. It is unclear whether meaningful delta restoration can be achieved without major effects on navigation and/or flood-risk management. However, the observations by Allison et al. (2013) near the Bonnette Carre Spillway during and after the 2011 flood suggest that shoaling related to flow diversion can be short-lived if the diversion can be regulated (i.e., closed). Closing the diversion would potentially allow any shoaling in the navigation channel to be flushed or reworked result in little to no impact to navigation (Allison et al. 2013).

The continued loss of coastal and delta lands will require reevaluation of flood risk to vulnerable areas such as New Orleans. Additional flood-mitigation measures will be needed to offset the losses in natural flood-mitigation benefits that lost wetlands once provided. Channel diversions for the promotion of land regeneration and enhanced structural flood-mitigation measures (i.e., larger levees, pumps, etc.) will require billions of dollars of new investments. These investments will be in addition to funds currently needed to maintain the Mississippi River's navigation and flood-mitigation infrastructure.

Summary and Conclusions

The Mississippi River is one of the most regulated and managed river systems in the world. The regulation and management of the Mississippi River and its tributaries for a select set of ecosystem services have provided substantial economic benefits to society. These benefits include transportation of goods, the mitigation of flood risk to both floodplain communities and agricultural lands, securing or increasing water supply, and reliable energy production. However, these benefits have come with externalities that counterbalance, at least in part, these economic gains. These externalities include altered hydrology, altered river hydraulics, decrease in sediment load, changes in the pattern of delta sedimentation, increased nutrient loads, reduced habitat diversity, decreased ecological diversity, and degraded water quality. The externalities coupled with resource constraints have resulted in several challenges for river management. Redressing many of these challenges going forward requires a shared, integrated, and system-wide management vision for the entire basin. Unfortunately, this system-wide integrated vision is currently absent in the MRB. If such a vision is attempted, it needs to be holistic and incorporate (and balance) economic, social, and ecological considerations. Historically, such a broad-scale vision has been left to government agencies such as the USACE. History has shown that one government agency's vision, as in the case with the "levees only" policy or the MR&T project, has been inadequate in meeting all the challenges that arise on a dynamically coupled human and natural system like the Mississippi River. Developing an integrated vision for the Mississippi River will require active participation from basin stake holders, policy makers, and government management agencies. Such a vision should also lead to commonly accepted priorities that serve as a guide for management across the MRB.

References

Alexander, J. S., R. C. Wilson, and W. R. Green. 2012. A brief history and summary of the effects of river engineering and dams on the Mississippi River system and delta. U.S. Geological Survey Circular 1375. Available: pubs.usgs.gov/circ/1375/C1375.pdf. (May 2014).

Alexander, R. B., R. A. Smith, G. E. Schwarz, E. W., Boyer, J. V. Nolan, and J. W. Brakebill. 2008. Differences in phosphorus and nitrogen delivery to the Gulf of Mexico from the Mississippi River basin. Environmental Science and Technology 42:822–830.

Allison, M. A., B. M. Vosburg, M. T. Ramirez, and E. A. Meselhe. 2013. Mississippi River channel response to the Bonnet Carre Spillway opening in the 2011 flood and its implications for the design and operation of river diversions. Journal of Hydrology 477:104–118.

Anfinson, J. O. 2003. The river we have wrought: a history of the upper Mississippi. University of Minnesota Press, Minneapolis.

Ashmore, P., and M. Church. 2001. The impact of climate change on rivers and river processes in Canada. Geological Survey of Canada, Bulletin 555, Ottawa. Available: www.publications.gc.ca/site/eng/93493/publication.html. (April 2015).

Baker, J. A., K. J. Killgore, and R. L. Kasual. 1991. Aquatic habitats and fish communities in the lower Mississippi River. Reviews in Aquatic Sciences 3:313–356.

Barry, J. M. 1997. Rising tide: the great Mississippi River flood of 1927 and how it changed America. Simon and Schuster, New York.

Bass, A. S., and R. E. Turner. 1997. Relationships between salt marsh loss and dredged canals in three Louisiana estuaries. Journal of Coastal Research 13:895–903.

Belt, C. B. 1975. The 1973 flood and man's constriction of the Mississippi River. Science 189:681–684.

Bernhardt, E. S., M. A. Palmer, J. D. Allen, G. Alexander, K. Barnas, S. Brooks, J. Carr, S. Clayton, C. Dahm, J. Follstad-Shah, D. Galat, S. Gloss, D. Goodwin, B. Hart, R. Hassett, S. Jenkinson, S. Katz, G. M. Kondolf, P. S. Lake, R. Lave, J. L. Meyer, T. K. Pangano, B. Powell, and E. Sudduth. 2005. Synthesizing U.S. river restoration efforts. Science 208:636–637.

Biedenharn, D. S., C. R. Thorpe, and C. C. Watson. 2000. Recent morphological evolution of the lower Mississippi River. Geomorphology 34:227–249.

Biedenharn, D. S., and C. C. Watson. 1997. Stage adjustment in the lower Mississippi River, USA. Regulated Rivers Research and Management 13:517–536.

Blum, M. D., and H. H. Roberts. 2012. The Mississippi delta region: past, present, and future. Annual Review of Earth and Planetary Sciences 40:655–683.

Blum, M. D., and H. H. Roberts. 2009. Drowning of the Mississippi delta due to insufficient sediment supply and global sea-level rise. Nature Geoscience 2:488–491.

Boesch, D. F., M. N. Josselyn, A. J., Mehta, J. T. Morris, W. K., Nuttle, C. A. Simestad, and D. J. P. Swift. 1994. Scientific assessment of coastal wetland loss, restoration and management in Louisiana. Journal of Coastal Research Special Issue 20.

Britsch, L. D., and J. B. Dunbar. 1993. Land loss rates: Louisiana coastal plain. Journal of Coastal Research 9:324–338.

Calvin, L., and P. Martin. 2010. The U.S. produce industry and labor: facing the future in a global economy. U.S. Department of

Agriculture, Economic Research Report Number 106, Washington, D.C. Available: www.ers.usda.gov/publications/err-economic-research-report/err106.aspx. (May 2014).

Camillo, C. A., and M. T. Pearcy. 2004. Upon their shoulders: a history of the Mississippi River Commission from its inception through the advent of the modern Mississippi River and tributaries project. Mississippi River Commission, Vicksburg.

Chick, J. H., B. S. Ickes, M. A. Pegg, V. A. Barko, R. A. Hrabik, and D. P. Herzog. 2005. Spatial structure and temporal variation of fish communities in the Upper Mississippi River System. U.S. Geological Survey, Upper Midwest Environmental Sciences Center, Long Term Resource Monitoring Program Technical Report 2005-T004, La Crosse, Wisconsin.

Chien, H., P. Yeh, and J. H. Knouft. 2013. Modelling the potential impacts of climate change on streamflow in agricultural watersheds of the Midwestern United States. Journal of Hydrology 491:73–88.

Chrzastowski, M. J., M. M. Killey, R. A. Bauer, P. B. DuMonetelle, A. L. Erdmann, B. L. Herzog, J. M. Masters, and L. R. Smith. 1994. The Great Flood of 1993: geological perspectives on the flooding along the Mississippi River and in tributaries in Illinois. Illinois Geologic Survey, Special Report 2, Champaign.

Clay, F. M. 1986. A century on the Mississippi: a history of the Memphis District U.S. Army Corps of Engineers, 1876–1981. U.S. Army Corps of Engineers, Memphis, Tennessee.

Coleman, J. M., H. H. Roberts, and G. W. Stone. 1998. Mississippi River delta: an overview. Journal of Coastal Research 14:698–716.

Couvillion, B. R., J. A. Barras, G. D., Steyer, W. Sleavin, M. Fisher, H. Beck, Trahan, N. B. Griffin, and D. Heckman. 2011. Land area change in coastal Louisiana from 1932 to 2010. U.S. Geological Survey Scientific Investigations Map 3164. Available: pubs.usgs.gov/sim/3164/. (May 2014).

Criss, R. E., and E. L. Shock. 2001. Flood enhancement through flood control. Geology 29:875–878.

Davinroy, R. D., E. J. Brauer, and G. Dyhouse. 2011. Environmental river engineering on the middle Mississippi River: St. Louis Harbor chevrons. Proceedings of the Mississippi River Research Consortium, volume 43. Available: http://m-r-r-c.org/2011/MRRC%202011%20Proceedings_final-1.pdf. (May 2014).

Davis, N. D. 1982. The father of waters: a Mississippi River chronicle. Sierra Club Books, San Francisco.

Day, J. W., J. D. Gunn, W. J. Folan, A. Yanez-Arancibia, and B. P. Horton. 2007. Emergence of complex societies after sea level stabilized. EOS, Transactions of the American Geophysical Union 88:169–170.

De Jager, N. R., and J. J. Rohweder. 2012. Spatial patterns of aquatic richness in the upper Mississippi River floodplain, USA. Ecological Indicators 13:275–283.

DeHaan, H., J. Stamper, and B. Walters. 2012. Mississippi River and tributaries system: 2011 post-flood report. U.S. Army Corps of Engineers, Mississippi Valley Division, Vicksburg, Mississippi.

Dobney, F. J. 1977. River engineers on the middle Mississippi: a history of the St. Louis District. U.S. Government Printing Office, Washington, D.C.

Dorothy, O. 2013. Restoring America's river: using the Water Resources Development Act to mend the upper Mississippi River. Izaak Walton League of America, Gaithersburg, Maryland. Available: www.iwla.org/index.php?ht=action/GetDocumentAction/i/72926. (May 2014).

DuBowy, P. J. 2013. Mississippi River ecohydrology: past, present and future. Ecohydrology and Hydrobiology 13:73–83.

Fennel, K., R. Hetland, Y. Feng, and S. DiMarco. 2011. A coupled physical-biological model of the northern Gulf of Mexico shelf: model description, validation and analysis of phytoplankton variability. Biogeoscience 8:1881–1899.

Fremling, C. R., J. L. Rasmussen, R. E. Sparks, S. P. Cobb, C. F. Bryan, and T. O. Claflin. 1989. Mississippi River fisheries: a case history. Pages 309–351 *in* D. P. Dodge, editor. Proceedings of the international large river symposium. Canadian Special Publication of Fisheries and Aquatic Sciences 106.

Galat, D. L., L. H. Fredrickson, D. D. Humburg, K. J. Bataille, J. R. Bodie, J. Dohrenwend,

G. T. Gelwicks, J. E. Havel, D. L. Helmers, J. B. Hooker, J. R. Jones, M. F. Knowlton, J. Kubisiak, J. Mzaourek, A. C. McColpin, R. B. Renken, and R. D. Semlitsch. 1998. Flooding to restore connectivity of regulated, large-river wetlands. BioScience 48:29–48.

Galloway, G. E. 2000. Three centuries of river management along the Mississippi River: engineering and hydrological aspects. Pages 51–64 *in* A. J. M Smits, P. H. Niehuis, and R. S. E. Leuven, editors. New approaches to river management. Backhuys, Leiden, Netherlands.

Garvey, J., B. Ickes, and S. Zigler. 2010. Challenges in merging fisheries research and management: the upper Mississippi River experience. Hydrobiologia 640:125–144.

Gibbons v. Ogden. 1824. 22 U.S. 1 (Wheat.)

Goolsby, D. A., W. A. Battaglin, G. B. Lawrence, R. S. Artz, B. T. Aulenbach, R .P. Hooper, D. R. Keeney, and G. J. Stensland. 1999. Flux and sources of nutrients in the Mississippi-Atchafalaya River basin. Topic 3 report for the integrated assessment on hypoxia in the Gulf of Mexico. National Oceanic and Atmospheric Administration, Coastal Ocean Program Decision Analysis Series No. 17, Silver Spring, Maryland.

Gordon, D. C. 2004. A remedy for a chronic dredging problem. Engineer: The Professional Bulletin for Army Engineers 34:32–36.

Grugbaugh, J. W., and R. V. Anderson. 1989. Long-term effects of navigation dams on a segment of the upper Mississippi River. Regulated Rivers Research and Management 4:97–104.

Hallegatte, S. 2009. Strategies to adapt to an uncertain climate change. Global Environmental Change 19:240–247.

Harmar, O. P., N. J., Clifford, C. R., Thorne, and D. S. Biedenharn. 2005. Morphological changes of the lower Mississippi River: geomorphologic response to engineering intervention. River Research and Applications 21:1107–1131.

Holmes, B. B. 1979. History of federal water resources programs and policies, 1961–1970. U.S. Department of Agriculture, Economics, Statistics, and Cooperatives Service, Miscellaneous Publications 1379, Washington, D.C.

Hoover, J. J., and K. J. Killgore. 1998. Fish communities. Pages 237–260 *in* M. G. Messina, and W. H. Conner, editors. Southern forested wetlands. Lewis Publishers, Boca Raton, Florida.

Humphreys, A. A., and H. L. Abbott. 1861. Report upon the physics and hydraulics of the Mississippi River; upon the protection of the alluvial region against overflow; upon the deepening of the mouths: based upon surveys and investigations made under the acts of Congress directing the topographical and hydrographical survey of the delta of the Mississippi River, with such investigations as might lead to determine the most practical plan for securing it from inundation, and the best mode of deepening the channels at the mouths of the river. Government Printing Office, Washington, D.C.

Hurley, K. L., R. J. Sheehan, R. C. Heidinger, P. S. Wills, and B. Clevenstine. 2004. Habitat use by middle Mississippi River Pallid Sturgeon. Transactions of the American Fisheries Society 133:1033–1041.

Huthoff, F., H. Barneveld, N. Pinter, J. Remo, H. Eerden, and A. Paarlberg. 2013. Optimizing design of river training works using 3-dimensional flow simulation. In P. Rigo and M. Wolters, editors. PIANC - Smart Rivers. Liege, Belgium. Available: www.pianc.org/downloads/sailingahead/Sailing%20Ahead%20June%202013/PIANC-SMART%20Rivers%20Booklet%20of%20Abstracts.pdf. (May 2014).

ITMS (Inland Marine Transportation Systems) Capital Investment Strategy Team. 2010. Inland Marine Transportation Systems (IMTS) capital projects business model. Inland Waterways Users Board, Final Report, Revision 1. Available: www.iwr.usace.army.mil/Portals/70/docs/Wood_doc/IMTS_Final_Report_13_April_2010_Rev_1.pdf. (May 2014).

Jemberie, A. A., N. Pinter, and J. W. F. Remo. 2008. Hydrologic history of the Mississippi and lower Missouri Rivers based upon a refined specific-gage approach. Hydrological Processes 22:4436–4447.

Jha, M., Z. Pan, E. S. Takle, and R. G. 2004. Impacts of climate change on stream flow in the upper Mississippi River basin: a regional climate model perspective. Journal of Geophysical Research 109:D09105.

Johnson, B. L., and K. H. Hagerty. 2008. Status and trends of selected resources of the Upper Mississippi River System. U.S. Geological Survey, Upper Midwest Environmental Sciences Center, La Crosse, Wisconsin. Available: http://pubs.usgs.gov/mis/LTRMP2008-T002. (May 2014).

Junk, W., P. B. Bayley, and R. E. Sparks. 1989. The flood pulse concept in river–floodplain systems. Pages 110–127 *in* D. P. Dodge, editor. Proceedings of the international large river symposium. Canadian Special Publication of Fisheries and Aquatic Sciences 106:110–127.

Kenney, M. A., B. F. Hobbs, D. Mohrig, H. Huang, J. A. Nittrouer, K. Wonsuck, and G. Parker. 2013. Cost analysis of water and sediment diversions to optimize land building in the Mississippi River delta. Water Resources Research 49:3388–3405.

Keown, M. P., E. A. Dardeau, and E. M. Causey. 1986. Historic trends in the sediment flow regime of the Mississippi River. Water Resource Research 22:1555–1564.

Kesel, R. H. 2003. Human modification to the sediment regime of the lower Mississippi River floodplain. Geomorphology 56:325–334.

Kesel, R. H., E. G. Yodis, and D. J. Craw. 1992. An approximation of the sediment budget of the lower Mississippi River prior to major human modification. Earth Surface Processes and Landforms 17:711–722.

Kim, W., A. Dai, T. Muto, and G. Parker. 2009. Delta progradation driven by an advancing sediment source: coupled theory and experiment describing the evolution of elongated deltas. Water Resources Research 45:W06428.

Klein, C. A., and S. B. Zellmer. 2007. Mississippi River stories: lessons from a century of unnatural disasters. Sothern Methodist Law Review 1471:1478–1480.

Knox, J. C. 1999. Long-term episodic changes in magnitudes and frequencies of floods in the Upper Mississippi River Valley. Pages 255–282 *in* A. G. Brown and A. G. Quine, editors. Fluvial processes and environmental change. Wiley, Chichester, UK.

Knox, J. C. 2003. North American paleofloods and future floods: responses to climatic change. Pages 143–164 *in* G. Benito and K. J. Gregory, editors. Paleohydrology: understanding global change. Wiley, Chichester, UK.

Knox, J. C. 2007. The Mississippi River System. Pages 145–182 *in* A. Gupta, editor. Large rivers: geomorphology and management. Wiley, Chichester, UK.

Knox, J. C. 2009. Evidence of Mississippi River flood history preserved in floodplain sedimentary deposits. Pages 28–33 *in* R. E. Criss and T. M. Kusky, editors. Finding the balance between floods, flood protection, and river navigation. Saint Louis University, Center for Environmental Sciences, St. Louis.

Koel, T. M. 2004. Spatial variation in fish species richness of the Upper Mississippi River System. Transactions of the American Fisheries Society 133:984–1003.

Kunkel, K. E., P. D. Bromirski, H. E. Brooks, T. Cavazos, A. V. Douglas, D. R. Easterling, K. A. Emanuel, P. Groisman, G. J. Holland, T. R. Knutson, J. P. Kossin, P. D. Komar, D. H. Levinson, and R. L. Smith. 2008. Observed changes in weather and climate extremes. in T. R.,Karl, G. A. Meehl, C. D. Miller, S. J. Hassol, A. M. Waple, and W. L. Murray, editors. Weather and climate extremes in a changing climate. Regions of focus: North America, Hawaii, Caribbean, and U.S. Pacific Islands. U.S. Climate Change Science Program and the Subcommittee on Global Change Research, Washington, D.C.

Lake, P. S., N. Bond, and P. Reich. 2007. Linking ecological theory with stream restoration. Freshwater Biology 52:597–615.

Langbein, W. B., and S. A. Schumm. 1958. Yield of sediment in relation to mean annual precipitation. Transactions of the American Geophysical Union 39:1076–1084.

Laurent, A., and K. Fennel. 2014. Simulated reduction of hypoxia in the northern Gulf of Mexico due to phosphorus limitation. Elementa: Science of the Anthropecene [online serial] 2:000022.

Laurent, A., K. Fennel, J. Hu, and R. Hetland. 2012. Simulating the effects of phosphorus limitation in the Mississippi and Atchafalaya River plumes. Biogeosciences 9:4707–4723.

Lohrenz, S. E., G. L. Fahnenstiel, D. G. Readalje, G. A. Lang, M. J. Dagg, T. E. Whitledge, and Q. Dortch. 1999. Enhanced primary production at the plume/oceanic interface of the Mississippi River plume. Continental Shelf Resources 19:1113–1141.

Mallman, E. P., and M. D. Zoback. 2007. Subsidence in the Louisiana coastal zone due to hydrocarbon production. Journal of Coastal Research Special Issue 50:443–449.

McIsaac, G. F, M. B. David, G. Z. Gertner, and D. A. Goolsby. 2002. Relating net nitrogen input in the Mississippi River basin to nitrate flux in the lower Mississippi River: a comparison of approaches. Journal of Environmental Quality 31:1610–1622.

McPhee, J. 1989. The control of nature. Farrar, Straus, and Giroux, New York.

Meade, R. H. 1995. Contaminants in the Mississippi River, 1987-92. U.S. Geological Survey Circular 1133.

Meade, R. H., and J. A. Moody. 2010. Causes for the decline of suspended-sediment discharge in the Mississippi River system, 1940–2007. Hydrological Processes 24:35–49.

Meade, R. H., and R. S. Parker. 1985. Sediment in rivers of the United States. Pages 49–60 *in* National water summary 1984. U.S. Geologic Survey Water-Supply Paper 2275.

Milly, P. C. D., J. Betancourt, M. Falkenmark, R. B. Hirsch, Z. W. Kundsewicz, D. P. Lenttenmaier, and R. J. Stouffer. 2008. Stationarity is dead: whither water management? Science 319:573–574.

Mishra, A. K., and V. P. Singh. 2010. A review of drought concepts. Journal of Hydrology 391:202–216.

Mississippi River Gulf of Mexico Watershed Nutrient Task Force. 2014. Reassessment 2013: assessing progress made since 2008. Available: water.epa.gov/type/watersheds/named/msbasin/upload/hypoxia_reassessment_508.pdf. (May 2014).

Moody, J. A., R. H. Meade, and D. R. Jones. 2003. Lewis and Clark's observations and measurements of geomorphology and hydrology, and changes with time. U.S. Geological Survey Circular 1246.

Moore, M. S. P. Romano, and T. Cook. 2010. Synthesis of Upper Mississippi River System submersed and emergent aquatic vegetation: past, present, and future. Hydrobiologia 640:103–114.

Moore, N.R. 1972. Improvement of the lower Mississippi River and tributaries, 1931-1972. U.S. Army Corps of Engineers, Mississippi River Commission, Vicksburg, Mississippi.

Morton, R. A., and J. C. Bernier. 2010. Recent subsidence-rate reductions in the Mississippi delta and their geological implications, Journal of Coastal Restoration 26:555–561.

Mossa, J. 1996. Sediment dynamics of the lowermost Mississippi River. Engineering Geology 45:457–479.

MRC (Mississippi River Commission). 2011. Mississippi River Commission 2011 MR&T flood report. Available: www.mvd.usace.army.mil/Portals/52/docs/MRC/MRC_2011_Flood_Report.pdf. (April 2015).

Niebling, W., J. Baker, L. Kasuri, S. Katz, and K. Smet. 2014. Challenge and response in the Mississippi River basin. Water Policy 16:87–116.

NCLS (National Committee on Levee Safety). 2009. Recommendation for a national levee safety program. A report to Congress from the National Committee on Levee Safety. Available: www.nfrmp.us/ncls/docs/NCLS-Recommendation-Report_012009_DRAFT.pdf. (May 2014).

NRC (National Research Council). 2013a. Corps of Engineers water resources infrastructure: deterioration, investment, or divestment? The National Academies Press, Washington, D.C.

NRC (National Research Council). 2013b. Levees and National Flood Insurance Program: improving policies and practices. The National Academies Press, Washington, D.C.

NRC (National Research Council). 2008. Mississippi River water quality and the Clean Water Act: progress, challenges, and opportunities. The National Academies Press, Washington, D.C.

O'Donnell, T. K., and D. L. Galat. 2007. River enhancement in the upper Mississippi River basin: approaches based on river uses, alterations and management agencies. Restoration Ecology15:538–549.

O'Neill, K. M. 2006. Rivers by design: state

power and origins of U.S. flood control. Duke University Press, Durham, North Carolina.

Paola, C. R. R. Twilley, D. A. Edmonds, W. Kim, D. Mohrig, D. G. Parker, E. Viparelli, and V. R. Voller. 2011. Natural processes in delta restoration: application to the Mississippi delta: Annual Review of Marine Science 3:67–91.

Parchure, T. M. 2005. Structural methods to reduce navigation channel shoaling. U.S. Army Corps of Engineer Research and Development Center, Coastal and Hydraulic Laboratory, Vicksburg, Mississippi.

Pflieger, W. L. 1975. The fishes of Missouri. Missouri Department of Conservation, Jefferson City.

Pinter, N., A. A. Jemberie, J. W. F. Remo, R. A. Heine, and B. S. Ickes. 2009. Flood trends and engineering on the Mississippi River system. Geophysical Research Letters 35:L23404.

Pinter, N., A. A. Jemberie, J. W. F. Remo, R. A. Heine, and B. S. Ickes. 2010. Cumulative impacts of river engineering, Mississippi and lower Missouri rivers. River Research and Applications 26:546–571.

Pinter, N., K. Miller, J. H. Wolsinski, and R. R. van der Ploeg.2004. Recurrent shoaling and channel dredging, middle and upper Mississippi River, USA. Journal of Hydrology 290:275–296.

Pinter, N., R. Thomas, and J. H. Wlosinski. 2001. Flood-hazard assessment on dynamic rivers. Eos, Transactions of the American Geophysical Union 82:333–339.

Quigg, A., J. B. Sylvan, A. B. Gustafson, T. R. Fisher, R. L. Oliver, S. Tozzi, and J. W. Ammerman. 2011. Going west: nutrient limitation of primary production in the Northern Gulf of Mexico and the importance of the Atchafalaya River. Aquatic Geochemistry 17:519–544.

Rabalais, N. N., and R. E. Turner, editors. 2001. Coastal hypoxia: consequences for living resources and ecosystems. American Geophysical Union, Coastal and Estuarine Studies 58, Washington, D.C.

Rabalais, N. N., and R. E. Turner. 2006. Oxygen depletion in the Gulf of Mexico adjacent to the Mississippi River. Pages 225–245 *in* L.N. Neretin, editor. Past and present marine water column anoxia. NATO Science Series: IV - Earth and Environmental Sciences. Kluwer, Dordrecht, Netherlands.

Rabalais, N. N., R. E. Turner, and D. Scavia. 2002. Beyond science into policy: Gulf of Mexico hypoxia and the Mississippi River. BioScience 52:129–142.

Rasmussen, J. L. 1994. Management of the upper Mississippi: a case history. Pages 441–463 *in* P. Calow and G. E. Petts, editors. The rivers handbook: hydrological and ecological principles, volume 2. Blackwell Scientific Publications, London.

Rebich, R. A., and D. K. Demcheck. 2007. Trends in nutrient and sediment concentrations and loads in major river basins of the south-central United States, 1993–2004. U.S. Geological Survey Scientific Investigations Report 2007-5090.

Rittenour, T. M., M. D. Blum, and R. J. Goble. 2007. Fluvial evolution of the lower Mississippi River valley during the last 100 k.y. glacial cycle: response to glaciation and sea-level change. Geological Society of America Bulletin 119:586–608.

Remo, J. W. F., M. Carlson, and N. Pinter. 2012. Hydraulic and flood-loss modeling of levee, floodplain, and river management strategies, middle Mississippi River, USA. Natural Hazards 61:551–575.

Remo, J. W. F., A. Khanal, and N. Pinter. 2013. Assessment of chevron dikes for the enhancement of physical-aquatic habitat within the middle Mississippi River, USA. Journal of Hydrology 501:146–162.

Remo, J. W. F., and N. Pinter. 2007. Retro-modeling the middle Mississippi River. Journal of Hydrology 337:421–435.

Remo, J. W. F., N. Pinter, and R. Heine, 2009. The use of retro- and scenario-modeling to assess effects of 100+ years river of engineering and land-cover change on middle and lower Mississippi River flood stages. Journal of Hydrology 376:403–416.

Roberts, H. H. 1997. Dynamic changes of the Holocene Mississippi River delta plain: the delta cycle. Journal of Coastal Research 13:605–627.

SAST (Scientific Assessment and Strategy Team). 1995. Metadata for levee coverage: upper Mississippi/Missouri River basin levees. Available: www.umesc.usgs.gov/

data_library/gis_data/umrs_floodplain/metadata/levees.html. (May 2014).

Schmidt, J. C., R. H. Webb, R. H., R. A. Valdez, G. R. Marzolf, and L. E. Stevens. 1998. Science and values in river restoration in the Grand Canyon. Bioscience 48:735-747.

Schramm, H. L., Jr. 2004. Status and management of fisheries in the Mississippi River. Pages 301-333 *in* R. Welcomme and T. Petr, editors. Proceedings of the second international symposium on river management of large rivers for fisheries, volume 1. Bangkok, Thailand, Food and Agriculture Organization of the United Nations, FAO Regional Office for Asia and the Pacific, RAP Publication 2004/16, 301-333, Bangkok, Thailand.

Schramm, H. L., M. S. Cox, T. E., Tietjen, and A. W. Ezell. 2009. Nutrient dynamics in the lower Mississippi River floodplain: comparing present and historic hydrologic conditions. Wetlands 29:476-487.

Schumm, S. A., I. D. Rutherfurd, and J. Brooks. 1994. Pre-cutoff morphology of the lower Mississippi River. Pages 13-44 *in* S. A. Schumm and B. R. Winkley, editors. The variability of large alluvial rivers. American Society of Civil Engineers, New York.

Sparks, R. E. 2010. Forty years of science and management on the upper Mississippi River: an analysis of the past and a view of the future. Hydrobiologia 640:3-15.

Sprague, L. A., R. M. Hirsch, and B. T. Aulenbach. 2011. Nitrate in the Mississippi River and its tributaries, 1980 to 2008: are we making progress? Environmental Science and Technology 45:7209-7216.

Stanley D., and A. Warne. 1997. Holocene sea level change and early human utilization of deltas. GSA Today 7(12):1-8.

State of Louisiana. 2012. Draft Louisiana 2012 master plan. Available: http://coastal.la.gov/a-common-vision/2012-coastal-master-plan/2012-coastal-master-plan-resources/. (May 2014).

Stevens, M. A., D. B. Simons, and S. A. Schumm. 1975. Man-induced changes of middle Mississippi River. Journal of the Waterways, Harbors, and Coastal Engineering Division 101:119-133.

Theiling, C. H. 1995. Habitat rehabilitation on the upper Mississippi River. Regulated Rivers Research and Management 11:227-238.

Theiling, C. H., and J. M. Nestler. 2010. River stage response to alteration of upper Mississippi River channels, floodplains, and watersheds. Hydrobiologia 640:17-47.

Trimble S. W. 1983. A sediment budget for Coon Creek basin in the Driftless Area, Wisconsin, 1853-1975. American Journal of Science 283: 454-474.

Trimble S. W. 1999. Decreased rate of alluvial sediment storage in the Coon Creek basin, Wisconsin, 1975-1993. Science 285:454-474.

Tucker, J. K., C. R. Dolan, J. T. Lamer, and E. A. Dustman. 2008. Climatic warming, red-eared sliders (*Trachemys scripta elegans*), and sex ratio in Illinois. Chelonian Conservation and Biology 7:60-69.

U.S. Floodplain Management Review Committee. 1994. Sharing the challenge: floodplain management into the 21st century. Report of the Interagency Floodplain Management Review Committee to the Administration Floodplain Management Task Force. Government Printing Office, Washington, D.C.

USACE (U.S. Army Corps of Engineers). 1994. Authorized and operating purposes of the Corps of Engineers reservoirs. USACE, Hydraulic Engineering Center, HEC Publication PR-19, Washington, D.C.

USACE (U.S. Army Corps of Engineers). 1995. Floodplain management assessment of the upper Mississippi River and lower Missouri River and tributaries. USACE, St. Paul District, St. Paul, Minnesota.

USACE (U.S. Army Corps of Engineers). 1998. Flood control, Mississippi River & tributaries; Mississippi River mainline levees enlargement and seepage control. Cape Girardeau, Missouri to Head of Passes, LA. USACE, Vicksburg, Mississippi.

USACE (U.S. Army Corps of Engineers). 2004. Upper Mississippi River System flow frequency study. Final Report. Available: www.mvr.usace.army.mil/Portals/48/docs/FRM/UpperMissFlowFreq/Upper%20Mississippi%20River%20System%20Flow%20Frequency%20Study%20Main%20Report.pdf. (May 2014).

USACE (U.S. Army Corps of Engineers). 2006a. Environmental design handbook. USACE, Rock Island District, Rock Island, Illinois. Available: www2.mvr.usace.army.mil/EMP/FY07%20EMP%20Design%20Handbook/EMP%20Environmental%20Design%20Handbook%20August%202006.pdf. (May 2014).

USACE (U.S. Army Corps of Engineers). 2006b. Levee owner's manual for non-federal flood control works: the rehabilitation and inspection program, Public Law 84-99. Available: http://media.swf.usace.army.mil/pubdata/ppmd/emermgt/pdf/leveeownersmanual.pdf. (May 2014).

USACE (U.S. Army Corps of Engineers). 2008. Upper Mississippi River comprehensive plan: main report. U.S. Army Corps of Engineers, Rock Island District, Rock Island, Illinois; St. Louis District, St. Louis; and St. Paul District, St. Paul, Minnesota. Available: www.mvr.usace.army.mil/Portals/48/docs/FRM/UpperMissCompPlan/UMRCPMainReport14Aug08update.pdf. (June 2014)

USACE (U.S. Army Corps of Engineers). 2009. Louisiana coastal protection and restoration: U.S. Army Corps of Engineers, New Orleans District, Final Technical Report and comment addendum, New Orleans. Available: www2.mvn.usace.army.mil/pd/projectslist/ProjectData/302/reports/LACPR%20Report%2014%20Aug%202009.pdf. (May 2014).

USACE (U.S. Army Corps of Engineers). 2011. Completed construction projects: Mosenthien Project. Applied River Engineering Center, St. Louis District, St. Louis.

USACE (U.S. Army Corps of Engineers). 2012. River training structures: innovative engineering keeps a dynamic river open for navigation. Available: www.army.mil/article/70627/. (May 2014).

USACE (U.S. Army Corps of Engineers). 2013. Conservation plan for the interior least tern, Pallid Sturgeon, and fat pocket mussel in the lower Mississippi River. USACE, Mississippi Valley Division/ Engineer Research and Development Center - Environmental Laboratory, Vicksburg, Mississippi.

USACE (U.S. Army Corps of Engineers). 2014a. Lock Performance Monitoring System. Available: corpslocks.usace.army.mil/lpwb/f?p=121:6:0. (May 2014).

USACE (U.S. Army Corps of Engineers). 2014b. National levee database. Available: nld.usace.army.mil/egis/f?p=471:1. (May 2014)

USFWS (U.S. Fish and Wildlife Service), 2000. Biological opinion for the operation and maintenance of the 9-foot navigation channel on the Upper Mississippi River System. USFWS, Bismarck, North Dakota.

USGS (U.S. Geological Survey). 1999. Ecological status and trends of the Upper Mississippi River System Long Term Resource Monitoring Program. U.S. Geological Survey, Upper Midwest Environmental Sciences Center, LTRMO 99-T001, La Crosse, Wisconsin.

Wiener, J. G., and M. B. Sandheinrich. 2010. Contaminants in the upper Mississippi River: historic trends, responses to regulatory controls and emerging concerns. Hydrobiology 640:49–70.

Wiener, J. G., C. R. Fremling, C. E. Korschgen, K. P. Kenow, E. M., Kirsch, S. J., Rogers, Y. Yin, and J. S. Sauer. 1998. Mississippi River. Pages 351–384 *in* M. J. Mac, P. A. Opler, C. E. Puckett-Haecker, and P. D. Doran, editors. Status and trends of the nation's biological resources, volume 1. U.S. Geological Survey, Reston, Virginia.

Wilcox, D. B., E. L. Stefanik, D. E. Kelner, M. Cornish, D. J. Johnson, I. J. Hodgins, S. J. Zigler, and B. L. Johnson. 2004. Improving fish passage through navigation dams on the Upper Mississippi River System. Interim report of the Upper Mississippi River-Illinois Waterway System navigation study. U.S. Army Corps of Engineers, Rock Island District, ENV Report 54, Rock Island, Illinois.

Winkley, B. R. 1977. Man-made cutoffs on the lower Mississippi River, conception, construction, and river response. U.S. Army Corps of Engineer, Potamology Investigations Report 300-2, Vicksburg District, Vicksburg, Mississippi.

Winkley, B. R. 1994. Response of the lower Mississippi River to flood control and navigation improvements. Pages 45–74 in S. A. Schumm and B. R. Winkley, editors. The variability of large alluvial rivers. American Society of Civil Engineers Press, New York.

Winkley, B. R., and J. J. Brooks. 1980. Navigation and flood control through geometric alignment. Potamology Investigation, Hydraulics Branch, U.S. Army Corps of Engineers, Hydraulics Branch, Potamology Investigations Report 300-3, Vicksburg District, Vicksburg, Mississippi.

Wuebbles, D. J., K. Hayhoe, and K. Cherkauer, 2009. Climate change and the upper Mississippi River basin. Pages 47–54 *in* R. E. Criss and T. M. Kusky, editors. Finding the balance between floods, flood protection, and river navigation. St. Louis University, Center for Environmental Sciences, St. Louis. Available: ces.slu.edu/annualreport/FloodForum_Book_final.pdf. (May 2014).

Wuebbles, D. J., and K. Hayhoe, 2004. Climate change projections for the United States Midwest. Mitigation and Adaptation Strategies for Global Change 9:335–363.

Yin, Y., and J. C. Nelson. 1996. Modifications of the upper Mississippi River and the effects on floodplain forests. Pages 29–40 *in* D. Galat and A.G. Frazier editors. Science for floodplain management into the 21st century, volume 3. U.S. Government Printing Office, Washington, D.C.

American Fisheries Society Symposium 84:235–262, 2016

The Lower Mississippi River Floodplain Ecosystem: Current Status and Future Potential

Michael A. Eggleton*
*Aquaculture/Fisheries Center, University of Arkansas at Pine Bluff
1200 North University Drive, Box 4912, Pine Bluff, Arkansas 71601, USA*

Quenton C. Fontenot
*Department of Biological Sciences, Nicholls State University
114 Gouaux Hall, Thibodaux, Louisiana 70310, USA*

John R. Jackson
*Department of Biological Sciences, Arkansas Tech University
1701 North Boulder Avenue, Russellville, Arkansas 72801, USA*

Abstract.—The lower Mississippi River encompasses the 1,535-km reach extending from the confluence with the Ohio River at Cairo, Illinois to the Gulf of Mexico. Waters of the lower Mississippi River have historically inundated vast areas of adjacent floodplain during spring flood pulses. Additionally, processes of land building within the river's deltaic plain supported vast forests and diverse freshwater and salt-marsh habitats. Flood pulses provided a mechanism of lateral exchange of energy and nutrients between the aquatic and terrestrial habitats, while sediment loads continually rebuilt and supported the deltaic plain. As human populations and agriculture expanded throughout the lower Mississippi Valley, construction of flood-protection levee systems and commercial navigational structures severely decreased the connectivity between the lower Mississippi River and its floodplain. The current lower Mississippi River floodplain is more than 90% reduced in area compared to historical conditions. Fluvial dynamics, which are the driving forces that stimulate floodplain function and create diverse habitats, appear to have been altered throughout approximately 80% of the river. As a result, the hydrograph, thermograph, sedimentation patterns, nutrient dynamics, and vegetation communities within the lower Mississippi River floodplain have experienced major changes through time, with many large alterations occurring during the past century. In addition, because most of the sediment load of the lower Mississippi River now enters the northern Gulf of Mexico, land building and associated processes are much reduced in the river's deltaic plain. This process has allowed intrusion of saltwater into coastal habitats, which has heavily impacted vegetation communities. This paper reviews the consequences of river modification to lower Mississippi River floodplain, current efforts towards restoring the floodplain and deltaic plain, and proposes future strategies towards restoring portions of the historical floodplain.

* Corresponding author: eggletonm@uapb.edu

Introduction—Overview of the Lower Mississippi River

The Mississippi River begins its 3,770-km journey to the Gulf of Mexico from Lake Itasca, Minnesota. The river's basin encompasses 4.7 million km^2, equivalent to 41% of the lower 48 United States and one-eighth of North America (Fremling et al. 1989). This large basin includes major tributaries such as the Illinois, Missouri, Ohio, White, and Arkansas rivers. Because the basin is so large and diverse throughout its length, the river is commonly classified into smaller units for comparative study. Schramm and Ickes (2016, this volume) divide the river into five ecological segments: the headwaters, the upper and lower impounded segments, and the upper and lower free-flowing segments. The headwaters are identified as the 824-km river segment between Lake Itasca and Coon Rapids, Minnesota (just upstream of Minneapolis, Minnesota). In this reach, the river is still somewhat natural in that floodplains are seasonally inundated, similar to historical conditions. The 1,069-km impounded segment downstream from Minneapolis to the confluence with the Missouri River north of St. Louis, Missouri contains 29 navigation locks and dams (Schramm 2004). These structures have created a series of run-of-the-river, lock-and-dam impoundments that have permanently inundated much of the historical floodplain. The upper and lower impounded segments are largely distinguished by differences in floodplain width and size, which generally increases in a downstream fashion (Fremling et al. 1989).

The free-flowing lower Mississippi River is the 1,884-km segment that begins at the confluence of the Missouri River and continues downstream to the Gulf of Mexico. The entire free-flowing segment is contained within a vast levee network except for a few upstream areas that are bounded by natural bluffs. The confluence of the Ohio River at Cairo, Illinois separates the upper and lower sections of the free-flowing segment. The upper section is commonly referred to as the middle Mississippi River (MMR) in the technical literature. In this section of the river, levees are less common, as much of the historical floodplain is bounded by natural bluffs. In contrast, the lower section of the free-flowing segment contains a wide, meandering channel flowing through a broad alluvial plain, though some natural bluffs do occur along the eastern bank in southwestern Mississippi. This segment is the largest at 1,535 km in length, comprising 81% of the free-flowing reach (Figure 1). Baker et al. (1991) further divides the lower free-flowing segment into two distinct reaches—one between Cairo, Illinois and Baton Rouge, Louisiana, with the other downstream of Baton Rouge. Below Baton Rouge, the river flows across a flat deltaic plain with a low slope, a deep, narrow channel, and high levees along both sides of the river channel. In this segment of the river, there is almost no existing floodplain (Figure 1).

Historical descriptions and discussion of contemporary anthropogenic modifications to the Mississippi River are well documented (e.g., Fremling et al. 1989; Baker et al. 1991; Schramm 2004; Sparks 2010; Alexander et al. 2012; Niebling et al. 2014; Schramm and Ickes 2016). While true that the lower Mississippi River receives less attention than the upper Mississippi River in the mainstream fisheries and aquatics literature, its floodplain is represented even less. Thus, this paper will focus on the floodplain of the lower free-flowing segment of the Mississippi River, which will be referred to herein as the LMR.

Three Centuries of Modifications to the Lower Mississippi River and Its Floodplain

Modifications of the LMR began almost three centuries ago for purposes of navigation, flood control, and/or bank stabilization. Because there are no dams on the LMR, development of hydroelectric power production has not occurred, though hydroelectric dams exist in many of its tributaries. The first levee network in the LMR was completed by the French colonial government to protect New Orleans around 1727 (Fremling et al.

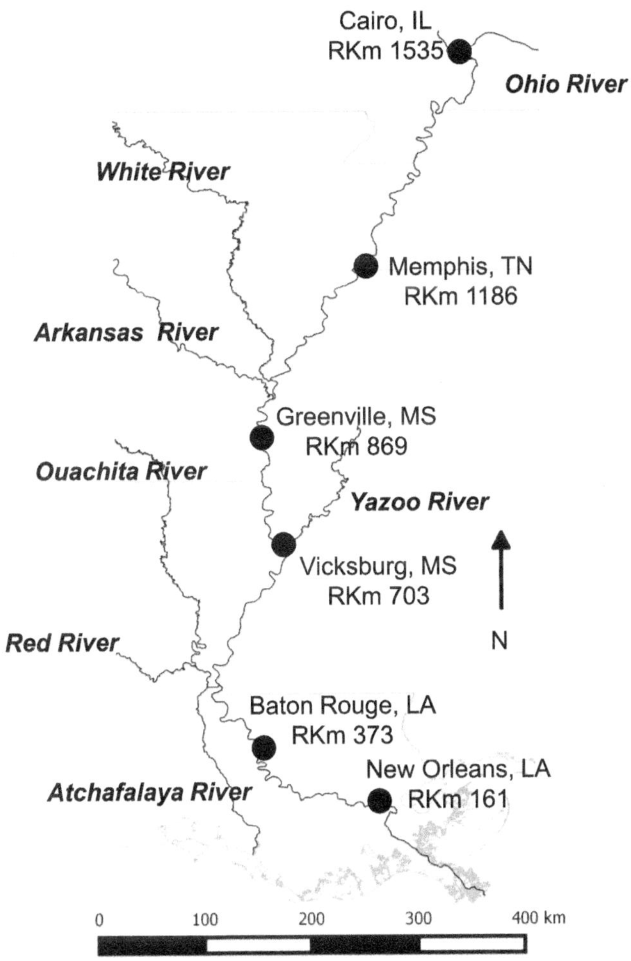

Figure 1. Map of the lower Mississippi River and its major tributaries. RKm = river kilometer.

1989). The French government considered levees essential to economic development such that landowners risked forfeiting their lands unless they constructed levees (Baker et al. 1991). As a result, many private levees were built and local levee districts evolved from the mid-1700s through the early 1900s. By 1844, the western bank of the LMR from just downstream of New Orleans north to the Arkansas River confluence was leveed to protect plantations and valuable farm lands, whereas the eastern bank was leveed upstream to Baton Rouge (Fremling et al. 1989). The U.S. Army Corps of Engineers (USACE) has been charged with maintaining navigation channels in the LMR for almost 200 years (Niebling et al. 2014). Although the USACE intermittently studied the potential of dredging to maintain a navigable main channel in the mid- to late 1800s, large-scale dredging was not common in the LMR until after 1900 (Baker et al. 1991).

Constant upgrading and expansion of levee systems throughout the LMR and its tributaries occurred throughout the 1800s (Pinter et al. 2010), leading to the initiation of the Mississippi River and Tributaries Project following the destructive 1927 Mississippi River flood (Baker et al. 1991). During the period 1929–1942, 14 neck cutoffs were constructed to shorten the river's path to the Gulf of Mexico (Biedenharn and Watson 1997). Although these modifications aided flood control, they were designed primarily to improve commercial navigation and fuel economic development in the region. In addition, three engineered floodways that allowed rechanneling of floodwaters through other river or lake basins were constructed in Louisiana (West Atchafalaya Floodway, Morganza Floodway, and Bonnet Carre Spillway; Baker et al. 1991). Another floodway exists in Missouri (Bird's Point-New Madrid Floodway) that allows temporary storage of Mississippi River floodwater for later discharge. Dikes and bank revetments to direct flow, maintain channel depth and alignment, and prevent bank erosion have been common throughout the LMR since the late 1800s (Alexander et al. 2012). The LMR contains approximately 125 dike fields constructed by the USACE, with fields containing 2–12 dikes each (Shields 1995). Since the 1960s, dikes have been constructed exclusively of limestone rock as opposed to wood pilings of previous decades (Baker et al. 1991). Bank revetments are used primarily to minimize bank erosion and collapse and to prevent the river from meandering across its floodplain (Sparks 2010). Since the 1940s, the USACE has used articulated concrete mattresses for bank revetments, as opposed to earlier periods when timber, brush fascines, wire gabions, rock riprap, asphalt paving, tires, and automobiles were commonly used to stabilize banks (Baker et al. 1991).

Modifications that have occurred throughout the LMR have undoubtedly affected the river's floodplain and, thus, river–floodplain dynamics (Schramm et al. 2000). Although comprehensive studies are rare and not sufficiently long term, it is unthinkable that fisheries production has not been directly and indirectly impacted (Schramm 2004; Schramm et al. 2015). The high levels of modification that have occurred in the LMR over the past 85 years to stimulate economic development through improved agricultural production and navigation, while minimizing flood risks, cannot be reversed due to economic, sociological, and political concerns. However, strategies may be possible to partly restore and maximize floodplain ecological function and, ostensibly, riverine fisheries production. The objectives of this paper are to (1) discuss how river modification in the LMR has affected its floodplain, (2) review current efforts towards restoring the LMR floodplain and deltaic plain, and (3) outline possible management that could be implemented to restore floodplain functionality and enhance fisheries production in the LMR.

Floodplain Effects Resulting from Modification of the LMR

Loss of Floodplain Area

The floodplain of the LMR is classified as largely decoupled from the main river channel (sensu Junk et al. 1989). Prior to the construction of the current levee system in the 1930s, the historical floodplain of the LMR encompassed approximately 10.1 million ha, or 101,000 km^2, which approximated the areal extent of a 100-year flood (see Table 1 in Schramm and Ickes 2016). In fact, 91% of the historical floodplain of the entire Mississippi River basin was located within the LMR (i.e., the lower free-flowing segment downstream of Cairo, Illinois). However, the present-day area of the LMR, including all aquatic and terrestrial habitats between the mainline levees, is estimated to be only 851,385 ha, equivalent to about 8% of its historical size (Schramm et al. 1999). The total area of current floodplain is estimated to be 651,839 ha, which includes about 77% of the total area between the levees (Schramm et al. 1999). Nearly all of this area is considered active LMR floodplain, in that it is periodically inundated by seasonal floods. The density of current LMR floodplain equates to only about 425 ha per river kilometer, which compares to almost 6,600 ha per river kilometer that existed previous to construction of the current levee system. Of the current 651,839 ha of active floodplain within the present-day LMR, 51,154 ha (8%) is contained within permanently inundated oxbow lakes and other floodplain lakes that do not seasonally dewater (Schramm et al. 1999).

Floodplain Inundation Patterns

The loss of historical LMR floodplain area is especially noticeable when floodplain inundation patterns are examined visually. Figure 2 contains a representative leveed 140-km reach of the LMR that encompasses river kilometers 703–843, located between Vicksburg, Mississippi and Eudora, Arkansas. Using data from the Vicksburg, Mississippi gauge, 53,343 ha of floodplain inundation occurs at the +0-m stage. This stage represents the low-water reference plane (LWRP) for this LMR location. The LWRP is equivalent to the river discharge that is exceeded 97% of the time, essentially reflecting base flow conditions for the river at this location. This figure includes the 51,154 ha of permanently inundated oxbow and other floodplain lakes mentioned above, which equals about 8% of total floodplain area in the LMR (Table 1; Figure 2). Thus, negligible floodplain inundation occurs at +0-m stage for this location. River stages of +6–10 m above LWRP represent typical flooding levels in the present-day LMR. At stage +4.6 m above LWRP, 70,583 ha of floodplain (11% of total floodplain area) becomes inundated, while 148,352 ha (23%) becomes inundated at +7.6-m stage (Table 1; Figure 2). At greater levels of flooding, 293,709 ha (45%) and 631,926 ha (97%) become inundated at stages of +9.1 and +13.7 m, respectively (Table 1; Figure 2; Schramm et al. 1999). Although extremely rare, river stages greater than +13.7 m represent nearly complete levee to levee flooding in the LMR. There are no records quantifying how much floodplain inundation occurred historically at different river stages, owing

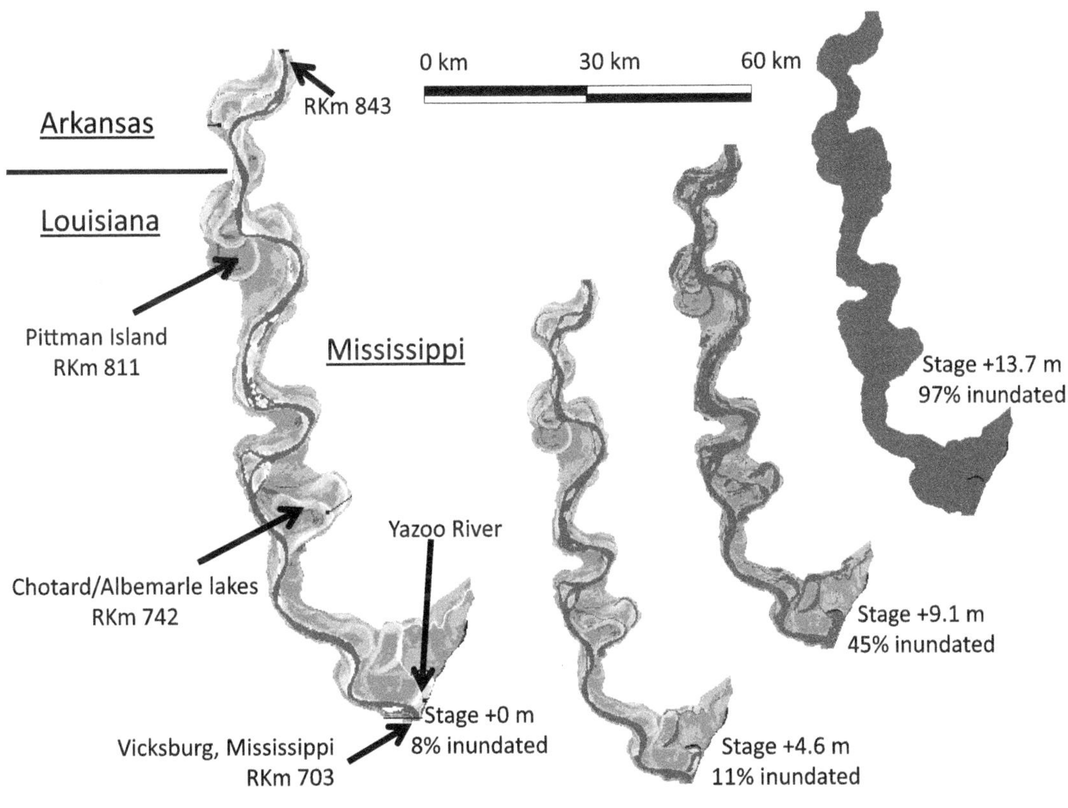

Figure 2. Floodplain inundation pattern within the 140- km reach of the lower Mississippi River above Vicksburg, Mississippi. Area shown on maps includes the area within the mainline levees of the present-day lower Mississippi River. Bank-full stage on the Vicksburg gauge is approximately +9.0 m above the low-water reference plane. The scale at the top refers only to the left map. Rkm = river kilometer.

Table 1. The lower Mississippi River classified as percentages of inundated main channel and floodplain area, floodplain connectance measures, and temporal characteristics of flooding versus river stages. LWRP = low-water reference plane.

River stage (m above LWRP)[a]	Percent total area inundated between levees[b]	Watered areas[b]		Percent of floodplain that is inundated	Floodplain connectance		Among of time ≥ stage during average year in weeks (% of year in parentheses)	
		Main channel (%)	Floodplain (%)		SDI[c]	FCI[d] (%)	Premodification (1900–1925)	Postmodification (1940–2010)
0.0	22.0	15.7	6.3	8.2	28.3	25.6	52 (100)	52 (100)
1.5	24.4	17.7	6.7	8.7	28.1	31.8	52 (100)	52 (100)
3.0	26.7	19.4	7.3	9.5	27.7	37.5	50 (96)	39 (75)
4.6	29.2	20.9	8.3	10.8	28.7	50.4	36 (69)	30 (58)
6.1	32.8	22.4	10.4	13.6	31.9	70.4	32 (62)	26 (50)
7.6	41.3	23.9	17.4	22.8	35.7	82.1	24 (46)	15 (29)
9.1	57.9	23.4	34.5	45.1	37.1	94.4	20 (38)	10 (19)
10.7	78.3	23.4	54.9	71.7	29.7	97.9	10 (19)	<1 (2)
12.2	92.2	23.4	68.7	89.8	20.2	99.7	2 (4)	<1 (2)
13.7	97.7	23.4	74.2	96.9	13.3	100	0 (0)	0 (0)
15.2	99.1	23.4	75.7	98.8	10.2	100	0 (0)	0 (0)
16.8	99.4	23.4	76.0	99.2	9.7	100	0 (0)	0 (0)
18.3	99.6	23.4	76.2	99.5	9.6	100	0 (0)	0 (0)
19.8	99.7	23.4	76.3	99.6	9.5	100	0 (0)	0 (0)
21.3	99.8	23.4	76.3	99.7	9.5	100	0 (0)	0 (0)

[a] Stages recorded from the Vicksburg, Mississippi stage (stage +9.0 m represents bank-full at this river location).

[b] Refers to the lower Mississippi River only during the postmodification period (i.e., the present day).

[c] SDI is the shoreline development index, which equals the total perimeter of inundated main channel and confluent floodplain water areas (L) divided by 2 times the square root of the corresponding area (A) multiplied by π, or SDI = $L/2(\pi A)^{0.5}$. A perfectly circular body of water has an SDI of 1.0.

[d] FCI is the floodplain connectance index, which equals the ratio of the inundated floodplain area connected to the main river channel minus areas of disconnected floodplain waters divided by the total floodplain area (Schramm et al. 1999).

to the lack of needed data sets. However, it can be assumed that the spatial extent was much greater compared to the present day, though the physical and temporal characteristics of that flooding may have been much different.

Stage Adjustment Resulting from Channel Incision

The LMR and its floodplain have been subject to impacts associated with channel incision for decades (Biedenharn and Watson 1997). Incision involves the down-cutting of a river channel into the landscape to the point that bed elevation and base water levels decrease over time (Galay 1983). Incision usually occurs along the thalweg and greatly changes river channel geometry (see conceptual model in Shields et al. 1998). Habitats are transformed over time and features relocate as the river attempts to achieve some minimum threshold gradient relative to its position above mean sea level (Biedenharn et al. 2000). Incision can be caused by a variety of activities, including reservoir construction, land-use changes, and occasionally natural events; however, construction of neck cutoffs and channelization practices (e.g., levees) are the most common sources (Galay 1983). Although incision impacts are more obvious in streams and smaller rivers (e.g., Shields et al. 1994), they have been documented in larger rivers such as the LMR (e.g., Biedenharn and Watson 1997).

Channel incision in the LMR has been largely attributed to past river training practices that have occurred. Biedenharn and Watson (1997) examined the response of the LMR to 16 neck cutoffs that occurred during the period 1929–1942 in response to the massive 1927 flood. Of these cutoffs, 14 were engineered while 2 were natural; the cutoffs shortened the river's course to the Gulf of Mexico by 245 km (Schramm et al. 2015). Prior to 1927, the LMR channel was considered to be in a state of dynamic equilibrium. Dynamic equilibrium implied that there was no significant widespread instability such as channel aggradation or degradation occurring, though an isolated area of natural aggradation was detected near Columbus, Kentucky prior to 1925 (Biedenharn and Watson 1997). In the present day, the LMR exists as a continuum of channel conditions, including reaches that are aggradational, degraditional, and transitional, with some reaches in states of dynamic equilibrium (Biedenharn and Watson 1997; Biedenharn et al. 2000). Within this continuum, however, river stage adjustments (i.e., a general lowering of base water levels) resulting from cutoff-induced channel incision are detectable at multiple locations between Columbus, Kentucky and Natchez, Mississippi during the postcutoff period (Biedenharn and Watson 1997).

Present-day river stages in the LMR have decreased resulting from channel incision processes associated with neck cutoffs constructed more than 80 years ago. Although bed materials have not become more coarse as predicted (Biedenharn et al. 2000), peak annual stages that typically occur in the LMR during mid- to late April are now (i.e., 1940–2010 or postmodification) approximately 3 m less compared to historical conditions (i.e., 1900–1925 or premodification) (Figure 3). Similarly, mean annual stages have declined by 1.5 m during the same period (+7.3 to +5.8 m), with stages falling below bank-full (i.e., stages that overtop the most pronounced stream banks and allow massive inundation of the adjacent floodplain; Armantrout 1998) by the end of May on average compared to late June or early July during historical conditions (Figure 3). There is some evidence that the resultant shortening of the LMR channel also has altered natural channel sinuosity (Schramm and Ickes 2016). By altering the channel's slope, some channel realignment has occurred that later triggered a massive slope recovery process known as "head-cutting" (i.e., channel degradation upstream, with channel aggradation downstream) in some areas of the LMR (Winkley and Brooks 1980; Schramm and Ickes 2016). The head-cutting phenomenon is somewhat diminished but still present in the LMR today (Pinter et al. 2010).

Altered Flood Pulse Characteristics

Recent studies indicate the current hydrograph of the LMR differs from the historical one in a variety of ways. Channelization practices implemented over the past 85 years (e.g., neck cutoffs, levees, and dike fields) that have shortened the LMR channel provide more efficient downstream transport of floodwaters to the Gulf of Mexico. Because faster downstream flows minimize stage increases during flood pulses, larger water volumes are needed to increase stages above bank-full and inundate the adjacent floodplain. As a result, the postmodification flood pulses in the present-day LMR are generally briefer and of less magnitude compared to historical or premodification conditions (Figure 3). Under postmodification (1940–2010) conditions, the LMR floodplain is inundated less often and for shorter time periods on average each year (Table 1), potentially altering many floodplain processes compared to historical conditions (Schramm and Eggleton 2006; Schramm and Ickes 2016). For example, at stages less than +2.5 m above LWRP, inundation duration patterns were similar during the pre- and postmodification times in the LMR, with these stages being maintained on average nearly year-round (Figure 3; Table 1). At stage +4.6 m above LWRP during the present day, stages persist at this level on average for about 30 weeks per year (58% of the year) compared to 36 weeks per year (69% of the year) during premodification times (Figure 3; Table 1). However, at +9.1 m above LWRP, which represents bank-full on the Vicksburg, Mississippi gauge, floodplain inundation duration was only one-half during postmodification conditions (i.e., 10 weeks or 19% of the year) compared to premodification conditions (i.e., 20 weeks or 38%

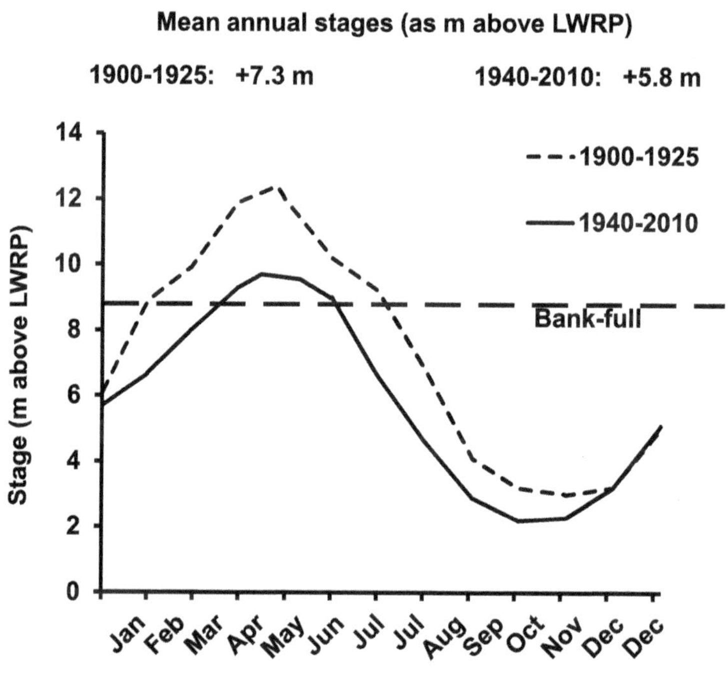

Figure 3. Mean annual river stages in the lower Mississippi River measured at the Vicksburg, Mississippi gauge before (1900–1925) and after (1940–2010) major modification occurred within the river. Modifications include construction of neck cutoffs that shortened the river's course to the Gulf of Mexico, dike fields to maintain channel depth, and tall levees to contain seasonal flooding. Horizontal grid represents bank-full stage on the Vicksburg gauge, which equals approximately +9.0 m above the low-water-reference plain (LWRP).

of the year) (Figure 3; Table 1). The significantly lower floodplain inundation durations during the postmodification period are consistent with lower river stages created by channel incision processes that have been occurring for decades (Biedenharn and Watson 1997; Biedenharn et al. 2000).

With respect to annual flood pulses, thermal characteristics appear to differ in the postmodification LMR compared to premodification times. The altered flood pulse described above has resulted in a general asynchrony between flooding and water temperatures optimal for the spawning and growth of many temperate riverine fishes (Schramm et al. 2000; Schramm and Eggleton 2006). In other words, water temperatures suitable for basic life history pursuits of many LMR fishes do not always occur during periods when the floodplain is inundated. This asynchrony occurs following river training practices (e.g., levee systems and dike fields) in which most of the annual flood pulse occurs prior to seasonal warming of water temperatures (Schramm and Eggleton 2006). Thus, poikilothermic, floodplain-dependent fishes accrue fewer benefits from seasonal access to floodplains because floodwaters contain suboptimal temperatures for active fish feeding, growth, and reproduction (Eggleton 2001). This thermal-hydrologic asynchrony or "thermal decoupling" (sensu Junk et al. 1989) has evolved through time due in part to the extensive river training practices. The phenomenon has produced earlier, briefer flood pulses in some rivers that also are colder due to the seasonal climates in temperate regions (Schramm and Eggleton 2006).

Recent studies from the LMR have indicated that a thermal-hydrologic asynchrony relative to annual flood pulses occurs during many years (Schramm and Eggleton 2006). Using data from Schramm et al. (2015) and 15°C as the threshold minimum temperature for feeding, growth, and reproduction of many LMR fishes, the river inundated its adjacent floodplain for only 3–4 weeks per year on average when water temperatures exceeded 15°C (Figure 4). The overlap be-

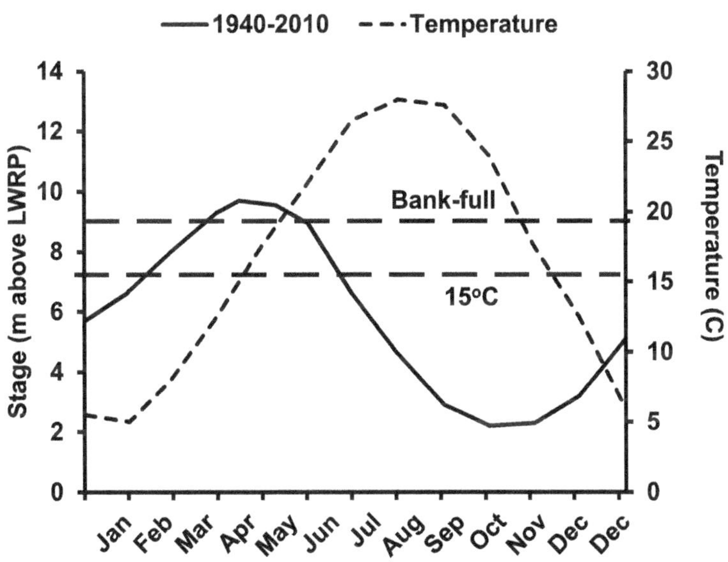

Figure 4 . Mean annual river stages (left axis) in the lower Mississippi River measured at the Vicksburg, Mississippi gauge after (1940–2010) major modification occurred within the river. Stage data obtained from the Vicksburg, Mississippi gauge. Water temperatures are plotted on the right axis. Top horizontal grid represents bankfull stage on the Vicksburg gauge (approximately +9.0 m above LWRP). Bottom horizontal grid represents 15°C water temperatures measured from the main river channel.

tween floodplain inundation and water temperatures exceeding 15°C is highly variable in the LMR from year to year. In examining 13 individual years (1990–2002), annual flooding coincident with water temperatures ≥15°C was substantial during some years (e.g., 1990, 1993, 1995, and 1998) but did not occur at all during other years (e.g., 1992, 2000, and 2001) (Figure 5). In addition, during some years when there was high overlap between floodplain inundation and water temperatures ≥15°C, much of the overlap period occurred when river stages were less than bank-full. Thus, energetic benefits accrued by LMR fishes from access to floodplain resources would have been bioenergetically less during those years (Schramm and Eggleton 2006). Because the LMR fish assemblage is largely warmwater, nearly all species resume active feeding and spawning at water temperatures greater than 15–18°C (Schramm and Eggleton 2006; Schramm et al. 2015). Thus, the hit-or-miss scenario that occurs in the LMR annually has the potential to impact seasonal feeding and growth, reproduction, and ultimately recruitment of many fish species. This type of thermal-hydrologic asynchrony has likely altered floodplain function throughout the LMR for decades, potentially having long-term impacts on fisheries production. In a bioenergetic analysis of two large-river catfishes, Eggleton (2001) reported greater estimated food consumption rates (measured as temperature-dependent caloric intake) during a year when floodplain inundation coincided with warmer water temperatures (1998) compared to another year where the spring flood pulse contained colder water temperatures (1997). Schramm and Eggleton (2006) further suggested that the flood-pulse concept proposed by Junk et al. (1989) was applicable to the LMR. However, models describing fish–floodplain relationships were much improved when thermal conditions of flood pulses were included in analyses. This particular topic concerning fish–floodplain interactions in highly modified rivers is of vital importance in understanding floodplain function and could serve as a template for river restoration. This particular topic is in need of further research.

Reductions in River–Floodplain Connectance

Contemporary theory on river–floodplain ecology infers a tight, energetic linkage between the river and floodplain, largely facilitated by the flood pulse that inundates the adjacent floodplains (Welcomme 1979; Junk et al. 1989; Bayley 1991; Sparks 1995a, 1995b; Lorenz et al. 1997). Junk et al. (1989) describes the flood pulse as a "moving littoral zone," in that shallow, shoreline habitats at the river's edge are fluid and ever changing and dependent on flooding magnitude. For example, shallow shoreline habitats during one week can be much deeper and far offshore the following week as floodwaters rise. Riverine fishes and other biota accrue significant energetic benefits from inundation of the floodplain. Flooding provides fishes with access to feeding, spawning, and nursery areas (Junk et al. 1989; Schramm and Ickes 2016). However, floodplain benefits are not limited to fishes. Floodplains contain shallow-water habitats that support resident and migratory waterfowl (Opperman et al. 2010) and support large stands of native forests (e.g., bottomland hardwoods, bald cypress *Taxodium distichum*, and tupelo gum *Nyssa aquatica*; Nelson et al. 2002). Floodplains also serve other vital ecosystem functions that support the river. Inundated floodplains supply nutrients and organic matter that stimulate primary and secondary production throughout the river (Junk et al. 1989; Bayley 1995). Floodplain soils serve as valuable sinks for nutrient runoff (Rabalais et al. 1996). In the LMR, floodplain soils have been purported as being a major sink for nitrogen being transported to the northern Gulf of Mexico, resultant from runoff from much of the central United States (Schramm et al. 2009). In particular, excessive nitrogen inputs from the Mississippi River basin have contributed to the well-documented hypoxic zone that has developed in the Gulf of Mexico in recent decades (Rabalais et al. 2002). Decreases in any of these functions typify re-

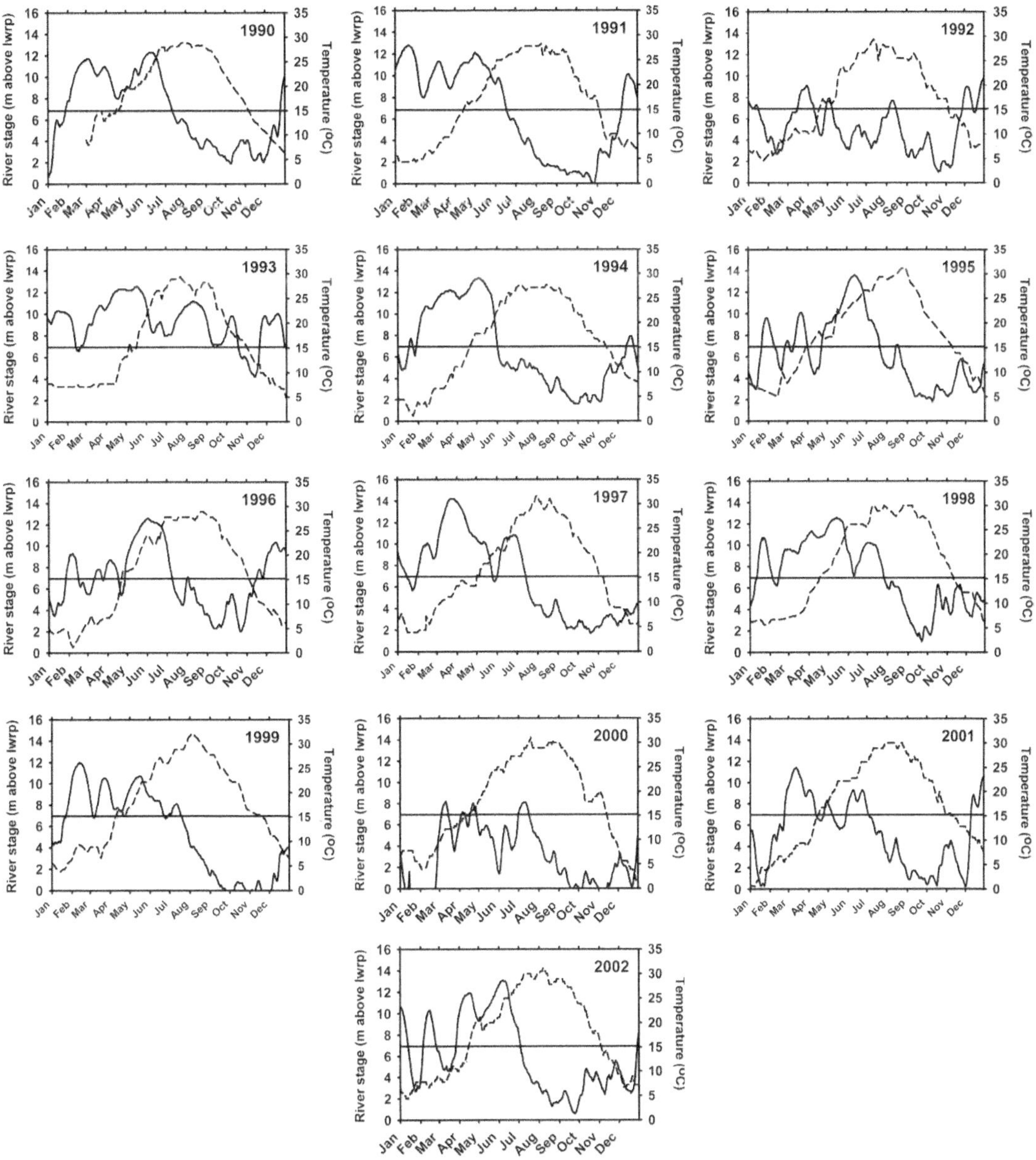

Figure 5. Stage (solid line) and water temperature (dashed line) of the lower Mississippi River measured at Vicksburg, Mississippi for calendar years 1990–2002. Horizontal grids represents water temperature of 15°C. River stages expressed as meters above low-water reference plane (LWRP), equivalent to the stage at which discharge is exceeded during 97% of the time. Bank-full stage on the Vicksburg gauge equals approximately +9.0 m above LWRP. Figure reprinted with permission from Schramm and Eggleton (2006).

ductions in interactions between rivers and their floodplains.

Modifications throughout the LMR during the past 80 years have primarily reduced floodplain area, induced channel incision, altered seasonal flood pulses, and induced thermal decoupling. These factors in combination have brought about a net reduction in what is termed river–floodplain "connectance" (sensu Ward and Stanford 1995; Galat et al. 1998), whereby the floodplain simply has less contact time with floodwaters due to the factors just described. Decreases in river–floodplain connectance are typified by several characteristics, including reduced spatial habitat heterogeneity, greater temporal instability, and diminished river–floodplain interactions (Ward and Stanford 1995; Shields et al. 1998; Eggleton 2001). In the LMR, mean annual stages have decreased approximately 1.5 m (20%) on average in recent decades (Figure 3), with annual flood pulses being briefer and of less magnitude (Table 1) and sometimes exhibiting thermal-hydrologic asynchrony (Figures 4–5). All of these factors reflect decreased river–floodplain connectance and have been occurring in the LMR for decades. Through time, it is likely that reduced river–floodplain connectance has altered the ecological function and processes within the LMR floodplain. For example, under the conditions above, LMR fishes on average have less time annually to use the floodplain for feeding, growth, and reproduction, while juvenile fishes have more variable conditions for hatching and less time for rearing.

Indices for floodplain connectance (i.e., shoreline development index and floodplain connectance index) can only be approximated for the present-day, postmodification LMR. Floodplain connectance index reflects the degree of physical or "watered" connection between the LMR main channel and its adjacent floodplain habitats, and infers high levels of river–floodplain interactions. Floodplain connectance index values increase linearly with river stages and peak at almost 100% when nearly all of the active floodplain is inundated (Figure 6; Table 1). Shoreline development index (SDI) values reflect habitat diversity, as indicated by the amount of inundated littoral-zone habitat. Shoreline development index values increase slightly with river stage, peaking at approximately bank-full stage (+9.0) as floodwaters begin massive inundation of the adjacent floodplain (Figure 6; Table 1). At river stages greater than +9.0 m, SDI values steadily decrease and stabilize at around 10, when floodplain inundation in almost levee to levee (Figure 6; Table 1). Using these con-

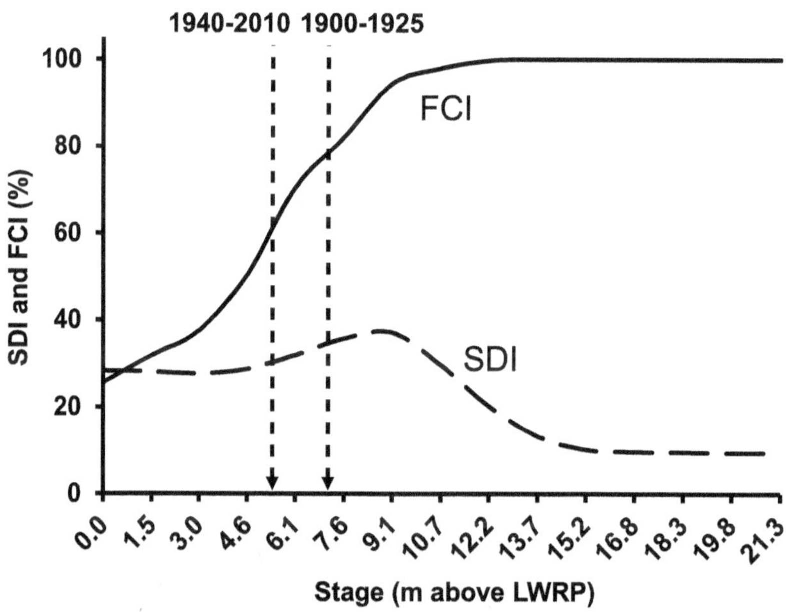

Figure 6. Floodplain connectance index (FCI, solid line), and shoreline development index (SDI, dashed line) versus river stages in the present-day lower Mississippi River. Stages expressed as meters above the low-water reference plain (LWRP) at the Vicksburg, Mississippi gauge. Vertical dashed lines represents mean annual stages at Vicksburg during the periods 1900–1925 and 1940–2010. FCI is a percentage metric, whereas SDI is unitless. Data compiled from Schramm et al. (1999).

nectance criteria, mean stage differences between the pre- and postmodification period in the LMR (+7.3 and +5.8 m, respectively) suggest that river–floodplain connectance is approximately 20% reduced during average hydrologic conditions (Figure 6; Table 1). However, the actual difference would be much greater owing to the fact that river–floodplain connectance was greater historically prior to river training practices when floodplain inundation duration was significantly longer and more widespread (Table 1). This finding is especially significant considering that the current LMR floodplain is only about 8% of its historical the size. Thus, 20% less connectance, which is a likely underestimate of the true value, with a floodplain that is more than 90% reduced in size has undoubtedly affected the LMR's fisheries, though this hypothesis has not been thoroughly researched. Shoreline development index values were not appreciably different at +7.3 and +5.8 m stages, and were similar to the greatest shoreline development index value that occurred at approximately bank-full stage (+9.1 m), though the physical and temporal characteristics of inundation were much different than in the present day (Figure 6).

River Modification Effects on Fishes

Fish diversity.—Losses of river–floodplain connectance are of special concern in large river systems because of their relatively high levels of biological diversity, including fishes (Ward et al. 1999). Comprehensive fish data sets do exist for the LMR, though compilation of them is challenging because of the river's large size, which necessitates monitoring by many state and federal agencies. Additionally, LMR data sets from the premodification period are rare. Baker et al. (1991) reported 91 fish species present throughout the LMR, though they cited earlier surveys indicating richness ranging from 110 to 121 species. Schramm and Eggleton (2002) reported 53 fish species collected from a series of main and secondary channel locations distributed throughout the LMR from Kentucky to southern Mississippi. Although this work excluded floodplain habitats, it demonstrated the ecological value of secondary river channels to LMR fishes. In later review papers, Schramm (2004) and Schramm and Ickes (2016) compiled species lists from several sources and reported that 109 fish species currently reside in all habitats of the LMR. This figure included at least 46 species classified as dependent on backwaters or floodplains (i.e., lentic habitats) during some portion of their life cycles. These species represent 42% of the present-day fauna and would be the mostly likely species affected by reduced river–floodplain connectance.

Despite obvious reductions in floodplain area and river–floodplain connectance, it is encouraging to see that there have been relatively few fish species extinctions within the LMR. For example, some sources consider Lake Sturgeon *Acipenser fulvescens* as a species that has disappeared from the LMR due to overfishing and high degrees of river training, as is postulated in the upper Mississippi River (Knights et al. 2002). However, because historical distributions of Lake Sturgeon are not well documented in the LMR, it may have been a peripheral species in this segment of the river and not prevalent historically. On the other hand, river training practices along with the associated floodplain losses and hydrologic alterations have probably impacted Alligator Gar *Atractosteus spatula* distributions throughout the LMR and its tributaries (Robison and Buchanan 1988; Pringle et al. 2000; O'Connell et al. 2007). A more likely effect of river training has been that reduced river–floodplain connectance in the LMR has produced shifts in fish assemblage structure, though these are not well documented and difficult to examine with historical data sets. In recent years, it has been widely reported that the exotic Silver Carp *Hypophthalmichthys molitrix* and Bighead Carp *H. nobilis* have become highly abundant in the LMR and many of its tributaries (Chick and Pegg 2001; Williamson and Garvey 2005; O'Connell et al. 2011). However, it has not been suggested that these invasions have in some way been facilitated by the extensive modifications in the LMR or the effects on its floodplain.

Fish-flooding relationships.—Recent studies suggest losses of river–floodplain connectance in the present-day LMR (Schramm et al. 2000; Schramm et al. 2015). Using the flood pulse concept (Junk et al. 1989) as a conceptual model describing river–floodplain interactions and their effects on LMR fishes, it would be expected that key fish population metrics, such as abundance, growth, and production, would be related to measures of flooding and floodplain inundation (e.g., Kwak 1988; Sparks 1995a, 1995b; Gutreuter et al. 1999; Eggleton and Schramm 2004). For example, growth or recruitment of a floodplain-dependent species might be expected to be enhanced during a year with high levels of flooding and floodplain inundation. Gutreuter et al. (1999) tested the relevance of the flood pulse concept in the highly modified upper Mississippi River. They reported a general concordance of the flood pulse concept predictions and demonstrated positive relationships between the growth of littoral-zone species (Largemouth Bass *Micropterus salmoides* and Bluegill *Lepomis macrochirus*) expected to utilize floodplain resources and flooding measures. Conversely, no relationship was detected between the growth of White Bass *Morone chrysops*, which is not a littoral-zone species nor expected to use floodplains, and the same flooding measures. Though Black Crappie *Pomoxis nigromaculatus* was considered a littoral-zone species, results were judged ambiguous for this species. Although the results of Gutreuter et al. (1999) were generally consistent with their predictions and supported the flood pulse concept, their data set was dominated, to an extent, by the massive 1993 Mississippi River flood. Similar fisheries effects were reported by Phelps et al. (2014) with respect to the temporary flooding of the Bird's Point-New Madrid floodway during the Mississippi River flood of 2011.

Studies from the LMR have generally reported inconsistent fish–flooding relationships, which contrast with studies from the upper Mississippi and other rivers. Rutherford et al. (1995) found no relationships between growth and abundance of young LMR fishes (Blue Catfish *Ictalurus furcatus*, Channel Catfish *I. punctatus*, Freshwater Drum *Aplodinotus grunniens*, and Gizzard Shad *Dorosoma cepedianum*) and measures of floodplain inundation. They attributed the lack of expected relationships to the large net reduction in area of the LMR floodplain due to levee construction, which is consistent to a reduction in river–floodplain connectance. Schramm and Mayo (1999) reported no relationships between the growth of age 3–5 Flathead Catfish *Pylodictis olivaris* and measures of flooding in the LMR. Although Flathead Catfish is not normally classified as a floodplain-dependent species, they are present in the LMR floodplain and the years being considered (1993–1997) contained a wide variety of flood pulse conditions in the LMR. Schramm and Eggleton (2006) failed to detect any positive relationships between annual growth increment and duration of floodplain inundation for Blue Catfish and Flathead Catfish in the LMR, although feeding and caloric intake differed across river habitats and years for these species (Eggleton 2001). However, annual growth increment of these species was positively related to days of floodplain inundation when main river channel water temperatures exceeded 15°C, which reflected warmer (18–20°C) water temperatures in adjacent floodplains. In this study, catfish growth in the LMR followed the flood pulse concept more closely when thermal conditions of the flood pulse were incorporated into flooding measures (e.g., degree-days above 15°C extrapolated over floodplain area; sensu Schramm et al. 2000). This finding would be consistent with the thermal-hydrologic asynchrony described above and may actually demonstrate the ecological drivers for fish use in floodplains of highly modified rivers such as the LMR.

One piece of information that is conspicuously lacking from studies of river–floodplain interactions is quantifying the spatial and temporal aspects of river–floodplain connectance needed to benefit fishes. During the period prior to the massive modifications that followed the 1927 flood, Schramm et al. (2015) indicated that although warmwater floodplain inundation averaged about 2 months per year, there appeared to be no "typical flood pulse" in the LMR. Several authors (e.g., Sparks et al. 1998; King et al. 2003; Janáč et al. 2010) have suggested that a minimum of 6 weeks of floodplain inundation is necessary for successful fish recruitment in unmodified large rivers. In the LMR, data sets suggested that floodplain inundation coincident with water temperatures greater than 18°C exceeded 6 weeks during at least 50% of years (refer to Figure 4 in Schramm et al. 2015). By comparison, however, floodplain inundation with water temperatures exceeding 18°C occurred for 3 weeks or less during half of the years in the postmodification LMR (Schramm et al. 2015). Although studies have been few, there is great potential for this altered hydrograph and thermal-hydrologic asynchrony to limit riverine fish production and perhaps modify floodplain function in the LMR. Additionally, shifts in fish assemblage structure would likely occur. Such fisheries impacts may be difficult to detect in the LMR without long-term data sets that include multistate cooperators. Some of these goals are being presently pursued by conservation organizations such as the Lower Mississippi River Conservation Committee.

Altered Sediment Dynamics in the Deltaic Plain

The LMR downstream of Baton Rouge, Louisiana is termed the "deltaic plain" and contains distinct characteristics and features not found elsewhere in the basin. River modifications have impacted this segment of the LMR much differently compared to areas upstream. The present-day LMR deltaic plain began forming approximately 7,500–8,000 years ago and has been shaped by the river changing course approximately every 1,000–1,500 years (Coleman et al. 1998). The cyclical development of deltaic headlands and subsequent delta abandonment has led to the formation of five major delta regions in the LMR deltaic plain over the previous 7,000 years (Blum and Roberts 2012). Extension of the delta lobe decreases the river channel elevation gradient and causes the river to change course and create a new delta headland approximately every 1,000 years. The abandoned delta loses its sediment source and is degraded through the natural processes of subsidence (i.e., decrease in elevation mostly due to compaction of newly deposited sediments), erosion, and sediment redistribution. Modern river training practices throughout the LMR prevent the river from changing course and, thus, depositing new sediments onto its historical floodplain. As a result, most of LMR's sediment load now enters the northern Gulf of Mexico and is deposited onto the continental shelf rather than being distributed throughout the deltaic plain (Kesel 2003).

Sediment delivered by overbank flooding and the distributary system of the LMR deltaic plain built the foundation for a vast network of bottomland hardwood forests, cypress-tupelo forests, and continua of freshwater to salt-marsh habitats along elevation and salinity gradients. Barrier islands and ridge habitats of the Mississippi River coastal region provided enough elevation to support maritime forests in otherwise low-lying areas (Braud et al. 2006). Nutrients and sediments delivered to floodplain habitats of the LMR deltaic plain sustained a diversity of habitats, and many organisms adapted a life-history strategy to take advantage of the predictable seasonal inundation of low-lying floodplain habitats during annual flood pulses. Levee construction and disconnection of distributaries has separated the LMR from approximately 93% of its historical floodplain (Schramm and Ickes 2016). This disconnection has altered hydrologic and sedimentation patterns throughout the LMR and its tributaries (Keown et al. 1986).

Vegetation Community Effects in the Deltaic Plain

Unlike river floodplains, water levels in estuaries are tidally influenced, though hydrology and sedimentation patterns in the Mississippi River estuary are influenced by water levels in the LMR. Sediment accretion throughout the Mississippi River deltaic plain maintained the elevation in the upland areas and ridges within the estuaries and increased the area of terrestrial areas within the estuary. However, river training occurring throughout the LMR has prevented river water and sediments from entering the historical delta floodplain; thus, elevation throughout this floodplain is decreasing mostly due to subsidence. For example, an elevation difference on the order of a few centimeters has greatly influenced the floodplain vegetative communities of the LMR deltaic plain (Rejmánek et al. 1987; Shaffer et al. 1992). Bottomland hardwood forests are found only at the highest elevations (i.e., upland areas and ridges) in the deltaic plain and typically experience the least amount of inundation among floodplain vegetative communities. Cypress-tupelo forests can withstand longer periods of flood inundation than bottomland hardwood forests and are, therefore, commonly associated with lower elevations than bottomland hardwood forests. As elevations continue to decrease due to subsidence, cypress-tupelo forest rapidly transition to marshes, which prevail at the lowest elevations (Nelson et al. 2002). Decreases in floodplain elevation typically coincide with increases in inundation period and cause shifts in floodplain vegetation communities (Shaffer et al. 1992).

The aforementioned river training effects have had noticeable impacts on the vegetative community within the Barataria Estuary in particular, which is the southernmost western floodplain of the LMR (Figure 7). Floodwaters of the LMR no longer inundate the Barataria Estuary; however, its historical floodplain does inundate when local bayous and canals overtop their banks during large rain events (Sklar and Conner 1979; Conner and Day 1992). The combination of rainfall and low elevation prevents most wetland areas within the Barataria Estuary from dewatering except during extended drought conditions (Conner and Day 1992). However, the loss of sedimentation from the LMR in the Barataria Estuary coupled with natural subsidence has decreased elevation and increased wetland area in the upper portion of the estuary (Nelson et al. 2002; Braud et al. 2006). Concurrently, approximately 38% of the bottomland hardwood forest in the upper Barataria Estuary was lost between 1972 and 1992 (Nelson et al. 2002). Although some of the lost bottomland hardwood forest was attributed to urbanization and logging, Nelson et al. (2002) concluded that increased inundation due to subsidence caused the majority of the bottomland hardwood forest to convert to cypress-tupelo forest.

In addition to losses from urbanization and logging, many stands of cypress-tupelo forests in the LMR deltaic plain are declining due to hydrologic alteration, nutrient deprivation, and saltwater intrusion (Nelson et al. 2002; Keddy et al. 2007; Shaffer et al. 2009). Gross primary production of cypress-tupelo forests is positively related to nutrient input (Brown 1981), with the sediment and nutrient deposits from flood pulses likely important for long-term forest sustainability (Myers et al. 1995; Keddy et al. 2007). Although adult bald cypress can withstand long periods of inundation, they experience increased mortality when the inundation is deeper and more long term (Eggler and Moore 1961; Harms et al. 1980; Brown and Lugo 1982). In particular, bald cypress seedling mortality increases significantly when inundated for more than 2 weeks (Brandt and Ewel 1989). Because of decreased elevation due to subsidence, some cypress-tupelo forests in the LMR deltaic plain have become permanently inundated. Thus, forests cannot naturally regenerate and are being gradually converted to marshes and, in some cases, open water (Nelson et al. 2002; Keddy et al. 2007). The disconnection of the Mississippi River from the estuary, the decreases in floodplain elevation, and the subsequent loss of coastal lands has allowed saltwater from the Gulf of Mexico to inundate historically inland freshwater systems. The resulting increases in salinity due to saltwater intrusion have killed entire stands of cypress-tupelo forests in close proximity to coastal areas (Penland and Ramsey 1990; Keddy at al. 2007; Shaffer et al. 2009).

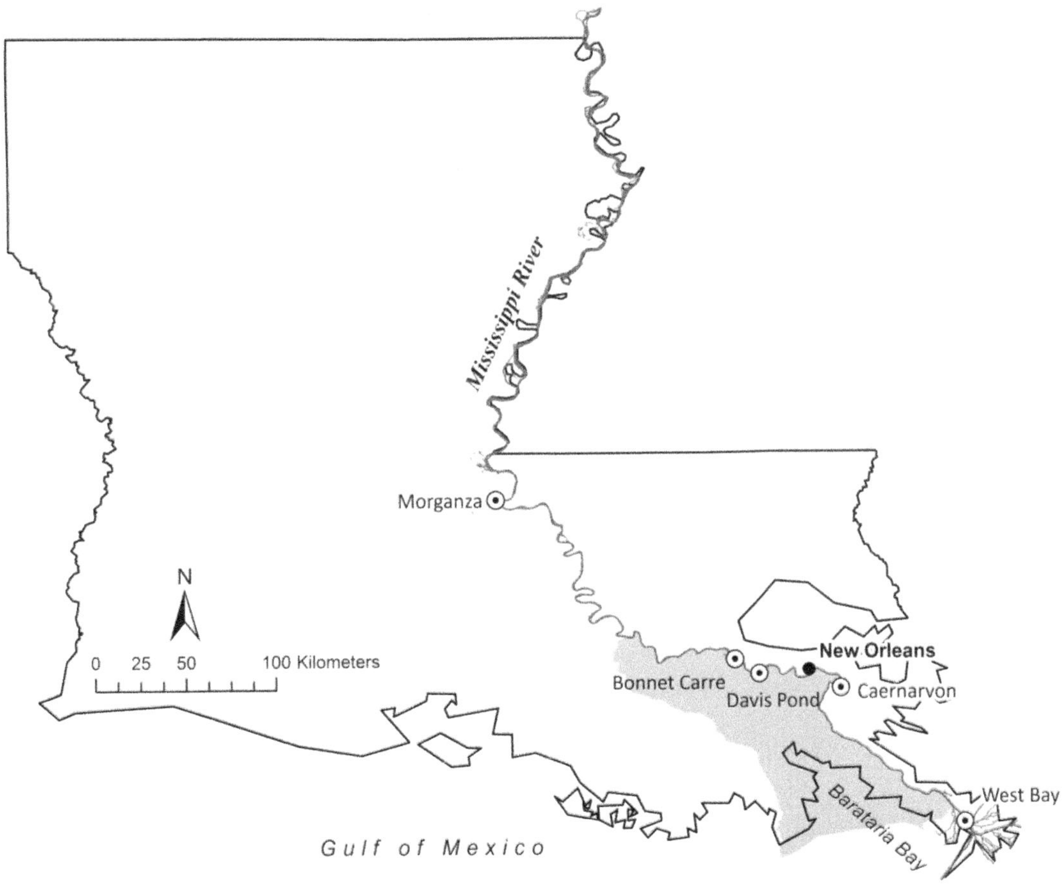

Figure 7. Map of the Barataria Estuary, Louisiana. This estuary represents the southernmost western floodplain of the lower Mississippi River.

Saltwater intrusion throughout habitats in the LMR deltaic plain has been exacerbated by the disconnection of the LMR from its floodplain, subsidence, sea level rise, and construction of navigational canals (Penland and Ramsey 1990; Ko and Day 2004; Poirrier 2013). Although bald cypress in Louisiana have a greater salinity tolerance than many other southeastern U.S. populations (Conner and Inabinette 2005), they are still highly susceptible to the combined effects of increased salinity and flooding (Allen et al. 1996). As a result, large stands of bald cypress have died in the LMR deltaic plain from the increased salinity. Increased water levels and salinity also have negatively impacted fresh and brackish marshes of the LMR deltaic plain (Keddy et al. 2007; Spalding and Hester 2007; Day et al. 2011). The synergistic effect of increased inundation and salinity has been identified as a primary factor causing coastal vegetation diebacks (Webb and Mendelssohn 1996). Over long-term time scales, this continual process leads to the eventual conversion of marshes and wetlands to open water (Turner and Rao 1990).

Dynamics of Land Losses in the Deltaic Plain

Subsidence and degradation of an abandoned delta is a natural process, but local anthropogenic factors such as training of the LMR channel, oil and gas extraction, and construction of navigational canals have contributed to the rates of land loss throughout the LMR deltaic plain (Reed 1995; Chan and Zoback

2007). In 1913, land-loss rates in the LMR deltaic plain were estimated to be 17 km^2/year (Gagliano et al. 1981). By 1980, several estimates indicated that loss rates had increased sixfold to 102 km^2/year (Gagliano et al. 1981; Boesch et al. 1994; Bernier et al. 2006). Awareness of this phenomenon and subsequent restoration efforts have decreased land loss rates by almost 60% to 43 km^2/year during the period 1985–2010 (Couvillion et al. 2011). Although this figure sounds encouraging, the percentage decrease is misleading because there is now much less land remaining to restore in some parts of the deltaic plain. In any event, despite restoration efforts during the past 30 years, more than 25% of the LMR deltaic wetlands have been completely lost during the past century (Couvillion et al. 2011). In addition to land losses, salinity increases also have transformed the vegetation communities in these areas. Tidwell (2003) reported that numerous citrus orchards had been converted to saline marshes in Lafourche Parish, Louisiana, which is south of New Orleans in an area of the deltaic plain that is permanently disconnected from the LMR. In fact, the LMR deltaic plain has been severely degraded due to the interaction of many different factors. Despite that strategies are in place to ameliorate some of the degradation, it has been speculated that it is unlikely that the deltaic plain can be restored to conditions that existed even as recently as 10 years ago.

Possible Restoration Strategies for the Lower Mississippi River Floodplain

Restoring and Enhancing Floodplain Function

In the present day, diverse stakeholders are debating future management of the LMR floodplain. Agricultural concerns favor raising levees to minimize flooding and protect valuable croplands reclaimed from the historic alluvial floodplain (Schramm et al. 2000; Schramm and Ickes 2016). Conversely, conservation agencies and organizations advocate levee modification to allow the LMR to reclaim at least parts of its historical floodplain, which is consistent with modern principles of river restoration (e.g., Bayley 1991; Gore and Shields 1995). The expense of minimizing and moderate flooding would include costs associated with raising the existing levee system and installing and operating water-control structures and pumping stations to manage floodwaters (Schramm et al. 2000). The cost of reconnecting the LMR to its original floodplain would include the purchase of large tracts of privately owned lands affected by flooding. Unlike the upper Mississippi River, where much of the floodplain is under state and federal ownership, the LMR floodplain is mostly privately owned (Schramm et al. 2015). Although potentially costing billions of U.S. dollars and requiring landowners willing to sell their lands or be granted conservation incentives, this management alternative is currently being pursued for the LMR floodplain by several conservation organizations (Schramm et al. 2000). However, it would be difficult to justify this enormous cost given that the LMR receives much less recreational use (especially angling) compared to other similar water bodies (Schramm et al. 2015).

A cheaper and more feasible alternative for the LMR might be manipulation of the existing floodplain to restore functionality. The reestablishment of the natural hydrologic regime has been argued as central and essential to river restoration (Bayley 1991, 1995; Poff et al. 1997). Implicit within this goal is restoration of river–floodplain connectance (Galat et al. 1998; Opperman et al. 2010). The LMR ecosystem is little studied and the ecological mechanisms and regulatory factors affecting fish production are not well understood (Eggleton 2001). However, it has been postulated that the altered hydrograph, variable thermal-hydrologic asynchrony, and subsequent reduced river–floodplain connectance have a high likelihood of adversely affecting riverine fish production on long-term time scales (Schramm et al. 2015). Thus, it is likely that long-term strategies designed to protect the lower Mississippi River Valley from flooding (e.g., tall levees, neck cutoffs, and floodways) and improve commercial navigation (e.g., dike fields and bank revetments) have had other biological and ecological consequences as described. Although natural solutions are the preferred approach to restore rivers (e.g., Poff et al. 1997; Galat et al. 1998), reengineering the LMR to reduce flood control and commercial navigation as priorities is not socially or economically feasible at this time (Schramm et al. 2000, 2015).

River-training strategies (e.g., levees and dike fields) and structural works (e.g., water-control structures, engineered floodways) have stabilized the alignment of the LMR main channel. As a result, the LMR channel no longer meanders, creating new channels and abandoning old ones (Schramm et al. 2000). Without intentionally creating aquatic habitats, the LMR and its floodplain will continue to experience progressive declines in the amount and complexity of aquatic and terrestrial habitat diversity. Creation or enhancement of existing habitats could include shallow-water areas that fill early during the flood pulse and remain isolated from additional inputs of cold floodwaters. These areas would warm earlier and more rapidly and provide conditions that should enhance growth, reproduction, and recruitment of riverine fishes, in effect mimicking historical conditions (Schramm et al. 2000, 2009). Some created habitats would need sufficient water depth to avoid dewatering and, thus, support fishes throughout the year. There are many natural (e.g., abandoned channels) and man-made (e.g., borrow-pit lakes from excavating materials for levee construction) water bodies within the LMR floodplain that retain water during all or part of the year (Schramm et al. 2000). However, some of these habitats are steadily being lost to sedimentation while others become anoxic and mostly inhospitable to fishes during midsummer (Sabo et al. 1991, 1999a, 1999b; Schramm et al. 2000). Sabo and Kelso (1991) suggested that LMR floodplain ponds with sinuous shorelines, higher volumes, and variable water depths maintained better water quality throughout the year (e.g., adequate dissolved oxygen and minimal pH fluctuations) and were more suitable fish nurseries. An approach using these guidelines as a template could be used to cost-effectively modify selected floodplain lakes within the LMR. This strategy would aid in partly restoring floodplain function.

Further strategies may exist for the LMR that would modify the present and connected floodplain through a series of shallow impoundments (Schramm et al. 2015). These impoundments would be designed to fill during seasonal flood pulses and retain floodwaters into the warmer seasons, thus mimicking historical floodplain conditions that should benefit fish production. Impoundments designed with different water-entry elevations would create diverse conditions for riverine fishes (Amoros and Bornette 2002) and be beneficial for many species in the LMR fish fauna (Schramm et al. 2015). Entrance of waters into the downstream ends of the impoundments also would mimic natural inundation of the floodplain and potentially minimize sedimentation (Sparks et al. 1998; Amoros and Bornette 2002). It has further been suggested that operation of these impoundments might be done to maximize benefits for native fishes while possibly suppressing nonnative fishes (e.g., Asian carps; Schramm et al. 2015). Sparks et al. (1998) proposed similar strategies for the upper Mississippi River, whereby the existing dam system was to be used to regulate or naturalize flooding, and aid in restoring floodplain function. Restoring the natural flood pulse as described above also could potentially be used to minimize the occurrence of hypoxia in these constructed floodplain impoundments by flushing organic matter from the system (Sabo et al. 1999a, 1999b). This strategy also would help restore floodplain function.

Another strategy involves establishing permanent reconnections of existing floodplain water bodies with the LMR main channel. Many of these lakes can support relatively high fish production (Cobb et al. 1984; Lowery et al. 1987) and contain high levels of energetically beneficial foods for riverine fishes (Eggleton 2001). However, in light of reduced stages from channel incision in the LMR, fishes may be largely confined to these habitats until the next flood cycle following the recession of floodwaters (Bonvillain et al. 2008). Extensive losses of fish production can occur in these lakes if anoxic conditions develop, if the lake dewaters, or if extensive predation occurs in the confined lake area (Sabo et al. 1999b; Bonvillain et al. 2008). This separation of floodplain lakes from the LMR is increasing, especially in the upper section of the LMR, due to channel incision processes and subsequent lowering of base water levels compared to several decades ago (Biedenharn and Watson 1997). As a result, floodplain lakes that were formerly connected to the LMR during summer base flows are now disconnected, with large portions of the lakes subject to dewatering. In other lakes, the downstream sections of lakes that connect to the LMR main channel (termed "chutes") are filled with sediments; thus, lakes become perched above the river because their water elevations are significantly greater (i.e., >8 m) than the LMR during large portions of the year (Schramm et al. 2000). These changes have occurred slowly over decadal time scales but have become obvious to recreational and commercial anglers that frequent the LMR. Excavating the downstream sections of floodplain lakes to reconnect them to the LMR and, if appropriate, managing water releases from these lakes through water-control structures, as described above, should have beneficial effects on riverine fish production (Schramm et al. 2000, 2015). These benefits for the LMR would center more on water from seasonal flood pulses being retained by the existing floodplain during more of the year while also providing more permanent connections between the main river channel and its floodplain habitats. More seasonal water retained in the floodplain increases feeding, growth, reproductive, and rearing windows for LMR fishes. More permanent connection of floodplain water bodies also would enhance habitat diversity at the ecosystem scale in the LMR and benefit floodplain-dependent fishes, which are nearly half of the current LMR fauna. It also is possible that these strategies would enhance recreational opportunities in the LMR.

Enhancing Floodplain Nutrient Cycling

In response to widespread concerns about the growing hypoxic zone in the northern Gulf of Mexico (e.g., Rabalais et al. 2002), the significance of the LMR floodplain as a nutrient sink has received increasing attention in recent years (Schramm et al. 2009). Although occupying only 8% of its historical size, the LMR active floodplain (i.e., the area inundated by spring floods, exclusive of permanent water bodies on the floodplain) is still approximately 650,000 ha in total area (Schramm et al. 1999; Schramm and Ickes 2016). Although the altered hydrograph and resultant thermal-hydrologic asynchrony may affect nitrogen removal, a substantial surface area still remains that could significantly affect nitrogen and phosphorus dynamics (Schramm et al. 2009). A restored and functional floodplain using the approaches described above also could potentially reduce excessive nitrogen inputs to the northern Gulf of Mexico. Using data collected during the period 1971–1994, Donner et al. (2004) estimated annual nitrate export from the Mississippi River basin into the northern Gulf of Mexico at 5.4 million kg/year, which was comparable to similar estimates generated by the U.S. Geological Survey (5.1 million kg/year).

Soils in the LMR basin account for most of the nitrogen removal, and denitrification rates are directly related to water temperature (Schramm et al. 2009). Thus, any floodplain management strategy in the LMR that retains water in floodplains during the warmer seasons will theoretically increase ecosystem-level denitrification and thus reduce excessive nitrogen inputs to the northern Gulf of Mexico (Schramm et al. 2015). Nitrogen cycling studies in the LMR are simple and based on fundamental soil chemistry and bioenergetics models that only considered water temperature and inundation duration (Schramm et al. 2009). Although other biotic, abiotic, and hydrologic factors play roles in denitrification processes (Kern et al. 1996), the shallow floodplain impoundments proposed by Schramm et al. (2015) as a restoration measure to benefit fish production should also increase denitrification. Further research is needed to provide better understanding of nitrogen and phosphorus dynamics in the LMR floodplain. Regardless, the role that the LMR floodplain performs in nutrient dynamics would only be enhanced under the restoration strategies proposed above that would produce more natural conditions (Schramm et al. 2009)

Restoring Natural Land-Building in the Deltaic Plain

The transfer and deposition of sediments from the LMR main stem to its degraded floodplain is key to restoring and preserving the dynamic deltaic plain. River diversions are an option to transport sediment-laden river water through the current flood protection levees into the adjacent floodplain. Diversions in the LMR were originally conceived to manipulate salinity levels to benefit oyster production (Chatry et al. 1983), though the potential benefit of sediment transport to degraded coastal wetlands has been more recently acknowledged (Gagliano et al. 1971; Day and Templet 1989; Allison and Meselhe 2010). The size of the diversion and sediment load in the river water column are important factors controlling the amount of sediment deposited onto the floodplain (Allison and Meselhe 2010; Kolker et al. 2012; Nittrouer et al. 2012).

The current LMR deltaic plain was built by sediments deposited from the Mississippi River water column; however, the suspended sediment load in the LMR has decreased by more than 70% since 1850 (Kesel 1988). Construction of dams and water-control structures on upstream tributaries and separation of the river from its floodplain are two main factors associated with the reduced sediment load in the LMR (Kesel 2003; Allison et al. 2013). The reduced sediment load may make small-scale river diversions (maximum discharge <1,420 m^3/s; Allison and Meselhe 2010) ineffective at depositing sufficient quantities of sediment in the adjacent floodplain to counter subsidence rates and maintain elevation (Mead et al. 2012). For example, the Davis Pond diversion project located 192 km above the LMR mouth at Head of Passes, Louisiana (Figure 7) is designed for a maximum discharge of 300 m^3/s. This diversion project is expected to reduce salinity in the Barataria Estuary (Das et al. 2012) while also depositing LMR sediments into the estuary (Kral et al. 2011). At present, these sediments appear to be stimulating primary productivity in the estuary (Poirrier et al. 2010); however, a significant accretion of land has not yet been documented. Another large diversion (discharge >1,420 m^3/s) project on the LMR is the West Bay diversion project located downstream of Venice, Louisiana (Figure 7; Allison and Meselhe 2010). With this diversion project, large deposits of sediments and creation of new land has been noticeable since the 2011 Mississippi River flood (Kolker at al. 2012), which was one of the largest floods on record (Allison et al. 2013).

The reduced amount of sediment in the LMR water column compared to historic levels indicates that water diversions alone will not provide enough sediment to substantially reverse coastal land loss in the deltaic plain unless many large diversions (discharge >1,420 m^3/s) are used (Allison and Meselhe 2010; Kolker et al. 2012). The amount of sediment transported by large diversions is dependent on seasonal river discharge. Construction of large diversions may alter shoaling patterns and could substantially impact instream sedimentation patterns in the adjacent downstream area (Wang et al. 2014), which may have implications for deep-draft river navigation (Allison et al. 2013). Pipeline transport of sediment dredged from the LMR bed can be used to rebuild degraded marsh (Ford et al. 1999; Slocum et al. 2005) and cypress forests (Middleton and Jiang 2013) in coastal Louisiana. The Louisiana Sediment Management Plan has been implemented to assist with restoration planning strategies and management of sediment resources

(Khalil and Finkl 2011). As opposed to being a permanent landscape feature like a large diversion, sediment pipelines can be moved to areas that need sediment. A combination of pipeline sediment transport coupled with small diversions operated seasonally to mimic the LMR flood pulse may be the most realistic strategy for long-term sustainability of the LMR deltaic plain.

Although physically external to the main river channel, two water-diversion projects located at the lower end of the LMR—Caernarvon (river kilometer 130; Figure 7) and Davis Pond (river kilometer 190)—are the only LMR projects designed to remove nitrogen from the water (Mitsch et al. 2005; Turner et al. 2007). These projects are operated to redirect LMR flow through the marshes of the deltaic plain. Although primarily intended to supply sediment to subsiding coastal marsh and reduce seawater intrusion, these systems also have the potential to reduce nitrogen loading to the northern Gulf of Mexico. Unfortunately, the combined diversion of river discharge through both projects represents only about 0.65% of the total annual discharge of the LMR and contains less than 1% of the river's total nitrogen load.

Lower Mississippi River Floodplain: Future Potential

For more than a century, navigation and flood control have been the predominate forces that have driven management of the LMR. Local governmental agencies, the federal government, conservation organization, and landowners have worked together to ensure navigation and flood protection. However, as a result of recent societal value changes, the costs of ongoing environmental harm to the ecological processes of the LMR, its floodplain, and the northern Gulf of Mexico are beginning to be addressed (Niebling et al. 2014). In response, new regulations and procedures involving greater public participation have been implemented to address these social value changes. Unfortunately, this has created complex and lengthy decision making and action processes that are often accompanied by extensive litigation (Niebling et al. 2014).

Multidisciplinary and interagency programs in the upper Mississippi River and tributaries have advanced the knowledge base of large, complex floodplain–river ecosystems through a combination of long-term monitoring, assessment, and restoration efforts (Sparks 2010). This same approach has been lacking for the lower Mississippi River, which has resulted in an incomplete understanding of its floodplain ecosystem (Schramm et al. 2015). However, new and adaptive approaches to restoration and management are emerging that provide opportunities to restore ecological function in the remaining LMR floodplain. One possible recent advance might be use of the EnviroFish model (Killgore et al. 2012), which integrates hydrology, hydraulics, land use, and empirical data on fish reproductive strategies in riverine floodplains to predict biological responses to different flooding scenarios. This model calculates the number of units of specific floodplain habitats that are available under different hydrologic and hydraulic scenarios. Each habitat is scored differently depending on its overall value for the spawning and rearing of riverine fishes. For example, flooded agricultural fields are scored lowest in terms of their value for fish spawning and rearing, whereas flooded bottomland hardwood forests and floodplain lakes contain the greatest scores. Such a holistic approach would be suitable for many federal planning processes and be more easily explainable to conservation organizations and other public groups. This modeling approach could also complement many of the approaches outlined above and could help guide floodplain restoration strategies.

Although it is not realistic to expect the LMR and its floodplain to be returned to their premodification states, there is ample reason to believe that at least portions of the historical floodplain can be restored to maintain some degree of natural floodplain structure and function. Although it is presently unclear how to quantify that degree, future studies can address this issue and better refine that number. Current studies in Louisiana have greatly enhanced understanding of land-building processes that will aid in restoring the LMR deltaic plain and, thus, inhibit further saltwater intrusion into coastal wetlands. These studies are currently documenting positive results from several water-diversion projects, with redistributed sediment deposition and reducing salinity levels being the primary focus. Another example would be Mitsch et al. (2005), who, using nitrogen-removal data from studies in Ohio and Louisiana wetlands, estimated that the entire Mississippi River basin would need 22,000 km^2 of created and/or restored wetlands in order to remove 40% of the nitrogen load discharged annually into the northern Gulf of Mexico. Although this study emphasized the obvious benefits of wetlands and their enormous nutrient absorption capacities, the enormous size of the remaining LMR floodplain and its numerous water bodies would seem a good starting point for this pursuit. Future studies are needed that consider the LMR floodplain ecosystem in this context. Additionally, future studies should be focused on improving understanding of fish–floodplain dynamics (including movement/migration studies), more detailed studies of LMR fishes with respect to population dynamics and bioenergetics, quantification of linkages between flooding dynamics and vegetation communities, and better estimates of floodplain inundation relative to seasonal flood pulses. All of these efforts combined would go far in developing a restoration template for the lower Mississippi River floodplain.

References

Alexander, J. S., R. C. Wilson, and W. R. Green. 2012. A brief history and summary of the effects of river engineering and dams on the Mississippi River system and delta. U.S. Geological Survey, Circular 1375, Reston, Virginia.

Allen, J. A., R. Pezeshki, and J. L. Chambers. 1996. Interaction of flooding and salinity stress on bald cypress (*Taxodium distichum*). Tree Physiology 16:307–313.

Allison, M. E., and E. A. Meselhe. 2010. The use of large water and sediment diversions in the lower Mississippi River (Louisiana) for coastal restoration. Journal of Hydrology 387:346-360.

Allison, M. E., B. M. Vosburg, M.T. Ramirez, and E. A. Meselhe. 2013. Mississippi River channel response to the Bonnet Carre Spillway opening in the 2011 flood and its implications for the design and operation of river diversions. Journal of Hydrology 477:104–118.

Amoros, C., and G. Bornette. 2002. Connectivity and biocomplexity in waterbodies of riverine floodplains. Freshwater Biology 47:761–776.

Armantrout, N. B., compiler. 1998. Glossary of the aquatic habitat inventory terminology. American Fisheries Society, Western Division, Bethesda, Maryland.

Baker, J. A., K. J. Killgore, and R. L. Kasul. 1991. Aquatic habitats and fish communities in the lower Mississippi River. Reviews in Aquatic Sciences 3:313–356.

Bayley, P. B. 1991. The flood pulse advantage and the restoration of river-floodplain ecosystems. Regulated Rivers: Research and Management 6:75–81.

Bayley, P. B. 1995. Understanding large river-floodplain ecosystems. BioScience 45:153–158.

Bernier, J. C., R. A. Morton, and J. A. Barras. 2006. Constraining rates and trends of historical wetland loss, Mississippi River delta plain, south-central Louisiana. Pages 371–382 *in* Y. J. Xu and V. P. Singh, editors. Coastal environment and water quality. Water Resources Publications, Highlands Ranch, Colorado.

Biedenharn, D. S., and C. C. Watson. 1997. Stage adjustment in the lower Mississippi River, USA. Regulated Rivers: Research and Management 13:517–536.

Biedenharn, D. S., C. R. Thorne, and C. C. Watson. 2000. Recent morphological evolution of the lower Mississippi River. Geomorphology 34:227–249.

Blum, M. D., and H. H. Roberts. 2012. The Mississippi delta region: past, present, and future. Annual Review of Earth and Planetary Sciences 40:655–683.

Boesch, D. F., M. N. Josselyn, A. J. Mehta, J. T. Morris, W. K. Nuttle, C. A. Simenstad, and D. J. P. Swift. 1994. Scientific assessment of coastal wetland loss, restoration and management in Louisiana. Journal of Coastal Research [online serial], Special Issue 20.

Bonvillain, C. P., A. M. Ferrara, and Q. C. Fontenot. 2008. Relative abundance and biomass estimate of a Spotted Gar population in a seasonally connected large river floodplain lake.

Proceedings of the Southeastern Association of Fish and Wildlife Agencies 62:177–182.

Brandt, K., and K. C. Ewel. 1989. Ecology and management of cypress swamps: a review. Florida Cooperative Extension Service Institute of Food and Agriculture Sciences, Bulletin 252. University of Florida, Gainesville.

Braud, D., A.J. Lewis, J. Sheehan, L. Foretich, A. Venuto, K. Fontenot, M. Landry, and S. M. Qadri. 2006. 2005 land use/land cover classification: Barataria basin. Louisiana Department of Environmental Quality, Baton Rouge.

Brown, S. 1981. A comparison of the structure, primary productivity, and transpiration of cypress ecosystems in Florida. Ecological Monographs 51:403–427.

Brown, S., and A. E. Lugo. 1982. A comparison of structural and functional characteristics of saltwater and freshwater forested wetlands. Pages 109–130 *in* B. Gopal, R. E. Turner, R. G. Wetzel, and D. F. Whigham, editors. Wetlands: ecology and management. International Scientific Publications, Jaipur, India.

Chan, M. W., and M. D. Zoback. 2007. The role of hydrocarbon production on land subsidence and fault reactivation in the Louisiana coastal zone. Journal of Coastal Research 23:771–786.

Chatry, M., R. J. Dugas, and K. A. Easley. 1983. Optimal salinity regime for oyster production on Louisiana's state seed grounds. Contributions in Marine Science 26:81–94.

Chick, J. H., and M. A. Pegg. 2001. Invasive carp in the Mississippi River basin. Science 292:2250–2251.

Cobb, S. P., C. H. Pennington, J. A. Baker, and J. E. Scott. 1984. Fishery and ecological investigations on main stem levee borrow pits along the lower Mississippi River. U.S. Army Corps Engineers, Lower Mississippi River Environmental Program, Report 1, Vicksburg, Mississippi.

Coleman, J. M., H. H. Roberts, and G. W. Stone. 1998. Mississippi River delta: an overview. Journal of Coastal Research 4:698–716.

Conner, W. H, and J. W. Day, Jr. 1992. Water level variability and litterfall productivity in forested freshwater wetlands in Louisiana. American Midland Naturalist 128:237–245.

Conner, W. H., and L. W. Inabinette. 2005. Identification of salt tolerant bald cypress *Taxodium distichum* (L.) for planting in coastal areas. New Forests 29:305–312.

Couvillion, B. R., J. A. Barras, G. D. Steyer, W. Sleavin, M. Fischer, H. Beck, N. Trahan, B. Griffin, and D. Heckman. 2011. Land area change in coastal Louisiana from 1932 to 2010. U.S. Geological Survey, Scientific Investigations Map 3164, Denver, Colorado.

Das, A., D. Justice, M. Inhoue, A. Hoda, H. Huang, and D. Park. 2012. Impacts of Mississippi River diversions on salinity gradients in a deltaic Louisiana estuary: ecological and management implications. Estuarine Coastal and Shelf Science 111:17–26.

Day, J. W., G. P. Kemp, D. J. Reed, D. R. Cahoon, R. M. Boumans, J. M. Suhayda, and R. Gambrell. 2011. Vegetation death and rapid loss of surface elevation in two contrasting Mississippi delta salt marshes: the role of sedimentation, autocompaction, and sea-level rise. Ecological Engineering 37:229–240.

Day, J. W., and P. H. Templet. 1989. Consequences of sea level rise: implications from the Mississippi delta. Coastal Management 17:241–257.

Eggler, W. A., and W. G. Moore. 1961. The vegetation of Lake Chicot, Louisiana, after eighteen years of impoundment. Southwestern Naturalist 6:175–183.

Donner, S. D., C. J. Kucharik, and J. A. Foley. 2004. Impact of changing land use practices on nitrate export by the Mississippi River. Global Biogeochemical Cycles 18:1–21.

Eggleton, M. A. 2001. Catfish feeding ecology and bioenergetics in a large river-floodplain ecosystem. Doctoral dissertation. Mississippi State University, Starkville.

Eggleton, M. A., and H. L. Schramm, Jr. 2004. Feeding ecology and energetic relationships with habitat of Blue Catfish *Ictalurus furcatus* and Flathead Catfish *Pylodictis olivaris* in the lower Mississippi River, USA. Environmental Biology of Fishes 70:107–121.

Ford, M. A., D. D. Cahoon, and J C. Lynch. 1999. Restoring marsh elevation in a rapidly subsiding salt marsh by thin layer deposition of dredged material. Ecological Engineering 12:189–205.

Fremling, C. R., J. L. Rasmussen, R. E. Sparks, S. P. Cobb, C. F. Bryan, and T. O. Claflin. 1989. Mississippi River fisheries: a case history. Pages 309–351 *in* D. P. Dodge, editor. Proceedings of the international large river symposium. Canadian Special Publication of Fisheries and Aquatic Sciences 106.

Gagliano, S. M., M. P. Light, and R. E. Becker. 1971. Controlled diversions in the Mississippi delta system: an approach to environmental management. Louisiana State University, Center for Wetland Resources, Hydrologic and Geologic Studies of Coastal Louisiana Report No. 8, Baton Rouge.

Gagliano, S. M., K. J. Meyer-Arendt, and K. M. Wicker. 1981. Land loss in the Mississippi River delta plain. Gulf Coast Association of Geological Societies Transactions 31:295-300.

Galat, D. L., L. H. Fredrickson, D. D. Humburg, K. J. Bataille, J. R. Bodie, J. Dohrenwend, G. T. Gelwicks, J. E. Havel, D. L., Helmers, J. B. Hooker, J. R. Jones, M. F. Knowlton, J. Kubisiak, J. Mazourek, A. C. McColpin, R. B. Renken, and R. D. Semlitsch. 1998. Flooding to restore connectivity of regulated, large-river wetlands. BioScience 48:721-733.

Galay, V. J. 1983. Causes of river bed degradation. Water Resources Research 19:1057-1090.

Gore, J. A., and F. D. Shields, Jr. 1995. Can large rivers be restored? BioScience 45:142-152.

Gutreuter, S., A. D. Bartels, K. Irons, and M. B. Sandheinrich. 1999. Evaluation of the flood-pulse concept based on growth of selected fishes of the upper Mississippi River system. Canadian Journal of Fisheries and Aquatic Sciences 56:2282-2291.

Harms, W. R., H. T. Schreuder, D.D. Hook, and C. L. Brown. 1980. The effects of flooding on the swamp forest in Lake Ocklawah, Florida. Ecology 61:1412-1421.

Janáč, M. M., M. Ondračková, P. Jurajda, Z. Valová, and M. Reichard. 2010. Flood duration determines the reproductive success of fish in artificial oxbows in a floodplain of a potamal river. Ecology of Freshwater Fish 19:1-12.

Junk, W. J., P. B. Bayley, and R. E. Sparks. 1989. The flood pulse concept in river-floodplain ecosystems. Pages 110-127 *in* D. P. Dodge, editor. Proceedings of the international large river symposium. Canadian Special Publication of Fisheries and Aquatic Sciences 106.

Keddy, P. A., D. Campbell, T. McFalls, G.P. Shaffer, R. Moreau, C. Dranguet, and R. Heleniak. 2007. The wetlands of Lakes Pontchartrain and Maurepas: past, present, and future. Environmental Reviews 15:43-77.

Keown, M. P., E. A. Dardeau, and E. M. Causey. 1986. Historic trends in sediment flow regime of the Mississippi River. Water Resources Research 22:1555-1564.

Kern, J., A. Darwich, K. Furch, and W. J. Junk. 1996. Seasonal denitrification in flooded and exposed sediments from the Amazon floodplain at Lago Camaleao. Microbial Ecology 32:47-57.

Kesel, R. H. 1988. The decline in the suspended load of the lower Mississippi River and its influence on adjacent wetlands. Environmental Geology and Water Sciences 11:271-281.

Kesel, R. H. 2003. Human modifications to the sediment regime of the lower Mississippi River flood plain. Geomorphology 56:325-334.

Khalil, S. M., and C. W. Finkl. 2011. Spoil or resource? Managing sediment for coastal restoration. Journal of Coastal Research SI 64:1433-1437.

Killgore, K. J., B. Bruchmann, R. Hunt, L.Y. Lin, J.J. Hoover, D. Johnson, D. Johnson, G. Young, K. Parrish, R. Goldman, and A. Casper. 2012. EnviroFish - user's manual, version 1.0. U.S. Army Engineer Research and Development Center, Vicksburg, Mississippi.

King, A. J., P. Humphries, and P. S. Lake. 2003. Fish recruitment on floodplains: the roles of patterns of flooding and life history characteristics. Canadian Journal of Fisheries and Aquatic Sciences 60:773-786.

Knights, B. C., J. M. Vallazza, S.J. Zigler, and M. R. Dewey. 2002. Habitat and movement of Lake Sturgeon in the upper Mississippi River, USA. Transactions of the American Fisheries Society 131:507-522.

Ko, J. Y., and J. W. Day. 2004. A review of ecological impacts of oil and gas development on coastal ecosystems in the Mississippi delta. Ocean and Coastal Management 47:597-623.

Kolker, A. S., M. D. Miner, and H. D. Weathers. 2012. Depositional dynamics in a river diversion receiving basin: the case of the West Bay Mississippi River diversion. Estuarine Coastal and Shelf Science 106:1-12.

Kral, F., R. Corstanje, J. R. White, and F. Veronesi. 2011. A geostatistical analysis of soil properties in the Davis Pond Mississippi freshwater diversion. Soil Science Society of America Journal 76:1107-1118.

Kwak, T. J. 1988. Lateral movement and use of floodplain habitat by fishes of the Kankakee River, Illinois. American Midland Naturalist 120:241-249.

Lorenz, C. M., G. M. Van Dijk, A.G.M. Van Hattum, and W. P. Cofino. 1997. Concepts in river ecology: implications for indicator development. Regulated Rivers: Research and Management 13:501-516.

Lowery, D. R., M. P. Taylor, R. L. Warden, and F. H. Taylor. 1987. Fish and benthic communities of eight lower Mississippi River floodplain lakes. U.S. Army Corps Engineers, Lower Mississippi River Environmental Program Report 6, Vicksburg, Mississippi.

Mead, A. A., C. R. Demas, B. A. Ebersole, B. A.

Kleiss, C. D. Little, E. A. Meselhe, N. J. Powell, T. C. Pratt, and B. M. Vosburg. 2012. A water and sediment budget for the lower Mississippi–Atchafalaya River in flood years 2008–2010: implications for sediment discharge to the oceans and coastal restoration in Louisiana. Journal of Hydrology 432–433:84–97.

Middleton, B. A., and M. Jiang. 2013. Use of sediment amendments to rehabilitate sinking coastal swamp forests in Louisiana. Ecological Engineering 54:183–191.

Mitsch, W. J., J. W. Day, L. Zhang, and R. R. Lane. 2005. Nitrate-nitrogen retention in wetlands in the Mississippi River basin. Ecological Engineering 24: 267–278.

Myers, R. S., G. P. Shaffer, and D. W. Llewellyn. 1995. Bald cypress (*Taxodium distichum* (L.) Rich.) restoration in southeast Louisiana: the relative effects of herbivory, flooding, competition, and macronutrients. Wetlands 15:141–148.

Niebling, W., J. A. Baker, L. Kasuri, S. Katz, and K. Smet. 2014. Challenge and response in the Mississippi River basin. Water Policy 16:87–116.

Nelson, S. A., P. A. Soranno, and J. Qi. 2002. Land-cover change in upper Barataria basin estuary, Louisiana, 1972–1992: increases in wetland area Environmental Management 29:716–727.

Nittrouer, J. A., J. L. Best, C. Brantley, R. W. Cash, M. Czapiga, P. Kumar, and G. Parker. 2012. Mitigating land loss in coastal Louisiana by controlled diversion of Mississippi River sand. Nature Geoscience 5:534–537.

O'Connell, M. T., A. U. O'Connell, and V. A. Barko. 2011. Occurrence and predicted dispersal of Bighead Carp in the Mississippi River system: development of a heuristic tool. Pages 51–71 *in* D. C. Chapman, and M. H. Hoff, editors. Invasive Asian carps in North America. American Fisheries Society, Symposium 74, Bethesda, Maryland.

O'Connell, M. T., T. D. Shepherd, A. M. U. O'Connell, and R. A. Myers. 2007. Long-term declines in two apex predators, Bull Sharks (*Carcharhinus leucas*) and Alligator Gar (*Atractosteus spatula*), in Lake Ponchartrain, an oligohaline estuary in southeastern Louisiana. Estuaries and Coasts 30:567–574.

Opperman, J. J., R. Luster, B.A. McKenney, M. Roberts, and A. W. Meadows. 2010. Ecologically functional floodplains: connectivity, flow regime, and scale. Journal of the American Water Resources Association 46:211–226.

Penland, S., and K. E. Ramsey. 1990. Relative sea-level rise in Louisiana and the Gulf of Mexico: 1908–1988. Journal of Coastal Research 6:323–342.

Phelps, Q. E., S. J. Tripp, D. P. Herzog, and J. E. Garvey. 2014. Temporary connectivity: the relative benefits of large river floodplain inundation in the lower Mississippi River. Restoration Ecology 23:53–56.

Pinter, N., A. A. Jemberie, J. W. F. Remo, R. A. Heine, and B. S. Ickes. 2010. Cumulative impacts of river engineering, Mississippi and lower Missouri rivers. River Research and Applications 26:546–571.

Poff, N. L., J. D. Allan, M. B. Bain, J. R. Karr, K. L. Prestegaard, B. D. Richter, R. E. Sparks, and J. C. Stromberg. 1997. The natural flow regime: a paradigm for river conservation and restoration. BioScience 47:769–784.

Poirrier, M. A. 2013. Effects of closure of the Mississippi River Gulf Outlet on saltwater intrusion and bottom water hypoxia in Lake Ponchartrain. Gulf and Caribbean Research 25:105–109.

Poirrier, M. A., K. Burt-Utley, J. F. Utley, and E. A. Spalding. 2010. Submersed aquatic vegetation of the Jean Lafitte National Historical Park and Preserve. Southeastern Naturalist 9:477–486.

Pringle, C. M., M. C. Freeman, and B. J. Freeman. 2000. Regional effects of hydrologic alterations on riverine macrobiota in the New World: tropical-temperate comparisons. BioScience 50:807–823.

Rabalais, N. N., R. E. Turner, Q. Dortch, D. Justic, V. J. Bierman, Jr., and W. J. Wiseman. 2002. Nutrient-enhanced productivity in the northern Gulf of Mexico: past, present and future. Hydrobiologia 475:39–63.

Rabalais, N. N., R. E. Turner, D. Justic, Q. Dortch, W.J. Wiseman, and B. K. S. Gupta. 1996. Nutrient changes in the Mississippi River and system responses on the adjacent continental shelf. Estuaries 19:386–407.

Reed, D. J., editor. 1995. Status and historical trends of hydrologic modification, reduction in sediment availability, and habitat loss/modification in the Barataria and Terrrebonne estuarine systems. Barataria-Terrebonne National Estuary Program, BTNEP Publication No. 20, Thibodaux, Louisiana.

Rejmánek, M., C. E. Sasser, and J. G. Gosselink.1987. Modeling of vegetation dynamics in the Mississippi River deltaic plain. Vegetatio 69:133–140.

Robison, H. W., and T. M. Buchanan. 1988. Fishes

of Arkansas. University of Arkansas Press, Fayetteville.

Rutherford, D. A., W. E. Kelso, C. F. Bryan, and G. C. Constant. 1995. Influence of physicochemical characteristics on annual growth increments of four fishes in the lower Mississippi River. Transactions of the American Fisheries Society 124:687–697.

Sabo, M. J., and W. E. Kelso. 1991. Relationship between the morphology of excavated floodplain ponds along the Mississippi River and their use as fish nurseries. Transactions of the American Fisheries Society 120:552–561.

Sabo, M. J., W. E. Kelso, C. F. Bryan, and D. A. Rutherford. 1991. Physicochemical factors affecting larval fish densities in Mississippi River floodplain ponds, Louisiana (USA). Regulated Rivers: Research and Management 6:109–116.

Sabo, M. J., W. E. Kelso, C. F. Bryan, and D. A. Rutherford. 1999a. Hydrology and aquatic habitat characteristics of a riverine swamp: I. Influence of flow on water temperature and chemistry. Regulated Rivers: Research and Management 15:505–523.

Sabo, M. J., W. E. Kelso, C.F. Bryan, and D. A. Rutherford. 1999b. Hydrology and aquatic habitat characteristics of a riverine swamp: II. Hydrology and the occurrence of chronic hypoxia. Regulated Rivers: Research and Management 15:525–542.

Schramm, H. L., Jr. 2004. Status and management of Mississippi River fisheries. Pages 301–333 *in* R. L. Welcomme and T. Petr, editors. Proceedings of the second international symposium on the management of large rivers for fisheries, volume I. Food and Agriculture Organization of the United Nations, Regional Office for Asia and the Pacific, RAP Publication 2004/16, Bangkok, Thailand.

Schramm, H. L., Jr., M. S. Cox, T. E. Tietjen, and A. W. Ezell. 2009. Nutrient dynamics in the lower Mississippi River floodplain: comparing present and historic hydrologic conditions. Wetlands 29:476–487.

Schramm, H. L., Jr., and M. A. Eggleton. 2002. Lower Mississippi River fisheries investigations. Final report submitted to the U.S. Army Corps of Engineers, Lower Mississippi Valley Division, Vicksburg, Mississippi.

Schramm, H. L., Jr., and M. A. Eggleton. 2006. Applicability of the flood-pulse concept to temperate floodplain river ecosystems: thermal and temporal components. Rivers Research and Applications 22:543–553.

Schramm, H. L., Jr., M. A. Eggleton, and R. M. Mayo. 2000. Habitat conservation and creation: invoking the flood-pulse concept to enhance fisheries in the lower Mississippi River. Polski Archiwum Hydrobiologii 47:45–62.

Schramm, H. L., Jr., M. A. Eggleton, and R. B. Minnis. 1999. Spatial analysis of floodplain habitat critical to lower Mississippi River fishes. Final report submitted to U.S. Fish and Wildlife Service and U.S. Geological Survey. Mississippi State University, Mississippi State, Mississippi.

Schramm, H. L., Jr., and B. S. Ickes. 2016. The Mississippi River: a place for fish. Pages 3–34 *in* Y. Chen, D. C. Chapman, J. R. Jackson, D. Chen, Z. Li, K. J. Killgore, Q. E. Phelps, and M. A. Eggleton, editors. Fishery resources, environment, and conservation in the Mississippi and Yangtze (Changjiang) River basins. American Fisheries Society, Symposium 84, Bethesda, Maryland.

Schramm, H. L. Jr., and R. M. Mayo. 1999. Growth of Flathead Catfish in the lower Mississippi River. Pages 121–124 *in* E. R. Irwin, W. A. Hubert, C. F. Rabeni, H. L. Schramm, Jr., and T. Coon, editors. Catfish 2000: proceedings of the international ictalurid symposium. American Fisheries Society, Symposium 24, Bethesda, Maryland.

Schramm, H. L., Jr., W. B. Richardson, and B. C. Knights. 2015. Managing the Mississippi River floodplain: achieving functional floodplain requires more than hydrological connection to the river. Pages 171–201 *in* P. F. Hudson and H. Middelkoop, editors. Geomorphic approaches to integrated floodplain management of lowland fluvial systems in North America and Europe. Springer Publishing, New York.

Shaffer, G. P., C. E. Sasser. J. G. Gosselink, and M. Rejmánek. 1992. Vegetation dynamics in the emerging Atchafalaya Delta, Louisiana, USA. Journal of Ecology 80:677–687.

Shaffer, G. P., W. B. Wood, S. S. Hoeppner, T. A. Perkins, J. Zoller, and D. Kandalepas. 2009. Degradation of bald cypress-water tupelo swamp to marsh and open-water in southeastern Louisiana, USA: an irreversible trajectory? Journal of Coastal Research 54:152–165.

Shields, F. D. 1995. Fate of the lower Mississippi River habitats associated with river training dikes. Aquatic Conservation Marine and Freshwater Ecosystems 5:97–108.

Shields, F. D., S. S. Knight, and C. M. Cooper. 1994. Effects of channel incision on base flow stream habitats and fishes. Environmental Management 18:43–57.

Shields, F. D., S. S. Knight, and C. M. Cooper. 1998. Rehabilitation of aquatic habitats in warmwater streams damaged by channel incision in Mississippi. Hydrobiologia 382:63–86.

Sklar, F. H., and F. H. Conner. 1979. Effects of altered hydrology on primary production and aquatic animal populations in a Louisiana swamp forest. Pages 191–210 *in* J. W. Day, D. D. Culley, R. E. Turner, and A. J. Mumphrey, editors. Proceedings of the third coastal marsh and estuary management symposium. Louisiana State University, Division of Continuing Education, Baton Rouge.

Slocum, M. G., I. A. Mendelssohn, and N. L. Kuhn. 2005. Effects of sediment slurry enrichment on salt marsh rehabilitation: plant and soil responses over seven years. Estuaries 28:519–528.

Spalding, E. A., and M. W. Hester. 2007. Interactive effects of hydrology and salinity on oligohaline plant species productivity: implications of relative sea-level rise. Estuaries and Coasts 30:214–225.

Sparks, R. E. 1995a. Maintaining and restoring the ecological integrity of the Mississippi River: importance of floodplains and floodpulses. Transactions of the North American Wildlife and Natural Resources Conference 60:90–97.

Sparks, R. E. 1995a. Need for ecosystem management of large rivers and their floodplains. BioScience 45:168–182.

Sparks, R. E. 2010. Forty years of science and management on the upper Mississippi River: an analysis of the past and a view of the future. Hydrobiologia 640:3–15.

Sparks, R. E., J. C. Nelson, and Y. Yin. 1998. Naturalization of the flood regime in regulated rivers: the case of the upper Mississippi River. BioScience 48:706–720.

Tidwell, M. 2003. Bayou farewell: the rich life and tragic death of Louisiana's Cajun coast, 1st edition. Pantheon Books, New York.

Turner, R. E., N. N. Rabalais, R. B. Alexander, G. McIsaac, and R. W. Howarth. 2007. Characterization of nutrient, organic carbon, and sediment loads and concentrations from the Mississippi River into the northern Gulf of Mexico. Estuaries and Coasts 30:773–790.

Turner, R. E., and Y. S. Rao. 1990. Relationship between wetland fragmentation and recent hydrologic changes in a deltaic coast. Estuaries 13:272–281.

Wang, H., G. D. Steyer, B. R. Couvillion, J. M. Rybczyk, H. J. Beck, W. J. Sleavin, E. A. Meselhe, M. A. Allison, R. G. Boustany, C. J. Fischenich, and V. H. Rivera-Monroy. 2014. Forecasting landscape effects of Mississippi River diversions on elevation and accretion in Louisiana deltaic wetlands under future environmental uncertainty scenarios. Estuarine Coastal and Shelf Science 138:57–68.

Ward, J. V., and J. A. Stanford. 1995. Ecological connectivity in alluvial river ecosystems and its disruption by flow regulation. Regulated Rivers: Research and Management 11:105–119.

Ward, J. V., K. Tockner, and F. Schiemer. 1999. Biodiversity of floodplain river ecosystems: ecotones and connectivity. Regulated Rivers Research and Management 15:125–139.

Webb, E. C., and I. A. Mendelssohn. 1996. Factors affecting vegetation dieback of an oligohaline marsh in coastal Louisiana: field manipulation of salinity and submergence. Journal of Botany 83:1429–1434.

Welcomme, R. L. 1979. Fisheries ecology of floodplain rivers. Longman Press, London.

Williamson, C. J., and J. E. Garvey. 2005. Growth, fecundity, and diets of newly established silver carp in the middle Mississippi River. Transactions of the American Fisheries Society 134:1423–1430.

Winkley, B. R., and J. H. Brooks. 1980. Navigation and flood control through geometric alignment. U.S. Army Corps of Engineers, Vicksburg District, Potamology Investigation Report 300-3, Vicksburg, Mississippi.

American Fisheries Society Symposium 84:263–291, 2016

The Yangtze River Floodplain: Threats and Rehabilitation

HONG-ZHU WANG*, XUE-QIN LIU, AND HAI-JUN WANG
State Key Laboratory of Freshwater Ecology and Biotechnology
Institute of Hydrobiology, Chinese Academy of Sciences
No. 7 Donghu South Road, Wuchang District, Wuhan 430072, P. R. China

Abstract.—The Yangtze (Changjiang) River floodplain is one of the most important ecosystems in China, as well as in the world, but is seriously threatened by multiple factors. Thus, it is crucial and urgent to rehabilitate the river floodplain. This paper reviews ecological studies conducted on the Yangtze River floodplain and presents suggestions for conservation and rehabilitation. First, the Yangtze River system is briefly introduced. Formed 23 million years ago, the Yangtze River is ca. 6,300 km in length with a mean annual runoff of 9.6×10^{11} m^3. Thousands of floodplain lakes are distributed along the mid-lower Yangtze River, and the total area remains 15,770 km^2 at present. Such a river-lake complex ecosystem holds a unique and diverse biota, with ca. 400 hydrophytes and hygrophytes, ca. 170 mollusks, ca. 200 fishes, ca. 400 water birds, and endangered dolphins and porpoises. Second, main threats to the Yangtze River floodplain ecosystem are identified: (1) habitat loss, including river channelization, sharp shrinkage of lake area (ca. 10,000 km^2 since the 1950s), degradation of lakeshore zones, and sand overmining; (2) alternations of hydrological regimes, including construction of ca. 47,000 reservoirs in the whole basin and disconnection of most floodplain lakes from the main stem; (3) water pollution, including eutrophication, heavy metals, and organic pollutants; and (4) overexploitation of biological resources, including overfishing and intensive pen culture. Third, effects of river–lake disconnection on lake ecosystems are summarized on the basis of our studies in the past 20 years. It was found that (1) disconnection is one of the main causes of lake eutrophication; (2) species diversity, biomass, and production of macrophytes and macrobenthos reach maxima at some levels of intermediate river connectivity; (3) disconnection greatly reduces fish species richness of each habitat guild, and natural fish larvae is severely depleted; and (4) disconnection simplifies macroinvertebrate food web structure, and the trophic basis of the simplified food web is more heavily dependent on detritus in disconnected lakes. Last, conservation strategies are proposed. Since the Yangtze River floodplain is a huge integrated system, the biodiversity conservation must be conducted on the whole basin scale. By establishing species–area models of fishes, the minimum protected area of Yangtze-connected lakes is estimated to be ca. 14,400 km^2. It means that at least 8,900 km^2 of disconnected lakes should be reconnected with the Yangtze main stem, and ecohydrological operation of dams and sluices is the feasible approach. Based upon our preliminary studies on environmental flow requirements, the following measures are suggested: (1) lower water levels during spring to improve germination of macrophytes, and control rising rates of water levels during spring–summer to ensure development of macrophytes;

* Corresponding author: wanghz@ihb.ac.cn

and (2) open sluice gates to restore migration routes for juveniles migrating into lakes during April-September and for adults migrating back to the Yangtze main stem during November-December.

Introduction

The Yangtze River floodplain is one of the most important ecosystems in China, as well as in the world, but is facing multiple stressors, such as habitat loss, alternations of hydrological regimes, pollution, and overexploitation. Therefore, it is crucial and urgent to rehabilitate the river floodplain. In this paper, we present a review of ecology and conservation of the Yangtze River floodplain. First, we briefly introduce the Yangtze River system. Second, we identify main ecological threats. Third, we summarize the effects of river–lake disconnection on lake ecosystems. Last, we propose holistic strategies for conservation and rehabilitation.

The History and Current Status of the Yangtze River Floodplain

Introduction to the Yangtze River Basin and Its Geological History

The Yangtze River is the longest river in Asia and the third longest in the world. Originating from Geladandong Peak, the main peak of the Tanggula Mountains on Qinghai–Tibet Plateau, the Yangtze River flows ca. 6,300 km from west to east and enters the East China Sea. The total catchment area is 1.8×10^6 km^2, comprising one-fifth of China's land territory. The terrain slopes downward stair-like from west to east, with a total fall of ca. 5,400 m. The upper reaches are the course above Yichang, being 4,504 km in length and 1×10^6 km^2 in catchment area; the middle reaches are between Yichang and Hukou, being 955 km in length and 6.8×10^5 km^2 in catchment area; and the lower reaches start from Hukou, being 938 km in length and 1.2×10^5 km^2 in catchment area (Yu and Lu 2005).

There are more than 10,000 tributaries in the Yangtze River system. Among them, there are 437 rivers, each with a catchment area of more than 1,000 km^2, and 22 rivers, each with a catchment area of more than 10,000 km^2. Numerous lakes are distributed on the plateaus and floodplains. Plateau lakes in the upper basin are mainly freshwater lakes, with a few saltwater ones. Floodplain lakes in the middle and lower basins are all freshwater ones (Yu and Lu 2005).

The mean annual runoff of the Yangtze River is 9.6×10^{11} m^3, constituting one-third of the total amount discharged into the sea by Chinese rivers and ranking forth in the world. Approximately 46% of the Yangtze runoff is from the upper basin, 18% from the Lake Dongtinghu drainage system, and 15% from the Lake Poyanghu system. Runoff of the main stem and tributaries is markedly different between flood and nonflood seasons within a year. The runoff of the main stem in the flood season (May–October) is 79–82% of the annual total in the upper basin, and 71–79% in the mid to lower basins (Yu and Lu 2005).

In the Early Cretaceous period, the Yanshan Movement turned the entire Yangtze River basin into land. However, the Paleo-Yangtze had not yet born until the Paleogene. From the Late Paleogene to the Early Neogene, the uplift of the Qinghai–Tibet Plateau altered the air circulation of lower atmosphere, turned the trade wind into the seasonal wind, and changed the climate from hot and dry to warm and humid. Because the precipitation and runoff sharply increased, the ancient Yangtze River started to develop in several regions. The water system of the upper reaches, including the Jinshajiang, Yalongjiang and Chuanjiang (Sichuan River) paleorivers, ran south into the South China Sea, and the rivers below Yichang drained into the Jianghan basin and others (Yu and Lu 2005). With regards to the age of the Yangtze River system, Zheng et al. (2013) argued that the connection through the Three Gorges must post-date 36.5 Ma (million years ago) and a river, much like the modern one, was in existence before 23 Ma.

During the Late Pleistocene (23–15 ka BP [thousand years before present]), the climate was dry and cold, and the level of the East China Sea was 130–160 m lower than the present level. The bed of the Yangtze River was intensely downcut, and headward erosion gradually moved up to Yichang, forming a deep narrow valley in the mid to lower reaches; compared to the present level, the riverbed was 5–25 m lower and the mean water level was 20–40 m lower. In the Early Holocene (13–12 ka BP), the climate became warmer, and the level of the East China Sea was uplifted to an altitude 50–100 m below the present level. In the Mid-Holocene (7.5–4 ka BP), temperature reached the highest level of the postglacial period; the mean annual temperature in the lower Yangtze basin was 3–4°C higher than the present one, and the sea level was 2–3 m higher. At that time, the Yangtze estuary was around Zhenjiang and Yangzhou. Due to large areas of forests and lakes, the amplitude of water level fluctuation in the river was relatively small (IGSNRRCAS et al. 1985; Yu and Lu 2005).

In the Late Holocene (4 ka BP to now), the climate fluctuated between cold-dry and hot-wet, the amplitude of temperature change was ca. 3.5°C (Zhu 1972), and the sea level was ±3 m around the present. Since the Western Jin Dynasty (265–317), large populations in the north were compelled to migrate to the south due to turmoil of wars, and they deforested for farming and reclaimed land from lakes. Due to climate change and aggravated soil water erosion, sand bars emerged in the river course and the Yangtze estuary kept expanding further east from Yangzhou, reaching somewhere near the present estuary in the Tang and Song dynasties (618–1279). For reasons such as the shifting of the main stem to the right and the construction of left banks, lakes along the left bank shrank gradually while lakes along the right expanded (IGSNRRCAS et al. 1985).

Distributions of Rivers and Lakes in the Mid to Lower Yangtze Plains

Downstream from the Three Gorges, the Yangtze River runs east through the Lake Dongtinghu Graben basin, Lake Poyanghu Graben basin, lower Yangtze Geosyncline belt, and Taihu–Wusong Graben basins. This series of fluvial-lacustrine systems in the mid to lower reaches of the Yangtze River were formed by sedimentation. The plains (108–122°E, 24–34°N) lie east-west for 1,800 km with a total area of ca. 1.6×10^5 km^2. The terrain is flat and low, with altitudes mostly ca. 50 m above sea level (ASL). Plains along the middle reaches consist of the Jianghan plain, Lake Dongtinghu plain, Edong plain, and Lake Poyanghu plain; those along the lower reaches comprise the Chaohu–Wanjiang plain and the Yangtze River delta among the Jiangsu and Zhejiang provinces and Shanghai (Figure 1; Zhao and Chen 1999).

The mid to lower Yangtze main stem is 1,893 km, with a sinuosity of ca. 1.6 (Figure 1). The main channel pattern is largely braided, comprising 70% of the mid to lower river in length. The river course is characterized by alternation of broad, braided reaches and narrow single channels, resembling lotus roots. At bank-full level, the mean river width (B) is 1,168–3,332 m and the mean depth (H) is 13–20.4 m, with $B^{0.5}/H$ of 1.88–4.45; maximum depth of the thalweg reaches 90.5 m. The longitudinal profile of the mid to lower course has a total fall of 56.4 m, with a mean slope of 0.34 (0.05–0.85) $\times 10^{-4}$ (IGSNRRCAS et al. 1985).

Numerous lakes are distributed along the mid to lower Yangtze River (Figure 1), with a total area of 15,770 km^2. They are divided into the following lake clusters: (1) The Jianghan–Dongtinghu plains lakes, distributed between Zhijiang and Wuxue. In the Mid-Holocene, this region was the vast waters of the paleo-Yunmeng Lake. Afterwards, great quantities of sand brought by the Yangtze River and its tributaries were deposited in the lake. As a consequence, land areas expanded while water areas shrank. The lake was thus divided into thousands of

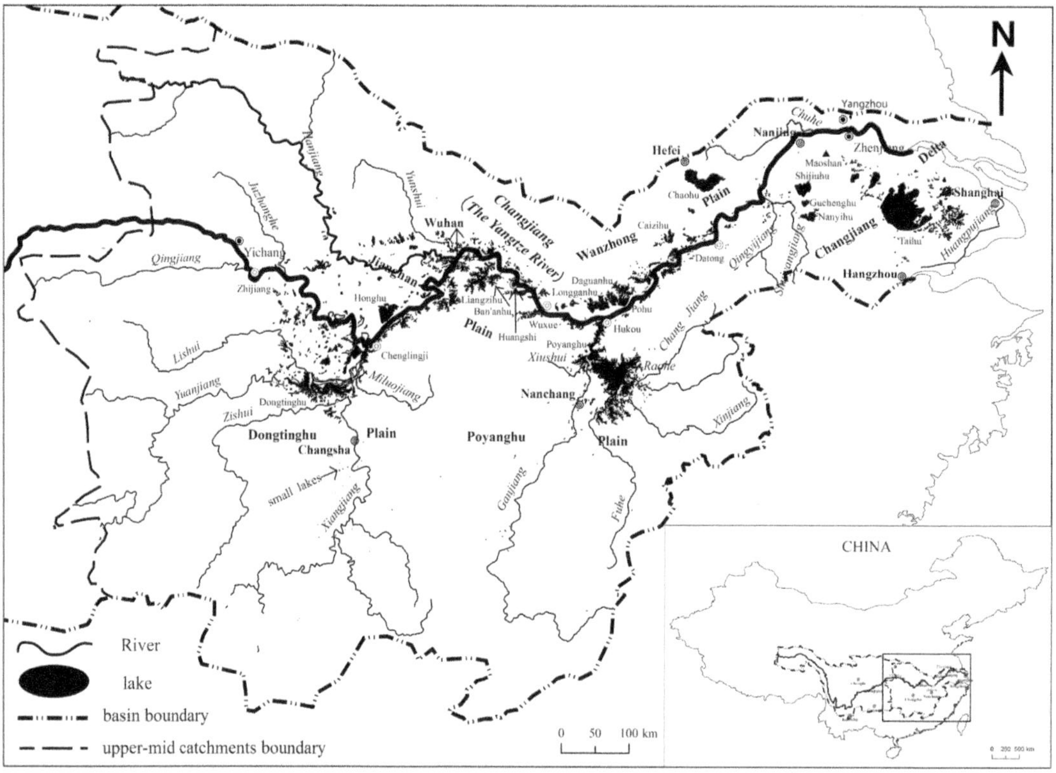

Figure 1. The mid to lower Yangtze River basin.

smaller lakes. At present, there remain more than 600 lakes, including Lake Dongtinghu, Lake Honghu, and Lake Liangzihu. (2) The Gan-Wan lakes, distributed between Wuxue and Datong. This group includes many lakes in eastern Hubei, southern Anhui, and the northern Jiangxi provinces, such as Lake Poyanghu, Lake Longganhu, Lake Bohu, and Lake Caizihu. (3) The Su-Wan lakes, distributed between Datong and Maoshan, including Lake Chaohu, Lake Nianyihu, Lake Shijiuhu, and Lake Guchenghu. (4) The Taihu plain lakes, distributed in the Yangtze River delta, including more than 200 lakes (Zhao and Chen 1999).

The origin of Yangtze lakes is roughly classified into five types (Rao 1956; Wang and Dou 1998): (1) Lakes formed by sedimentation in graben/depression basins. Because sedimentation rates are higher than subsidence rates, the lakes become gradually shallower. Typical ones are the Jianghan lakes, Lake Dongtinghu, Lake Poyanghu, and Lake Chaohu. (2) Lakes formed in abandoned channels. One type is formed by the main-stem migration, including the oxbow lakes along the middle reaches and the Yangtze-side lakes between Huangshi and Datong. Another type is irregular long lakes formed by obstruction of tributaries. (3) Lateral levee lakes, formed by overflowing narrow lowlands or valleys along floodplain boundaries. The outlets were blocked by sedimentation. Typical ones are Lake Luhu and Lake Liangzihu around Wuhan. (4) Scour lakes, deep pools scoured by floods after banks burst. (5) Lakes formed by river action on lagoons, such as Lake Taihu.

Climate and Hydrologic Features of the Mid to Lower Yangtze River Floodplain

The mid to lower Yangtze basin mostly belongs to the subtropical monsoon zone. The annual mean temperature is 13–20°C and the annual accumulated temperature (≥10°C) is

4,000–6,500°C. The mean temperature of the coldest month (January) is above 0°C, being 0–2°C in areas north of the main stem, 2–10°C in areas south of the main stem, and 10–12°C in the Nanling mountainous areas. The freezing duration is short, and the frost-free season lasts for ca. 285 d. The annual precipitation is 800–1,600 mm. It decreases from southeast to northwest, being 1,200–1,800 mm in east hill areas, ca. 1,500 mm in mountainous areas south of the Yangtze main stem, and 1,000–1,200 mm in the Yangtze plains. Seasonally, precipitation is greatest during summer (June–August), which constitutes 40% of the annual total. Spring and fall periods contain modest levels of precipitation, with the least (<10% of annual total) occurring during the winter (Zhao and Chen 1999).

In the mid to lower Yangtze main stem, the multiyear mean discharge is from 13,900 m^3/s at Yichang to 28,700 m^3/s at Datong, and the mean annual runoff is from 4.38×10^{11} m^3 at Yichang to 9.60×10^{11} m^3 in the estuary. The runoff in May–October is more than 70% of the annual total, with a peak in July. The multiyear mean water level is from 43.8 m ASL at Yichang to 8.7 m ASL at Datong. The water level is highest in July–September and lowest in December–February, with annual fluctuation amplitude of 13.5–18.4 m. The water table slope is $(0.193–0.579) \times 10^{-4}$ in the middle main stem and $(0.097–0.0203) \times 10^{-4}$ in the lower. The sediment content of water is moderate. The multiyear mean sediment concentration is from 1.14 kg/m^3 at Yichang to 0.48 kg/m^3 at Datong, and the mean annual sediment runoff is from 5.0×10^8 metric tons (mt) at Yichang to 4.3×10^8 mt at Datong (Yu and Lu 2005).

As a consequence of human disturbance, hydrological regimes of the Yangtze floodplain lakes have been altered to various extents. According to water level fluctuations (WLFs), three lake types can be identified (Table 1). The first WLF type is quasi-natural. It is similar to the WLF of the Yangtze main stem, with large fluctuation amplitude. This type includes all Yangtze-connected lakes and some disconnected ones. The second WLF type is intermittent. This type includes lakes connected to rivers or Yangtze-connected lakes only during the high-water level period, with moderate fluctuation amplitude. Water levels are low and stable during the low-water level period. The third WLF type is reservoir-like. It mainly includes Yangtze-disconnected lakes, with small fluctuation amplitude. Before the flood season, lakes are drained to the lowest water levels for flood control. After the flood season, sluice gates are shut to store water and lakes are kept at high water levels.

Environments and Biological Resources of the Yangtze River Floodplain

The Yangtze floodplain lakes are characterized by their shallow, flat, and large basins. The mean water depth of these lakes is generally ca. 2 m and rarely over 4 m. There are 18 lakes each with an area larger than 100 km^2. Because of these characteristics, interaction between water and sediments happens frequently in the lakes under the influence of monsoon climate. Moreover, the lakes are characterized by high levels of nutrients. Thus, the transparency is quite low (Table 2). In the main stem, total nitrogen was lower, but total phosphorus was higher (Table 2).

The nutrient levels of the lakes were found to be high (ca. 50 μg/L) historically, according to a palaeolimnological research (Yang et al. 2006). With a combination of high nutrients and suitable water-thermal conditions, a unique and diverse biota has evolved in the Yangtze River and its adjoining lakes. There are ca. 400 recorded species of hydrophytes and hygrophytes (Zhang 2013), 167 mollusk species with 92 endemic to China (Shu et al. 2009), ca. 200 fish species with 50% endemic to China (Yu et al. 2005; Liu and Wang 2010), ca. 50 reptile species, ca. 400 water bird species, and more than 60 mammal species, including endangered river dolphin *Neophocaena phocaenoides* and finless dolphin *Lipotes vexillifer* (Fang et al. 2006a).

The abundances of plankton, macrophytes, and macroinvertebrates are rather high in the lakes, but very low in the main stem (Table 2). Fishes are abundant in the

Table 1. Hydrological features of typical lakes in the Yangtze floodplain.

Lake	Connectivity with Yangtze	Water level fluctuation (WLF) type	Lake area (km^2)	Catchment area (km^2)	Mean depth (m)	Maximum depth (m)	WLF amplitude (m)	Lake volume (m^3)	Annual water turnover
Dongtinghu	Connected	Quasi-natural	2,432.5	257,000	6.39	23.50	13.35	178.00	18.30
Honghu	Disconnected	Quasi-natural	344.4	10,352	1.91	2.20	2.13	6.58	1.92
Liangzihu	Disconnected	Reservoir-like	304.3	3,265	4.16	6.20	2.82	12.65	1.17
Poyanghu	Connected	Quasi-natural	2,933.0	162,000	5.10	29.19	13.31	249.00	6.40
Dahuchi (sublake of Poyanghu)	Seasonally connected	Intermittent	36.7		4.55	6.00	3.50	1.67	
Shahuchi (sublake of Poyanghu)	Seasonally connected	Intermittent	14.0		0.45	3.96	3.40		
Shengjinhu	Disconnected	Quasi-natural	78.5	1,554	1.26	3.50	6.98	0.99	
Chaohu	Disconnected	Reservoir-like	769.6	9,258	2.69	3.77	2.46	19.00	2.17
Taihu	Disconnected	Reservoir-like	2,338.1	36,895	1.90	2.60	1.26	44.28	1.18

Table 2. Basic limnological characteristics of lakes and the Yangtze main stem in the mid to lower Yangtze River basin (reanalyzed from Wang et al. 1999, 2005, 2006, 2007, 2008, 2014; Feng 2005; Pan et al. 2009, 2011a, 2011b, 2014; Wang and Wang 2009; Zhao 2010). Ind = individuals; "-" = no data.

Parameters	Lakes				Yangtze main stem
	Mean	Median	Min	Max	(mean)
Secchi depth, m	1.1	0.9	0.3	3.3	0.41
Total nitrogen of water, mg/L	2.02	0.94	0.07	13.7	0.66
Total phosphorus of water, mg/L	0.11	0.04	0.005	0.76	0.13
Phytoplankton chlorophyll *a*, µg/L	16.5	4.4	0.8	139	1.69
Density of phytoplankton, 10^4 ind/L	947	361	16	6,615	-
Biomass of phytoplankton, mg/L	6.3	2.3	0.2	35.3	-
Density of zooplankton, ind/L	2,032	1,518	233	16,145	17.5
Biomass of zooplankton, mg/L	2.2	1.5	0.4	8.3	0.07
Biomass of submerged macrophytes, wet mass, g/m^2	931	295	0	9,484	Rare
Biomass of emergent macrophytes, wet mass, g/m^2	406	82	0.4	1,728	-
Density of macroinvertebrates, ind/m^2	864	259	8	23,194	91
Biomass of macroinvertebrates (mollusks with shells), wet mass, g/m^2	30.0	16.1	0.005	313	0.17

river floodplain, and the fish yield of natural lakes was 75–135 kg/ha (Wu and Jao 1958).

Main Threats to the Yangtze River Floodplain Ecosystem

Habitat Loss

River channelization.—Embankments along the mid to lower Yangtze River was initiated between 345 and 356 during the Eastern Jin Dynasty and began to take shape during the Ming and Qing dynasties (1368–1911). Discrete dikes had been gradually connected together by that time, and the total length was ca. 3,600 km, forming a huge system (Wang 1959). Prior to 1950, the river course was changed frequently due to the weakness of embankments. Since then, embankment engineering along the Yangtze main stem has been conducted on a large scale and the river regime has been stabilized, resulting in serious channelization. The work includes three major engineering projects (Yu and Lu 2005). The first one is dike reinforcement, especially along the reaches with bank caving. During the past 50 years, 8.9×10^7 m^3 of bank enrockments have been constructed, 4.9×10^6 m^2 of sunken fascine mattresses have been installed, 685 spur dikes have been built, and 20 km of longitudinal dikes have been added. The total length of reinforced bank is ca. 1,400 km. The second one is cut-off of meanders. During 1966–1972, three meanders were cut off from the lower Jingjiang in the middle Yangtze. The river length was shortened by 78 km, with sinuosity reduced from 2.83 to 1.93, and the slope was increased. As a result, the riverbed was severely eroded and the water level was lowered, reducing river water flowing into Lake Dongtinghu. The third one is blockage of secondary channels. Since the 1970s, secondary channels in 7 of 41 braided reaches were blocked. After blockage, flow regimes were simplified and main channels were deepened. With regard to ecological effects of channelization along the mid to lower Yangtze, no systematic studies have been conducted. The probable impacts are habitat loss, environmental heterogeneity decrease, and biodiversity decline.

Sharp shrinkage of lake area.—Since the Western Jin Dynasty (265–317), large fluxes of northern Chinese have migrated to the Yangtze basin many times. Due to land reclamation from lakes and sedimentation caused by deforestation, the area of Yangtze lakes began to decrease (IGSNRRCAS et al. 1985). Land reclamation reached a considerable scale in the Southern Song Dynasty (1127–1279), rapidly increased during the Ming Dynasty (1368–1644), and peaked in the middle of the Qing Dynasty (1644–1911) (Du et al. 2011). In 1949, the total area of lakes was 25, 828 km^2 (CWRC 1999). From the 1950s to 1970s, the reclaimed area sharply increased, and the area of remaining lakes is only 15,770 km^2 at present.

In the Jianghan and Dongtinghu plains of the middle basin, the total lake area was reduced by 5.2% between the 1930s and 1950s, by 46.4% between the 1950s and 1970s, and by 17.5% between the 1970s and 2000 (Du et al. 2011; Figure 2). The major causes of these reductions in lake surface area are land reclamation and, to a lesser extent, sedimentation. Historically, forest coverage of the Yangtze basin was 50–60%, but it decreased to 30–40% in the 1950s and to 10% in 1986 (cited from Wei et al. 2007). This resulted in increased soil erosion, and the rate of sediment accumulation reached 1.6–3.4 mm/year in three lakes of the Jianghan plain, and 10–30 mm/year in Lake Dongtinghu (cited from Du et al. 2011). The siltation in lakes further stimulated the reclamation process.

Degradation of lakeshore zones.—In many Yangtze lakes, shore zones have degraded or even disappeared due to human activities such as land reclamation, urbanization, and road construction. Natural shorelines have been artificially replaced by hard banks

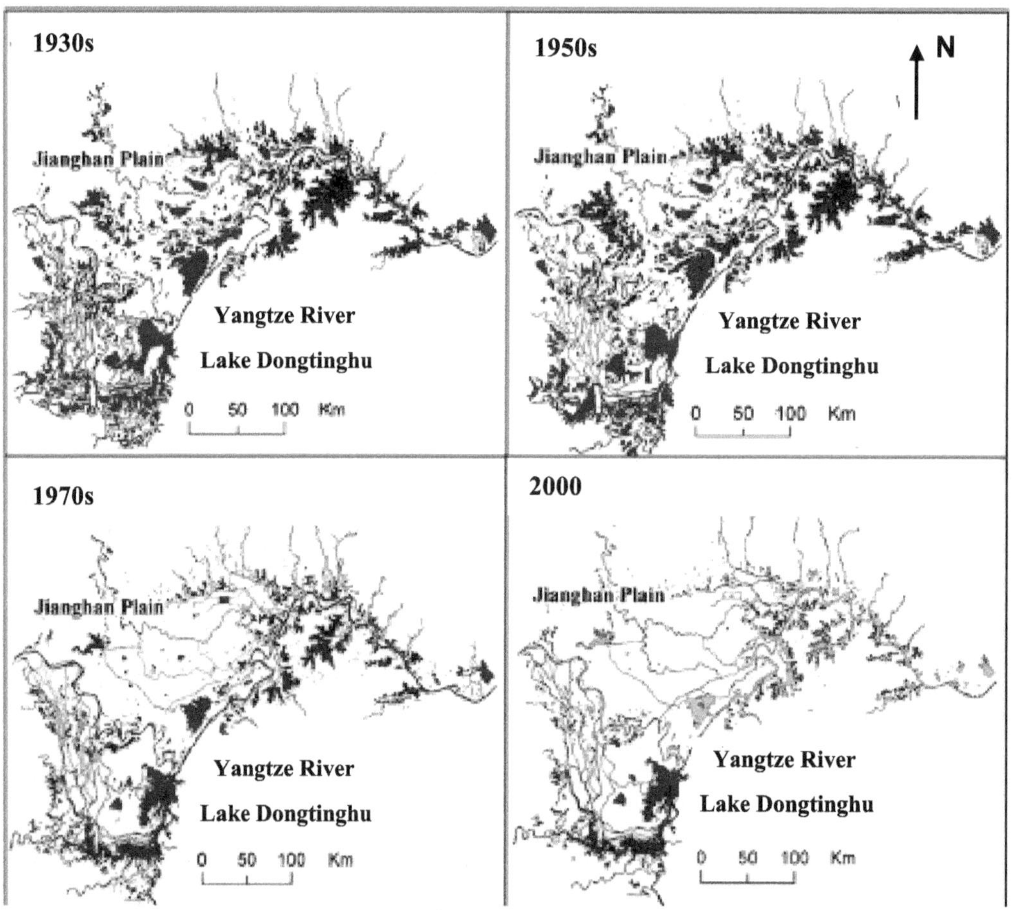

Figure 2. Lakes from the Jianghan plain and the Dongtinghu plain in the 1930s, 1950s, 1970s, and 2000. (From Du et al. 2011.)

(rocks and concrete). For example, in Lake the lower reaches at Lake Chaohu, stone bank is 129.0 km long, cement revetment is 40.7 km long, in total constituting 90% of the whole shoreline length (Wang et al. 2012). Meanwhile, a vast area of the riparian zone has become impermeable ground. Degradation of lakeshores results in deterioration of aquatic and terrestrial vegetation along shorelines, diminution of capabilities to intercept external pollution and to reduce internal loading, and loss of lake-ecosystem integrity.

Sand overmining.—Sand mining in the Yangtze main stem began in the 1950s, and limited amounts of cobbles were collected from the Jinjiang reaches during low-water periods. Since the 1980s, the amount and extent of sand mining has gradually increased, though most of these operations were unauthorized. Concurrently, ship traffic increased into the thousands of vessels. Due to serious damage of sand overexploitation, the State Council enacted a regulation on sand mining in the Yangtze River, which has been in force since 2002. At present, sand mining in the main stem has been controlled (Wu and Li 2010). However, large-scale sand mining moved quickly to Lake Poyanghu and Lake Dongtinghu, and their tributaries following enactment of regulations. For example, in Lake Poyanghu, mining ships increased from 140 in 2001 to 450 in 2005; after 2010, mining area enlarged from the north to the middle, and more than 90 additional ships were added (Cui et al. 2013). Recently, local governments began to issue relevant rules to control the serious situations of unauthorized sand mining.

Sand overmining should have very serious impacts on river floodplain ecosystems, but no systematic studies have been conducted. According to some reports (Zhang and Huang 2008; Wu and Cui 2008; Ma and Xu 2013), the negative effects may include five aspects. First, large amounts of sediment are pumped out and habitats of benthic animals and aquatic plants are destroyed. Second, suspended substance dramatically increases. For example, in northern Lake Poyanghu, suspended substance reached 0.35 kg/m^3, and Secchi depth in the summer dropped from 1.5 to 0.1–0.5 m. Turbid water affects plankton directly and inhibits plant growth. Third, pollution from wastewater, oil, garbage, noise, light, and so forth from mining ships is produced, and probably release of sediment pollution due to disturbance is enhanced. Fourth, feeding, breeding, migration, communication, and other activities of fishes, finless porpoise, water birds, and so forth are affected. Fifth, hydrogeomorphology and related processes are changed, resulting in bank collapse and decrease of lake and wetland area.

Alternations of Hydrological Regimes

Cascade reservoirs in the main stem and tributaries.—By 2009, ca. 47,000 reservoirs had been built in the Yangtze basin. The total storage capacity was 2.5×10^{11} m^3, and the usable capacity was 1.2×10^{11} m^3, being 25% and 2% of the multiyear mean surface runoff of the Yangtze River. Among them, there were 166 large reservoirs, with 38 in the upper basin, 114 in the middle basin, and 14 in the lower basin (Chen 2012). Moreover, many new reservoirs are being constructed or planned, and the South–North Water Diversion Project is being implemented.

Due to multiple factors, such as hydraulic projects, water use for industry and agriculture, climate changes, and land-use changes, hydrological regimes of the mid to lower Yangtze main stem and floodplain lakes have been greatly altered. Next, we take the changes after impoundment of the Three Gorges Reservoir (TGR) as an example. First, the mean annual runoff below the dam after the impoundment was reduced by 6–10% during 2003–2007 (CAE 2010). The water release from TGR was decreased greatly from mid-September to November to store water, but increased from January to June; during the major flood season, it was reduced only at risk of serious flooding (Wang and Chen 2010; Zhang et al. 2013). Second, the mean annual sediment discharge below the dam after the impoundment was greatly reduced by 63–86% during 2003–2007, and the sediment released from TGR became much finer. This resulted in riverbed erosion, even down to the Datong reach in the lower basin, and the water level at equal discharge during the flood season was decreased by 0.1–1 m. The most serious erosion occurred in the Yichang–Chenglingji reach, with scour depth in the thalweg up to more than 10 m. The amount of sediment flowing into Lake Dongtinghu from the Yangtze main stem was greatly reduced, and the mean annual sedimentation rate in 2003–2006 was reduced by 90.7%, thus prolonging the lifetime of Lake Dongtinghu (CAE 2010). Third, water storage in TGR and riverbed scour downstream changed water levels of the main stem and Yangtze-connected lakes and their fluctuation rhythms. It was estimated that the water level in the middle reaches decreased by 2–3 m while storing water and increased by 0.6–0.8 m in January–March (Wang and Chen 2010). During 2003–2007, the water level of Lake Dongtinghu (at Lujiao station) increased by 0.5–1 m in January–March but deceased by 1.2–1.4 m in July–August and by 1.5–2.2 m in October–November; the water level of Lake Poyanghu (at Hukou station) increased by 0.9–1.4 m in January–March but decreased by 0.1–0.2 m in July–August and by 1.3–1.5 m in October–November (Shi et al. 2011). Fourth, water temperature was changed. During 2003–2008, the water temperature at Yichang gauging station of the main stem increased by 0.5–2.3°C in October–January but decreased by 0.3–1.8°C in March–May (Li et al. 2013).

The alterations of hydrological regimes exert great impacts on the river-floodplain ecosystem along the mid to lower Yangtze River. First, the drop of water levels in the Yangtze main stem during summer and autumn blocks the exchange of organic matter and energy with the floodplain, decreasing floodplain area, habitat heterogeneity, and species diversity (Poff and Hart 2002; Shi et al. 2011). Second, alternations of water level fluctuation rhythms affect wetland vegetation. The rise of water levels in spring can retard germination of hygrophytes and emergent macrophytes, and the drop of water levels in autumn can kill submerged macrophytes, and invertebrates such as bivalves due to dewatering (Xu and Chen 2013; Zhang 2013; Zhang et al. 2014). Because vegetation provides important habitats and food sources for many types of fishes and waterbirds, these species were seriously affected (CAE 2010). Third, the lowering of flood peaks and water temperature from spring to early summer retards fish spawning, and fish resources seriously decline. After the impoundment of TGR, the initiation of the spawning season was delayed ca. 1 month, early growth and development was suppressed, and the eggs and larvae were reduced by 90% (Xie et al. 2007; Zhang et al. 2011). Additionally, the earlier drop of water levels in autumn may disturb fish migration (Ru and Liu 2013).

River–lake disconnection.—From the mid-1950s to 1960s, most lakes were disconnected from the Yangtze main stem by embankments and sluice gates, leaving few connected lakes, including Lake Dongtinghu, Lake Poyanghu, and Lake Shijiuhu, at present. After disconnection, water level fluctuations in the lakes became reservoir-like. For example, prior to damming in Lake Chaohu (777 km^2), water levels reached annual maxima during summer and minima during spring, with annual amplitudes exceeding 2 m. Following damming, amplitude declines to about 1 m, with water levels reaching minima before the flood season (Figure 3). Ecological effects of river–lake disconnection will be discussed in detail later in this paper.

Water pollution.—The discharge of wastewater and sewage in the Yangtze River basin has been continuously increasing since the 1970s. The total amount increased from 9.5×10^9 mt/year in the late 1970s through 1.5×10^{10} mt/year in the late 1980s and 2.0×10^{10} mt/year in the mid to late 1990s to 4.4×10^{10} mt/year in 2012 (Figure 4; MWR 2004–2013). More than 80% of the wastewater is discharged into the mid to lower basins.

Lake eutrophication is one of the severe impacts of excessive wastewater and sewage. More than 40% of the lakes were in eutrophic–hypertrophic states (Figure 5). Two of the four largest lakes herein, Lake Taihu and Lake Chaohu, have suffered very severe cyanobacteria blooms for decades (Zhu 1996; Lin 2003; Xie 2009).

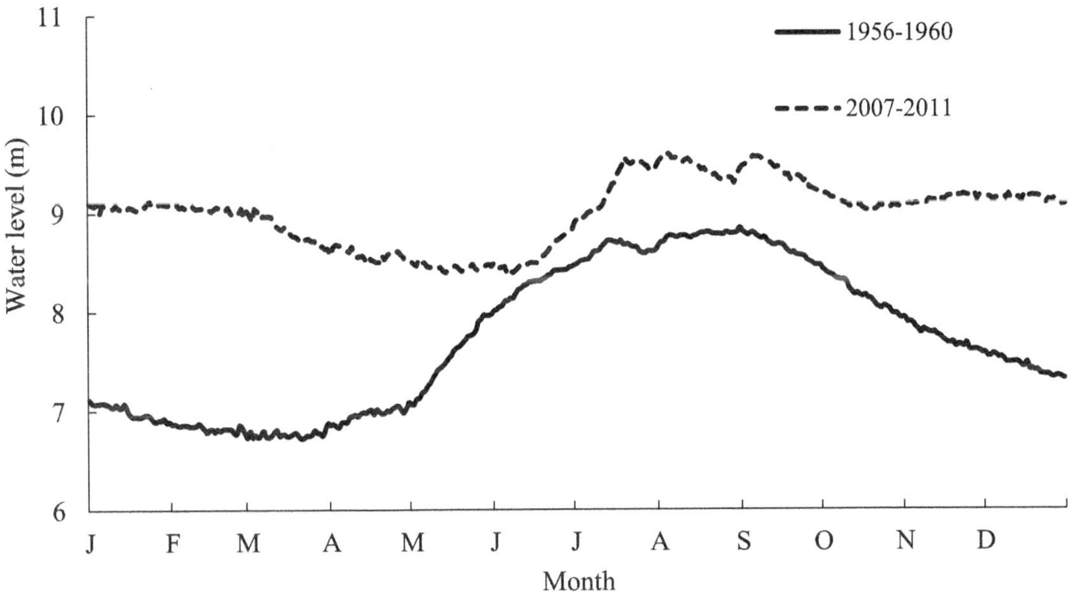

Figure 3. Alteration of water levels before (1956–1960) and after (2007–2011) building sluices in Lake Chaohu. (From Zhang et al. 2014.)

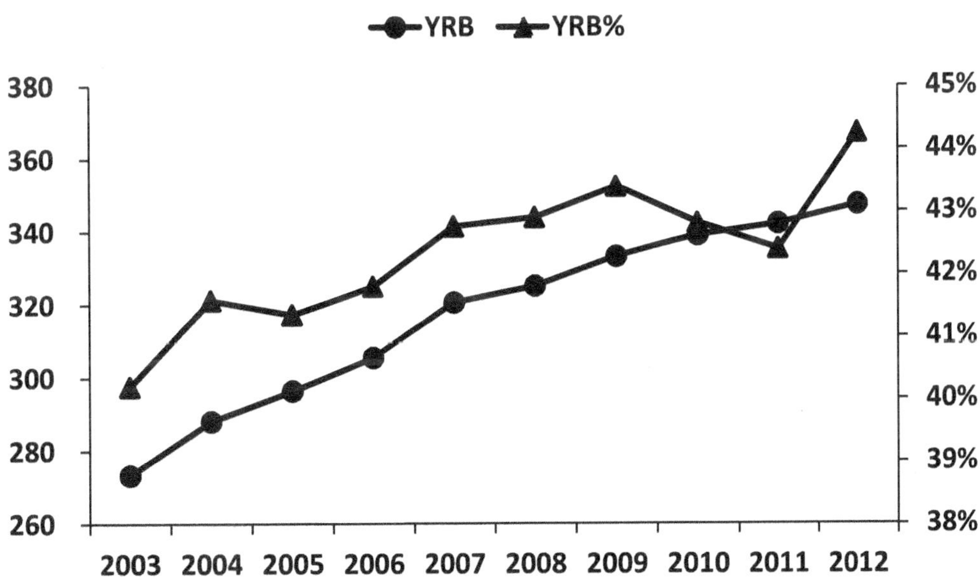

Figure 4. Amount of wastewater and sewage discharged in the Yangtze River basin (YRB) between 2003 and 2012 (10^8 metric tons) and its percentage of the total amount in China (YRB%). (Data from MWR 2004–2013.)

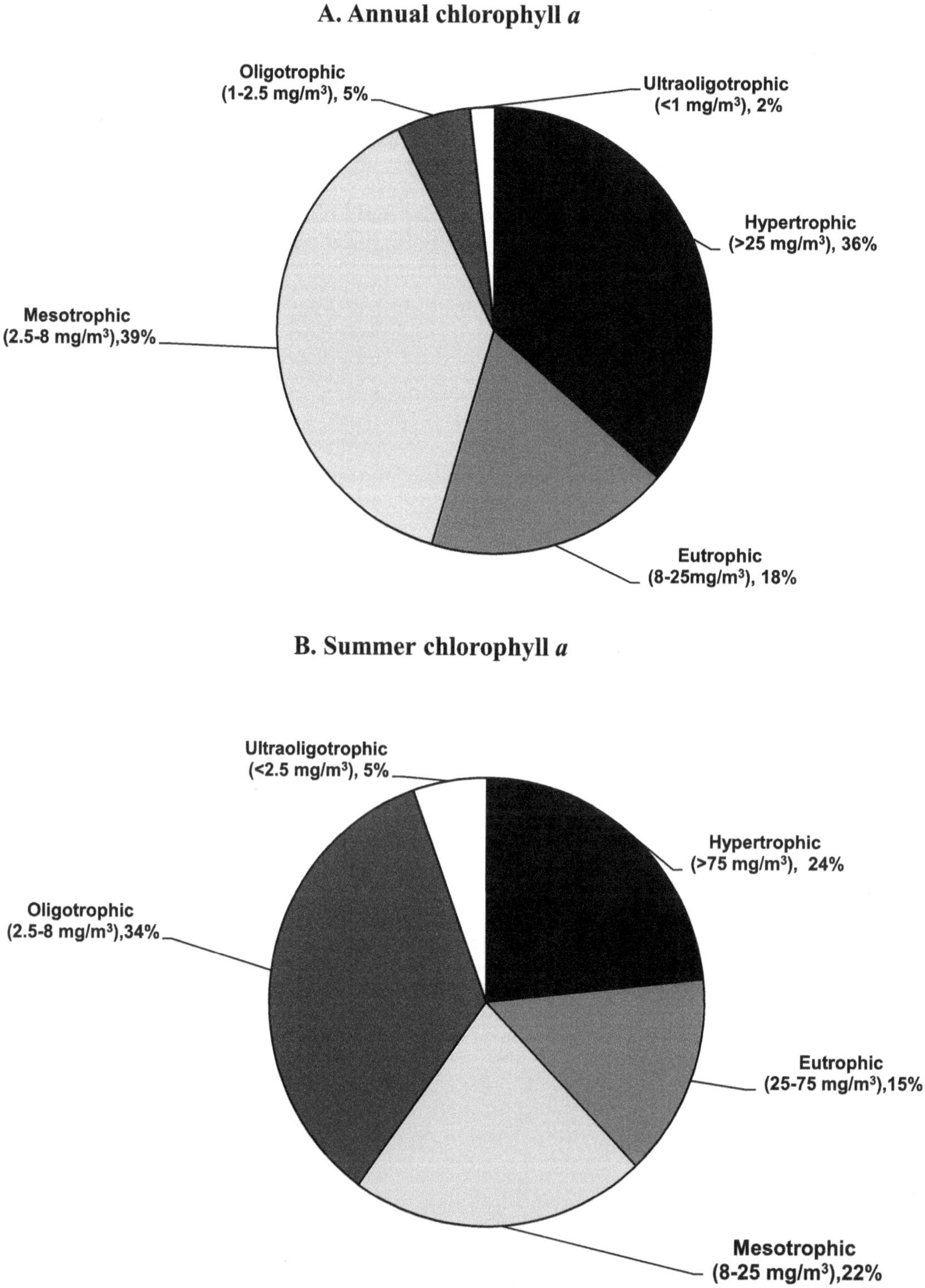

Figure 5. Trophic states of Yangtze lakes according to the fixed boundary classification system (OECD 2006). (From Wang 2007.)

Pollution of heavy metals and organic pollutants are among other impacts on the lakes. In the main stem (Wuhan, Hubei), ca. 64% of sediment sites were polluted by heavy metals to a medium level of ecological hazard; contents of eight metal species were higher than the background soil levels (Shen et al. 2008). In lakes (Wuhan, Hubei), heavy metals pollution in the sediment was more severe. The contents of four out of eight metal species were more than twice as much as the background levels, and the average content of zinc was as high as 297 μg/g in Lake Donghu, Lake Longyanghu, and Lake Moshuihu (Shen et al. 2008). Concentrations of persistent organic pollutants were higher than the standards in many lakes (Wuhan); 19 species of organo-chlorine pesticides also were detected in Lake Honghu (Fang et al. 2006b).

Overexploitation of biological resources.—Biological resources in the Yangtze River floodplain have been seriously overexploited. In the main stem and connected lakes, electrofishing boats and fine-meshed (0.5–1 cm) fish traps are the main direct threats to fish resources (Cao 2011). Due to overfishing and reservoir construction, river–lake disconnection, and pollution, fish yield in the whole river-lake system has decreased from 4.3×10^5 to 1.0×10^5 mt/year between the 1950s and 2000s (cited from Liu and Wang 2010). The famous Reeves Shad *Tenualosa reevesii* has disappeared from the Yangtze River; Tapertail Anchovy *Coilia ectenes* and Obscure Pufferfish *Takifugu obscurus* have become rare, and natural populations of the four major domestic Chinese carps have sharply decreased (Cao 2008, 2011). Meanwhile, mollusks and other organisms also have been overused as human and fish food or for other purposes.

In disconnected lakes, pen culture has become the dominant fishery model in the past 20 years, overstocking Chinese mitten crab *Eriocheir sinensis* in most cases. Although useful for production and management, pen culture can have severe negative effects on ecosystems. In Lake Biandantang (Hubei), after years of continuous intensive crab culture, submerged macrophytes declined from coverage of ca. 90% in 1991 to less than 5% in 2002 and disappeared in the end; Secchi depth was only 1.1 m, being around half of the original state. Under the stocking density of 1 kg/ha of crab larvae or juveniles, the density and production of macroinvertebrates were reduced by more than 60% (Xu et al. 2003). Fish culture using chemical fertilizers also is common, especially in small to medium-sized lakes, thus worsening eutrophication in this region.

Ecological Effects of River–Lake Disconnection on Yangtze Floodplain Lakes and the Mechanisms

We focused upon this issue in the past 20 years, and the results are summarized as follows.

Plankton

The mean species number of phytoplankton in Yangtze-connected lakes was 25% greater than that of disconnected lakes while the total density of the former was 42% lower (Wang and Wang 2009). The dominant taxon of connected lakes was cryptophytes while that of disconnected lakes was cyanobacteria. Densities of cyanobacteria, chlorophytes, and bacillariophytes in connected lakes were obviously greater than those of disconnected lakes, while those of cryptophytes and euglenophytes were in reverse. The mean species number of zooplankton (rotifers, cladocerans, and copepods) in connected lakes was ca. 30% lower than that of disconnected lakes, and densities of the total and each group in the former were much lower.

In terms of phytoplankton chlorophyll *a* (Chl *a*), disconnected lakes had the greatest concentrations (30 μg/L), followed by lentic areas of connected lakes (5.0 μg/L) and lotic areas of connected lakes (2.1 μg/L). On the other hand, total nitrogen (TN) and total phosphorus (TP) ranked differently across these lake classifications. Specifically, TN concentrations were 2.6, 1.2, and 1.5 mg/L, TP concentrations were 0.2, 0.06, and 0.13 mg/L in disconnected lakes, lentic areas of connected lakes, and lotic areas of connected lakes, respectively (Wang and Wang 2009). This means that the greatest abundance of phytoplankton in disconnected lakes cannot be explained by nutrients alone. A further analysis showed that TP is the limiting factor of Chl *a* in disconnected lakes as well as in lentic areas of connected lakes, but the slope of the TP-Chl *a* regression in the former was about twice as great as that in the latter (Figure 6). This disparity may be attributed to the different residence times. In lotic areas of connected lakes, the limiting factor of Chl *a* is water current. Chl *a* had a parabolic relationship with surface velocity, peaking at a velocity of 0.12 m/s. When current is slow, Chl *a* tends to increase with velocity, due to the fact that greater flow is closely linked to greater input of external nutrient sources. When current is fast, Chl *a* tends to decline, suggesting that current inhibition has prevailed over nutrient effects on growth of algae. Fast current may cause high concentrations of suspended substance, resulting in light limitation on algal growth. The above analyses demonstrate that river–lake disconnection makes lake water more still, which allows phytoplankton to propagate in large numbers. Similarly, due to stable environment and plentiful food of phytoplankton, zooplankton develops well in disconnected lakes.

Macrophytes

There was no significant difference in the mean species number of the total as well as floating and submerged species between Yangtze-connected and disconnected lakes (Wang and Wang 2009). The species number of emergent macrophytes in disconnected lakes was 40% greater than that of connected lakes while that of hygrophytes was 8% lower in the former.

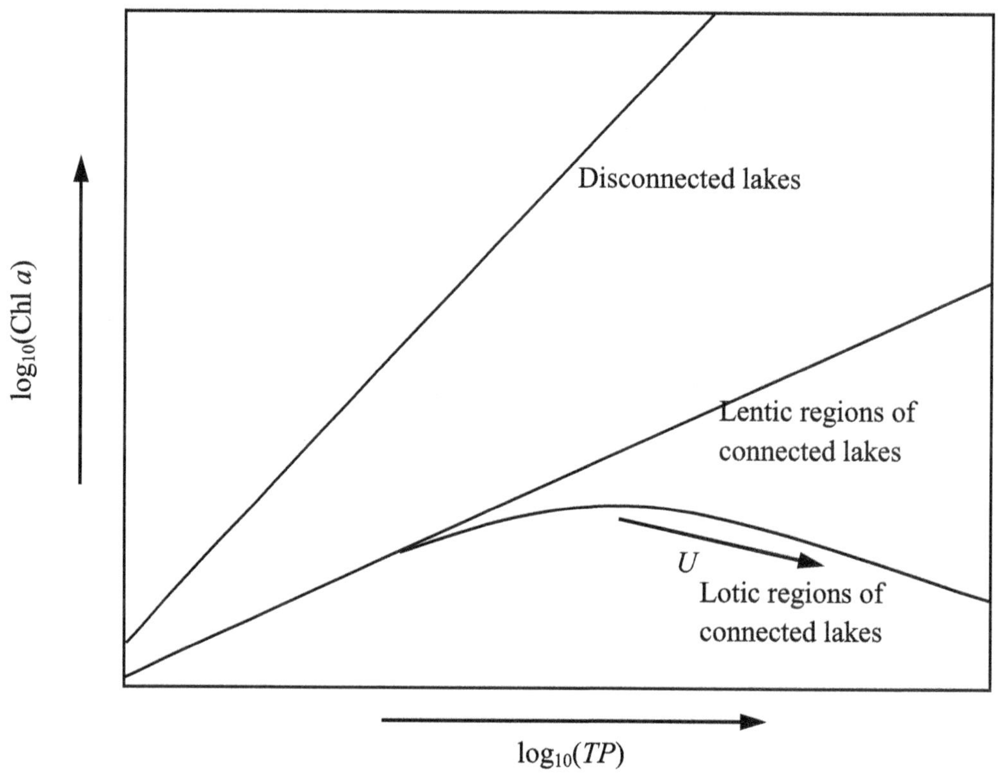

Figure 6. Comparison of chlorophyll *a*–phosphorus regression lines from Yangtze-connected and disconnected lakes. TP, total phosphorus of lake water; Chl *a*, phytoplankton chlorophyll *a*; and *U*, water velocity. (From Pan et al. 2009.)

Hygrophytes develop very well in the Yangtze-connected lakes, with coverage up to 15% and biomass up to 1,500 g/m^2, but develop poorly in disconnected lakes. Emergent macrophytes can grow well in connected and disconnected lakes, with biomass up to 2,000 g/m^2 but coverage lower than 10%. Freely floating macrophytes occur more frequently in disconnected lakes, but coverage and biomass are very low. Floating-leaved macrophytes may cover large areas in the two lake types, but biomass is relatively low. Submerged macrophytes develop better in disconnected lakes than in connected ones, with coverage up to 80% and 40%, respectively, and biomass up to 2,600 and 1,200 g/m^2, respectively (Wang and Wang 2009).

In terms of dominant species, hygrophytes are dominated by the genus *Carex* in connected lakes and *Cynodon dactylon* in disconnected lakes. Emergent macrophytes are dominated by *Phragmites australis* and *Triarrhena lutarioriparia* in connected lakes and by *P. australis* and *Zizania latifolia* in disconnected lakes. For submerged macrophytes, the dominant species in connected lakes are *Potamogeton malaianus, Hydrilla verticillata, Ceratophyllum dermersum,* and *Vallisneria spiralis.* Among them, *P. malaianus* prefers waters with larger water level fluctuation amplitude while the others prefer areas with lower connectivity. The dominant submerged species in disconnected lakes are *P. maackianus, H. verticillata, C. dermersum,* and *V. spiralis.* Among them, *P. maackianus* dominates the climax community. The dominant submerged species usually shifts from *P. malaianus* to *P. maackianus* after river–lake disconnection.

There are two mechanisms for disconnection to change macrophytes. First, water level fluctuations (WLFs) are altered (e.g., Figure 3). Water level fluctuation amplitude is closely correlated with diversity and biomass of all macrophytes. Species richness maximizes at an intermittent level of fluctuation amplitude (ca. 5 m; Figure 7). Biomass of hygrophytes increases with fluctuation amplitude while that of submerged macrophytes decreases (Figure 7; Zhang 2013). The fluctuation amplitude is decreased in disconnected lakes, thus resulting in decreased coverage and biomass of hygrophytes and increased submerged macrophyte biomass. After disconnection, the reduced rate of water-level change also is harmful to hygrophyte development. If WLF type is changed as reservoir-like, this can lead to degradation of all vegetation. For example, vegetation coverage in Lake Chaohu has decreased from 30% before disconnection to less than 1% at present (Zhang et al. 2014). Second, other habitat conditions are changed. In disconnected lakes, water velocities decrease so sediment concentrations of water decrease, and the lake bottom is covered with more silt, thus favoring freely floating and submerged macrophytes.

Macrobenthos

Compared with those of the Yangtze main stem and disconnected lakes, macrozoobenthos of the Yangtze-connected lakes are characterized by greater diversity (Wang et al. 1999, 2007; Pan et al. 2011a). This pattern of species richness conforms to the theory that α diversity of macrobenthos in floodplain waters reaches a maximum at an intermediate level of connectivity (Amoros and Bornette 2002). Mollusks in connected lakes were especially species-rich, and the total number of species was about four times as high as that in disconnected lakes; the total number of insect species in connected lakes was about 1.5 times as high as that in connected lakes; the total number of oligochaete species and other animals in connected lakes was about twice as high as that in connected lakes. This result seems attributable to habitat heterogeneity within such a large lotic-to-lentic area. For example, bottom structures of connected lakes range from silt to sand and to gravel, but in disconnected lakes, sediment is primarily with silt. Additionally, river–lake disconnection can indirectly affect macrobenthos by changing vegetation types.

River–lake disconnection affects standing crops of macrobenthos to a considerable extent (Pan et al. 2011a). Although total

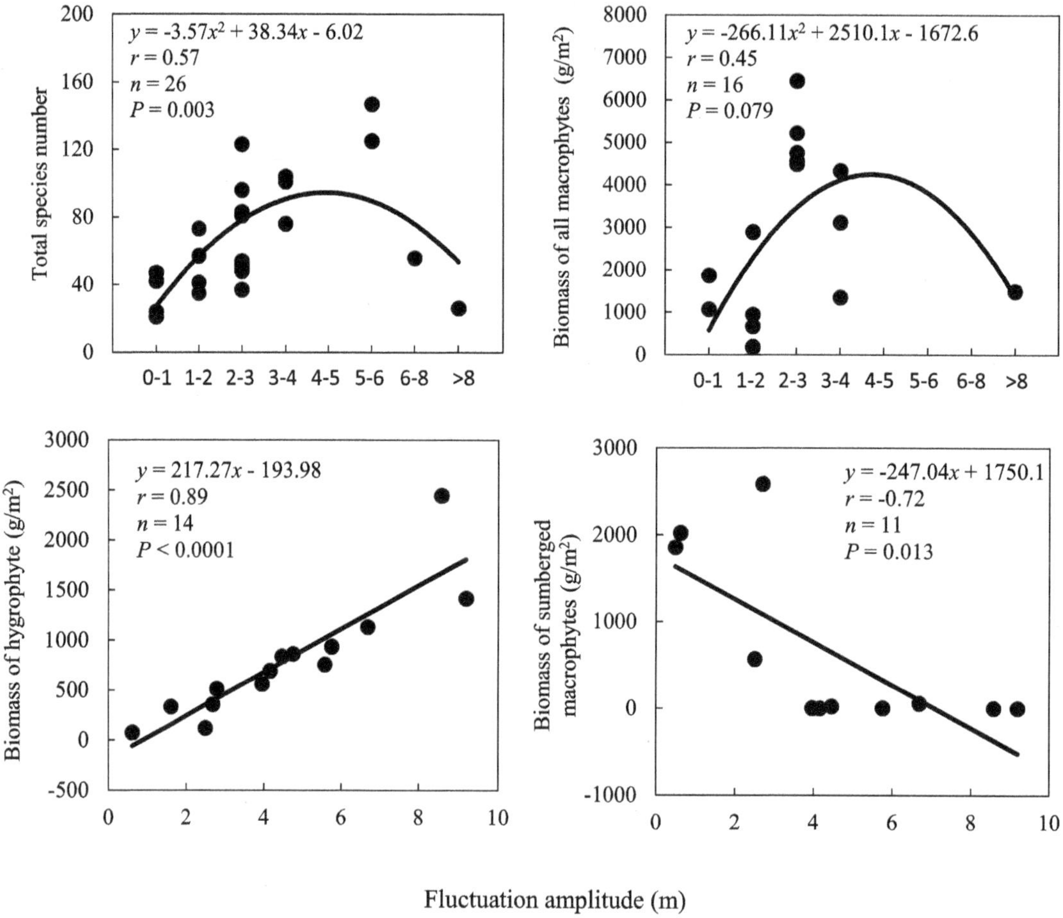

Figure 7. Relationships between water level fluctuation amplitude and species richness and biomass of macrophytes in Yangtze floodplain lakes. (From Zhang 2013.)

densities had no much difference, connected lakes had abundant bivalve filterers such as the genus *Corbicula*, which was more than eight times denser than in disconnected lakes. Oligochaetes and insects were more abundant in disconnected lakes, and the density of each was 4.6 times and 4.5 times as high as that of connected lakes, respectively.

Along the river connectivity gradient, the change patterns of functional groups of macrobenthos in Yangtze lakes could be classified into three types (Pan et al. 2011a). First, density, biomass, and production of shredders, collector-filterers, and predators show unimodal changes. Shredders (e.g., the genus *Stictochironomus*) feed on coarse particulate organic matter, such as decaying leaves. Therefore, their abundance should increase due to more external organic matter brought by increased river inflow but may be inhibited in case of overdisturbance. At an intermediate level of river connectivity, the substrate is hard and sandy in general (Figure 8C), and suspended organic particles are relatively abundant (Figure 8C). Thus, the substrate favors filterers (mainly bivalves). At a high level of river connectivity, the substrate becomes unstable and water is full of inorganic particles so that filterers would not be supported. The increase of predators (e.g., the family Dytiscidae) at the initial stage may be the result of reduction in macrophytes because plants can provide refuges for zoobenthos and impede predation. The decrease of predators at later stages seems to be caused by high turbidity and habitat instability. High turbidity may prevent predators from seeing prey (Figure 8C). Second, density, biomass, and production decrease when river connectivity increases. This pattern is represented by collector-gatherers (mainly the families Tubificidae and Chironomidae). Collector-gatherers mainly consume fine particulate organic matter, so their decrease probably is caused by the fact that the organic-rich silt diminishes with increasing river connectivity (Figure 8C). Third, density decreases with increasing river connectivity, whereas biomass and production change unimodally. Scrapers exhibit this trend. The decrease of scrapers density could be ascribed to the reduction of macrophytes, which small-sized gastropods prefer (Figure 8C). The increase of biomass and production at the initial stage could be ascribed to the replacement of small-sized epiphytic gastropods (mainly the family Bithyniidae) by large-sized bottom dwellers (mainly the genus *Bellamya*), and the decrease at later stages may result from the increase of sand fractions. Because of instability and detritus shortage, sandy bottoms in the Yangtze-connected lakes should not be suitable for scrapers, which feed mainly on detritus there. In terms of total benthos, density decreases with increasing river connectivity (Figure 8A), showing the same pattern with the dominant groups in density (i.e., scrapers and collector-gatherers). Biomass and production change unimodally (Figure 8B), showing the same pattern with the dominant groups in biomass (i.e., scrapers and collector-filters).

Fishes

By comparing species–area relationships of fishes in Yangtze-connected and disconnected lakes (Figure 9), our study shows that river-lake disconnection reduced the total species number in lakes on the Yangtze River floodplain by 38.1% on average. In terms of habitat guilds in disconnected lakes, river-sea migratory fishes was reduced by 87.5%, riverine fishes by 71.7%, river–lake migratory fishes by 40.6%, and lake resident fishes by 25.4% (Figure 9A–D). The reduction in fish diversity after river-lake disconnection should be attributed mainly to the blockage of migration routes, the loss of fluvial environments in which some species spend parts of their life cycles, and decreased habitat heterogeneity.

The resource of eggs and larvae of four domestic Chinese carps in the Yangtze main stem was estimated to be 1.0×10^{11} in 1964 (Yi et al. 1988) but was reduced to only 1.7×10^{10} in 1981 (STSGDFCR 1982). River–lake disconnection is one of the main causes for the observed decline in fish larvae.

Food Webs

As stated above, river–lake disconnection changes lake communities completely, thus affecting the food web structure. For example, macroinvertebrate food web structure differs greatly between Yangtze-connected and disconnected lakes, and the food web of the latter is much simplified (Figure 10; Liu et al. 2006; Liu and Wang 2008). The numbers of total species, macroinvertebrate species, functional groups, predators, and total links are much higher in connected lakes than in disconnected ones. By contrast, connectance and connectivity are much lower in connected lakes, indicating that the species interaction strength is weaker in these lakes than that of disconnected ones. Connected lakes have more heterogeneous habitats and wider range of food resources, and this can decrease the interaction strength among species. As a result, the greater seasonal fluctuations of physical environments such as water levels can decrease the frequency that animal species meet each other and, thus, the interaction strength. Although the trophic basis of macroinvertebrate food web is detritus in both connected and disconnected lakes, the latter is more heavily based on detritus (70%) than the former (40%). In terms of energy flow, the total energy flowing through the food web is much greater (by orders of magnitude) in connected lakes.

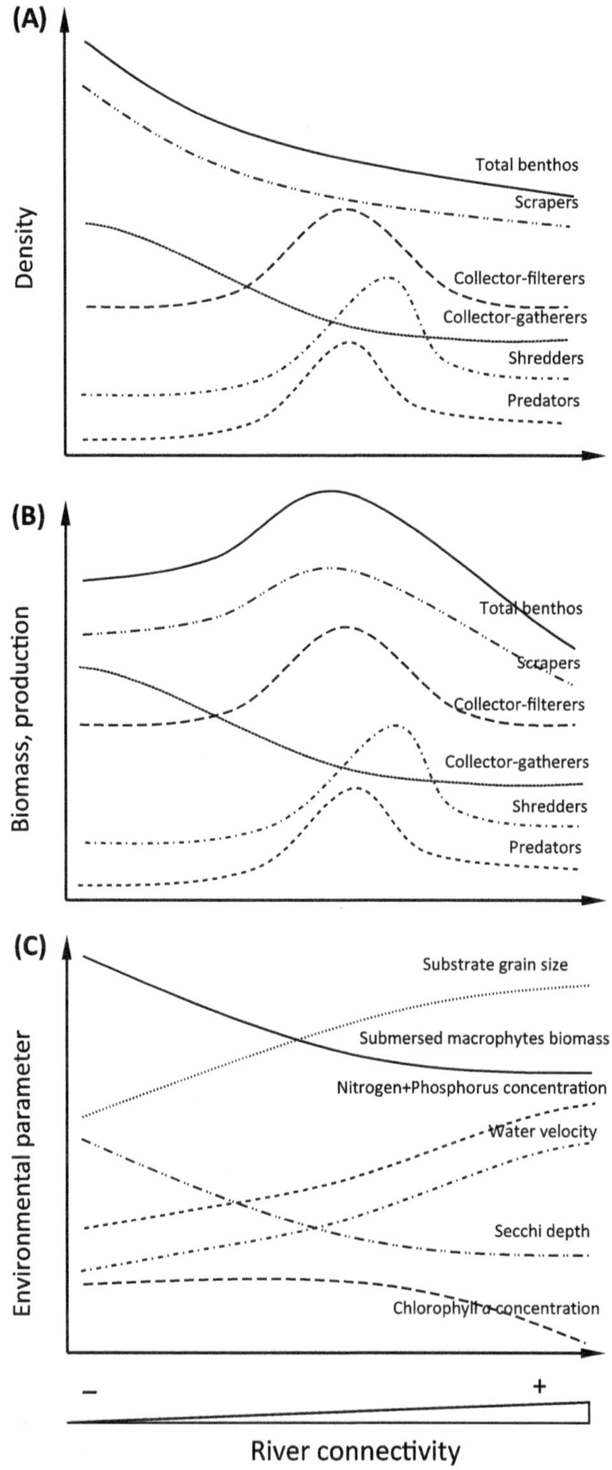

Figure 8. Tendency diagrams to illustrate responses of **(A)** density of different functional feeding groups, **(B)** biomass and production of different functional feeding groups, and **(C)** environmental parameters to river connectivity in the Yangtze floodplain. (From Pan et al. 2011b.)

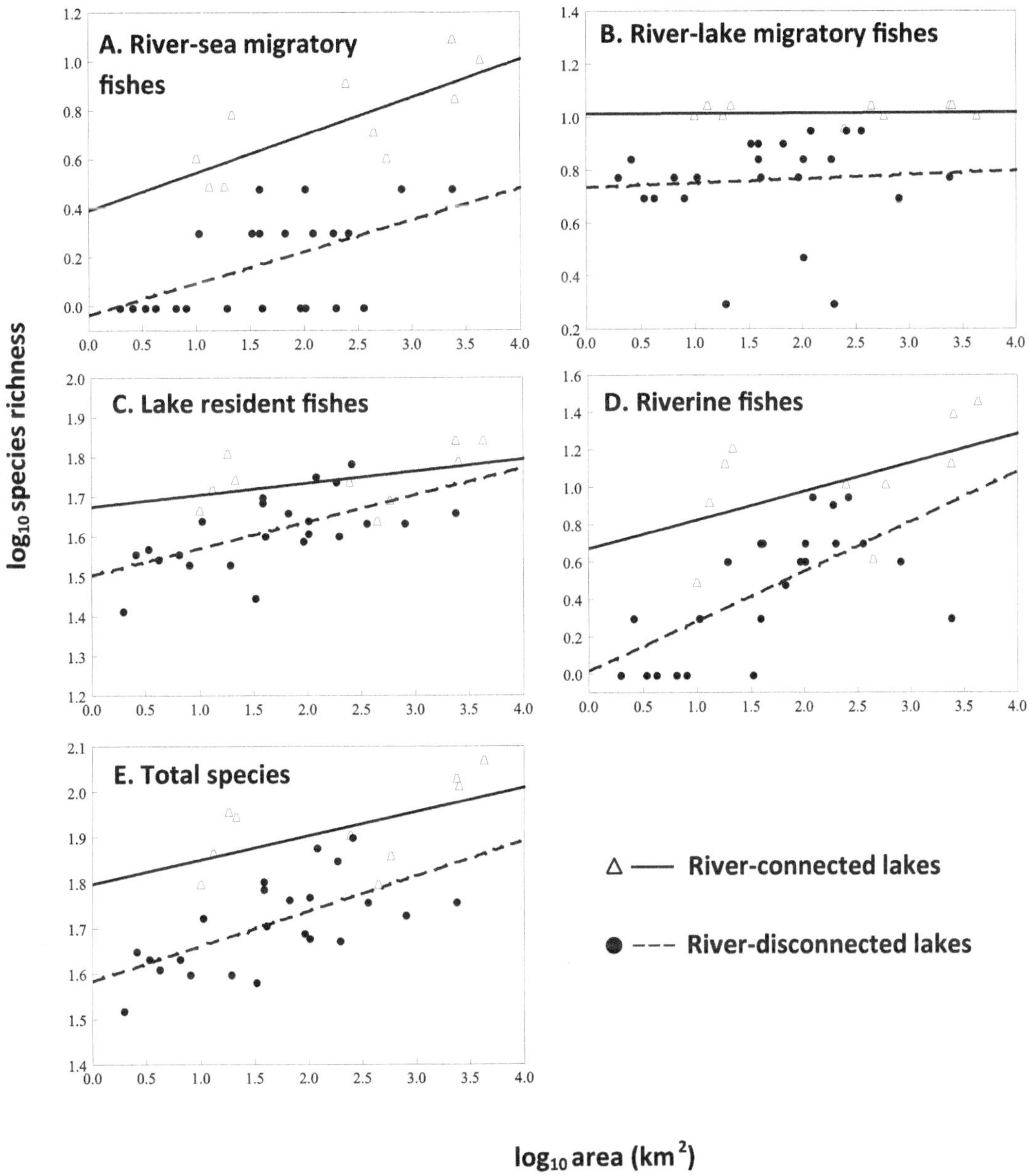

Figure 9. Species–area relationships of fishes in Yangtze-connected and disconnected lakes in the Yangtze floodplain. (From Liu and Wang 2010.)

Mechanisms for Effects of River–Lake Disconnection on Lake Ecosystems

We have systematically revealed the relationships between river connectivity and lake ecosystems (Figure 11). The main findings are as follows: (1) river–lake disconnection is one of the main causes of lake eutrophication, increasing the conversion efficiency of nutrient into phytoplankton biomass to a large extent; (2) species diversity, biomass, and production of macrophytes and macrobenthos reach maxima at some levels of intermediate river connectivity; (3) river–lake disconnection greatly reduces fish species richness of each habi-

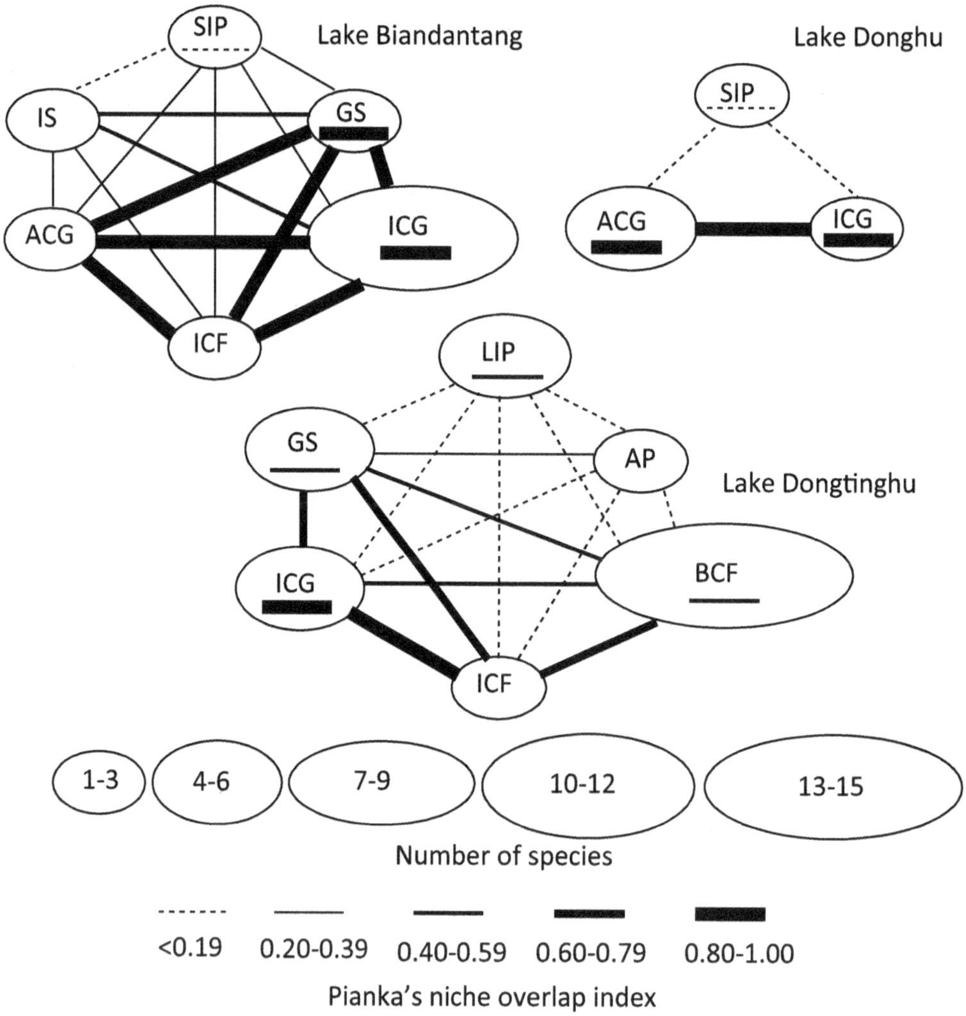

Figure 10. Comparison of niche overlap web (summary web) of Yangtze-connected Lake Dongtinghu with two disconnected lakes. IS, insect shredders; GS, gastropod scrapers; ACG, annelid collector-gatherers; ICG, insect collector-gatherers; BCF, bivalve collector-filterers; ICF, insect collector-filterers; SIP, small insect predators (mainly chironomids); and LIP, large insect predators (mainly odonates). The line thickness in each ellipse shows the dietary overlap of species within the functional feeding group. (From Liu and Wang 2008.)

tat guild, and natural resource of fish larvae is severely depleted; and (4) river–lake disconnection simplifies macroinvertebrate food web structure in disconnected lakes and increases the connectance and degree of dietary overlaps, and trophic basis is more dependent on detritus.

River–lake disconnection disturbs the natural flow regimes, decreasing the species diversity of lakes sharply and changing the community structure substantially. It has been revealed that natural flow regimes are the main mechanisms by which river floodplain ecosystems have quite rich species and very high production (Mahoney and Rood 1998; Amoros and Bornette 2002; Lytle and Poff 2004; Junk 2005; Rood et al. 2005). Over evolutionary time, all species have developed life history strategies adapting to predictability of water regimes, and thus, hydrological alternations will exert great effects on any organisms. In summary, obstruction of

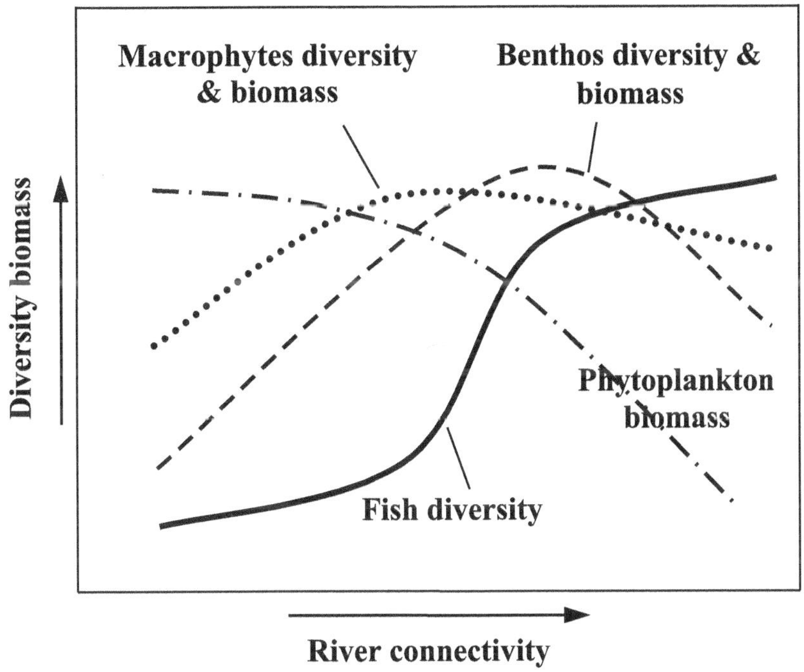

Figure 11. Relationships between hydrological connectivity and different ecological groups in the Yangtze floodplain.

river–lake hydrological connectivity affects Yangtze lake ecosystems via three mechanisms (Ward 1997; Amoros and Bornette 2002; Ward et al. 2002). First, block the exchange of water, sediment, nutrients, and organic matter between the Yangtze main stem and lakes, disturb periodicity of hydrological fluctuations, and hinder/delay growth and development of organisms. Free connection is of particular importance to organisms with habitat shifts in life histories. Second, lessen hydrogeomorphological dynamics, and decrease spatiotemporal heterogeneity of habitats. Hydrological processes create various landscape elements and environmental gradients, which are changing over time. Third, stop or weaken natural disturbance by hydrological dynamics, which interrupts successions of various communities and keeps them in different successional stages, thus forming mosaics of community patches.

According to situations in the Yangtze floodplain, Figure 12 shows how river–lake connection works on lake ecosystems:

In winter (Figure 12A), water levels in the Yangtze main stem and lakes are the lowest and some areas are dry. Hygrophytes on lakeshore and some submerged macrophytes in shallow waters begin to germinate, invertebrates such as mollusks winter in sediment and other habitats, and river–lake migratory fishes stay in deep pools of the main stem.

In spring (Figure 12B), the main stem floods, and river water containing large amounts of sediment and nutrients flows into lakes. Hygrophytes and hydrophytes grow well, invertebrates revive and breed, and migratory and resident fishes spawn and then larvae migrate/drift into lakes.

In summer (Figure 12C), the water level in the main stem reaches a maximum. Lakeshore is submerged and then hygrophytes degenerate gradually, but hydrophytes flourish; fishes and invertebrates assemble and forage in vegetation.

In autumn (Figure 12D), the water level in the main stem goes down and lake water flows back into the river. Lakeshore emerg-

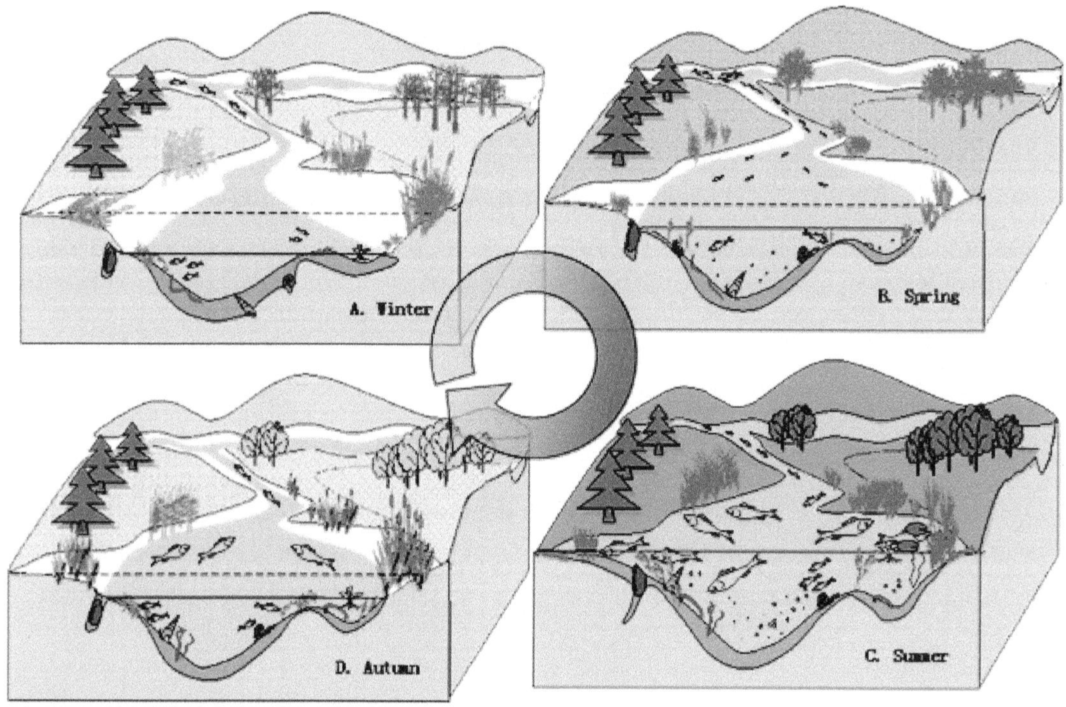

Figure 12. Diagram of ecosystem dynamics of the Yangtze River floodplain. (From Wang and Wang 2009.)

es from water and hydrophytes decline; migratory fishes swim to the main stem.

Holistic Conservation of the Yangtze River Floodplain Ecosystem and Strategies of Ecohydrological Rehabilitation

Conservation Principles and Key Scientific Questions

At present, species diversity in the Yangtze floodplain is continually decreasing and water quality is continually degrading. Therefore, the ecosystem protection should be oriented to resolve these two problems. Because pollution control has been discussed extensively, we focus upon biodiversity conservation in this paper. According to our studies and related literature, we propose the following two principles concerning biodiversity conservation in the Yangtze River floodplain.

The principle of holistic conservation.—Since the Yangtze River is a huge integrated system formed before 23 Ma, biodiversity conservation must be conducted on the whole water system scale. The Yangtze River floodplain covers a large geographical region and holds diverse subsystems such as the main stem, tributaries, lakes, ponds and ditches, hyporheic zones, riparian zones, and estuary. All the subsystems are interconnected and interact with each other. Apparently, any single/local reserve is not enough to realize biodiversity conservation of the whole river floodplain. Therefore, it is of great importance to conserve the huge ecosystem from a holistic perspective.

The principle of free hydrological connectivity.—The principle concerns free water exchange among water bodies of the river floodplain. There are four dimensions of hydrological connectivity (i.e., longitudinal, lateral, vertical, and temporal). Free connection can increase temporal-spatial heterogeneity of habitats, and thus species diversity, and can help to prevent overgrowth of some species and thus reduce ecological disasters (e.g., blue-green algal blooms). Migratory species shift habitats in different stages of life history, and connection is of particular importance to them. To avoid habitat fragmentation, it is also necessary to maintain internal connection in a single water body.

To implement the above principles, two scientific questions should be answered. First, what is the minimum protected area of Yangtze-connected lakes for biodiversity conservation of the Yangtze River floodplain? Estimating reserve size is a key step for effective conservation of biodiversity. At present, only a few connected lakes maintain free connection with the Yangtze main stem. Do they meet the requirement of biodiversity conservation for the entire floodplain? Second, what are the environmental flow requirements of the Yangtze River floodplain? The hydrological regimes of the floodplain waters have been altered by river damming and river–lake disconnection, leading to ecosystem degradation to various degrees. Therefore, it is urgent to understand the hydrological requirements of these ecosystems so as to provide the scientific basis for ecohydrological operation of dams and sluices.

Minimum Protected Area of Yangtze-Connected Lakes

Species–area relationship is regarded as one of the general laws in ecology and can be used to calculate the minimum protected area. Targeting fish conservation, we estimated the minimum protected area of the Yangtze-connected lakes. The reasons to choose fish are (1) fish species are sensitive to hydrological connectivity, especially migratory ones; (2) fishes are K-strategists and position at higher trophic levels in general, and it is reasonable to assume that most organisms can be conserved if fishes are under protection; and (3) fish data are most abundant.

Based on collected fish diversity data of the past 50 years, the cumulative species–area model has been constructed (Liu and Wang 2010; Figure 13). Given that the total number of fish species is 173 in this region, the minimum protected area is estimated to be 14,400 km^2 according to the model. Since

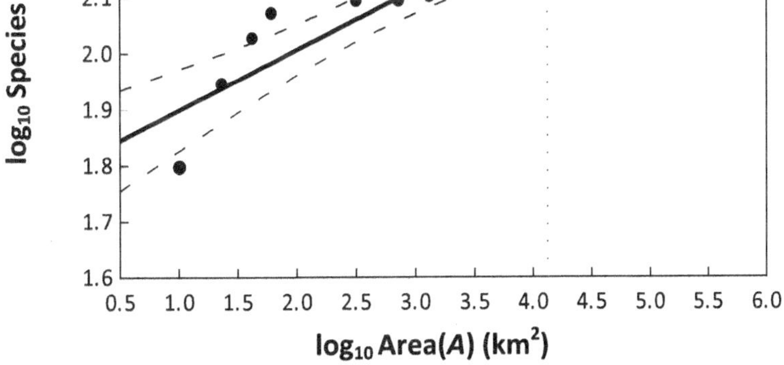

Figure 13. Estimation of minimum protected area of Yangtze-connected lakes for fish diversity conservation in the Yangtze floodplain. (From Liu and Wang 2010.)

the total area of existing Yangtze-connected lakes is 5,500 km^2, at least 8,900 km^2 of disconnected lakes need to be reconnected.

Ecohydrological Operation of Dams and Sluices

Our research shows that the existing Yangtze-connected lakes do not meet the area requirement of biodiversity conservation in the Yangtze floodplain. Therefore, we must reconnect enough disconnected lakes with the main stem. The feasible approach is to operate dams and sluices in ecohydrological ways, and this should be based on the assessment of environmental flow requirements (EFRs).

In terms of macrophytes EFRs, we focused on WLFs. Macrophyte distributions show clear zonation along elevation gradients, and the life histories are closely correlated with lake WLFs (Keddy and Reznicek 1986). Hygrophytes are distributed at the greatest elevations and adapt to environments with large fluctuation amplitude. For example, biomass of *Carex* in lakes with large amplitude (e.g., Lake Dahuchi) is about 10 times as much as those with small amplitude (e.g., Lake Yandonghu) (Zhang 2013). Hygrophytes germinate near lakeshore in low water level periods, and the seedlings cannot tolerate submergence. During growth, hygrophytes prefer stable water levels and cannot survive if water levels rise too fast. Most hygrophytes mature before June, and thus the suitable submergence time is June–July. Emergent macrophytes grow at lower elevations than hygrophytes and are distributed at a wider range of fluctuation amplitude. In low water periods, they germinate in places from 1 to 7 m above lake surfaces to shallows. Taking the common reed *Phragmites australis* as an example, root-sprouting is obviously prevented at 20 cm under water and absolutely fails beyond 30 cm (Cao 2007). The sprouts can resist submergence to a certain degree, but fast rising of water levels beyond the growth rate has detrimental effects on their growth, even leading to death. The suitable submergence time is July–August, and biomass is much higher in areas submerged in July–August than in June (Zhang 2013). Submerged macrophytes prefer lakes with small fluctuation amplitude. Underwater light is the key factor affecting their growth, and the ratio of Secchi depth to mean depth should be more than 0.5 during March–June in order to enable a normal growth (Wang et al. 2005). A certain increase in water levels can promote the growth of macrophytes (Zhang et al. 2013), but fast rising of water levels can inhibit the growth or kill the plants (e.g., Keddy and Reznicek 1986).

With regards to fishes, we analyzed the migration pattern of river-lake migratory fishes (the four Chinese carps, i.e., Black Carp *Mylopharyngodon piceus*, Grass Carp *Ctenopharyngodon idella*, Silver Carp *Hypophthalmichthys molitrix*, and Bighead Carp *Hypophthalmichthys nobilis*; Figure 14; Ru and Liu 2013). Between late spring and early summer, the adults breed in the main stem when the water level rises sharply, and the eggs drift downstream and develop in the main stem. After spawning, adults will migrate into floodplain lakes to feed. Some larvae and juveniles then move passively into lakes along with the flow. Others feed in backwaters of the river and migrate later into floodplain lakes. They grow and mature in the lakes for 3–4 years. When the water recedes in late autumn, the adults again return to the main stem. The subadults also may leave if the lake water is not deep enough.

According to the above results, we propose the following ways to operate dams and sluices to meet ecohydrological requirements: (1) lower water levels during spring to improve germination of macrophytes, and control rising rates of water levels during spring–summer to ensure development; and (2) open sluice gates to restore migration routes for juvenile fishes migrating into lakes during April–September and for adults migrating into the Yangtze main stem during November–December.

Strategies for Holistic Conservation

Based on conservation principles and our researches, we propose the following strategies to conserve the Yangtze biodiversity.

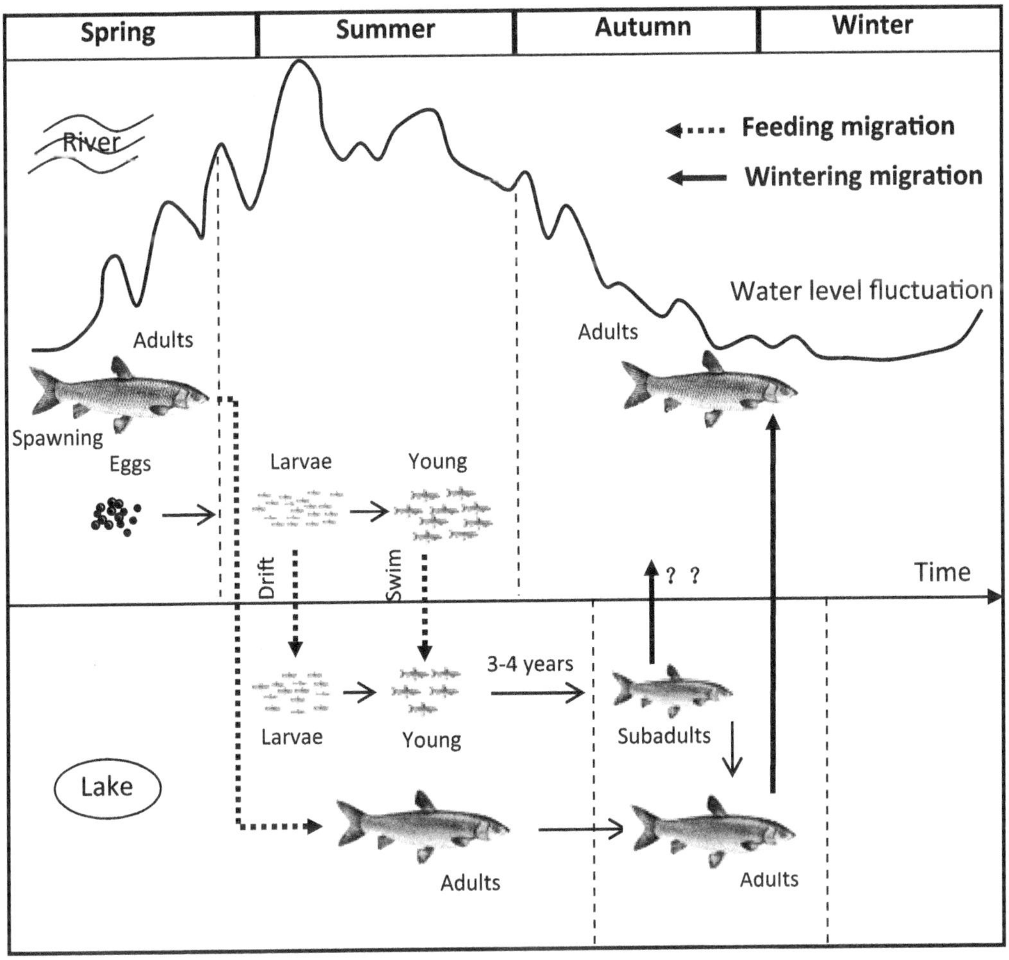

Figure 14. Migration diagram of four major Chinese carps in the Yangtze floodplain. (From Ru and Liu 2013.)

First, establish a holistic nature reserve to protect the entire Yangtze River ecosystem. In the mid to lower river floodplain, the reserve should include (1) the main stem and connected lakes, and (2) the rehabilitation of at least 8,900 km^2 of disconnected lakes.

Second, strengthen conservation actions in the main stem and connected lakes. The existing connected lakes include only two large-sized lakes, Lake Poyanghu and Lake Dongtinghu, and one medium-sized lake, Lake Shijiuhu (see Figure 1). They are biodiversity hotspots. However, fishes and other organisms in these lakes are facing multiple threats, such as overfishing, sand overmining, water pollution, and unusually low water levels in dry seasons because of flow regulations by upstream reservoirs and climate change. Although the Chinese central government has banned fishing in spring in the Yangtze main stem and connected lakes since 2002, the protection achieved appears to be rather limited. Numerous juvenile fish are being caught during open fishing seasons with illegal fishing gears. We propose that commercial fishing in the Yangtze main stem and connected lakes be banned all year round, and a core protected area where all human activities are forbidden should be demarcated. In addition, it is necessary to reoperate the upstream dams to meet the ecohydrological requirements and connectivity of the connected lakes.

Third, rehabilitate disconnected lakes. Those with good conservation potentials are of first consideration, such as oxbows, Honghu and Liangzihu lakes along the middle reaches, and Anqing lakes along the lower reaches. Ecohydrological operation of dams and sluices had better be based on EFR assessments of key organisms (e.g., macrophytes, benthos, fishes, water birds, and dolphins) and ecosystem services (e.g., clean water, food production, and scenery). As stated above, we have done some works in this aspect, but more studies are needed. To meet the urgent need of conservation at present, ecohydrological operation must be implemented as much as possible, even if knowledge is limited. Additionally, enclosures for aquaculture spread over many disconnected lakes, leading to vegetation degradation and then eutrophication. Therefore, these enclosures should be removed completely or only confined to small areas.

Acknowledgments

This work is financially supported by the State Key Laboratory of Freshwater Ecology and Biotechnology (2011FBZ14), the National Natural Science Foundation of China (30900194 and 41371054), the Chinese Academy of Sciences (KZCX1-SW-12), 973 Programs (2008CB418006 and 2002CB412309), and WWF China. The map and area data of Yangtze lakes are from the Scientific Data Sharing Platform for Lake and Watershed made by the Nanjing Institute of Geography and Limnology, Chinese Academy of Sciences. We thank Shu-Ran Wang, Ya-Jing He, Sai-Bo Yuan, Yu Peng, Qing Yu, Xiao-Peng Shi, and Xiao-Ke Zhang for their help.

References

Amoros, C., and G. Bornette. 2002. Connectivity and biocomplexity in waterbodies of riverine floodplains. Freshwater Biology 47:761–776.

CAE (Chinese Academy of Engineering). 2010. Stage assessment of the Three Gorges projects. China Water and Power Press, Beijing. (In Chinese.)

Cao, W. X. 2008. Some issues concerning protecting fish resources in the Yangtze River basin. Resources and Environment in the Yangtze basin 2:163–164. (In Chinese.)

Cao, W. X. 2011. The present situation and protection measures of fish resources in the Yangtze River. Jiangxi Fishery Sciences and Technology 2: 1–4. (In Chinese.)

Cao, Y. 2007. Study on impact factor and technique of vegetation restoration for flood beaches wetlands along Yangtze River. Doctoral thesis. Nanjing Normal University, Nanjing, Jiangsu, China. (In Chinese.)

Chen, J. 2012. Evolution and water resource utilization of the Yangtze River. Changjiang Press, Wuhan, Hubei, China. (In Chinese.)

Cui, L. J., Y. F. Zhai, and G. F. Wu. 2013. Dredging being moved southward enlarges the impacted region in Poyang Lake: the evidences from multi-remote sensing images. Acta Ecologica Sinica 33:3520–3525. (In Chinese.)

CWRC (Changjiang Water Resources Commission of the Ministry of Water Resources). 1999. Atlas of the Changjiang River basin. SinoMaps Press, Beijing. (In Chinese.)

Du, Y., H. P. Xue, S. J. Wu, F. Ling, F. Xiao, and X. H. Wei. 2011. Lake area changes in the middle Yangtze region of China over the 20th century. Journal of Environmental Management 92:1248–1255.

Fang, J. Y., Z. H. Wang, Y. Li, X. W. Li, A. P. Chen, C. Y. Zheng, S. Q. Zhao, S. Rao, Z. R. Tang, S. L. Xu, H. T. Cui, H. Zeng, and F. Jiang. 2006a. Ecological basis of regional biodiversity conservation of wetlands in the central Yangtze. Higher Education Press, Beijing. (In Chinese.)

Fang, M., S. H. Qi, C. X. Wu, L. S. Hu, and Q. K. Su. 2006b. High resolution sedimentary record of organochlorine pesticides in sediments core from Honghu Lake. Geological Science and Technology Information. 25:89–92. (In Chinese.)

Feng, W. S. 2005. Resource and environmental characters of a large river-joining lake, zooplankton. Pages 134–142 *in* Y. B. Cui and Z. J. Li, editors. Fishery resources and conservation of environment in lakes of the Changjiang River basin. Science Press, Beijing. (In Chinese.)

IGSNRRCAS (Institute of Geographic Sciences and Natural Resources Research, Chinese Academy of Sciences), Changjiang Institute of Water Resources and Hydropower Research), and Changjiang Waterway Planning Design and Research Institute. 1985. The mid–lower reaches of the Yangtze River channel characteristics and their evolution. Science Press, Beijing. (In Chinese.)

Junk, W. J. 2005. Flood pulsing and the linkages between terrestrial, aquatic, and wetland systems. Internationale Vereinigung fur Theoretische und Angewandte Limnologie Verhandlungen 29:11-17.

Keddy, P. A., and A. A. Reznicek. 1986. Great lake vegetation dynamics: the effects of fluctuating water levels and buried seeds. Journal of Great Lakes Research 12:25-36.

Li, J., Z. Q. Xia, H. C. Dai, W. Yi. 2013. Effect of the Three Gorges Reservoir initial filling on downstream habitat suitability of the typical fishes. Journal of Hydraulic Engineering 44:892-900. (In Chinese.)

Lin, Z. X. 2003. Analysis of water environmental change in Taihu watershed. Journal of Lake Sciences 14:97-103. (In Chinese.)

Liu, X. Q., and H. Z. Wang. 2008. Food web of benthic macroinvertebrates in a large Yangtze River-connected lake: the role of flood disturbance. Fundamental and Applied Limnology 171:297-309.

Liu, X. Q., and H. Z. Wang. 2010. Estimation of minimum area requirement of river-connected lakes for fish diversity conservation in the Yangtze River floodplain. Diversity and Distributions 16:932-940.

Liu, X. Q., H. Z. Wang, and X. M. Liang. 2006. Food web of macroinvertebrate community in a Yangtze shallow lake: trophic basis and pathways. Hydrobiologia 571:283-295.

Lytle, D. A., and N. L. Poff. 2004. Adaptation to natural flow regimes. Trends in Ecology and Evolution 19:94-100.

Ma, S. S., and Q. Q. Xu. 2013. Key technical issues in sand excavation management in Yangtze River basin. China Water Resources 10:15-18. (In Chinese.)

Mahoney, J. M., and S. B. Rood. 1998. Streamflow requirements for cottonwood seedling recruitment: an integrative model. Wetlands 18:634-645.

MWR (Ministry of Water Resources of China). 2004-2013. China water resources bulletin (2003-2012). China Water and Power Press, Beijing. (In Chinese.)

OECD (Organization for Economic Co-operation and Development). 2006. Eutrophication of waters: monitoring, assessment and control. Prepared by Soil & Water Conservation Society of Metro Halifax. Available: http://lakes.chebucto.org/TPMODELS/OECD/oecd.html. (May 2015).

Pan, B. Z., H. J. Wang, X. M. Liang, and H. Z. Wang. 2009. Phytoplankton chlorophyll a in the Yangtze-connected lakes: present state, influencing factors and empirical relationships. Fresenius Environmental Bulletin 18:1894-1900.

Pan, B. Z., H. J. Wang, X. M. Liang, and H. Z. Wang. 2011a. Macrozoobenthos in Yangtze floodplain lakes: patterns of density, biomass, and production in relation to river connectivity. Journal of the North American Benthological Society 30:589-602.

Pan, B. Z., H. J. Wang, H. Z. Wang, and Z. Y. Wang. 2011b. Macrozoobenthic assemblages in relation to environments of the Yangtze-isolated lakes. Frontiers of Environmental Science and Engineering 6:246-254.

Pan, B. Z., H. Z. Wang, and H. J. Wang. 2014. A floodplain-scale lake classification based on characteristics of macroinvertebrate assemblages and corresponding environmental properties. Limnologica 49:10-17.

Poff, N. L., and D. D. Hart. 2002. How dams vary and why it matters for the emerging science of dam removal an ecological classification of dams is needed to characterize how the tremendous variation in the size, operational mode, age, and number of dams in a river basin influences the potential for restoring regulated rivers via dam removal. BioScience 52:659-668.

Rao, Q. Z., editor. 1956. Basic knowledge of lake survey. Science Press, Beijing. (In Chinese.)

Rood, S. B., G. M. Samuelson, J. H. Braatne, C. R. Gourley, F. M. Hughes, and J. M. Mahoney. 2005. Managing river flows to restore floodplain forests. Frontiers in Ecology and the Environment 3:193-201.

Ru, H. J., and X. Q. Liu. 2013. River-lake migration of fishes in the Dongting Lake area of the Yangtze floodplain. Journal of Applied Ichthyology 29:594-601.

Shen, Z. Y., J. F. Niu, J. Qi, L. L. Wang, Z. W. Tang, and J. L. Feng, editors. 2008. Characteristics and ecological risks of water pollution in typical segments of the mid-Yangtze River. China Environmental Science Press, Beijing. (In Chinese.)

Shi, Y., D. W. Yang, L. Ning, and J. A. Ma. 2011. Assessment of effects of the operation of Three Gorges Dam on the flood control in the mid-lower Changjiang River. Changjiang Press, Wuhan, Hubei, China. (In Chinese.)

Shu, F. Y., H. J. Wang, L. X. Liu, and H. Z. Wang, 2009. Assessment of species status of Mollusca in the mid-lower Yangtze lakes. Acta Hydrobiologia Sinica 33:84–91. (In Chinese.)

STSGDFCR (Survey Team of Spawning Grounds of Domestic Fishes in Changjiang River). 1982. A survey on the spawning grounds of the "four famous Chinese carps" in the Changjiang River after dammed by the key water control project at Gezhouba. Journal of Fisheries of China 6:287–304. (In Chinese.)

Wang, H. J. 2007. Predictive limnological researches on small- to medium-sized lakes along the mid-lower Yangtze River. Doctoral dissertation. University of Chinese Academy of Sciences, Beijing. (In Chinese.)

Wang, H. J., X. M. Liang, P. H. Jiang, J. Wang, S. K. Wu, and H. Z. Wang. 2008. TN:TP ratio and planktivorous fish do not affect nutrient-chlorophyll relationships in shallow lakes. Freshwater Biology 53:935–944.

Wang, H. J., H. Z. Wang, X. M. Liang, and S. K. Wu. 2014. Total phosphorus thresholds for regime shifts are nearly equal in subtropical and temperate shallow lakes with moderate depths and areas. Freshwater Biology 59:1659–1671.

Wang, H. Z., C. L. Song, X. Q. Liu, and K. Li. 2012. Lakeshore overview of Lake Chaohu and ecological rehabilitation schemes for shoreline and littoral zones. Resources and Environment in the Yangtze basin 21(Z2):62–68. (In Chinese.)

Wang, H. Z., and H. J. Wang. 2009. Ecological effects of river-lake disconnection and restoration strategies in the mid-lower Yangtze basin. Pages 379-396 *in* Z. Y. Wang, D. G. Shao, X. J. Shao, Y. J. Lu, and X. H. Duan, editors. Integrated management of water, sediment and ecology in the Yangtze River basin. Science Press, Beijing. (In Chinese.)

Wang H. Z., Z. C. Xie, X. P. Wu, and Y. L. Liang. 1999. A preliminary study of zoobenthos in the Poyang Lake, the largest freshwater lake of China, and its adjoining reaches of Changjiang River. Acta Hydrobiologica Sinica 23(Supplement):132–138. (In Chinese.)

Wang, H. Z., H. J. Wang, X. M. Liang, and Y. D. Cui. 2006. Stocking models of Chinese mitten crab (*Eriocheir japonica sinensis*) in Yangtze lakes. Aquaculture 255:456–465.

Wang, H. Z., H. J. Wang, X. M. Liang, L. Y. Ni, X. Q. Liu, and Y. D. Cui. 2005. Empirical modelling of submersed macrophytes in Yangtze lakes. Ecological Modelling 188:483–491.

Wang, H. Z., Q. Q. Xu, Y. D. Cui, and Y. L. Liang. 2007. Macrozoobenthic community of Poyang Lake, the largest freshwater lake of China, in the Yangtze floodplain. Limnology 8:65–71.

Wang, J., and H. Y. Chen. 2010. Water regime variation of the middle and low reaches of Yangtze River during the water storage period of Three Gorges Reservoir and its countermeasures. China Water Resources 19:15–17. (In Chinese.)

Wang, J. T. 1959. The embankment project of the Yangtze River. Yangtze River 5:61–64.

Wang, S. M., and H. S. Dou. 1998. Lakes of China. Science Press, Beijing. (In Chinese.)

Ward, J. V. 1997. An expansive perspective of riverine landscapes: pattern and process across scales. GAIA - Ecological Perspectives for Science and Society 6:52–60.

Ward, J. V., Malard, F., and K. Tockner. 2002. Landscape ecology: a framework for integrating pattern and process in river corridors. Landscape Ecology 17:35–45.

Wei, X., Y. Du, S. Cai, T. Zhang, and T. Liu. 2007. On the evolution of lakes in Hubei Province and countermeasures. Journal of Lake Sciences 19:530–536. (In Chinese.)

Wu, G. F., and L. J. Cui. 2008. Remote sense-based analysis of the sand dredging impact on water clarity in Poyang Lake. Acta Ecologica Sinica 28:6113–6120. (In Chinese.)

Wu, H. W., and C. C. Jao. 1958. Fish farming in the shallow lakes in China. Verhandlungen des Internationalen Verein Limnologie 13:765–769.

Wu, Z. G., and J. Li. 2010. Seeking truth and being pragmatic for reform and innovation to develop a new situation of sand mining management of Yangtze River. Yangtze River 41:31–36. (In Chinese.)

Xie, P. 2009. Reading about the histories of cyanobacteria, eutrophication and geological evolution in Lake Chaohu. Science Press, Beijing. (In Chinese.)

Xie, S. G., Z. J. Li, J. S. Liu, S. Q. Xie, H. Z. Wang, and B. R. Murphy. 2007. Fisheries of the Yangtze River show immediate impacts of the Three Gorges Dam. Fisheries 32:343–344.

Xu, J. J., and J. Chen. 2013. Study on the impact of Three Gorges Reservoir on Poyang Lake and some proposals. Journal of Hydraulic Engineering 7:3. (In Chinese.)

Xu, Q. Q., H. Z. Wang, and S. P. Zhang. 2003. The impact of overstocking of mitten crab, *Eriocheir sinensis*, on lacustrine zoobenthic community. Acta Hydrobiologica Sinica 27:41–46. (In Chinese.)

Yang, X. D., J. Shen, X. H. Dong, and E. F. Liu.

2006. Nutrition state evolution and lake ecological response of shallow lakes in history period in mid-lower Yangtze River—taking Lake Long'gan and Lake Tai'bai for example. Science in China Series D: Earth Sciences 35(Supplement II):45–54. (In Chinese.)

Yi, B. L., Z. T. Yu, and Z. S. Liang. 1988. Gezhouba water control project and four famous fish in Yangtze River. Hubei Science and Technology Press, Wuhan, Hubei, China. (In Chinese.)

Yu, W. C., and J. Y. Lu. 2005. The evolution and regulation of the Yangtze River. China Water and Power Press, Beijing. (In Chinese.)

Yu, X. D., T. H. Luo, and H. Z. Zhou. 2005. Large-scale patterns in species diversity of fishes in the Yangtze River basin. Biodiversity Science 13:473–495. (In Chinese.)

Zhang, D. B., S. P. He, G. X. Hu, and G. P. Hu. 2013. Preliminary analysis on runoff variation of middle Yangtze River after impoundment of Three Gorges Reservoir. Yangtze River 1: 1–3. (In Chinese.)

Zhang, G., L. Wu, H. T. Li, M. Liu, F. Cheng, B. R. Murphy, and S. G. Xie. 2011. Preliminary evidence of delayed spawning and suppressed larval growth and condition of the major carps in the Yangtze River below the Three Gorges Dam. Environmental Biology of Fishes 93:439–447.

Zhang, X. K. 2013. Water level fluctuation requirements of plants in the Yangtze floodplain lakes. Doctoral dissertation, University of Chinese Academy of Sciences, Beijing. (In Chinese.)

Zhang, X. K., X. Q. Liu, and H. Z. Wang. 2014. Developing water level regulation strategies for macrophytes restoration of a large Yangtze-disconnected lake, China. Ecological Engineering 68:25–31.

Zhang, Z. L., and L. Z. Huang. 2008. Influence of quarrying in Poyang Lake in the ecological environment. Jiangxi Hydraulic Science and Technology 34:7–10. (In Chinese.)

Zhao, J., and C. K. Chen, editors. 1999. China's geography. Higher Education Press, Beijing. (In Chinese.)

Zhao, W. H. 2010. Macroecological patterns of macrozoobenthos in rivers of China and environmental flow requirements in lower reaches of the Yellow River. Doctoral dissertation. University of Chinese Academy of Sciences, Beijing. (In Chinese.)

Zheng, H., P. D. Clift, P. Wang, R. Tada, J. Jia, M. He, and F. Jourdan. 2013. Pre-Miocene birth of the Yangtze River. Proceedings of the National Academy of Sciences of the United States of America 110:7556–7561.

Zhu, K. Z. 1972. A preliminary study of the climatic changes in China since recent 5000 years. Acta Archaeological Sinica 1:15–38. (In Chinese.)

Zhu, M. 1996. Variation trend and protection steps of water quality in Taihu Lake. Journal of Lake Sciences 8:133–138. (In Chinese.)

American Fisheries Society Symposium 84:293–309, 2016

Agriculture in the Mississippi River Basin: Effects on Water Quality, Aquatic Biota, and Watershed Conservation

Yushun Chen*[,1]
Aquaculture and Fisheries Center, University of Arkansas-Pine Bluff
1200 North University Drive, Pine Bluff, Arkansas 71601, USA

Mike Daniels
Division of Agriculture Cooperative Extension Service, University of Arkansas
2301 South University Avenue, Little Rock, Arkansas 72203, USA

Michele Reba
USDA Agricultural Research Service, National Sedimentation Laboratory
504 University Loop East, Jonesboro, Arkansas 72401, USA

Jennifer Bouldin
Department of Biological Sciences, Arkansas State University
Post Office Box 599, Jonesboro, Arkansas 72467, USA

Chris Henry
Rice Research and Extension Center, University of Arkansas
2900 Highway 130 East, Stuttgart, Arkansas 72160, USA

Pearl Daniel
Division of Agriculture Cooperative Extension Service, University of Arkansas
2301 South University Avenue, Little Rock, Arkansas 72203, USA

Sagar Shrestha and Jack Farrelly
Aquaculture and Fisheries Center, University of Arkansas-Pine Bluff
1200 North University Drive, Pine Bluff, Arkansas 71601, USA

Dennis Frame
University of Wisconsin-Madison Discovery Farms
40195 Winsand Drive, Pigeon Falls, Wisconsin 54760, USA

Dennis Busch
University of Wisconsin-Platteville Pioneer Farm
1 University Plaza, Platteville, Wisconsin 53818, USA

Amber Radatz
University of Wisconsin-Madison Discovery Farms
40195 Winsand Drive, Pigeon Falls, Wisconsin 54760, USA

* Corresponding author: yushunchen@ihb.ac.cn
[1] Current address: Institute of Hydrobiology, Chinese Academy of Sciences, 7 Donghu South Road, Wuhan, Hubei 430072, China

Eric Cooley
University of Wisconsin-Madison Discovery Farms
4319 Expo Drive, Manitowoc, Wisconsin 54221, USA

Beatrix Haggard[2]
LSU AgCenter
212A Macon Ridge Road, Winnsboro, Louisiana 71295, USA

James Hendrix and Donnie Miller
LSU AgCenter
107-A Arts Drive, St. Joseph, Louisiana 71366, USA

Abstract.—Agriculture has been identified as a potential leading source of nutrients (nitrogen and phosphorus) and sediment enrichment of water bodies within the Mississippi River basin (MRB) and contributes to impaired water quality and biological resources in the MRB and the northern Gulf of Mexico (GOM). This study reviewed agriculture, impacts on water quality and biological resources, and a brief introduction of watershed conservation programs in the MRB. Agriculture has increased nutrients and sediment loads to the Mississippi River and the northern GOM since the 1950s. Fish and macroinvertebrate communities have shifted, and low oxygen and high-turbidity-tolerant groups became dominant. In addition to existing conservation practices such as the Conservation Reserve Program through the 1985 farm bill and other related programs (e.g., the Wetlands Reserve Program), a recent basin-wide conservation initiative—the Mississippi River Basin Healthy Watersheds Initiative (MRBI)—was launched by U.S. Department of Agriculture Natural Resources Conservation Service in 2010. The MRBI provides financial incentives (more than US$222 million) to producers and landowners in 640 watersheds of 13 states to implement voluntary conservation practices that improve water quality, restore wetlands, enhance wildlife habitat, and sustain agricultural profitability. Edge-of-field and watershed monitoring have been initiated through the MRBI and related agricultural conservation programs such as Section 319 of the U.S. Environmental Protection Agency Clean Water Act and new initiatives such as the Discovery Farms program in Arkansas, Wisconsin, Minnesota, and North Dakota, Pioneer Farm in Wisconsin, the Louisiana Master Farmer Program in Louisiana, and others in the MRB states. These efforts will greatly improve downstream watershed ecosystem health by avoiding, controlling, and trapping nutrient and sediment runoff from agricultural fields to the Mississippi River and GOM. Although there continues to be problems with nutrient transport, sedimentation, and depleted groundwater supplies, agriculture will likely have less influence on the future ecological health condition of the Mississippi River and GOM. Future restoration programs need to focus more on state or regional coordination by classifying restoration projects and standardizing the geographic scale and evaluation methods across the whole MRB.

[2] Current address: Department of Plant and Soil Sciences, Oklahoma State University, 371 Agricultural Hall, Stillwater, Oklahoma 74078, USA

Introduction

The Mississippi River basin (MRB) is the largest river basin in North America and the third largest river basin in the world, covering roughly 3.2 million km^2 (Van der Leeden et al. 1990; Foley et al. 2004; Figure 1). About 70 million people live within the MRB, which includes all or parts of 30 U.S. states (Goolsby et al. 1999). The main stem of the Mississippi River originates in northern Minnesota and flows southward more than 3,700 km to the Gulf of Mexico (GOM) (Goolsby et al. 1999; Figure 1). At St. Louis, Missouri, the average flow of the Mississippi River is almost doubled from about 3,530 to 6,790 m^3/s downstream of its confluence with the Missouri and Illinois rivers (Goolsby et al. 1999). About 320 km downstream from St. Louis, its average flow is more than doubled again to about 15,340 m^3/s below the confluence with the Ohio River. From its confluence with the Ohio River to 860 river kilometers downstream at Vicksburg, Mississippi, its average flow increases to 17,400 m^3/s (or 541 $km^3/year$), with most of this increase coming from the Arkansas and White rivers (Goolsby et al. 1999). The Atchafalaya River is a major distributary of the Mississippi and discharges to the west of the Mississippi River, and carries approximately 34% of the total Mississippi–Atchafalaya system flow (Dagg and Breed 2003).

Agriculture in the Mississippi River Basin

The MRB is one of the most productive farming regions in the world, producing the majority of the corn, soybeans, wheat, rice, cattle, and hogs grown in the United States, and about 58% of the basin is cropland (Goolsby et al. 1999; USDA NRCS 2013). During the 1950s to 1970s, soybean prices reached all-time highs, and the combination of publicly funded flood-control measures and the prevailing farm economies further facilitated widespread forest clearing in the Mississippi River alluvial valley for agriculture (King et al. 2006). The United States produces approximately 40% of the world's corn, with the majority coming from the MRB (USDA NRCS 2013). Corn, soybeans, and wheat dominate the MRB (Foley et al. 2004). In the eastern half of the MRB, rain-fed corn and soybean cropping systems, grasslands, pastures, and forests dominate (Foley et al. 2004). In the western portion of the basin, rain-fed wheat, irrigated (corn, soybean) cropping systems, grasslands, and shrublands dominate (Foley et al. 2004). One of the most significant changes in the MRB land cover from 1950 to present was the expansion in soybean and corn cultivation. Most of this change occurred within the central United States and the lower Mississippi valley where soybeans were cultivated in rotation with nitrogen-demanding corn (Donner 2003; Foley et al. 2004). Thus, currently the traditional Corn Belt, which is from Nebraska east to Ohio (Foley et al. 2004), is expanding to the lower Mississippi valley. Moreover, in the Mississippi alluvial plain ecoregion (Omernik and Griffith 2012; Figure 2), wetlands accounted for more than 50% of the land use in the past but dropped to 6%, and the majority of land use (80%) in this ecoregion is agriculture (Caskey 2003). Floodplain habitat resources, hydrology, and water quality conditions are transformed by agricultural development. The agricultural development includes tile drainage, precision land leveling, stream channelization, filling of wetlands, excavation of ditches, flood control, and water table manipulation (Shields et al. 2013). In addition to the land-use changes, crop yields in the MRB have increased dramatically, which are caused by changes both in farm management and climate (Foley et al. 2004). One of the most important changes in farm management has been the dramatic increase in industrial fertilizer use (Foley et al. 2004). For instance, corn has stimulated a nearly 20-fold increase in fertilizer application rates since 1950 (Foley et al. 2004).

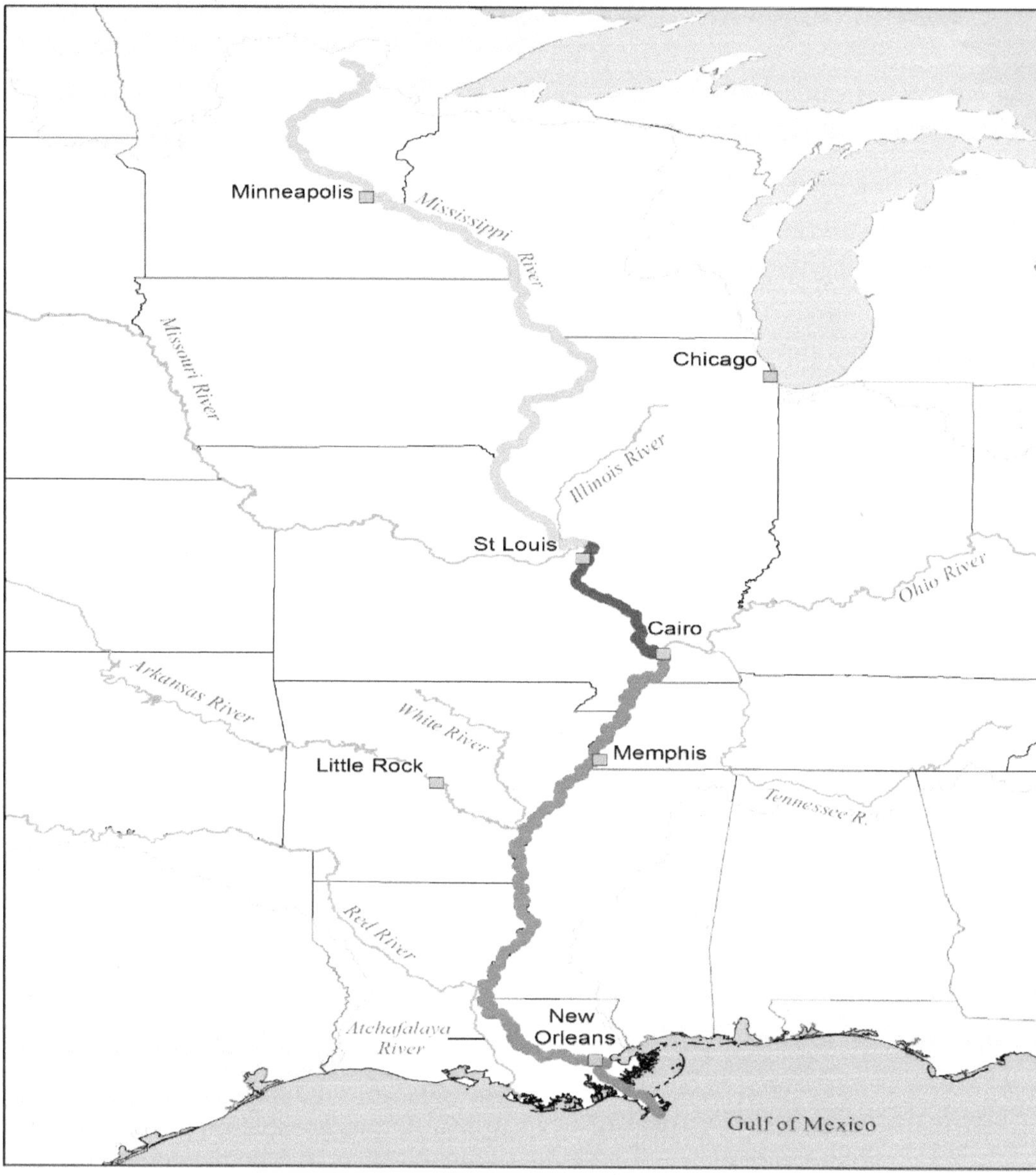

Figure 1. Location of the Mississippi River basin and the Gulf of Mexico.

Agricultural Pollution and Its Impacts on Water Quality and Biota in the Mississippi River Basin and Northern Gulf of Mexico

Agricultural Pollution in the Mississippi River Basin

The majority of all the fertilizers (with nutrients of N and P) and pesticides used in the United States are applied to cropland within the MRB because of the large-scale agriculture (Goolsby et al. 1999). These applied fertilizers are potential sources of nutrient (N and P) pollution from agriculture in the MRB. Agricultural sources of N and P dominate inputs in the MRB to affect the North Atlantic Ocean (Howarth et al. 1996; Goolsby et al. 1999). The Mississippi River delivers approximately 1.82×10^9 kg N/

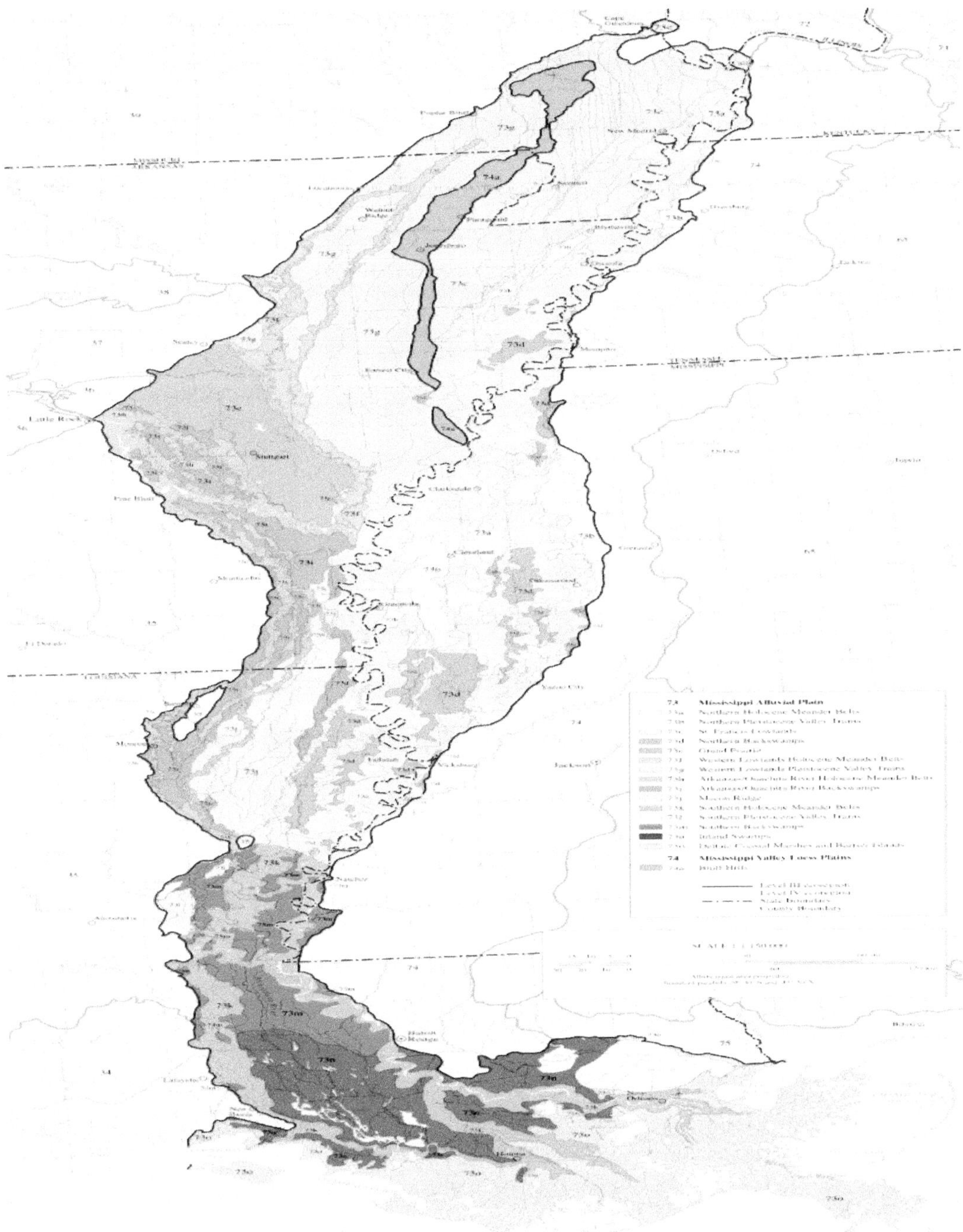

Figure 2. Location of the Mississippi alluvial plain ecoregion (Omernik and Griffith 2012).

year (1.82 million metric tons N/year or 1.3 × 10^{11} mol N/year) and 0.107 × 10^9 kg P/year (3.4 × 10^{11} mol P/year) to the northern GOM (Howarth et al. 1996). The total N discharged is about 61% nitrate, 37% organic N, and 2% ammonium (Table 1; Goolsby et al. 1999). On a yearly average, the total dissolved nitrogen is comprised of 53% nitrate,

Table 1. Discharge of nutrient and sediment from the Mississippi River to the Gulf of Mexico. (Sources: Turner and Rabalais 1991; Meade 1995; Howarth et al. 1996; Goolsby et al. 1999; Dagg and Breed 2003.)

Category	Total	Nitrate	Ammonia	Organic
Nitrogen (10^9 kg N/year)	1.82	1.11	0.67	0.04
Phosphorus (10^9 kg P/year)	0.107			
Sediment (10^9 kg/year)	150			

43% organic N, and 4% ammonium (Turner and Rabalais 1991; Dagg and Breed 2003).

The total N loading to the GOM was doubled from 1954–1959 to 1980–1985 (Turner and Rabalais 1991). For example, nitrate flux from the MRB to the GOM has averaged nearly 1 million metric tons per year since 1980 and is about three times larger than it was in the early 1950s (Goolsby et al. 1999). From 1981 to 1987, annual N and P loading averaged 130.3 million kg atoms N and 3.44 million kg atoms P to the GOM, respectively (Goolsby et al. 1999). This large input dominates the biological processes of the region and stimulates high levels of phytoplankton production, which in turn support high rates of bacterial production, protozoan and metazoan grazing, and fisheries production (Dagg and Breed 2003).

The principal source areas for the N that discharges to the GOM are watersheds draining intensively cultivated regions in southern Minnesota, Iowa, Illinois, Indiana, and Ohio. These regions contribute several times more N per unit area than other areas in the MRB (Goolsby et al. 1999). For example, streams draining Iowa and Illinois contribute as much as 35% of the total N flux of the Mississippi River during years of average rainfall, and much more during years of high rainfall. However, these two states comprise only about 9% of the area of the Mississippi–Atchafalaya basin (Goolsby et al. 1999).

In addition to nutrient exports, soil erosion from agricultural activities contributes to sedimentation in the downstream water bodies. The Mississippi River discharges an average of 1.5×10^{11} kg/year (150 million metric tons/year) sediment into the northern GOM (Table 1; Meade 1995). Prior to discharging into the gulf, nitrate concentration in the Mississippi River is usually more than 100 μM. But high concentrations of suspended particulate materials absorb and scatter light, and their mixing minimizes the time any phytoplankton cell will be in a surface, lighted environment (Dagg and Breed 2003).

Impacts on Habitat and Water Quality

Habitat

In Wisconsin, stream habitat quality was negatively correlated with the amount of agricultural land in both an entire watershed and a 100-m-wide buffer along the stream in 103 streams (Wang et al. 1997). Model results also showed that stream habitat quality started to decline when watershed agricultural land use exceeded 50% in those Wisconsin streams (Wang et al. 1997). In the Yazoo River basin of northwestern Mississippi, associated floodplain habitats exhibit long-term degradation due to upstream impoundment, channelization, and floodplain cultivation in one studied reach of the Coldwater River (Shields et al. 2013). For instance, water depth in floodplain lakes was one of the habitat parameters affected by agricultural disturbance in the Yazoo River basin (Miranda 2011). In the Mississippi Delta streams, floodplain deposits are rich in organic material and stream channels are largely altered by intense agricultural activities (Killgore et al. 2008). In southwestern Louisiana, land and water-use practices associated with agricultural activities have influenced habitat conditions by modifying natural channel geomorphology (e.g., aggradation of sediment in streams, channelization, and clearing of riparian corridors), simplifying stream microhabitat diversity, and reducing the availability of riparian food resources (Mize et al. 2008).

Overall water quality.—Agriculture practices can reduce the ability of ecosystems to provide goods and services. For example, high applications of fertilizers and pesticides can increase nutrients and toxins in groundwater and surface waters, increase health and water purification costs, and decrease fishery and recreational values (Tilman et al. 2002). The intensification of cultivation of three primary crops (corn, soybean, and wheat) has altered land cover and water quality across the United States since 1950 (Donner 2003). One serious consequence of agricultural activities in the MRB is increased nutrients in groundwater systems where it poses a significant human health threat and in surface aquatic ecosystems where it can cause eutrophication and hypoxia (Foley et al. 2004). In the delta area of Arkansas (i.e., the Arkansas part of the Mississippi alluvial plain ecoregion, Figure 2), for example, water quality is significantly influenced by non-point-source runoff from its highly agriculturalized areas (ADEQ 2012). In northwestern Mississippi, the majority of streams are on the 303(d) list of water-quality-limited waters, and agricultural effects on streams have reduced the number of unimpaired streams for water quality comparisons (Bryson et al. 2007). In addition to surface water-quality issues, the groundwater in the agricultural region can be affected as well. For instance, because of the intense irrigation in the Mississippi alluvial plain ecoregion, both the quality and quantity of groundwater could be affected when the groundwater pumping rates exceed the aquifer recharge rates (Czarnecki et al. 2002).

Nutrients (nitrogen and phosphorus).—Model simulations indicate that agricultural sources in watersheds contribute more than 70% of the delivered N and P in the MRB (Alexander et al. 2008). The high present-day rate of fertilizer use on corn and the expansion of soybean cultivation could be largely responsible for the change in nutrient ratios (e.g., N:P) in the Mississippi River since 1950 (Donner 2003). About 42% of the N fertilizer and 37% of the P fertilizer used annually in the United States from 1981 to 1985 was applied in states that are partially or completely in the Mississippi River watershed, where applications equal 4.2 million metric tons of N and 0.53 million metric tons of P (Turner and Rabalais 1991). It was estimated that a maximum 44% of the applied N and 28% of the applied P fertilizers in these states may have made its way to the GOM via the Mississippi and Atchafalaya rivers (Table 2). The largest contribution of N (52%) to the GOM is from the cultivation of corn and soybeans while P originates primarily from animal manure on pasture and rangelands (37%) (Alexander et al. 2008).

In a watershed, the area of corn was strongly related to the concentration of total N and the ratio of dissolved inorganic N and organic N at the watershed outlet (Donner 2003). Even with recommended fertilizer and land-conservation measures, losses of 10–40 kg/ha/year were measured (Table 2; Simpson et al. 2008). As a result of increased corn acres, N and P loss to water increased by 37% (117 million kg) and 25% (9 million kg), respectively (Simpson et al. 2008). Average loss of P in runoff from corn is 2–15 kg/ha/year (Table 2; Carpenter et al. 1998; Sharpley et al. 2001; Simpson et al. 2008). Model results show that the increase in corn cultivation required to meet the goal of $1.5–3.6 \times 10^{10}$ gallons of renewable fuels by the year of 2022 suggested by a recent U.S. Senate energy policy would increase the annual average flux

Table 2. Loss of nutrient from agricultural watersheds of the Mississippi River basin. (Sources: Carpenter et al. 1998; David and Gentry 2000; Sharpley et al. 2001; Simpson et al. 2008)

Category	Nitrogen	Phosphorus
Percent applied fertilizer loss (%)	44	28
Nutrient loss from corn fields (kg/ha/year)	10–40	2–15
Percent nutrient contribution in agriculture dominated watersheds (%)	86	67

of dissolved inorganic N export by the Mississippi and Atchafalaya rivers by 10–34% (Donner and Kucharik 2008).

Moreover, the form of N and P fertilizers used has changed during the past several decades. In 1929, N fertilizer was 48% ammonia and 19% nitrate and the balance mostly organic N. By 1949, the proportion was 79.5% ammonia and 12% nitrate, while the percent ammonia has remained the same or increased slightly since then (Turner and Rabalais 1991).

Average nitrate concentration in the lower Mississippi River and N fertilizer use in the United States is strongly related after 1960, when N fertilizer use increased beyond 2 million metric tons per year (Turner and Rabalais 1991). Nitrate concentrations in the Mississippi River and some of its tributaries in the upper Midwest have increased two- to fivefold in the 1990s (Goolsby et al. 1999). Highly fertilized croplands in the central Corn Belt and along the lower Missouri and Mississippi rivers now act as hot spots of nitrate leaching, particularly during the late spring (Foley et al. 2004). Nitrate leaching has contaminated the well water in many areas of the Midwestern United States and in other parts of the world (Schlesinger et al. 2006). In Iowa, nitrate-nitrogen concentrations were directly related to the percentage of row crop in both large (1,002–2,774 km^2 with row crop percentage from 24 to >87%) and small (47–775 km^2 with row crop percentage from 28% to 87%) watersheds (Schilling and Libra 2000).

Linear regression models suggested that average annual surface water nitrate concentrations in Iowa, and possibly similar agricultural areas in the Midwestern United States, can be approximated by multiplying a watershed's row crop percentage by 0.1 (Schilling and Libra 2000). In Illinois, large inputs of N were found beginning about 1965, coinciding with increased N fertilizer applications (about 800,000 Mg N/year). The N input (about 400,000 Mg N/year) was 8.6 million Mg N for the 1979 to 1996 crop years, with a corresponding riverine flux of 4.4 million Mg N (51% of net anthropogenic inputs discharged by rivers) (David and Gentry 2000). In less-populated areas dominated by agriculture, such as northwestern, central, and southern Illinois, it was estimated that about 86% of the N and 67% of the total P were from agricultural surface runoff and subsurface tile drainage flow (David and Gentry 2000). In Wisconsin, soil-test P level was related to dissolved P concentration in runoff for most of the rainfall simulation tests in corn fields (Andraski and Bundy 2003). The higher intensity and volume of rain events from climate change combined with increased flow from other rivers may be some of the main factors of the higher loading of sediment and some nutrients to the northern GOM. It has also been documented that the majority of the total nitrate load of the GOM originated north of the mouth of the Ohio River (Rabalais et al. 2002a).

Nutrients leaching and transport mechanism.—There was a rapid decline in the average first-order rate of N loss with channel size from 0.45/d in small streams to 0.005/d in the Mississippi River (Alexander et al. 2000). The proximity of sources to large streams and rivers was an important determinant of N delivery to the estuary in the MRB (Alexander et al. 2000). Model simulations demonstrated that three factors led to the doubling of nitrate export by the Mississippi River since 1960: (1) an increase in fertilizer application rates, particularly on corn; (2) an increase in runoff across the basin; and (3) the expansion of soybean cultivation (Donner et al. 2004). Increasing base flow and base flow percentage over the second half of the 20th century in Iowa has contributed to increasing nitrate-nitrogen concentrations measured in Iowa rivers because nitrate was primarily delivered to streams as base flow and tile drainage (Schilling 2005). Results from 11 Iowa rivers and their watersheds indicated that increasing base flow in Iowa's rivers was significantly related to increasing row crop intensity (Schilling 2005). A 13–52% increase in row crop percentage in many Iowa watersheds has contributed to an increase of 33–135 mm in base flow and a 7–31% increase in base flow percentage (Schilling 2005). A variety of factors can influence the rate of leaching of N from agricultural systems into surface waters, including water runoff, rate of fertilizer application, type and texture of soil, and the type of land use (Howarth et al. 1996).

First, drainage systems play a big role in nutrient transport. In the Midwestern United States, drainage water from fields is typically conveyed through a network of tile lines, generally located approximately 0.6–1.0 m below the soil surface, to managed drainage ditches. The ditches then convey the water to natural streams or rivers (Smith and Pappas 2007). The greatest density of subsurface drainage occurs in the Midwestern United States, which coincides with the general region with the most potentially productive soils in the nation (Smith and Pappas 2007). In Illinois watersheds, artificial drainage through underfield tiles was the primary mechanism for nitrate export (Royer et al. 2006). Based on an average effective drainage area of 30 ha in the Embarras River watershed of Illinois, 38 and 64 kg N/ha leached out of the watershed through tile drainage for a total of 3.1 Mg N for the 1995 and 1996 water years, respectively (Gentry et al. 1998). Differences in annual N export for each tile were the result of a combination of factors, including timing and area of N fertilization, amount and distribution of precipitation, crop uptake of soil-derived N, and inorganic N pools remaining after harvest (Gentry et al. 1998). Transport of P species showed different routes. For instance, tile drainage and overland flow contributed to dissolved reactive phosphorus export, whereas export of particulate P was almost exclusively from overland flow (Royer et al. 2006). Dredging can also affect nutrient removal in agricultural watersheds. In the Midwestern United States, bed material collected from drainage ditches prior to dredging was better able to remove nitrate, ammonium, and soluble phosphorus from water than material collected from the bed of the ditches after dredging (Smith and Pappas 2007), as the latter may have less interaction between water column and the bed material.

Discharge is another factor affecting nutrient export. For instance, tile N export from the watershed was greatest during high-flow events when there concurrently existed large pools of soil inorganic N in the form of nitrate (Gentry et al. 1998). Analyses also revealed that nearly all nutrient export occurred when discharge was larger than the median discharge, and extreme discharges (≥90th percentile) were responsible for more than 50% of the nitrate export and more than 80% of the P export (Royer et al. 2006). Positive relationships between suspended sediment concentration and water discharge have been found in a small (98 km^2) agricultural watershed in east central Illinois (Borah et al. 2003). Positive relationships of phosphate, atrazine, and metolachlor concentrations with water discharge were also found in different sizes (98, 622, and 932 km^2) of agricultural watersheds in east central Illinois (Borah et al. 2003). A related factor is storm size. For instance, a significant relationship was found between stormflow dissolved reactive phosphorus concentration and storm size (Sharpley et al. 2008).

Crops can also affect nutrient discharge. In the MRB, fields with more than 50% of total N additions from legume N sources and fields with complex crop rotations that included both annual and perennial species were approximately in N balance (3.7 and 5.7 kg N/ha/year, respectively) compared to fertilizer-based practices in corn–soybean rotations with average annual surpluses near 35 kg N/ha/year (Blesh and Drinkwater 2013). Nutrient exports have a specific time period during the agriculture operation. For instance, in Illinois and other intensively drained and farmed watersheds of the upper Midwest, nutrient exports occur annually beginning in mid-January and continue through June (Royer et al. 2006). However, prior to entering the GOM, large particles settle out from the water column in the lower Mississippi River, which was supported by observed declines in particulate organic C (42%), particulate N (57%), total suspended matter (43%), and chlorophyll-*a*-containing particles greater than 20 μm (39%) (Dagg et al. 2005). In addition, nutrient sequestration in lower Mississippi River backwaters could be one reason for declines of nutrients in the lower Mississippi River main stem.

Organic materials and sediments.—Agricultural land use has increased sediment loading in rivers in the MRB (Foley et al. 2004). Deep sediments in streams and lakes are one of the major impairments that caused by agriculture. For instance, in the Yazoo River basin, agricultural activities have reduced water depth of lakes, which eventually affected the aquatic communities (Miranda 2011). The sediment load of the Mississippi River to the GOM has been reduced by more than 50% (Meade 1996). In southeastern Arkansas where row crop agriculture is dominant, the majority of the waters in this area have been severely altered by channelization, ditching, and rerouting the drainage patterns. Some of the stream reaches have been evaluated as not meeting their respective turbidity and mineral standards (ADEQ 2012). Organic carbon from the Mississippi and Atchafalaya rivers was one of the major contributing carbon sources in northern GOM (Rabalais et al. 2002a).

Runoff and floods.—Landscape sensitivity to precipitation and running water can be accelerated by agricultural land use because the protective effects of land cover are depleted compared with pre-agriculture (Knox 2001). High-resolution floodplain stratigraphy of the past two centuries showed that accelerated runoff associated with agricultural land use has increased the magnitudes of floods across a wide range of recurrence frequencies (Knox 2001). Large floods have been particularly important to the movement and storage of sediment in the floodplains of the upper Mississippi valley, where tiling and channelization have increased drainage efficiency and probably have contributed in part to the occurrence of large floods (Knox 2001). The evidence from the upper Mississippi valley showed that agricultural land use increased peak discharges from high-frequency floods by about 200–400% above their pre-agricultural magnitudes in small watersheds (Knox 2001). Artificial drainage was identified as the largest driver of increased flow. For instance, the majority of the increase in flow in the upper MRB watersheds was attributed to changes in water residence time on the landscape and subsequent reductions in evapotranspiration resulting from installation of artificial drainage networks (Schottler et al. 2014). In the lower MRB, agriculture plays less of a role in runoff and floods. For instance, in 2011, the lower MRB experienced one of the largest floods in recorded history but was mostly contained by levees (Turnipseed et al. 2011). Flooding in the tributaries is mostly controlled by reservoirs. Agriculture in the lower MRB contributed less to these flood events compared to those in the upper MRB.

Hypoxia.—Seasonally severe and persistent hypoxia, or low dissolved oxygen concentration (DO < 2 mg/L), occurs on the inner- to mid-Louisiana continental shelf (depths of 5 to 60 m) to the west of the Mississippi River and Atchafalaya River deltas (Rabalais et al. 2001). It is the second largest zone of coastal hypoxia in the world (Rabalais et al. 2002b). Hypoxia occurs below the pycnocline from as early as late February through early October but is most widespread, persistent, and severe in June, July, and August (Rabalais et al. 2001). Depending on the depth of the water, hypoxia may encompass from 10% to more than 80% of the total water column. In cases of the higher percentages above, hypoxia may reach to within 2 m of the surface in the 10-m water column or to within 6 m of the surface in a 20-m water column (Rabalais et al. 2001). The general consensus is that hypoxia in the northern GOM is caused primarily by algal production stimulated by excess N delivered from the Mississippi–Atchafalaya River basin and seasonal vertical stratification of incoming stream flow and Gulf waters, which restricts replenishment of oxygen from the atmosphere (Alexander et al. 2000; Foley et al. 2004; Donner and Kucharik 2008; Bianchi et al. 2010). The higher intensity and volume of rain events from climate change combined with increased flow from other tributaries could be some of the main factors of the higher loading of sediment and some nutrients (e.g., from agriculture or other sources in the MRB) could contribute to the hypoxia of GOM (Turnipseed et al. 2011). An analysis of diatoms, foraminiferans, and carbon accumulation in the sedimentary record provides evidence of increased eutrophication, and hypoxia in the Mississippi River delta might be coincident with changes in N loading (Rabalais et al. 2001). Model results showed an increasing trend in hypoxic thickness and significant increases in hypoxic volume and area from 1985 to 2011 in the northern GOM (Obenour et al. 2013).

Impacts on Aquatic Biota

Aquatic ecosystems have been substantially modified and impacted by land-use, hydrologic, and water-quality changes associated with the extensive development of agricultural surface and subsurface drainage. The cumulative effects of changes occurring over the past century have been as follows: widespread declines in many intolerant species, dramatic shifts in the composition of aquatic communities, and homogenization of aquatic faunal assemblages toward more tolerant, generalist species (Blann et al. 2009). In a nationwide structural equation model, cropland affected benthic communities by both altering structural habitats and imposing water-quality-related stresses (Riseng et al. 2011). In an agricultural landscape, low water depth and deep sediments in streams and rivers are main stressors for aquatic biota (Killgore et al. 2008).

Upper Mississippi River basin.—In Wisconsin, fish index of biotic integrity (IBI) scores were negatively correlated with the amount of agricultural land in the entire watershed and in a 100-m-wide buffer along the stream in 103 studied streams (Wang et al. 1997). Also, fish IBI scores started to decline when watershed agricultural land use exceeded 50% (Wang et al. 1997). In northern Midwestern agricultural streams, instream habitat and physical conditions explained 14–50% of the variability in fish community composition (Talmage et al. 2002). When the dominant substrate type varied from silt, fish community descriptors (e.g., species richness, community diversity, and total abundance) in the Red River basin in northwestern Minnesota and eastern North Dakota and the Minnesota River basin in southern and southwestern Minnesota responded positively with higher IBI scores (Talmage et al. 2002). For instance, woody debris has a positive effect on various fish community descriptors. The woody debris and overhanging vegetation can provide fish communities with cover, temperature stabilization, a food source, and reduced fine sediment (Talmage et al. 2002).

Lower Mississippi River basin.—In the lower MRB, most delta streams (i.e., streams formed in the floodplains of large river systems) have been altered through intense agricultural activities and flood control measures (Killgore et al. 2008). Total suspended solids from typical agricultural runoff mixtures played a very important role in contributing to periphyton changes in the Mississippi alluvial plain streams (Murdock et al. 2013). In the Mississippi alluvial plain ecoregion, percent of deciduous forest was positively correlated with fish community Shannon diversity ($r = 0.4694$) while percent corn was positively correlated with the number of intolerant taxa ($r = 0.5382$), the number of minnow taxa ($r = 0.4749$), and the relative abundance of insectivores ($r = 0.6114$) (Caskey 2003). In the Arkansas delta ecoregion, the fishery is impacted and in an altered form in the majority of these waters, as they are within the agriculture landscape (ADEQ 2012). Once a natural stream has been channelized, only those organisms that do not require instream cover and can exist in highly turbid waters will survive (ADEQ 2012). Within these systems, the fishable goal of the Clean Water Act is still being met, even though the aquatic life communities have been substantially altered (ADEQ 2012). In the upper Steele Bayou system of west central Mississippi, Killgore et al. (2008) found that managing the sediments (e.g., removal, prevention from drop pipes) without restoring perennial flow conditions could hamper the recovery of ichthyofauna. In southwestern Louisiana streams, the abundance and taxa richness of macroinvertebrate communities declined significantly with increases in concentrations of fipronil (a broad-use insecticide and chemical from the phenylpyrazole family) compounds and rice-cultivation land-use intensity (Mize et al. 2008). Macroinvertebrate community tolerance scores also increased linearly with increases in the percentage of rice cultivation in the basins (Mize et al. 2008). As rice-cultivation intensity and concentration of fipronil compounds increased, the most notable response in macroinvertebrate-community structure was a shift in dominance from insects (midges, mayflies, and caddisflies) to noninsects (scuds, snails, and worms) (Mize et al. 2008). In streams of the Pine Hills region of Louisiana, agricultural land use in riparian corridors and associated reduced water quality resulted in disturbances that reduced mussel diversity and abundance (Daniel and Brown 2013).

Gulf of Mexico.—The hypoxia zone or dead zone refers to when the oxygen concentration falls below a critical level in water near the seabed and results in the failure to capture fish, shrimp, and crabs in bottom-dragging trawls (Renaud 1986; Rabalais et al. 2002b; Figure 3). The severely stressed seasonal hypoxic and anoxic zone of the Louisiana inner shelf in the northern GOM loses many higher taxa during the peak of hypoxia (Rabalais and Turner 2001). A hypoxia-stressed benthos is typified by short-lived, smaller surface deposit-feeding polychaetes and the absence of marine invertebrates such as pericaridean crustaceans, bivalves, gastropods, and ophiuroids (Rabalais et al. 2002b). Nitrogen loading to estuaries and coastal waters is responsible for overenrichment, eutrophication, and low-oxygen conditions that endanger fisheries (Tilman et al. 2002). Important fisheries were variably affected by increased or decreased food supplies, mortality, forced migration, reduction in suitable habitat, increased susceptibility to predation, and disruption of life cycles (Rabalais et al. 2002b). The numbers of stressed or dying benthic infaunal organisms within the sediments increased substantially when the oxygen levels remained low for prolonged periods. In addition, biodiversity was diminished and community structure and ecosystem functions were altered (Rabalais and Turner 2001; Rabalais et al. 2002b). These

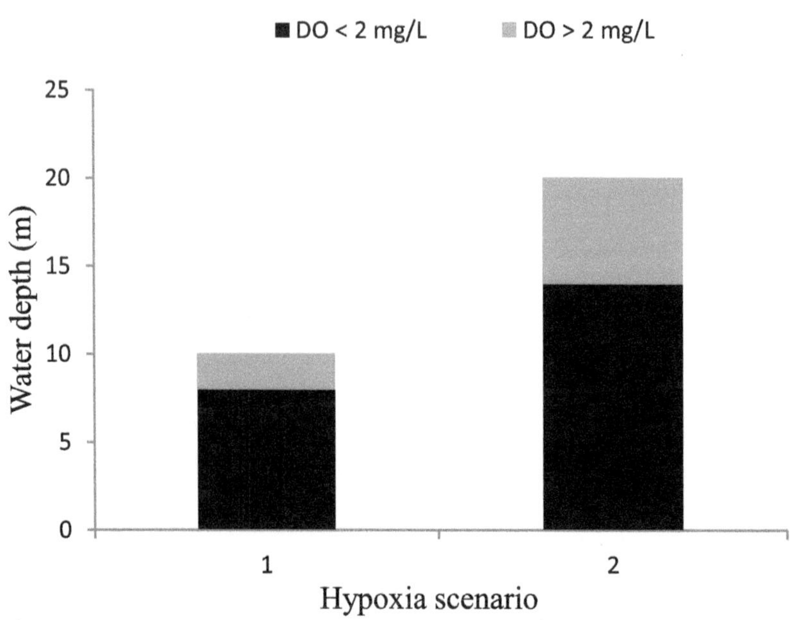

Figure 3. Hypoxia scenarios and depth profiles (Rabalais et al. 2001). Gray is the high oxygen surface water and black is the hypoxic bottom water.

stresses can cause behavioral and physiological responses in marine organisms. Behavioral responses include active avoidance, a response only available for swimming organisms versus those attached to substrates or associated with the bottom, and reduced feeding. Physiological adaptations include dormancy, reduction in molting and growth rates, increased ventilation rate, increased circulation of blood past respiratory organs, increased production of hemoglobin or other binding proteins, and anaerobic metabolism (Figure 4; Rabalais et al. 2002b).

Watershed Conservation Efforts

To reduce effects of agricultural activities on downstream water quality and biological resources, the U.S. federal government has been very proactive in providing technical and financial support to farm operators for a wide range of voluntary conservation practices in the MRB (Lambert et al. 2006). For instance, the 1985 farm bill established the Conservation Reserve Program to take marginal and erodible crop land out of production and place it into permanent cover (Food Security Act of 1985). Other similar programs include the Wetlands Reserve Program through the 1990 farm bill, the Wildlife Habitat Incentives Program and the Environmental Quality Incentives Program in 1996, the Conservation Reserve Enhancement Program in 1997, and the Farmable Wetlands Program under the 2002 farm bill. Maintenance and restoration of small streams become a high priority of land management, both rural and suburban (Schlesinger et al. 2006).

Since 2010, the U.S. Department of Agriculture Natural Resources Conservation

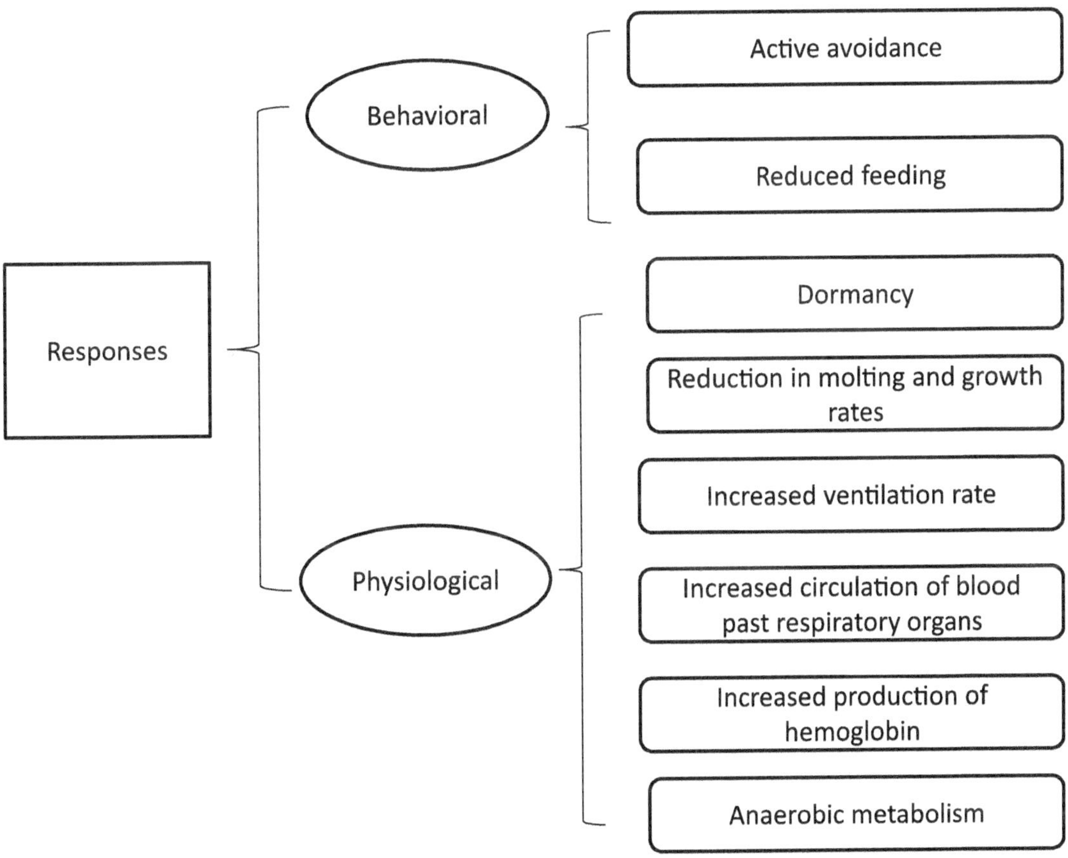

Figure 4. Behavioral and physiological responses of marine organisms to hypoxia (Rabalais et al. 2002b).

Service (NRCS) and partners have worked with producers and landowners through the Mississippi River Basin Healthy Watersheds Initiative (MRBI) to implement voluntary conservation practices that improve water quality, restore wetlands, enhance wildlife habitat, and sustain agricultural profitability in the MRB (USDA NRCS 2010). To achieve these MRBI goals, NRCS works with producers to avoid, control, and trap nutrient and sediment runoff and to maintain or improve agricultural productivity (USDA NRCS 2010). The MRBI is implemented by NRCS through the Conservation Partnership Initiative, the Wetlands Reserve Enhancement Program, Conservation Innovation Grants, and other programs (USDA NRCS 2010). Since 2010, NRCS has invested more than US$222 million in financial and technical assistance to support 123 MRBI partnership projects covering 640 watersheds and nearly 233,504 ha of targeted conservation planning and implementation in 13 MRB states (USDA NRCS 2013). Edge-of-field and watershed monitoring efforts have been initiated through the MRBI and related agricultural conservation programs such as EPA 319 grants and new initiatives such as the Discovery Farms program in Arkansas, Wisconsin, Minnesota, and North Dakota; Pioneer Farm in Wisconsin, the Louisiana Master Farmer Program in Louisiana, and others in the MRB states. In 2012, the Lower Mississippi River Conservation Committee began working with the nonprofit Mississippi River Trust to reforest frequently flooded cleared land within the lower Mississippi River floodplain in six states through the Wetlands Reserve Enhancement Program (LMRCC 2012). More than 4,000 ha were enrolled or accepted for enrollment in this program. Reforestation of these batture lands will help (1) lessen the amount of excess nutrients entering the river and the GOM, (2) reduce flooding of farmland, and (3) expand habitat for bears, migratory birds, and other wildlife (LMRCC 2012). More recently, the NRCS works with farmers and ranchers in selected small watersheds in the MRB to implement conservation practices through the National Water Quality Initiative (USDA NRCS 2014). These efforts will greatly improve downstream watershed ecosystem health by avoiding, controlling, and trapping nutrient and sediments runoff from agricultural fields in the MRB.

Conclusions

Agriculture has increased nutrient and sediment loads and reduced dissolved oxygen in the Mississippi River and the northern GOM. Due to agricultural development, macroinvertebrates, fish, and other biological communities have been impacted resulting in dominant groups tolerant of low oxygen and high turbidity. The Mississippi River Basin Healthy Watersheds Initiative and related conservation programs and efforts will greatly improve downstream watershed ecosystem health by avoiding, controlling, and trapping nutrient and sediments runoff from agricultural fields in the MRB. These conservation programs and better farming practices overall are reducing agricultural impacts to aquatic systems in the MRB. Although there continues to be problems with nutrient transport, sedimentation, and depleted groundwater supplies, agriculture will likely have less influence on the future ecological health condition of the Mississippi River and GOM. But uncertainties exist as other human-related stressors (e.g., urbanization or industry) and climate events (e.g., extreme droughts or floods) can change the overall trend. Future restoration programs may need to focus more on state or regional coordination by classifying restoration projects and standardizing the geographic scale and evaluation methods across the whole MRB.

Acknowledgements

This research is partially funded by a USDA Natural Resources Conservation Service (NRCS) Conservation Innovation Grant (award #69-3A75-12-227), the NRCS Mississippi River Basin Healthy Watersheds Initiative, and National Water Quality Initiative programs. We thank Andrew Sharpley for constructive comments on improving the manuscript.

References

ADEQ (Arkansas Department of Environmental Quality). 2012. Integrated water quality monitoring and assessment report. ADEQ, North Little Rock, Arkansas. Available: www2.adeq.state.ar.us/water/pdfs/draft_2012_integrated_water_quality_monitoring_and_assessment_report.pdf. (November 2013).

Alexander, R. B., R. A. Smith, and G. E. Schwarz. 2000. Effect of stream channel size on the delivery of nitrogen to the Gulf of Mexico. Nature 403:758–761.

Alexander, R. B., R. A. Smith, G. E. Schwarz, E. W. Boyer, J. V. Nolan, and J. W. Brakebill. 2008. Differences in phosphorus and nitrogen delivery to the Gulf of Mexico from the Mississippi River basin. Environmental Science and Technology 42:822–830.

Andraski, T. W., and L. G. Bundy. 2003. Relationships between phosphorus levels in soil and in runoff from corn production systems. Journal of Environmental Quality 32:310–316.

Bianchi, T. S., S. F. DiMarco, J. H. Cowan, Jr., R. D. Hetland, P. Chapman, J. W. Day, and M. A. Allison. 2010. The science of hypoxia in the northern Gulf of Mexico: a review. Science of the Total Environment 408:1471–1484.

Blann, K. L., J. L. Anderson, G. R. Sands, and B. Vondracek. 2009. Effects of agricultural drainage on aquatic ecosystems: a review. Critical Reviews in Environmental Science and Technology 39:909–1001.

Blesh, J., and L. E. Drinkwater. 2013. The impact of nitrogen source and crop rotation on nitrogen mass balances in the Mississippi River basin. Ecological Applications 23:1017–1035.

Borah, D. K., M. Bera, and S. Shaw. 2003. Water, sediment, nutrient, and pesticide measurements in an agricultural watershed in Illinois during storm events. Transactions of the ASAE 46:657–674.

Bryson, J. R., R. H. Coupe, and M. A. Manning. 2007. Characterization of water quality in unmonitored streams in the Mississippi alluvial plain, northwestern Mississippi, May–June 2006. U.S. Geological Survey, Scientific Investigations Report 2007-5116, Reston, Virginia. Available: http://pubs.usgs.gov/sir/2007/5116/pdf/SIR07-5116_508.pdf. (October 2013).

Carpenter, S. R., N. F. Caraco, D. L. Correll, R. W. Howarth, A. N. Sharpley, and V. H. Smith. 1998. Nonpoint pollution of surface waters with phosphorus and nitrogen. Ecological Applications 8:559–568.

Caskey, B. J. 2003. Relations of stream fish communities to physical and chemical parameters and land use in the Mississippi alluvial plain ecoregion. U.S. Geological Survey, National Water Quality Assessment Program, Open File Report 03-31, Jackson, Mississippi. Available: http://pubs.usgs.gov/of/2003/0031/report.pdf. (December 2015).

Czarnecki, J. B., P. D. Hays, and P. W. McKee. 2002. The Mississippi River valley alluvial aquifer in Arkansas: a sustainable water resource? U.S. Geological Survey, USGS Fact Sheet 041-02, Little Rock, Arkansas. Available: http://ar.water.usgs.gov/LOCAL_REPORTS/FS-041-02.pdf. (April 2015).

Dagg, M. J., T. S. Bianchi, G. A. Breed, W.-J. Cai, S. Duan, H. Liu, B. A. McKee, R. T. Powell, and C. M. Stewart. 2005. Biogeochemical characteristics of the lower Mississippi River, USA, during June 2003. Estuaries 28:664–674.

Dagg, M. J., and G. A. Breed. 2003. Biological effects of Mississippi River nitrogen on the northern gulf of Mexico—a review and synthesis. Journal of Marine Systems 43:133–152.

Daniel, W. M., and K. M. Brown. 2013. Multifactorial model of habitat, host fish, and landscape effects on Louisiana freshwater mussels. Freshwater Science 32:193–203.

David, M. B., and L. E. Gentry. 2000. Anthropogenic inputs of nitrogen and phosphorus and riverine export for Illinois, USA. Journal of Environmental Quality 29:494–508.

Donner, S. D. 2003. The distribution of the primary crops in the U.S. since 1950 and the relationship to river nutrient levels. Global Ecology and Biogeography 12:341–355.

Donner, S. D., and C. J. Kucharik. 2008. Corn-based ethanol production compromises goal of reducing nitrogen export by the Mississippi River. Proceedings of the National Academy of Sciences of the United States of America 105:4513–4518.

Foley, J. A., C. J. Kucharik, T. E. Twine, M. T. Coe, and S. D. Donner. 2004. Land use, land cover, and climate change across the Mississippi basin: impacts on selected land and water resources. Pages 249-261 in R. S. Defries, G. P. Asner, and R. A. Houghton, editors. Ecosystems and land use change. American Geophysical Union, Geophysical Monograph Series 153, Washington, D.C.

Food Security Act of 1985. U.S. Code, volume 16, sections 3801–3862.

Gentry, L. E., M. B. David, K. M. Smith, and D. A. Kovacic. 1998. Nitrogen cycling and tile drainage nitrate loss in a corn/soybean watershed. Agriculture, Ecosystems and Environment 68:85–97.

Goolsby, D. A., W. A. Battaglin, G. B. Lawrence, R. S. Artz, B. T. Aulenbach, R. P. Hooper, D. R. Keeney, and G. J. Stensland. 1999. Flux and sources of nutrients in the Mississippi–Atchafalaya River basin. Topic 3: report for the integrated assessment on hypoxia in the Gulf of Mexico. National Oceanic and Atmospheric Administration, National Ocean Service, Coastal Ocean Program, Decision Analysis Series No. 17, Silver Spring, Maryland. Available: http://oceanservice.noaa.gov/products/hypox_t3final.pdf. (October 2013).

Howarth, R. W., G. Billen, D. Swaney, A. Townsend, N. Jaworski, K. Lajtha, J. A. Downing, R., Elmgren, N. Caraco, T. Jordan, F. Berendse, J. Freney, V. Kudeyarov, P. Murdoch, and Z. Zhu. 1996. Regional nitrogen budgets and riverine N and P fluxes for drainages to the North Atlantic Ocean: natural and human influences. Biogeochemistry 35:75–139.

Killgore, K. J., Hoover, J. J., Murphy, C. E., Parrish, K. D., Johnson, D. R., and K. F. Myers. 2008. Restoration of delta streams: a case history and conceptual model. U.S. Army Engineer Research and Development Center, Ecosystem Management and Restoration Research Program, EMRRP-ER-08, Vicksburg, Mississippi.

King, S. L., D. J. Twedt, and R. R. Wilson. 2006. The role of the wetland reserve program in conservation efforts in the Mississippi River alluvial valley. Wildlife Society Bulletin 34:914–920.

Knox, J. C. 2001. Agricultural influence on landscape sensitivity in the upper Mississippi River valley. Catena 42:193–224.

Lambert, D., P. Sullivan, R. Claassen, and L. Foreman. 2006. Conservation-compatible practices and programs: who participates? U.S. Department of Agriculture, Economic Research Service, Economic Research Report Number 14. Available: http://ageconsearch.umn.edu/bitstream/7255/2/er060014.pdf. (October 2013).

LMRCC (Lower Mississippi River Conservation Committee). 2012. Reforestation of frequently flooded land. Available: http://www.lmrcc.org/success-stories/. (April 2015).

Meade, R. H. 1995. Setting: geology, hydrology, sediments and engineering of the Mississippi River. Pages 13–30 in R. H. Meade, editor. Contaminants in the Mississippi River, 1987–92. U.S. Geological Circular 1133.

Miranda, L. E. 2011. Depth as an organizer of fish assemblages in floodplain lakes. Aquatic Sciences 73:211–221.

Mize, S. V., S. D. Porter, and D. K. Demcheck. 2008. Influence of fipronil compounds and rice-cultivation land-use intensity on macroinvertebrate communities in streams of southwestern Louisiana, USA. Environmental Pollution 152:491–503.

Murdock, J. N., F. D. Shields, Jr., and R. E. Lizotte, Jr. 2013. Periphyton responses to nutrient and atrazine mixtures introduced through agricultural runoff. Ecotoxicology 22:215–230.

Obenour, D. R., D. Scavia, N. N. Rabalais, R. E. Turner, and A. M. Michalak. 2013. Retrospective analysis of midsummer hypoxic area and volume in the northern Gulf of Mexico, 1985–2011. Environmental Science and Technology 47:9808–9815.

Omernik, J., and G. Griffith. 2012. Ecoregions of the Mississippi alluvial plain (EPA). The encyclopedia of Earth. Available: www.eoearth.org/view/article/51cbed8b7896bb431f692c04. (April 2015).

Rabalais, N. N., and R. E. Turner, editors. 2001. Coastal hypoxia: consequences for living resources and ecosystems. Coastal and Estuarine Studies 58.

Rabalais, N. N., R. E. Turner, and W. J. Wiseman, Jr. 2001. Hypoxia in the Gulf of Mexico. Journal of Environmental Quality 30:320–329.

Rabalais, N. N., R. E. Turner, and D. Scavia. 2002a. Beyond science into policy: Gulf of Mexico hypoxia and the Mississippi River. BioScience 52(2):129–142.

Rabalais, N. N., R. E. Turner, and W. J. Wiseman, Jr. 2002b. Gulf of Mexico hypoxia, a.k.a. "the dead zone." Annual Review of Ecology and Systematics 33:235–263.

Renaud, M. 1986. Hypoxia in Louisiana coastal waters during 1983: implications for fisheries. Fishery Bulletin 84:19–26.

Riseng, C. M, M. J., Wiley, R. W. Black, and M. D. Munn. 2011. Impacts of agricultural land use on biological integrity: a causal analysis. Ecological Applications 21:3128–3146.

Royer, T. V., M. B. David, and L. E. Gentry. 2006. Timing of riverine export of nitrate and phosphorus from agricultural watersheds in Illinois: implications for reducing nutrient loading to the Mississippi River. Environmental Science and Technology 40:4126–4131.

Schilling, K. E. 2005. Relation of baseflow to row crop intensity in Iowa. Agriculture, Ecosystems and Environment 105:433–438.

Schilling, K. E., and R. D. Libra. 2000. The relationship of nitrate concentrations in streams to row crop land use in Iowa. Journal of Environmental Quality 29:1846–1851.

Schlesinger, W. H., K. H. Reckhow, and E. S. Bernhardt. 2006. Global change: the nitrogen cycle and rivers. Water Resources Research [online serial] 42:W03S06.

Schottler, S. P., J. Ulrich, P. Belmont, R. Moore, J. W. Lauer, D. R. Engstrom, and J. Almendinger. 2014. Twentieth century agricultural drainage creates more erosive rivers. Hydrological Processes 28:1951–1961.

Sharpley, A. N., P. J. A. Kleinman, A. L. Heathwaite, W. J. Gburek, G. J. Folmar, and J. P. Schmidt. 2008. Phosphorus loss from an agricultural watershed as a function of storm size. Journal of Environmental Quality 37:362–368.

Sharpley, A. N., R. W. McDowell, and P. J. A. Kleinman. 2001. Phosphorus loss from land and water: integrating agricultural and environmental management. Plant and Soil 237:287–307.

Shields, F. D., Jr., R. E. Lizotte, Jr., and S. S. Knight. 2013. Spatial and temporal water quality variability in aquatic habitats of a cultivated floodplain. River Research and Application 29:313–329.

Simpson, T. W., A. N. Sharpley, R. W. Howarth, H. W. Paerl, and K. R. Mankin. 2008. The new gold rush: fueling ethanol production while protecting water quality. Journal of Environmental Quality 37:318–324.

Smith, D. R., and E. A. Pappas. 2007. Effect of ditch dredging on the fate of nutrients in deep drainage ditches of the Midwestern United States. Journal of Soil and Water Conservation 62:252–261.

Talmage, P. J., J. A. Perry, and R. M. Goldstein. 2002. Relation of instream habitat and physical conditions to fish communities of agricultural streams in the northern Midwest. North American Journal of Fisheries Management 22:825–833.

Tilman, D., K. G. Cassman, P. A. Matson, R. Naylor, and S. Polasky. 2002. Agricultural sustainability and intensive production practices. Nature 418:671–677.

Turner, R. E., and N. N. Rabalais. 1991. Changes in Mississippi River water quality this century implications for coastal food webs. BioScience 41(3):140–147.

Turnipseed, D. P., Y. C. Allen, B. R. Couvillion, K. L. McKee, and W. C. Vervaeke. 2011. Ecosystem effects in the lower Mississippi River basin. U.S. Geological Survey, Professional Paper 1798-L, Reston, Virginia. Available: http://pubs.usgs.gov/pp/1798l/pdf/pp1798l.pdf. (April 2015).

USDA NRCS (U.S. Department of Agriculture National Resources Conservation Service). 2010. The Mississippi River basin healthy watersheds initiative: conservation beyond boundaries. Available: www.nrcs.usda.gov/wps/portal/nrcs/detailfull/national/home/?cid=STELPRDB1048200. (December 2015).

USDA NRCS (U.S. Department of Agriculture National Resources Conservation Service). 2013. Agricultural statistics 2013. Available: www.nass.usda.gov/Publications/Ag_Statistics/2013/Agricultural_Statistics_2013.pdf. (December 2015).

USDA NRCS (U.S. Department of Agriculture National Resources Conservation Service). 2013. The Mississippi River basin healthy watersheds initiative 2012 progress report. Available: www.nrcs.usda.gov/wps/portal/nrcs/detailfull/national/home/?cid=STELPRDB1048200. (December 2015).

USDA NRCS (U.S. Department of Agriculture National Resources Conservation Service). 2014. The national water quality initiative. Available: www.nrcs.usda.gov/wps/portal/nrcs/detail/national/programs/financial/eqip/?cid=stelprdb1047761. (December 2015).

Van der Leeden, F., F. L. Troise, and D. K. Dodd. 1990. The water encyclopedia. Lewis Publishers, Chelsea, Michigan.

Wang, L., J. Lyons, P. Kanehl, and R. Gatti. 1997. Influences of watershed land use on habitat quality and biotic integrity in Wisconsin streams. Fisheries 22(6):6–12.

American Fisheries Society Symposium 84:311–350, 2016

Mississippi River Ecosystem Restoration: The Past Forty-Plus Years

Gretchen L. Benjamin*
The Nature Conservancy, Missouri Chapter Field Office
Post Office Box 440400, St. Louis, Missouri 63144, USA

Angeline J. Rodgers
U.S. Fish and Wildlife Service
6578 Dogwood View Parkway, Suite A, Jackson, Mississippi 39213, USA

K. Jack Killgore
U.S. Army Corps of Engineers, Engineer Research and Development Center
Environmental Laboratory
3909 Halls Ferry Road, Vicksburg, Mississippi 39180, USA

Abstract.—Ecosystem restoration of the Mississippi River main stem has been ongoing since the early 1970s. After the passage of environmental laws in the late 1960s to the early 1970s, private citizens and state and federal natural resource agency managers began to seek programs and funding for restoration and conservation that eventually resulted in mitigation measures of adverse impacts. Environmental-type actions that include the Great River Environmental Action Team, the Avoid and Minimize program, the middle Mississippi River biological opinion, and the lower Mississippi River conservation plan and biological opinion originated from laws or legal action. The Upper Mississippi River Restoration, Navigation and Ecosystem Sustainability Program, Restoring America's Greatest River, and Operation and Maintenance activities, which support system ecological restoration measures, are, to a large extent, done in a cooperative setting to improve the river for multiple benefits. This coalition of agencies and professions has resulted in the application of hundreds of different types of measures to restore form and function to the third largest river in the world. Over the years, dredging and disposal practices have improved in an effort to minimize the impacts from these activities. Lost floodplain islands have been replaced, backwater lakes and channel depths have been recovered, active river flow has been reintroduced to backwaters, and microhabitats for special concern species have been restored, all to recreate broad functional floodplain habitat. Wing-dike and side-channel closure structures have been shortened, notched, or removed to recover flow along the main-channel border and side channels, increasing hydraulic residence time and recovering valuable habitat along with restoring nutrient and sediment assimilation processes the floodplain provides. Field monitoring has shown positive responses from endangered and threatened species, migratory and resident aquatic and wildlife species, abiotic conditions like water quality, and increased use by humans enjoying the benefit of a restored river system. Collectively, this work is some of the most extensive large river restoration in the world, but it only represents a small contribution to what is necessary to maintain a diverse and resilient Mississippi River. The informa-

* Corresponding author: gbenjamin@tnc.org

tion provided in this chapter provides a basis for continuing restoration efforts that should become a routine part of Mississippi River management.

Introduction

From Lake Itasca in Minnesota to Head of Passes in Louisiana, the Mississippi River main stem flows over 3,700 km from north to south (Fremling 2005). The river is divided into three sections with distinctly different geographies, flow rates, and management characteristics (Figure 1). The upper Mississippi River (UMR) is described as the portion of the river from its headwaters to the confluence of the Missouri River at St. Louis, Missouri (Anfinson 2003). The middle Mississippi River (MMR) is a relatively short section of river from the Missouri River to the confluence of the Ohio River at Cairo, Illinois and is typically included in the description of the UMR. The lower Mississippi River (LMR) extends from the Ohio River to the Gulf of Mexico.

Beginning in the headwaters, the river meanders 793 km through Minnesota. Finding the main thread of the river can be difficult as islands and wetlands mask the channel (Fremling 2005). St. Anthony Falls in Minnesota is the northern extent of the UMR, continuing to Cairo, Illinois. In this 1,394-km stretch, large tributaries enter the main stem, including the Minnesota, Wisconsin, Iowa, Illinois, and Missouri rivers (USACE 2011). Below the confluence with the Missouri River, the river flow is more than 5,578 m^3/s, 75% of the time to the confluence of the Ohio River (USGS 2014). This 314-km stretch of river between the Missouri and Ohio rivers confluences is often referred to as the MMR but is within the UMR basin. In all, the headwaters and the upper Mississippi River main stem reaches cover about 2,187 km.

The lower Mississippi River alluvial valley encompasses 13 million ha, is 40–200 km wide, and starts at the mouth of the Ohio River, running 1,535 river kilometers to the Gulf of Mexico. River volume enlarges substantially after the Ohio River, as it contributes the greatest volume of any tributary. Mean annual discharge at Vicksburg, Mississippi (river kilometer [rkm] 704, measuring from Head of Passes), midway between the mouth of the Ohio River and the Gulf of Mexico, is 15,637 m^3/s (Baker et al. 1991). The maximum discharge that occurred during the 2011 flood, the largest flood on record, was 65,695 m^3/s (DeHaan et al. 2012). Although the entire valley is an alluvial delta, the distributaries of the Mississippi River delta are generally south of Baton Rouge and have changed course over the past 5,000 years (Saucier 1994).

Channel Modifications on the Mississippi River

The Mississippi River main stem has been a transportation corridor for human habitants, but with European settlement it became a major route to move people and eventually bulk commodities in and out of the middle portion of the United States. The General Survey Act of 1824 provided the establishment of the Rivers and Harbors Act and authorized US$75,000 to clear sandbars and remove debris (mostly trees) in the LMR and portions of the Ohio River (Anfinson 2003). By 1878, a 1.4-m-deep channel was authorized, and in 1907, a 2-m-deep channel was authorized for the entire main stem of the Mississippi River. Initial channel modification activities included removing snags, removing rapids and water falls, installing channel training structures, and dredging sand shoals (Anfinson 2003). In 1930, the U.S. Congress authorized a 2.7-m-deep navigation channel, and a variety of engineering techniques were used along the length of the Mississippi River to accommodate this depth. With the continued need to deepen the river, channel engineers included more channel training structures to focus river flow to one main thread (thalweg) in order to use the power of the river to scour the channel. Most notable, to maintain the 2.7-m channel depth on the UMR, a series of locks and dams were built to impound the river.

The only provisions for navigation in the area known as the headwaters are the six res-

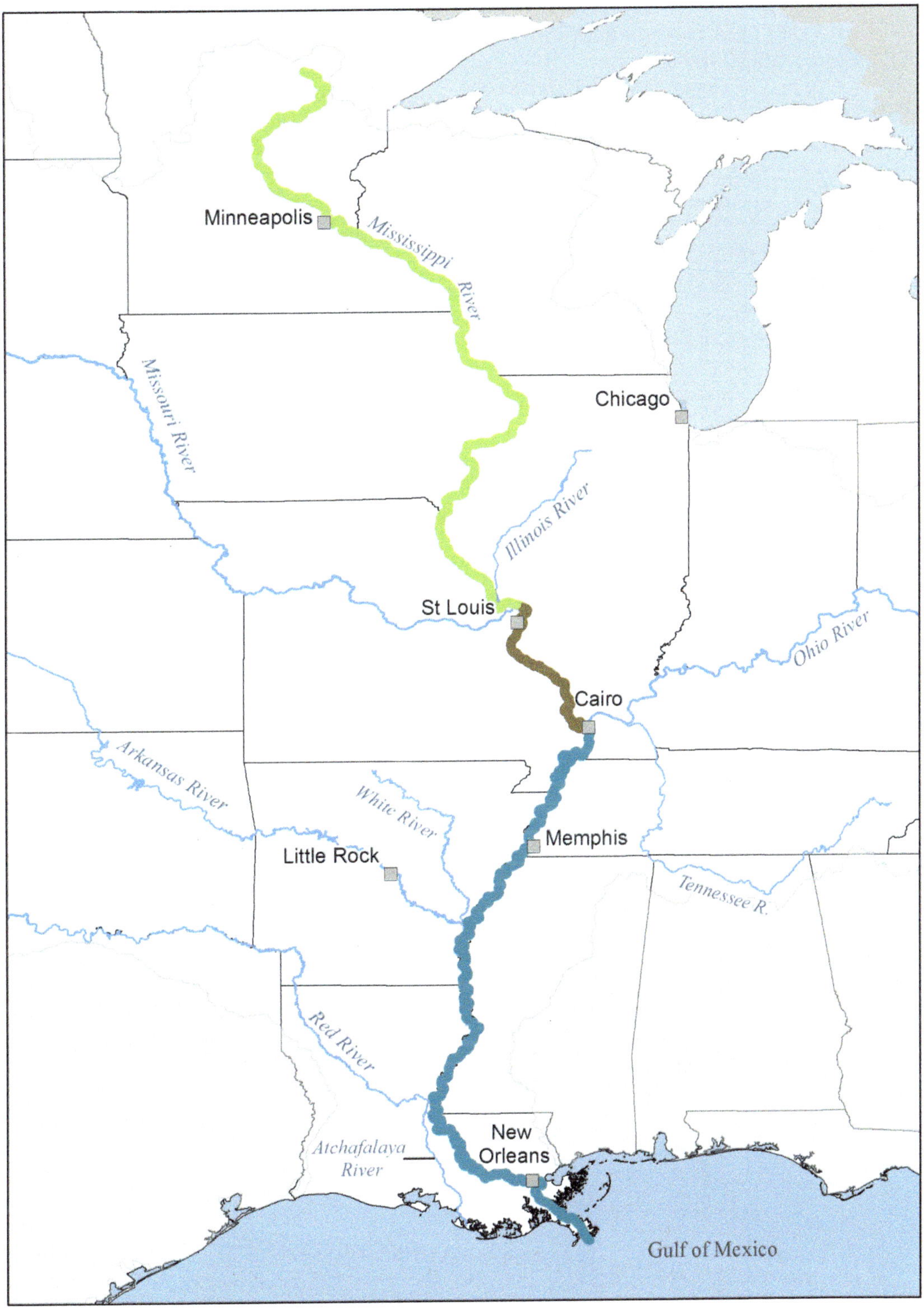

Figure 1. Map of the main-stem Mississippi River: green—upper, brown—middle, and blue—lower.

ervoirs in Minnesota (Winnibigoshish, Leech Lake, Pokegama, Sandy Lake, Pine River, and Gull Lake) used to augment flows during the navigation season with a collective flow of 17.6 m^3/s (Rivers and Harbors Act of 1888; USOFR 2011). Beyond these relatively minor alterations, the 793 river kilometers of the headwaters are primarily without levees or other flow-altering infrastructure.

In contrast, the UMR navigation system is a complex series of stair step reservoirs created by navigational locks and dams that begins north of Minneapolis, Minnesota and extends to St. Louis, Missouri (Figure 2). The Illinois River is also an important element of the UMR navigation system that connects the Mississippi River with Chicago, Illinois and ultimately the Great Lakes with eight locks and dams. Altogether, there are a total of 37 navigational locks and dams on the UMR navigation system, including those on the Illinois River (UMRBC 1982).

Below the confluence of the Missouri River to Cairo, Illinois, the river is not impounded longitudinally but is contained laterally within levees. Known as the MMR, it is managed for commercial navigation as part of the UMR, but hydrologic and hydraulic alterations are more comparable to the LMR. This transitional area between the upper and lower Mississippi River contains unique rock formations, and Lock 27 provides navigation traffic the means to avoid the treacherous portion of channel known as the Chain of Rocks just north of St. Louis, Missouri. The MMR does not have the flow volume as provided by the Ohio River to the LMR, so it requires more channel training structures per river kilometer to maintain the navigational corridor.

The LMR (1,535 river kilometers) is authorized as a 3.7-m-deep navigational channel by the 1944 Flood Control Act but is maintained as a 2.7-m-deep navigational channel down to Baton Rouge, Louisiana (rkm 378) to match the draft depth of the UMR (Camillo 2012). However, stabilization efforts together with dredging provide a 3.7-m-deep channel through most of this reach at low river stag-

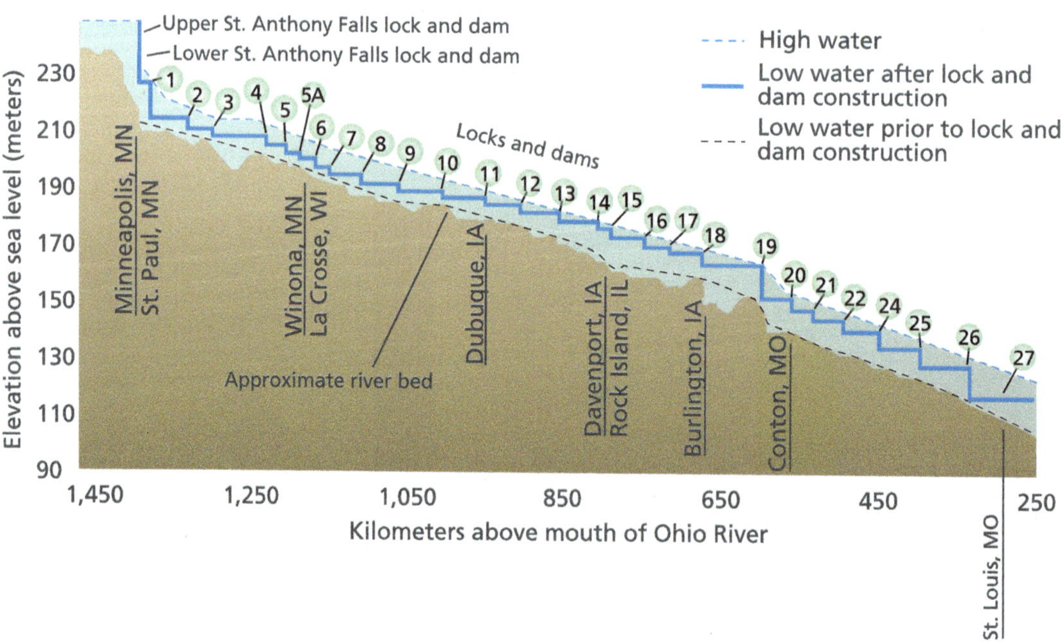

Figure 2. The upper Mississippi River navigation system is a series of stair step pools created by navigational locks and dams that begins north of Minneapolis, Minnesota and extends to St. Louis, Missouri. Diagram courtesy of the Integration and Application Network (IAN.umces.edu), University of Maryland Center for Environmental Science.

es. Below Baton Rouge to the Gulf of Mexico, the navigation channel is maintained at 12 m for ocean going ships.

The problems for navigation in the unimpounded reaches are primarily in the bendways where point bars extend into the channel, creating a narrow navigation lane during low water. In addition to being one-way traffic during these conditions, towboats moving downstream are at greater risk of colliding into the concave bank. Lateral erosion along the concave bank can also threaten the integrity of levees and possibly alter the river course. Therefore, channel training structures were primarily designed to manage the morphometry and stability of bendways. For almost 150 years (since 1877), channel training structures have been constructed in the MMR and LMR to confine the channel and create self-scouring conditions from increased velocities to maintain navigation depth. Traditional structures have different names and different impacts depending on location along the 3,200+ river kilometers, but in general, the structures are

1. Dikes (wing dams, wing dikes, transverses, and spurs), which are mostly perpendicular to the primary flow path;
2. Closure dams and closure dikes, which eliminate most if not all flow to off-channel areas during low stages; and
3. Revetment, which armors the main-channel bank line.

Today, the main channel border of the Mississippi River, from Minneapolis to Baton Rouge, is captured by thousands of dikes, secondary channel closure dams, and

Figure 3. Dike field along Keyes Point bendway in the lower Mississippi River, river kilometer 1268. Sediment accumulates between the dikes. Photo credit: Jack Killgore, U.S. Army Corps of Engineers, Engineer Research and Development Center.

variations of these original structures (Figure 3). Shoreline stabilization such as riprap and articulated concrete mattress occurs throughout the entire system, interspersed with natural banks and sandbars. River volume and depth ease the management of the navigation channel, but risk of widespread flooding requires significant flood control measures. Since the 1927 flood and the subsequent 1928 Flood Control Act, the LMR has been part of a comprehensive flood control system created for the "project design flood" (Camillo 2012). The Mississippi River and Tributaries Project consists of a total of 5,998 km of levees (3,566 km are mainline Mississippi River levees), a series of floodways for use during large flooding events, and water storage capacity in key tributary systems (Kentucky and Barkley Lakes in the Tennessee River basin; four reservoirs in the Yazoo Basin, Mississippi; Camillo 2012). Additionally, the river was shortened by more than 241 km between 1929 and 1960 using bendway cutoffs (Baker et al. 1991).

Impacts of River Modifications

Beginning in the 1820s, the Mississippi River was modified by snag removal, dredging and disposal, channel training structures, and the removal of rock rapids and falls (Anfinson 2003). These changes, as well as significant pollution from industrializing cities and land-use alterations in the early to mid-20th century within the river basin, diminished the quality of river habitat (Anfinson 2003). Channel training structures caused filling between the structures and off-channel areas and focused the majority of below bank-full flows to the main channel for navigation.

Lock and dam construction on the upper river created impoundments, which initially increased shallow/deep wetlands, channels, and backwater features of the river floodplain. However, within 20 years of impoundment, aquatic habitats began to degrade. Increased water levels permanently inundated low-elevation floodplain lands or higher-elevation lands became islands. Newly inundated islands and shoreline became more vulnerable to erosion from wave action and released sediment to the water column. With time, the suspended sediment fell out in low-velocity environments, filling important habitats such as backwater wetlands and lakes. Today, impounded areas above the dams are windswept open-water expanses with elevated turbidity, little to no aquatic vegetation, and reduced habitat value for many species (UMRCC 1993).

Confounding the issue of habitat loss, migratory fish movement was hampered by the lock and dam structures, which only provide passage when river flows are elevated and dam gates are partially or completely lifted to manage river flow. River flow sufficient to require gate openings is different at every lock and dam structure and may not happen every year or at the right time of the year for fish movement when it does happen.

The upper most navigable portion of the UMR is only about 3% leveed, due in part to preservation ascribed by the Upper Mississippi River National Fish and Wildlife Refuge, which was established in 1924 (USFWS 2006). Downstream of the refuge, and in the MMR and LMR, levees reduced the lateral connection to the functional floodplain by more than 80–90%, limiting the natural assimilation of nutrients and sediment, water retention and infiltration, and off-channel fish and wildlife habitat, and dampening primary production (Baker et al. 1991; Camillo 2012). Dikes and revetment reduced hydraulic connections between the channel and floodplain and caused sediment aggradation between dike structures, resulting in a loss of aquatic habitat. Since 1960, there has been a gradual loss of secondary channels due to closure structures that divert water to the main channel during low to moderate river stages (Guntren et al., in press). Collectively, the alterations to hydrology and hydraulics of the lower 1,535 km of the Mississippi River, particularly levees, confine the active-flowing portion of the river from a 13-million-ha alluvial valley to a 0.6-million-ha river channel and floodplain referred to as the batture (Baker et al. 1991).

Ecosystem Restoration Programs

The continued degradation of the Mississippi River eventually led to practices and legislation to restore natural river floodplain features. Over the past 40 years, there have been five active and one federally authorized, but so far not appropriated, ecosystem restoration programs that merit explanation for main-stem restoration of the Mississippi River. These programs are typically set up within the UMR, MMR, or LMR segments. On the UMR, GREAT (Great River Environmental Action Team) I, II, and III (Water Resources Development Act of 1976); Upper Mississippi River Restoration-Environmental Management Program (1986); and Avoid and Minimize (1988) have provided or are active programs that provide funding for ecosystem restoration. The Navigation and Ecosystem Sustainability Program (2007) is authorized but has not received funds for ecological restoration. On the MMR, restoration programs have been funded as part of the Avoid and Minimize program that extends down to the mouth of the Ohio River, in addition to compliance with jeopardy biological opinions of the Federal Endangered Species Act. On the LMR, the Restoring America's Greatest River plan developed by the Lower Mississippi River Conservation Committee laid out priorities and provided a template for ecological restoration. As the result of unique partnerships, the U.S. Army Corps of Engineers (USACE) completed a conservation plan for the Channel Improvement Program under Section 7(a)(1) of the Endangered Species Act for the Pallid Sturgeon *Scaphirhynchus albus*, fat pocketbook mussel *Potamilus capax*, and interior least tern *Sterna antillarum athalassos* (Killgore et al. 2014). The recent U.S. Fish and Wildlife Service (USFWS) Biological Opinion on the Channel Improvement Program recognized the conservation benefits of USACE's efforts to incorporate restoration designs in channel training structures and work with USFWS to restore flow into secondary channels. However, there is no formal ecological restoration program in USACE with associated funding on the LMR; all efforts to date have been the result of cooperative partnerships.

Upper Mississippi River Restoration Programs

GREAT I, II, III.—The states and USFWS had a long history of disputes with USACE about dredging and disposal in the UMR. These tensions ignited with the passage of the National Environmental Policy Act in 1969 and the Clean Water Act in 1972. The state of Wisconsin (Wisconsin v. U.S. Army Corps of Engineers et al. 1973) used these newly established laws to push the federal government to find an environmentally acceptable method to manage the disposal of dredge material for maintaining the 2.7-m navigation channel (GREAT 1980).

In 1974, under the leadership of the USACE and USFWS, an interagency team was organized to identify and assess the problems associated with multipurpose use of the river and develop recommendations for improved management of the river. This effort developed into the GREAT study (UMRBC 1982), which was formally authorized by Section 117 of the Water Resources Development Act of 1976. The effort began in the St. Paul District of USACE and came to be called GREAT I (Water Resources Development Act of 1976). The study reach of GREAT I included 348 navigational river kilometers.

The solutions were multipronged. River hydraulic engineers studied actual dredging depth practices and determined that it would be more efficient to maintain the dredging depth at 3.6 m rather than the historic 3.9 m (Barges and tows actually draft 2.7 m, but the channel is managed with maneuvering depths; USACE 1997). This overdraft dredging was actually causing head cutting of the channels, resulting in more dredging and ultimately less channel reliability. Along with reduced draft dredging, the USACE determined that the optimum channel width for both the straight channel lengths and channel bends would be 107 and 152 m, respectively. By doing this, USACE was able to reduce dredging amounts by about

382,277 m^3 annually while maintaining a safe and reliable channel for commercial navigation. This solution provided economic and environmental wins; it saved dredging costs and it reduced the amount of area needed for dredge material disposal (USACE 1997).

Another solution was to define the chronic dredge cuts (areas that are dredged on a regular basis) and find environmentally acceptable sites to place the dredge material. This comprehensive effort took several years and was the most contentious. In the end, about 90 disposal sites were established for more than 100 chronic dredge cuts and saved more than 405 ha of wetlands in the process (USACE 1997). The capacity of these disposal sites is renewed by providing free use of the dredge material for construction, road sanding, and other applications.

Following along with the success of GREAT I, GREAT II in the Rock Island District of USACE and GREAT III in the St. Louis District attempted to follow similar patterns for managing dredging and disposal on the UMR. Both programs provided important results but were not as comprehensive as GREAT I. Perhaps most valuable from this entire process is the fact that all the teams (River Resources Forum - St. Paul, River Resources Coordinating Committee - Rock Island, and River Resources Action Team - St. Louis) established in the GREAT processes still work to resolve river-related issues in a manner that allows for multiple river uses while conserving the natural resources of the UMR.

Other important outcomes from these programs:

- USACE conducts sediment contaminants studies and monitors native mussel populations within the footprint of chronic dredging sites. In both cases, tremendous data sets were established and the information documents the decline in contaminants in river sediments and describes population trends and recruitment of native mussels in these areas.
- The teams systematically evaluated channel training structures to determine if dredging requirements of the navigation channel can be minimized. This delicate balance between structures that channelize the river and minimizing dredging often requires important consideration of the environmental and economic costs associated with channel maintenance. Economically, the considerations include dredging cost, disposal site availability, and channel reliability. Environmental considerations include disturbance of sensitive species such as native mussels and fish habitat, maintaining ecological connections between the flowing and backwater portions of the river floodplain, and site specific tradeoffs between impacts from dredging and building of channel training structures.
- Since most of the dredging and disposal issues were resolved, river managers worked on other important issues associated with balancing the multiple uses on the river. These same groups are now working to optimize dam operations, determine channel structure placement and removal, promote recreational opportunities, and continue to adapt the dredging and disposal activities to balance commercial navigation needs with ecological conditions to maintain a healthy river system.

Upper Mississippi River Restoration-Environmental Management Program.—The U.S. Congress authorized the Environmental Management Program (recently renamed the Upper Mississippi River Restoration Program [UMRR]) in the 1986 Water Resources Development Act to help address ecological needs on the Upper Mississippi River System. Program elements were identified and designed in the Comprehensive Master Plan for the management of the Upper Mississippi River System and became the framework for initiating the program (Water Resources Development Act of 1986). Subsequent amendments have helped refine program policy, implementation, and establish continuous funding (Water Resources Development Acts of 1990, 1992, 1999, 2007). The Upper Mississippi River Restoration Program is the most comprehensive, long-term restoration program in the history of the Mississippi River. There are two major components of UMRR: the habitat rehabilitation and enhancement projects (HREPs) and the Long Term Resource Monitoring Program (LTRMP). Together, HREPs and LTRMP are designed to improve the ecological health of the Upper Mississippi River System and increase our understanding of its natural resources in the 1,931 river kilometers of the Mississippi and Illinois rivers above Cairo, Illinois (USACE 2010).

Through HREP implementation, 55 projects are complete, including more than 43,301 ha of improved fish and wildlife habitat (EMPCC 2012). Planning is underway for an additional 34 projects with an estimated 32,780 ha of habitat benefits. These numbers do not represent the totality of projects necessary to restore the UMR, but simply represent the next set of priorities for the program. In 2000, the Habitat Needs Assessment identified at least 54,633 additional ha of essential habitat restoration within five major habitats and also identified ecological functions and processes that would need to be restored (USACE 2000). Off-channel and backwater complexes are the primary focus of HREPs within the impounded portion of the UMR floodplain (Figure 4). Typical restoration projects average 202–405 ha of habitat directly restored and seek to restore and rehabilitate an entire mosaic of habitats to include topographically diverse islands, off-channel and deep-water diversity, protection of desirable floodplain features, and rebuilding critical habitat features for vulnerable species. Design and construction work to reestablish these essential habitats requires several techniques that are combined to produce the desired results. With the tremendous innovation and lessons learned from HREPs, there are direct benefits to the Upper

Figure 4. A backwater complex within the impounded portion of the upper Mississippi River. Photo credit: Jeff Janvrin, Wisconsin Department of Natural Resources.

Mississippi River System and indirect benefits to other large-river restoration, nationally and internationally, which can use this experience to inform their projects.

The Upper Mississippi River Restoration Program was uniquely structured to include the LTRMP for key river factors: water quality, fisheries, aquatic plants, land cover/land use, and periodic updates for topographic and bathymetric data to provide river managers with the knowledge they need to make sound restoration decisions. It was established as a partnership program between the USACE, U.S. Geological Survey, USFWS, U.S. Environmental Protection Agency, and the states of Illinois, Iowa, Minnesota, Missouri, and Wisconsin (Water Resources Development Act of 1986, Section 1103).

The Long Term Resource Monitoring Program was established to develop a better understanding of the UMR and its resource problems. Key to the program success is managing the substantial database and providing analyses for data that has been gathered in some cases for more than 25 years. Collectively, this program component provides knowledge to inform river managers of the status and trends of these river factors to improve ecological management of the Upper Mississippi River System.

Habitat rehabilitation and enhancement projects and LTRMP have numerous ancillary benefits for the UMR and other large rivers. In 2012, the updated Environmental Design Handbook (USACE 2012) detailed the program knowledge for large river restoration, focusing on habitat project planning, design, and implementation for large river restoration specialists, nationally and internationally. Continual reassessment of this work will ensure that project designs improve in both efficiency and effectiveness. Using LTRMP data, numerous models have been developed to aid project design to identify the best placement and layout of restoration features (Johnson and Hagerty 2008; Rohweder et al. 2012). Finally, the HREP Planning and Sequencing Framework has been developed to help determine project sequencing for a comprehensive planning approach and could serve to help other large systems establish restoration plans (USACE 2003).

Each floodplain area within the UMR requires extensive knowledge of local ecological factors to determine the best tools to apply for riverine floodplain restoration. In all cases, a strong partnership of state and federal agency representatives with expertise in hydraulic engineering, river ecology, fish and wildlife management, and local habitat conditions is essential to design and implement these projects. After 28 years, the UMRR continues to thrive and in recent years has received funding close to the full authorized amount of US$33.187 million. This funding level reflects the successive progress in habitat recovery, the cornerstone of this program since its inception. Each project builds upon knowledge gained from previous work, resulting in a state of the art program for river recovery. While the resulting impacts of 150 years of intense river alteration is still prevalent on the UMR, the cumulative recovery of more than 40,469 ha of habitat within the main channel border, floodplain islands, side channels, and backwater features like shallow to deep wetlands, sloughs, oxbows, and deep channels is evident from floral and faunal response (USACE 2010).

Navigation Ecosystem Sustainability Program.—The Feasibility Study for the Upper Mississippi River-Illinois Waterway System Navigation Study began in 1993 after the completion of the reconnaissance phase documented a federal interest in investigating improvements for navigation on the Upper Mississippi River System. Initial study goals were to assess the need for improving the 37 locks and dams on the system and determine the incremental environmental impact of additional traffic on the river if system upgrades were warranted. In 1998, the study was halted due to issues with the economic portion of the study and was not restarted until August 6, 2001. The restructured study required the USACE to address the ongoing cumulative effects of navigation along with the incremental increase of traffic induced by navigation improvements, with an overall goal of attaining a sustainable river system for navigation and the ecosystem. The final integrated feasibility report (USACE 2004) outlines the 15-year implementation details of the 50-year plan scoped out in the study process.

The authorized study included the following recommendations: 2.2 billion navigation infrastructure investment for small-scale navigational measures, new 366-m lock chambers at seven existing lock sites on the upper Mississippi and Illinois rivers, mitigation for site-specific impacts, and continued review of options for navigational efficiency (USACE 2004). The environmental portion of the authorized study outlined a 2.0 billion increment of the 50-year UMR restoration plan. Restoration would include fish passage, dam reoperation for navigation and environmental purposes, island construction, off-channel reconnection, channel structure alterations for enhanced aquatic connections, and protection of existing valuable natural features.

The environmental portion of the Navigation Ecosystem Sustainability Program (NESP) and the Upper Mississippi River Restoration-Environmental Management Program (UMRR-EMP) established a joint vision statement, overarching ecological goals, and system-wide reach and project-scale objectives for the UMR. River managers and policy experts are cautious about any transition from UMRR-EMP to NESP due to concern over the potential loss of UMRR construction and monitoring efficiencies and unproven implementation of NESP. In 2007, the Upper Mississippi River Basin Association (UMRBA) addressed this issue with the guidance document entitled, "Integrating NESP and EMP: A UMRBA Vision for the Future of Ecosystem Restoration on the Upper Mississippi River System" (UMRBA 2007). The key provisions in the document outline transition actions to ensure that the capabilities of UMRR-EMP are not lost in the transition to NESP.

Middle and Lower Mississippi River Programs

Avoid and Minimize.—In October 1992, the St. Louis District of USACE issued "Design Memorandum No. 24, Avoid and Minimize Measures, Melvin Price Locks and Dam, Upper Mississippi River - Missouri and Illinois" (USACE 1992). These measures applied to both the UMR and MMR. The document was developed as a result of a commitment made in the 1988 Record of Decision associated with the environmental impact statement for the second lock (Lock 26) at Melvin Price Locks and Dam (USACE 1988). The intent of the program was to reduce the possible environmental impacts of increased navigation traffic associated with construction of the second lock.

The actions associated with the Avoid and Minimize program were streamlined to implement and have numerous positive elements, including habitat restoration primarily through modification to existing channel training structures, innovative design for channel training structures, environmental pool management, and modifications to dredging and disposal. In addition, lock mooring cells were installed to reduce navigation impacts as tows wait for lock passage and a portion of the program is set up to monitor benefits from habitat improvements and model different alternatives that benefit the ecosystem as well as navigation (USACE 1992–2010/2011). Full-scale implementation of the program began in 1996. Program direction is coordinated through the River Resources Action Team, which consists of state, federal, and private partners in both natural resources and river industry.

Middle Mississippi River biological opinion.—The Endangered Species Act of 1973 (ESA) requires federal agencies to consult with the USFWS or the National Marine Fisheries Service (NMFS) if any project occurs in the habitat of a listed species. Under Section 7(a) (2) of the act, the services will undergo formal consultation with the action agency to determine if the project will jeopardize listed species. In 2000, the USFWS issued a draft jeopardy biological opinion for the operation and maintenance of the 2.7-m navigation channel on the UMR, including the MMR (USFWS 2004). The USFWS concluded that navigation channel operation and maintenance was detrimental to the continued existence of the Pallid Sturgeon. Consequently, USACE began a habitat restoration program in the MMR that included island construction, side-channel restoration, wing-dike notching, and chevron-dike construction to diversify habitat for the Pallid Sturgeon and other species. The U.S. Army Corps of Engineers has a commitment to continue this program until the listed species are fully recovered based on long-term monitoring of the populations.

Restoring America's greatest river.—The LMR does not have the benefit of a federally authorized and funded restoration program but has relied on a unique partnership to restore valuable habitat for numerous species, including three federally endangered species: interior least tern, Pallid Sturgeon, and fat pocketbook mussel. The Lower Mississippi River Conservation Committee (LMRCC) was formed in 1994 as a coalition of natural resource conservation and environmental quality agencies from the states of Arkansas, Kentucky, Louisiana, Mississippi, Missouri, and Tennessee. There are several cooperating federal agencies, including USFWS, USACE, U.S. Geological Survey, U.S. Environmental Protection Agency, and U.S. Department of Agriculture Natural Resource Conservation Service. The U.S. Fish and Wildlife Service provides a full-time coordinator to the LMRCC.

In the mid-2000s, the LMRCC held state-level planning meetings to identify habitat restoration projects for the LMR that were compiled into Restoring America's Greatest River plan (LMRCC 2015). Projects were selected to enhance habitat complexity in the main channel, but also to restore floodplain hydrology and connectivity between the river and its floodplain. Restoring America's Greatest River provided a template for restoration efforts on the lower river. Restoration of secondary channels was prioritized as a restoration focus for the lower river, and the USACE Engineer Research and Development Center developed a ranking system for proposed projects through a priority index that considers habitat quality and economy of restoration indices (Boysen et al. 2012; Killgore et al. 2012). A unique partnership between LMRCC, USFWS, and USACE allowed for cost-effective rehabilitation of these vital habitats, many of which had been altered or lost due to Channel Improvement Program features of the Mississippi River and Tributaries Project such as dikes. Through this partnership, 14 secondary channel restoration projects, totaling over 90 km and almost 2,833 ha, have been completed since 2006 by notching dikes to provide more permanent flow during low river stages (LMRCC 2015). Additional dike notching has been conducted by USACE districts through maintenance or new construction.

Lower Mississippi River conservation plan and biological opinion.—Endangered species issues in the LMR were similar to the MMR since both reaches comprised the 1,931-km free-flowing Mississippi River within the range of most listed species. For more than a decade, the USACE worked with the USFWS and state conservation agencies to identify and resolve endangered species and ecosystem management issues associated with USACE civil works projects to provide flood risk management infrastructure and to facilitate navigation in the LMR. These programs that have significantly affected the river are now potentially the most important and cost-effective tools to maintain, enhance, and restore its ecological functions. This was accomplished by considering and incorporating ecological engineering opportunities during the design phase of channel improvement and channel maintenance projects. Early consideration of conservation designs can result in localized improvements in habitat function and value, with little to no effect on flood risk management, navigation, or project cost.

In the case of the LMR, a conservation plan (Endangered Species Act of 1973, section 7(a)(1)) for the interior least tern, Pallid Sturgeon, and fat pocketbook mussel, which were the three listed species in the lower river, was completed by USACE prior to formal consultation (Killgore et al. 2014). Section 7(a)(1) requires all federal agencies to use their authorities as appropriate to carry out programs for the conservation (i.e., recovery) of endangered and threatened species. The conservation plan outlined the programmatic mechanisms by which the Channel Improvement Program of the Mississippi River and Tributaries Project administered by USACE is being utilized to implement conservation measures that maintain and improve habitat values within the LMR for recovery of endangered and other trust species inhabiting the river channel and floodplain. U.S. Army Corps of Engineers then submitted the conservation plan as a biological assessment (i.e., Endangered Species Act of 1973, section 7 consultation) for the Channel Improvement Program and requested formal consultation with the USFWS (USFWS 2013). The conservation plan outlined conservation measures USACE has implemented and will continue to implement to benefit the listed species: avoid adverse impacts directly associated with river engineering practices; develop and implement channel construction and maintenance operation guidelines that conserve and restore LMR habitats such as gravel bars and secondary channels; develop and implement cost-effective monitoring programs to document species response to channel operations; and share restoration, research, and monitoring responsibilities and costs by maintaining partnerships with federal agencies, state agencies, and nongovernmental organizations. As of 2012, almost 30% (230) of the dikes in the LMR (774) have been constructed or retrofitted with notches to diversify channel border, increase scour holes, and reconnect seasonally flooded habitat (Killgore et al. 2014). Additionally, USACE has collaborated with LMRCC and have cost-efficiently rehabilitated 14 secondary channels, as noted previously.

Based upon the information and conservation measures in the conservation plan, the USFWS issued a biological opinion in December 2013 stating that the Channel Improvement Program is not likely to jeopardize the continued existence of the interior least tern, Pallid Sturgeon, and fat pocketbook mussel (USFWS 2013). Since secondary channels were important to the recruitment and survival of the three listed species, the USFWS used secondary channel abundance and condition as a surrogate for take (i.e., harass, harm, or kill) of all three species (USFWS 2013). The U.S. Army Corps of Engineers recently revitalized the Mississippi River Geomorphology and Potamology Program, which will aid in quantifying secondary channel abundance and condition, among other studies.

Tools for Restoration

The use and benefit of restoration tools vary among the three reaches of the Mississippi River due to differences in geomorphic and hydrologic conditions and differences in human alterations to the floodplain. The water volume of the UMR is less than the MMR and LMR, so the navigation channel is managed using a combination of four techniques: impoundment by low lift locks and dams, dredging and disposal, channel training structures, and bank stabilization. The 1930s dams impounded the UMR raising water levels so channel training structures are completely submersed in most areas. Aquatic restoration typically incorporates recreating bathymetric diversity with dredging, flow modifications to increase or decrease connectivity to a more natural balance, and island construction to replace terrestrial floodplain habitats.

Increased water volume in the MMR between the confluences of the Missouri and Ohio rivers is usually adequate to provide navigation. However, more than 80% of the floodplain is leveed and an extensive system of channel training structures, bank stabilization, and dredging and disposal are necessary to maintain the 2.7-m navigation channel. Here, alterations to channel training structures can recreate diverse habitat in the main channel border and reconnect side channels

Below the mouth of the Ohio River, the LMR is typically a high water volume system. A network of levees and floodways contain the lateral extent and active river area as a part of the Mississippi River and Tributaries Project, which is almost 90% complete

(Camillo 2012). Within this river segment, the 2.7-m-deep channel and flood risk reduction is managed by an extensive system of channel training structures, bank stabilization, and dredging and disposal coupled with a levee system along the length of the LMR. The batture, or the land between the levees, most of which is forested, is confined within 0.6 million ha of what was once the 13 million ha alluvial valley of the Mississippi River (Baker et al. 1991).

Biodiversity of large floodplain rivers is arrayed along intermediate levels of a disturbance/connectivity gradient (Ward et al. 1999). Ecological restoration in the Mississippi River attempts to recreate this type of gradient using various tools to manage bathymetry, water velocity regimes, and sediment deposition or scouring at specific locations. The goal is to move the river towards the least degraded and most ecologically dynamic state possible, given the regional context (Palmer et al. 2005). There are at least 12 major categories of tools used to restore and diversify riverine habitats in the Mississippi River impacted by channel engineering (Tables 1–4). In many cases, the river engineering programs that tried to harness the river are now being used to restore, create, or preserve habitats in the river and floodplain. The following section describes these tools and how they are applied in restoration programs.

Description of Restoration Tools

Modified channel training structures.— There are two general categories of river training structures (Pokrefke 2012): (1) redirective, such as dikes to scour the navigation channel, which uses the river energy to prevent sediment accretion, and (2) resistive, such as riprap to protect shorelines, which attempts to maintain the system as status quo. There are at least eight modifications of channel training structures used today to restore habitat (Table 1). According to Soar et al. (2005), the applicability of a particular type of structure to any given situation depends on a number of factors such as hydrologic conditions, sediment size and loading, channel morphology, floodplain and valley characteristics, availability of construction materials, and project objectives, as well as the inevitable time and funding constraints. The successful use of a particular type of structure in one situation does not necessarily ensure that it will be effective in another.

Notching transverse/longitudinal dikes.—In recognition of the potential loss of habitat resulting from sedimentation in dike fields, notches have been constructed in the dikes to increase bathymetric diversity, including deep scour holes below the notch. Notching is by far the most widespread restoration activity in the Mississippi River (Figure 5). A notch is a weir section in the dike designed to maintain flow through dike fields at low river stages. Notches are also placed through longitudinal dikes to provide fish and boat access. According to Pokrefke (2012), the size of the weir section depends on local site conditions, but typically the top width varies from about 30–91 m, with a weir invert 1.5–4.6 m below the crown of the dike. Not all dikes are subject to notching, either because of location, potential impacts to integrity of the dike or shoreline, or minimal environmental benefits gained. Rootless dikes, which have an offset typically 30 m or more from the riverbank, create similar bathymetric diversity as notching (Pokrefke 2012).

Hubbard et al. (2006) conducted a limited analysis of pre- and postnotching surveys of transverse dikes within the Memphis and Vicksburg USACE districts. The study showed that the spatial and temporal scour and fill trends in the dike fields were extremely dynamic. The response to notching was observed to be site specific, with locations of scour and fill varying from year to year within the dike fields. These findings are important because they document that dike fields are not static environments, but rather temporally and spatially highly dynamic systems.

Today, all six districts of the USACE Mississippi Valley Division pursue opportunities to remove, notch, or shorten channel training structures for ecological benefits. Notches

Table 1. Type and purpose of the more common river restoration techniques in the free-flowing Mississippi River. Number of applications for each tool is through 2012 for the lower Mississippi River (LMR) and 2015 for the middle Mississippi River (MMR). U.S. Army Corps of Engineers (USACE) districts responsible for most restoration actions in each reach are St. Paul (UMR); St. Louis (MMR); Memphis, Vicksburg, and New Orleans District (LMR). Descriptions of actions were based on USACE (2012a), Davinroy (1996), Pokrefke (2013), Franco (1967), and Fenwick (1969). Number of applications in the LMR were from Killgore et al. (2015) unless otherwise stated and Kip Runyon (USACE, St. Louis District for the MMR, personal communication.)

Type of structure or action	Description of action	Expected habitat benefits	Number of applications
Notching or shortening transverse (also called wing and spur) dikes	A notch is constructed or created through one or more dikes. Typical top width varies from about 31 to 91 km, with a notch invert 1.5–4.6 km below crown elevation of dike.	Designed to maintain flow through the dike fields at low river stages and create scour holes for bathymetric diversity.	• LMR – 225 of 774 dikes • MMR – 177 of 978 dikes
Notching closure structures	Same as transverse dikes	Designed to maintain flow, wetted perimeter, and connectivity through the upstream reach of secondary channels.	• LMR – 14 secondary channels; 91.3 km of in-channel habitat (LMRCC 2015)
Modified revetment	Three types: longitudinal grooves created in articulated concrete mat, woody debris from bank clearing placed along the bankline, and offshore revetment constructed 1.5–4.6 km riverside of existing bankline.	Grooves in articulated concrete mat increase surface area for attachment of macroinvertebrates, woody debis provides velocity refugia and attachment sites for invertebrates, and offshore revetment creates a long ecotone between swift and slackwater.	• LMR – 1,598 km
Stepped-up dikes	In a dike field, the crown elevation of each dike in a downstream sequence is constructed at a slightly higher elevation than the previous one.	Promotes scour between successive dikes in a dike field rather than deposition that occurs in stepped down or dikes with constant elevations.	Unknown
Chevron dikes	Constructed as an ellipse-shaped head oriented upstream and two legs that extend downstream, parallel to the flow.	Scour holes form in the interior of the dike and along the outside of the legs, and a small island may form downstream of the chevron.	• LMR – 7 • MMR – 31
Trail dikes	Extensions attached to a transverse dike. Extensions can be added upstream perpendicular to the flow to form a J-hook or L-head dike, or added both upstream and downstream forming a T-head dike.	Scour holes form at the extensions and slackwater forms between the bank and trail dike.	Unknown
Hard points	In lieu of revetment, hard points consist of stone fills spaced along an eroding bankline, protruding only short distances into the channel.	Compared to revetment, hardpoints conserve natural river bank, diversify riverine habitat through deposition and scour, and provide velocity refugia for aquatic organisms.	• LMR – 227
Roundpoints	Alternating rows of rock mounds are placed within the footprint downstream of a typical dike.	Spacing of the rock mounds are a function of the structure height to create flow and bathymetric diversity within a dike field.	• MMR – 1 set, 11 piles
Island building	Flexible dredge ppe used to build low-elevation islands.	Intended to promote nesting habitat for least terns.	• MMR – 4

Table 2. Island design and building includes the actions and benefits for river restoration listed below for the impounded portion of the upper Mississippi River (UMR). Without exception, islands are built in a complex using the multiple techniques listed below. MMR = middle Mississippi River; LMR = lower Mississippi River.

Type of structure or action	River reach	Description of action	Expected habitat benefits	Number of applications
Island design—natural levee restoration	UMR	Islands are placed along the main channel border to provide the historical border between the fast-moving current of the main thalweg and slowing currents of secondary, tertiary, and backwater complexes. Design elements replicate natural floodplain geomorphology, but also factor in the altered conditions of the current engineered system. In all cases, island design complements the natural river dynamics, including seasonal water-level variations to ensure long-lasting restoration benefits.	Replacing the lost topographic diversity across the floodplain provides diversified flow patterns, reduces wind fetch, increases littoral zones complete with sandbars and mudflats, and promotes terrain variability necessary to support natural aquatic and terrestrial vegetation.	UMR—31 of 54 habitat restoration projects include islands (USACE 2012).
Island design—wind fetch reduction	UMR	Within larger complexes of island, additional islands are strategically placed to break up wind from the prevailing direction(s). These islands are typically small, require extra armoring, or are constructed entirely of rock to assure stability against substantial wind fetch. Although orientation and height may be altered slightly, these islands are often reconstructed based off the platform of historial images and maps.	As wind fetch is reduced, the calm water in the shadow of these islands provides prime conditions for aquatic plant growth, waterfowl brood cover, spawning and juvenile fish habitat, rest areas for migratory waterbirds, and overwintering areas for backwater (lentic) fish species.	UMR—31 of the 54 habitat restoration projects include islands (USACE 2012).
Island design—altered flow distribution	UMR	Along with recreating natural levees and breaking up wind fetch, river managers are using islands to redirect flow to promote channel scour. Strategically placed islands recreate bathymetric diversity to direct flow and transport sediment in a manner that promotes self-perpetuating channel diversity.	Islands design redirects flow paths, scours accumulated substrates, and increases flow heterogeneity. Riverine fishes flourish in these improved flow conditions, and as a consequence, dependent native mussel species are also found in these renewed habitats.	UMR—31 of the 54 habitat restoration projects include islands (USACE 2012).
Island design—dredge material placement	UMR	Maintenance of the channel for commercial navigation requires the placement of dredge material in an environmentally reasonable manner to meet state and federal regulatory policy. In collaboration, staff responsible for channel maintenance and river resource managers work to find	Dredge material can be beneficially used to create habitat islands when properly coordinated with natural resource managers and channel maintenance engineers. When building these islands, they follow the guidance necessary to safeguard	UMR—Pools 4, 5, 5a, 8

Table 2. Continued.

Type of structure or action	River reach	Description of action	Expected habitat benefits	Number of applications
Island design—dredge material placement (continued)		beneficial locations for dredge material and construct islands within the floodplain for the benefit of habitat and channel maintenance.	natural resources, recreate lost habitat, and provide a cost saving for environmental programs as channel maintenance pays for dredging and placement.	
Island design—recreate sand flats	UMR MMR LMR	Mechanical or hydraulic dredge material placement is used to specifically restore critical sand flats necessary for the life history of many riverine species but especially the endangered interior least tern. In the LMR, these sand flats are generally isolated from large islands or riverbanks to reduce predation for terrestrial species. In the UMR, these features are often a component of another island construction project.	Historically, sandbar/sand flats were scattered across the river floodplain. Channelization has diminished and degraded the quality of bare sandbar island habitat throughout the system. Restoration of these critical habitats is necesary for interior least tern nesting, other nonendangered shore birds, reptile basking and nesting, and invertebrate habitat.	UMR—Pools 4, 5a, 8, 9, 10 LMR—84 km of reconnected side channel develop sandbars habitat.
Island design—recreate mud flats (emergent wetlands)	UMR	Mechanical or hydraulic dredge material placement is used to regenerate natural mudflats at an elevation similar to average summer and fall water surface elevation. and fall water surface elevation. Typically a special feature of larger islands, a berm is constructed around the periphery of the mudflat and filled with fine sediments. As the sediment dries, the berm is breached to provide ingress and egress of mobile aquatic species to this valuable habitat. Hydraulic dredge placement requires the movement of pipe to assure an even distribution of fine sediment across the delineated mud flat.	Similar to sandbars and sand flats, mud flats are essential off-channel habitat. A variety of plants and aquatic organisms rely on mudflats to complete one or more of their life requisites, including native emergent and floating leaf aquatic plants, invertebrates, waterbirds, and juvenile habitat for a variety of ducks and fish.	UMR—Pools 5, 5a, 6, 8, 9

Table 3. Backwater/off-channel restoration dredging for river restoration is done in both the impounded (upper Mississippi River [UMR]) and free-flowing (middle Mississippi River [MMR] and lower Mississippi River [LMR]) portions of the Mississippi River.

Type of structure or action	River reach	Description of action	Expected habitat benefits	Number of applications
Backwater dredging – hydraulic	UMR	Dredging to regain bathymetric diversity and critical water depth in off channel areas has become a vital component of river restoration. Dredge cut considerations include typical water levels, long-term desired depth, and projected sedimentation rate. In northern climates, the depth of ice is also a consideration. Hydraulic dredging of the backwaters is possible if the dredging volumes are high, typically more than 38,000 m^3, and there is sufficient depth to allow equipment access.	Bathymetric diversity has been lost to impoundment, sedimentation from upland channelization and island erosion. Regaining off-channel deep holes, flowing channels, and varying gradient of shallow and deepwater marsh is imperative to system restoration. Restored depth benefits habitat for aquatic plants, invertebrates, water birds, herptiles, fish, and mammals.	UMR – 29 of 54 habitat projects that have dredging as a part of the project. Hydraulic versus mechanical not differentiated (USACE 2012).
Backwater dredging – mechanical	UMR	Mechanical dredging has the same considerations and benefits as hydraulic dredging but it is more agile in the backwater situations. In many cases, it is the best option for creating pilot channels when side channels are reopened and reestablishing flows is an important factor. Mechanical dredging is the preferred method of dredging when dredging quantities are relatively low, generally less than 38,000 m^3, access depth is restrictive, and precision cuts are desired.	Bathymetric diversity has been lost to impoundment, sedimentation from upland, channelization, and island erosion. Regaining off-channel deep holes, flowing channels, and varying gradient of shallow- and deepwater marsh is imperative to system restoration. Restored depth benefits habitat for aquatic plants, invertebrates, water birds, herptiles, fish and mammals.	UMR – 29 of 54 habitat projects that have dredging as part of the project. Hydraulic versus mechanical not differentiated (USACE 2012).

Table 4. Water level management for river restoration is done primarily on the impounded (upper Mississippi River [UMR]) portion of the Mississippi River and includes two different techniques listed below. Hydrologic conditions dictate which technique is most effective in different segments of the impounded portion of the river.

Type of structure or action	River reach	Description of action	Expected habitat benefits	Number of applications
Water level control with managed connections	UMR	Although not the preferred option, developing areas that can be sequestered from dynamically fluctuating river water levels is often the only answer to recover more natural seasonal variability. This requires the construction of some form of structure or the modification of an existing levee to keep these fluctuations out of the floodplain complex for desired conditions. Stoplog structures, gated structures, pump stations or a combination are typically built within the levee system to manage the water levels. These projects do not generally provide ingress and egress of fish species and can be costly to build, maintain, and operate.	Managing water levels closer to natural seasonal variability promotes the biogeochemical processes necessary for sediment drying, seed germination, restored native aquatic plant vigor, and nutrient assimilation.	UMR – 24 of the 55 habitat restoration projects include water level control as a project component (USACE 2012).
Water level management with dam reoperation	UMR	Large-scale water level reductions can be managed by the dam structures in the upper Mississippi River. With careful consideration for all river uses, including navigation, water supply, and recreation, minimal summer drawdowns can, on average, expose from 400 to 800 ha of river bottom for sediment drying and seed germination and restore aquatic plant vigor. Each dam requires different discharge parameters and dredging requirements to maintain the channel at each dam and upstream pool.	Restoring more natural seasonal water levels promotes favorable conditions for seed germination of annual and perennial emergent, floating leafed and submersed aquatic vegetation. Water-level reductions can also promote beneficial conditions for sediment drying and consolidation. Increased plant density and diversity increases water clarity, habitat for aquatic life from invertebrates to fish, and nutrient cycling and may reduce harmful algal blooms.	UMR Pools 5, 6, 8, 24, 25, 26. One attempt on Pool 13 was abandoned soon after implementation (USACE 1996, 1999, 2007).

Figure 5. A notch constructed in a lower Mississippi River dike that creates scour holes, eddies, and enhanced connectivity. Photo credit: Jan Hoover, U.S. Army Corps of Engineers, Engineer Research and Development Center.

are often incorporated during routine channel maintenance activities, thus reducing or eliminating overall costs of the restoration measure. The St. Paul, St. Louis, Memphis, and Vicksburg districts have been the most active in actually altering these structures. Scour holes (or plunge pools) created by notched dikes are important habitat targets for LMR recreational anglers, particularly for the three species of catfish found in the river: Blue Catfish *Ictalurus furcatus*, Channel Catfish *I. punctatus*, and Flathead Catfish *Pylodictis olivaris* (LMRCC 2012).

Notching closure structures in secondary channels.—Secondary channels are referred to by various names: chutes, abandoned channels, side channels, and sloughs. Geomorphically, they form islands as the secondary channel cuts through the floodplain before rejoining the main channel. Closure dikes are constructed in secondary channels to shunt more water into the main channel during low flows. A notch, similar to that described for transverse dikes, is constructed to restore flow through the secondary channel without compromising navigable depths in the main channel (Figure 6). Multiple dikes in a secondary channel are usually notched to increase flow through the upper to lower reaches.

The Lower Mississippi River Conservation Committee partners with USACE and USFWS in the LMR to construct notches in closure structures to rehabilitate secondary channels, provide more permanent flow to these areas at lower river stages, and provide habitat for riverine fishes including the federally endangered Pallid Sturgeon. The LMRCC coordinates with USACE to determine potential secondary channel candidates with high priority index values (Killgore et al. 2012) that are logical from engineering and restoration perspectives. Issues for consideration include overall bed elevation of

Figure 6. Closing dike along the lower reach of a secondary channel in the lower Mississippi River that was notched to restore a flow-through system at lower river stages, Lookout Dike Field, river kilometer 1241. Photo credits: Jack Killgore, U.S. Army Corps of Engineers, Engineer Research and Development Center.

sandbars within a dike field, potential impacts to navigation, and potential threats to the structural integrity of the dike. To create hydrological connection between the secondary channel and main channel, it is often necessary to notch every dike in a dike field, where practicable. Engineering plans and design are developed, but often, site-specific conditions dictate the extent to which work can be accomplished in the field. Invert elevations of notches range from 2 to 3 m below the crest of the dike, although field conditions may not permit reaching that depth. The LMRCC is responsible for hiring the construction contractor, coordinating the work between organizations, and securing all necessary permits. The USFWS provides funding for project construction through the Fish Passage Program. Average cost per dike is approximately US$25,000, which only includes construction costs. There is considerable work-in-kind contributions, in addition to private matching funds used to complete these projects in a cost-efficient manner.

Bank stabilization and modified revetment.—Shoreline and bank stabilization are used in a number of different applications. In one scenario, stabilization is used to protect existing habitat subject to erosion and under threat of complete loss if preventative action is not taken. Typically, this is an island or shoreline along a main flowing portion of the river or an area with extensive wind fetch where the land mass is vulnerable to excessive erosion. Conversely, when islands are built in the

UMR, the new structures need to be stabilized to assure that they remain in place long enough to mature and naturalize with the dynamics of the river processes.

Innovative designs have transformed the composition of stabilization from traditional linear rock structures to designs that incorporate living plants, wood, sacrificial berms, and hydraulically reshaped structures that capture or funnel flow to create the desired habitat outcome. These changes and the lessons learned from in-river application are well documented in the environmental design handbook that should serve as a reference guide for engineering design for river floodplain restoration (USACE 2012).

Actual stabilization techniques depend on erosive forces exerted on the newly created or existing features, but in all cases, minimal intervention is the construction target. Three categories of stabilization are constructed: riprap/revetment, vegetative, and biogeostabilization to provide the appropriate protection.

Revetments are placed on riverbanks to arrest bank caving and protect levees and other structures and to maintain an efficient channel alignment. Several modifications are used to minimize impacts and restore some level of habitat complexity.

- Articulated concrete mattress grooves—In the 1980s, articulated concrete mattress design was modified for environmental enhancement in the LMR (Figure 7). Articulated concrete mattress is now constructed with longitudinal grooves over the surface of each block to reduce current velocity and increase surface area for the attachment of macroinvertebrates (Way et al. 1995).
- Off-bankline revetment—Pokrefke (2012) described off-bankline revetments as an environmental enhancement alternative, providing slack water habitat compared to traditional on-bankline revetments. According to Pokrefke (2012), the off-bankline revetment is a row of stone, typically "A" stone, and is placed 1.5 to 4.6 m riverside of the existing bankline at an elevation of the existing bank height. Notches are typically left in the revetment to allow fish and boat access to the slack water areas.

Newly constructed river features that are expected to sculpt with river dynamics require engineered but typically less substantial armoring to allow feature development. Vegetative stabilization or biogeostructures are most commonly used in the UMR. Vegetative stabilization is typically the planting of willow *Salix* spp. cuts planted manually or with a machine, which fluidizes the substrate allowing easy insertion. Biogeostructures are often a combination rock groins, sacrificial berms, and planted flood-tolerant woody vegetation like willows (USACE 2012). The use of sacrificial berms promotes a gradual slope with extensive littoral zone and a final island slope approaching 10:1. Over time, these features are extremely resilient and stable to river dynamics in pooled reaches due to the long slope and likelihood to promote aquatic plant growth in the shadows of the feature.

Other modifications to river training structures.—New and innovative river training structures or designs have been developed to redirect flow to the main channel and protect shorelines from erosion while providing depth and velocity diversity for habitat enhancement. A brief description of these structures are provided below and summarized in Table 1.

- Stepped-up dikes—Dikes comprising a system may be stepped-up (i.e., dike crown elevations increase downriver to promote scour between successive dikes; Franco 1967; Fenwick 1969). The river's current, moving over each submerged dike, allows the sediment buildup to be redistributed back into the main channel and carried downstream (USACE 2012).
- Chevron dikes—These structures are constructed with an ellipse-shaped head oriented upstream and two legs that extend downstream, parallel to the flow (Figure 8). Scour holes form in the interior of the dike and along the outside of the

Figure 7. U.S. Army Corps of Engineers Matt Sinking Unit laying out articulated concrete mattress along a bank in the lower Mississippi River. Inset shows longitudinal grooves to increase surface area for benthic macroinvertebrate attachment. Photo credits: Jack Killgore, U.S. Army Corps of Engineers, Engineer Research and Development Center.

legs, and a small island may form downstream of the chevron. Chevrons were originally used by the St. Louis District in the MMR as a river training structure for erosion control at the upstream end of islands and habitat restoration tool (Davinroy et al. 1996; Johnson 2003). Chevrons are typically used in wider reaches of the river where a flow split is desired (Davinroy et al. 1996). W dikes are similar except shaped like the letter W, with the apex of two legs facing upstream forming two scour holes and one depositional bar upstream.

- Trail dike—An extension of a transverse dike can be added upstream perpendicular to the flow to form a J-hook or L-head dike, or added both upstream and downstream forming a T-head dike. Scour holes form at the extensions and slack water forms between the bank and trail dike.
- Hardpoints—This technique is a cost-effective alternative to revetment consisting of stone fills spaced along an eroding bank line, protruding only short distances into the channel (Figure 9). Groins are similar rock structures extending perpendicular to the shoreline to reduce erosion from wind driven wave action. Over time, the area between hardpoints and groins will create a scalloped beach effect, further enhancing the bank environment (USACE 2012). Compared

Figure 8. Aerial view of chevrons. Photo credit: Jack Killgore, U.S. Army Corps of Engineers, Engineer Research and Development Center.

Figure 9. A series of hardpoints along an eroding bank at Delta Point Secondary Channel, river kilometer 562. Photo credit: Jack Killgore, U.S. Army Corps of Engineers, Engineer Research and Development Center.

to revetment, hardpoints conserve natural riverbank, diversify riverine habitat through deposition and scour, and provide velocity refugia for aquatic organisms.

- Roundpoints—These are alternating rows of rock mounds within the footprint of a typical dike with the spacing of the rock mounds a function of the structure height to create flow and bathymetric diversity within a dike field (Pokrefke 2012).

Island design and building.—Historically, islands were integral to the physical structure and function of the UMR floodplain. Without islands, shallow littoral zones are diminished for aquatic plant growth, wide expanses of open waters become subject to erosional processes from wind and boat waves, biogeochemical processes are severely limited, and habitats for invertebrates, herptiles, fish, wildlife, and birds become degraded or nonexistent. Restoring islands within the UMR plays a pivotal role in the restoration of the mosaic of habitats and functions that were historically present in the system (USACE 2012).

Basic riverine dynamics dictate island morphology. Islands adjacent to the main flowing portion of the river naturally have a higher elevation because sediment-laden water leaves the high energy thalweg and drops its sediment. In the broader floodplain, islands in the secondary and tertiary channels are often lower and stable because they capture occasional flood waters and finer sediment that remain longer in suspension. River engineers and ecologists attempt to replicate the natural diversity within the confines of an altered, impounded system. To ensure lasting benefits from restoration, island design must consider seasonal variation and compliment natural river dynamics.

Through the Environmental Management Program, island reconstruction has developed based on several factors: site-specific needs, including current conditions; desired outcomes together with funds available to construct the optimum feature(s); and lessons learned from constructing islands in a dynamic riverine environment (USACE 2012). Careful consideration must be given to cumulative changes that occur to the system hydraulics as a result of the completed project. Therefore, each structure and complex area is often a combination of science and art that considers historic maps and photographs, along with current site conditions to determine the best island location, elevation, and size (Figure 10).

- Throughout the history of HREPs, most island construction serves to reduce fetch in impounded portions of a pool and emulate a more natural flow regime by reducing below bank-full connectivity between the active river flow and the slack backwater areas. In the 27 years of UMRR-EMP, a few basic standards have been established creating a set of baseline factors to consider (Table 2). Islands should be built in water less than 1.8 m to reduce erosive forces that will continually destabilize the structure. Island construction should start with a course-grain sediment base to provide optimum stability (Figure 11).
- A top layer of fine material (0.2–2.0 m thickness) is incorporated into the island footprint to promote terrestrial plant growth and secondary island stabilization. Fine material can often be excavated from the river bottom adjacent to the island to minimize material movement and create more depth in the area (USACE 2012).
- Islands are designed to flood from downstream to upstream to reduce hydraulic forces from eroding the island. Designing islands to flood from downstream to upstream reduces the hydraulic jump and resultant erosive forces that could jeopardize the structure during flooding conditions (USACE 2012).
- Island stabilization is necessary to ensure the longevity of the constructed islands within the floodplain. Depending on river currents, wind fetch, and other erosive forces, rock riprap, rock groins, and biological stabilization are used strategi-

Figure 10. Multiple designs of island construction and rehabilitation are used along Pool 8 of the upper Mississippi River. Photo credit: John Nelson, Wisconsin Department of Natural Resources.

Figure 11. Construction of an island in Pool 8 of the upper Mississippi River. Photo credit: Sharrone Baylor, U.S. Fish and Wildlife Service.

Figure 12. Construction of rock groins in Pool 9 of the upper Mississippi River. Photo credit: Jeff Janvrin, Wisconsin Department of Natural Resources.

cally to protect the island footprint (Figure 12).

Habitat specific features have been designed within the footprint of large islands over the 27-year program (USACE 2012):

- Rock sills (armored low section of an island) and wide, low elevation earthen islands provide an area for overtopping during fairly routine flood events, typically every 2 years, to minimize local effects on 100-year flood stages. These features have also been used to enhance natural scour during flood events to maintain backwater channels and, when combined with a notch, maintains dissolved oxygen levels year round.
- Island sand points and flats are island features designed to promote bare or near-bare sand persistence needed for shorebirds, reptiles, and amphibians. Through careful consideration of water levels, these sand areas remain in the unvegetated condition indefinitely.
- Mudflats and isolated wetlands are features that have been incorporated into constructed islands to diversify habitat benefits. Fine sediments are carefully deposited into an area within a containment berm (that may or may not be breached after construction) to create the proper conditions for the desirable aquatic plant growth.
- Woody debris is occasionally incorporated as a feature for island stabilization to create an alternative to a rock structure and creates natural microhabitats missing from the river with intense river management (Figure 13).

More recently, the St. Louis District was the first agency in the Mississippi River to use flexible dredge pipes to create sandbar habitat (Figure 14). Traditional dredging within the St. Louis District utilized a rigid metal pipe that discharged dredged material parallel to the dredge cut resulting in long, narrow disposal bars limited in size,

Figure 13. Woody debris has been incorporated as a feature for island stabilization using groins in Spring Lake, Pool 5 of the upper Mississippi River. Photo credit: Jeff Janvrin, Wisconsin Department of Natural Resources.

elevation, location, and habitat diversity (Cox et al. 2011). The flexible dredge pipe can target discharge sites, create islands and sandbars, and ultimately restore habitats impacted by traditional dredging and river training structures.

Backwater dredging.—Another consequence of impoundment in the UMR is the loss of bathymetric diversity, especially directly upstream of the dams. While difficult to observe, the combination of island loss from erosion and watershed inputs of sediment is filling the deep channels and covering other water features, particularly in slack water above the dam structures (Johnson and Hagerty 2008). Dredging to regain critical depth has become an integral part of backwater habitat restoration; however, there are several challenges: determining where to dredge, quantities of material to dredge, and contours of the dredge cuts. If cuts are too deep, they create slumping and headcutting, while too shallow can create conditions that fill quickly and need to be maintained on a regular basis. Understanding sedimentation rates and patterns in the areas can be advantageous to determine appropriate dimensions but are not always readily available (USACE 2012).

The cumulative experience from restoration dredging in the UMR can provide reasonable guidelines for projects elsewhere. Dredge-cut dimensions must include consideration of typical water levels, long-term desired depth, and projected sedimentation rate. The depth of ice will also be a consideration in northern climates. Dredging can either be hydraulic or mechanical depending on quantities, depth of access to get equipment into dredging locations, and ability to meet water quality standards. In many cases, contract specifications provide the contractor with the option to determine which method of dredging is possible and cost effective (Table 3).

Dredging depth for channel restoration varies based on habitat objectives, usually associated with habitat for backwater fish-

Figure 14. Flexible dredge pipe being used to construct islands in the middle Mississippi River. Photo credits: U.S. Army Corps of Engineers, St. Louis District.

es. Backwater dredging depths to provide overwintering habitat for fishes have ranged from 1.2 to 4.0 m (USACE 2012). The depth is determined based on a balance of disposal capacity (usually a feature associated with an island), site-specific habitat needs, forecasted sedimentation rates, and cost. More recent projects have incorporated a variety of dredging depths to increase diversity within the backwater complexes (J. Janvrin, Wisconsin Department of Natural Resources, personal communication). Access channels for construction are typically 1.8 m deep to accommodate draft of equipment. When possible, access channels are located to provide channel or backwater habitat benefits upon completion of the project. The channel slope is dependent on the site-specific conditions but is commonly built in the range of a 2:1 to 4:1 slope (USACE 2012). Last, deeper holes can provide important habitat and are typically 6 m deep with varying dimensions based on site conditions and habitat needs. Site-specific knowledge of sediment rates is valuable to assure that the deeper water will not simply create a sediment trap.

Dredging is often done in close proximity to island construction and dredge material is typically suitable as fill for newly constructed islands, generating cost savings for the overall ecosystem restoration project. Grain size and the potential for contaminants are carefully considered before dredge material placement.

Water-level management.—The Mississippi River historically experienced seasonal variation in flow and water levels, often referred to as the flood pulse. Late winter/early spring was marked by high water from snowmelt and rains that would trail off until midsummer when flows and water levels typically reached lower levels. Late summer and early fall often came with additional rain, resulting in a slight raise in flow and water levels. Early winter flows and stages are typically the lowest within the annual hydrologic cycle. This pattern has persisted since the glacial retreat (Fremling 2005), resulting in multiple ecological benefits but especially conducive to the growth of native floodplain and aquatic vegetation communities.

Locks and dams were constructed on the UMR from 1930 to 1940 to maintain the 2.8-m navigation channel. Managed water levels at the locks and dams function to eliminate the historic seasonal low water levels, but higher flood waters are allowed to pass through the lock and dam system. The loss of seasonal (summer predominantly) lower water levels has led to a change in the distribution and abundance, significant interannual variability, and, in some cases, elimination of native aquatic plant communities (USACE 1996).

Natural resource river managers are using two different techniques to manage water levels and simulate a more natural hydrograph to restore essential ecosystem components. In river reaches with highly variable flow and water levels, floodplain areas can be sequestered from the adjacent river by levees or dikes and managed using stop logs, gates, or pump stations to regain more-natural seasonal water levels. In the large impounded portions of the river, lock and dam systems have been used to reduce summer water levels to improve ecological conditions within the constraints of the annual hydrology and the need to maintain the navigation channel. Both methods are important restoration tools requiring careful planning to minimize challenges associated with implementation (Table 4).

Human activities influence river hydrology that subsequently creates complex situations for river managers to replicate seasonal hydrographs. Although not ideal, using existing sequestered floodplain areas or developing new areas is often the only answer to recover more natural seasonal variability. This requires some form of structure or the modification of an existing levee to maintain desired habitat conditions. Stoplog structures, gated structures, pump stations, or a combination are typically incorporated to provide the mechanisms to manage water levels within the area for multiple benefits that include plants, birds, reptiles, amphibians, and fishes. In general, these projects restore vital ecological conditions but may impede ingress and egress of fishes and are more costly to build, maintain, and operate (USACE 2012).

Within UMRR, 27 site-specific water-management projects have developed moist-soil management units or isolated backwater lakes (USACE 2012). Both are managed for native aquatic vegetation production, but moist-soil management units are managed with persistent low water until early fall primarily for waterfowl use while isolated backwater lakes are managed to replicate more natural seasonal variability for multiple fauna. In both cases, water-level control is essential to meet the habitat and restoration needs of the area and is the best way to contend with dramatically fluctuating river water levels. If the adjacent river hydrograph were to return to more historic flow patterns and associated water levels, these areas could be reopened to the flowing river.

Without a specific program or funding designation, state and federal agencies began cooperatively investigating the possibility of manipulating water levels at dams to create pool-wide summer drawdowns. River managers understood that they could not impact commercial navigation with reduced water levels, but with careful planning, the locks and dams could be used to draw down water levels to mimic more natural summer conditions.

Each of the 29 lock and dam structures on the UMR have different operating conditions for managing water levels. In Pools 24, 25, and 26 the operational range allows water reductions below full pool without changing operation procedures. The commercial channel, ports, marinas, and water intakes are designed for a wide range of water levels, so reducing water levels does not require burdensome considerations to implement. Most of the pools in the upper portion of the UMR have a much narrower range of water level operation and require careful consideration of all river users. Any changes to operations must be approved by USACE. Pool-scale drawdown in these areas requires extensive planning to manage the reduced water levels for commercial navigation, recreational access, water intake and discharge, and biological benefits. These differences have resulted in two different implementation scenarios; one that is relatively simple to implement and has been underway for 20 years and one that requires years of planning and has only been done five times in 19 years.

Working cooperatively, USACE along with local, state, and federal natural resource managers planned and implemented the first pool level drawdown on the UMR in 1994. The Pool 25 experiment consisted of holding the pool elevation approximately 0.6 m lower for a period of about 30 d during the beginning of the growing season. At the completion of the drawdown, the pool was slowly returned to normal water level operations. Exposed river mudflats became covered with moist soil plants that provided valuable fish and wildlife habitat and arguably fostered sediment and nutrient assimilation in proximity to the restored plant beds (Figure 15).

With the 1994 success in Pool 25, environmental pool management was expanded to Pools 24, 25, and 26, or the southern-most impounded portion of the UMR. Environmental Pool Management goals change each year based on expected and actual flow conditions and input from the natural resource managers. This flexibility allows consideration for the ecological needs to enhance conditions for resident or migrating biota along with the need to manage the commercial navigation channel.

Environmental pool management exemplifies how successful demonstrations can become routine operation by blending management for multiple river benefits. Currently, river managers are considering additional modifications to extend the drawdowns from 30–45 d to 45–90 d of the growing season to potentially gain a response from perennial as well as annual aquatic vegetation and provide additional opportunities to dry and consolidate sediment. Many factors will have to be considered, including economic, recreational, and biological impacts, to test these adaptive approaches.

On the northern portion of the UMR, a USACE planning process identified Pools 5,

Figure 15. Moist soil plants growing along mudflats in Pool 25 of the upper Mississippi River after a 30-d drawdown that exposed mud flats. Photo credit: Kenneth Dalrymple, U.S. Fish and Wildlife Service.

7, 8, and 9 as the best candidates for the first pool-scale demonstration drawdown (USACE 1996). Through an exhaustive process that included significant public involvement, UMR Pool 8 was selected as the initial study pool because of a wealth of existing data, including bathymetry, sediment-type distribution, and expected benefits to aquatic vegetation. The drawdown was a 0.46-m reduction, with water levels managed at both the dam structure and the La Crosse, Wisconsin stage gauge, depending on river flow (USACE 1999).

The Pool 8 demonstration drawdown was conducted in 2001 from June 30 through September 15, with an estimated 791 ha of substrate exposed. This demonstration was repeated in 2002 with almost ideal conditions to conduct the second year drawdown. Monitoring in Pool 8 documented both annual and perennial emergent plants production in the exposed sediments. Plant species composition found on sample transects during the drawdown matched the results of 1999 premonitoring seedbank study done for Pool 8. At the conclusion of the second year drawdown, transect sampling documented plant species shifted from annual to perennial, and in the case of broadleaf arrowhead *Sagittaria latifolia*, tuber production increased 16-fold (3.4 g/m^2 versus 55.3 g/m^2) across all transects during the second year (Nissen 2014).

Today, three demonstration drawdowns have been conducted and a fourth drawdown is planned for Pool 3 (Figure 16). Abiotic and biotic responses associated with the Pools 5, 6, and 8 drawdowns have been documented in a number of published (Cavanaugh et al.

Figure 16. Pool 5 of the upper Mississippi River before (left) and after (right) drawdown enhancing aquatic and marsh plant growth. Photo credits: Tim Schlagenhaft, Minnesota Department of Natural Resources.

2006; Custer et al. 2007; Kenow and Lyon 2009) and unpublished reports (Kenow et al. 2007; USACE 2007). Preliminary assessments indicated that drawdowns enhanced deep and shallow marsh perennial, rooted floating aquatic, and shallow marsh annual plant communities. More than 70 species of moist-soil, perennial emergent, and aquatic species grew on the exposed substrate, including rice cutgrass *Leersia oryzoides*, broadleaf arrowhead, flatsedge *Cyperus* sp., and smartweed *Polygonum* sp. These changes provided positive effects on fish spawning and nursery areas, improved forage production and availability for migrating waterfowl and shorebirds, and had no measureable adverse effects on fish populations, water quality, or contaminant bioavailability (Kenow et al. 2015).

Diversions

Freshwater diversions have been constructed in the lowermost 193 km of the LMR for habitat restoration, reduction of saltwater intrusion, sediment introduction and land building, and flood control. Land loss due to subsidence, sea level rise, and hurricanes is motivating plans for more and larger freshwater diversions in southern Louisiana (CPRA 2012). Five freshwater diversions are currently operating as restoration tools but differ in physical structure, capacity, and operation (Davis Pond Diversion, the Violet Siphon, the Caernarvon Diversion, the White Ditch Siphon, and the Naomi Siphon; Figure 17). Diversions have ranged from siphon pipes with a maximum capacity of 7 m^3/s (White Ditch) to concrete box culverts with a maximum discharge of 302 m^3/s (Davis Pond). Larger diversions are planned that may exceed 2,100 m^3/s to move more sediment into the marshes for land-building purposes. Controversy surrounds the predictions that diversions will result in measurable land building and not impact important recreational and commercial fish and shellfish.

Reforestation.—The lower Mississippi River alluvial valley was almost entirely covered by bottomland hardwood forests and cypress-tupelo swamps before European settlement. Estimates were 10 million ha of forested floodplain prior to clearing and levee development, with an estimated 2.8 million ha today (King et al. 2006). According to the U.S. Forest Service Forest Inventory and Analysis National Program, the majority of forested lands in the lower Mississippi River corridor (Industrial Economics 2014) are privately owned (84%). Within the active floodplain (batture), there are more than 300,000 ha of agricultural lands (2008 National Agricultural Statistics Service data). Much of the reforestation efforts in the lower Mississippi alluvial valley, to date, have been accom-

Figure 17. Davis Pond Diversion in the lower Mississippi River operated to restore coastal marshes in Louisiana. Photo credit: U.S. Army Corps of Engineers, New Orleans District.

plished through cooperative efforts among the U.S. Department of Agriculture Natural Resources Conservation Service and nongovernmental partners.

Typically, reforestation in the LMR and other riverine systems must consider the following criteria (Allen et al. 2001):

- Choosing the planting site: Plantings should take place primarily on lands least suitable for agriculture and include edges of rivers, streams, borrow pits, drainage channels, farm ponds, roadsides, and fencerows.
- Choosing the species: Base selection of tree species on several factors, including which species are capable of growing on the site, the availability of planting stock, and personal objectives for reforestation.
- Choosing the planting method: The two major planting methods are direct seeding and planting seedlings. Both can be done either by hand or with machines.

Due to logging, agriculture and urbanization, the UMR floodplain forest represents only a small portion of the presettlement range. Today, floodplain forests are often confined by levees and must adapt to elevated water levels in the impounded portion of the UMR. Island erosion and increased water levels have reduced topographic diversity, making conditions less suitable for forest species that are less tolerant of continual inundation. As trees die and forest canopy opens up, the typical regenerate of pioneer to climax species is replaced by herbaceous vegetation or grasses like reed canary grass *Phalaris arundinacea* that favor current moisture regimes of the island and shoreline (Guyon et al. 2012). Impoundment also exacerbates erosional forces along islands and shoreline areas, making trees more vulnerable to uprooting, especially during flooding or other episodical weather events.

River ecologists are concerned with these trends and developed the Upper Mississippi

River Systemic Forest Stewardship Plan to address the management necessary to maintain healthy forests in the UMR (Guyon et al. 2012). The restoration tools are management intensive due to the altered conditions and the magnitude of the issues associated with forest succession. Landscape actions include prescribed burning, elevation modification, and water level management. Site specific work includes stand improvement, site preparation with weed inhibitors, and tree plantings that entails containerized saplings, bare root seedlings, and direct seeding. The U.S. Army Corps of Engineers and USFWS, along with other natural resource partners, coordinate activities to ensure that this work moves forward on federal land in this area.

Conclusions and Recommendations

The environmental movement of the 1960s brought about change for natural resource protection, conservation, and restoration. Ecosystem restoration and set-aside conservation areas on the Mississippi River came out of the changes that resulted for people caring more about clean water, clean air, and healthy fish and wildlife populations. More than 40 years later, the legislation that provided for these opportunities has resulted in a healthier ecosystem and a better appreciation of the diverse biota that inhabit the Mississippi River (Figure 18).

In most cases, river managers, the public, or both had to fight for opportunities to improve river conditions for something other than human uses. These struggles have resulted in federal authorizations and appropriations such as GREAT, the UMRR-EMP, and biological opinions for the recovery of endangered species. These programs allowed brain trusts to form among hydraulic engineers, river managers, and scientists to create and implement the actions necessary to restore and recover degraded habitats. Large river restoration was virtually unknown at the beginning of these programs, but the combination of river expertise on a site specific level and knowledge of river dynamics provided a successful laboratory to find, test, and refine solutions. The adaptive component of all these programs continues today.

A number of restoration measures have been part of innovative demonstrations, multiple phase project complexes, and systematic retooling of the river. In all cases, the projects provide ecological benefits, along with meeting the needs of commercial navigation, recreation, energy use, flood risk management, and water supply. Successful partnerships between state and federal government, nongovernmental organizations, and the public have provided the potency to move this important work forward. The innovative work and the partnerships will be vital for continued success.

For all that has been accomplished, we have merely scratched the surface of what needs to be done to restore the Mississippi River system as a whole. The main stem is a sum of its parts, and many watersheds that enter the Mississippi River are severely degraded. Watershed programs help to improve these conditions, but current and legacy inputs of sediment, nutrients, and pollutants will take time to change. Additionally, the hydrologic delivery from these systems is vastly different due to increased urban impermeable surfaces, agricultural drainage, aquifer depletion, and altered or complete elimination of floodplain and riparian areas. With these conditions, maintaining current river system health can be challenging.

In the main stem of the UMR, we have only begun to restore the degraded habitat. In the 27 years of UMRR-EMP, it is estimated that the program has directly restored about 3% of the floodplain habitat mosaic, and with indirect benefits, that number could likely be doubled. The UMR is only a portion of the system, with the obvious implication that there is still much work to be done. Similar needs occur in the MMR and LMR to continue and expand modifications to channel training structures that minimize impacts (e.g., sedimentation on gravel bars) and diversify habitats (e.g., plunge pools). Conservation is also important in the free-flowing

Figure 18. Braided floodplain habitat in mid-pool 8, upper Mississippi River. Photo credit: ©Robert Hurt.

LMR to preserve the large tracts of natural riverine habitat that still exist today.

The future will have to include structural remedies for lost habitat, some reverse engineering of the river, and at least some ability to allow the river to seasonally function like a river-floodplain system. Structurally, the islands, along with dredging and flow management, will be essential to maintain and regain these elements in the floodplain, especially in the impounded reaches of the river. In all reaches of the river, it will be necessary to conduct strategic reverse engineering, such as reconnecting side channels to the active river flow, notching structures to improve main channel border habitat, and reconnecting the portion of the floodplain to adjacent river flow. Restoring seasonal variation to water levels and flows that more closely replicate historic conditions is also a key to returning river processes and function. In the impounded section of the river, this means managing water levels at the locks and dams for more than just commercial navigation, as has been demonstrated by summer growing season drawdown in six UMR pools.

In all cases, restoration measures have been used and adapted and continue to improve. It has taken decades to develop and refine restoration tools that are effective and compatible with the multiple uses of the river. Meaningful changes will only come from a long-term commitment to incorporate ecological engineering practices into all water resource planning and construction. Creating a more resilient Mississippi River is a lofty goal but certainly within reach with the experience of the past and the drive to continue to look for new solutions beyond current knowledge to recover a more ecologically sustainable Mississippi River.

Acknowledgments

The authors would like to recognize the numerous basin partners who provided key knowledge and fiscal support throughout the preparation of this chapter. We would like to acknowledge

- U.S. Army Corps of Engineers, Mississippi Valley Division, Mississippi River and Tributaries Project, Mississippi River Geomorphology and Potamology Program, and the Ecosystem Management and Restoration Research Program at ERDC-EL who supported chapter preparation and provided sponsorship.
- Members of the Lower Mississippi River Conservation Committee, past and present, who developed the Restoring America's Greatest River plan for the lower Mississippi River and supported preparation of this chapter.
- U.S. Army Corps of Engineers staff as well as other federal and state partners, who worked to implement, innovate. and adapt the Environmental Management Program (now called Upper Mississippi River Restoration) and supported preparation of this chapter.
- U.S. Army Corps of Engineers Channel Improvement Programs, Mississippi Valley Division, who supported innovation channel management and improved dam operations.
- The Nature Conservancy, North America Water program, who supported the writing and publication of this chapter.

References

Allen, J. A., B. D. Keeland, J. A. Stanturf, A. F. Clewell, and H. E. Kennedy, Jr. 2001. A guide to bottomland hardwood restoration. U.S. Geological Survey, Biological Resources Division Information and Technology Report 2000-0011, Reston, Virginia. U.S. Forest Service, Southern Research Station, General Technical Report SRS-40.

Anfinson, J. O. 2003. The river we have wrought: a history of the upper Mississippi. University of Minnesota Press, Minneapolis.

Baker, J. A., K. J. Killgore, and R. L. Kasul. 1991. Aquatic habitats and fish communities in the lower Mississippi River. Reviews in Aquatic Sciences 3:313–256.

Boysen, K. A., K. J. Killgore, and J. J. Hoover. 2012. Ranking restoration alternatives for the lower Mississippi River: application of multi-criteria decision analysis. U.S. Army Corps of Engineers, Research and Development Center, ERDC TN-EMRRP-ER-17, Vicksburg Mississippi.

Camillo, C. A. 2012. Divine providence: the 2011 flood in the Mississippi River and Tributaries Project. U.S. Army Corps of Engineers, Omaha District, Paper 142, Omaha, Nebraska.

Cavanaugh, J. C., W. B. Richardson, E. A. Strauss, and L. A. Bartsch. 2006. Nitrogen dynamics in sediment during water level manipulation on the upper Mississippi River. River Research and Applications 22:651–666.

Cox, A. N., J. L. Brown, R. D. Davinroy, and E. Brauer. 2011. Engineering considerations for island and sandbar creation using flexible floating dredge disposal pipe: middle Mississippi River, miles 200.0–0.0. U.S. Army Corps of Engineers, Applied River Engineering Center, St. Louis District, Report M57, St. Louis.

CPRA (Coastal Protection and Restoration Authority). 2012. Louisiana's comprehensive master plan for a sustainable coast. CPRA, Baton, Rouge, Louisiana. Available: www.lacpra.org/assets/docs/2012 Master Plan/Final Plan/2012 Coastal Master Plan.pdf. (May 2015).

Custer, T. W., P. M. Dummer, C. M. Custer, A. U. Li, D. Warburton, M. J. Melancon, D. J. Hoffman, C. W. Matson, and J. W. Bickham. 2007. Water level management and contaminant exposure to tree swallows nesting on the upper Mississippi River. Environmental Monitoring and Assessment 133:335–345.

Davinroy, R. D., S. L. Redington, C. N. Strauser. 1996. Design of blunt nosed chevrons in the Mississippi River for sediment transport. Pages X-86–X-93 *in* Proceedings of the sixth federal interagency sedimentation conference, volume 2. Interagency Advisory Committee on Water Data, Subcommittee on Sedimentation. U.S. Government Printing Office, Washington, D.C.

DeHaan, H., J. Stamper, and B. Walters. 2012. Mississippi River and Tributaries System 2011 postflood report. U.S. Army Corps of Engineers, Mississippi Valley Division, Vicksburg, Mississippi.

EMPCC (Environmental Management Program Coordinating Committee). 2012. Plan to transition work under the Upper Mississippi River Restoration-Environmental Management Program to the Navigation and Ecosystem Sustainability Program, April 2012. Available: www.mvr.usace.army.mil/Portals/48/docs/Environmental/EMP/Key%20 Docs/2013_UMRR_EMP_to_NESP_Transition_Plan_3May.pdf. (May 2015).

Endangered Species Act of 1973. U.S. Code, volume 16, sections 1531 et seq.

Fenwick, G. B. 1969. State of knowledge of channel stabilization in major alluvial rivers. U.S. Army Corps of Engineers, Committee on Channel Stabilization, Technical report no. 7, Vicksburg, Mississippi.

Franco, J. J. 1967. Research for river regulation dike design. ASCE Journal of the Waterways and Harbors Division 93:71–87.

Fremling, C. R. 2005. Immortal river: the upper Mississippi River and modern times. University of Wisconsin Press, Madison.

GREAT (Great River Environmental Action Team). 1980. A study of the upper Mississippi River. U.S. Army Corps of Engineers, St. Paul District, St. Paul, Minnesota.

Guntren, E., A. Oliver, T. Keevin, and D. Williams. In press. Changes in lower Mississippi River chutes: a compendium of bathymetric and photographic data. U.S. Army Corps of Engineers, Research and Development Center, MRG&P Report, Vicksburg, Mississippi.

Guyon, L., C. Deutsch, J. Lundh, and R. Urich. 2012. Upper Mississippi River Systemic Forest Stewardship Plan. U.S. Army Corps of Engineers, St. Paul District, St. Paul, Minnesota.

Hubbard, L. C., D. S. Biedenharn, and P. Hoffman. 2006. Investigation of the effects of notched dikes on pool habitat. Draft report submitted to U.S. Army Corps of Engineers, Mississippi Valley Division, Vicksburg, Mississippi.

Industrial Economics. 2014. The economic profile of the lower Mississippi River: an update. Final Report. Prepared for Lower Mississippi River Conservation Committee, Vicksburg, Mississippi by Industrial Economics, Cambridge, Massachusetts.

Johnson, B. 2003. Biological monitoring report: over-wintering fish use of chevron dikes—Pool 25, Mississippi River. U.S. Army Corps of Engineers, Avoid and Minimize Program, St. Louis District, St. Louis.

Johnson, B. L., and K. H. Hagerty, editors. 2008. Status and trends of selected resources of the upper Mississippi River System. U.S. Geological Survey, Upper Midwest Environmental Sciences Center, La Crosse, Wisconsin.

Kenow, K. P., G. L. Benjamin, T. W. Schlagenhaft, R. A. Nissen, M. Stefanski, G. J.Wege, S. A. Jutila and T. J. Newton. 2015. Process, policy, and implementation of pool-wide drawdowns on the upper Mississippi River: a promising approach for ecological restoration of large impounded rivers. River Research and Applications. DOI: 10.1002/rra.2857.

Kenow, K. P., and J. E. Lyon. 2009. Composition of the seed bank in drawdown areas of navigation pool 8 of the upper Mississippi River. River Research and Applications 25:194–207.

Kenow K. P., J. T. Rogala and P. J. Boma 2007. Evaluation of 2006 vegetation response on areas exposed during the 2005 drawdown of Navigation Pool 5, upper Mississippi River. U.S. Army Corps of Engineers, Navigation and Ecosystem Sustainability Program, NESP ENV Report 5, Rock Island District, Rock Island, Illinois; St. Louis District, St. Louis; and St. Paul District, St. Paul, Minnesota. Available: http://www2.mvr.usace.army.mil/UMRS/NESP/Documents/Evaluation%20of%202006%20vegetation%20on%20Pool%205%20-%20MH%20edit.pdf. (May 2015).

Killgore, K. J., P. Hartfield, R. A. Fischer, W. T. Slack, J. J. Hoover, A. B. Harrison, D. Biedenharn, and B. Kleiss. 2014. Conservation plan for the interior least tern, Pallid Sturgeon, and fat pocketbook mussel in the lower Mississippi River (Endangered Species Act, Section 7(a)(1)). U.S. Army Engineer Research and Development Center, Mississippi River Geomorphology and Potomology Program Report 4, Vicksburg, Mississippi.

Killgore, K. J., J. J. Hoover, and B. R. Lewis. 2012. Ranking secondary channels for restoration using an index approach. U.S. Army Engineer Research and Development Center, Ecosystem Management and Restoration Research Program, ERDC TN-EMRRP-ER-15, Vicksburg, Mississippi.

King, S. L., D. J. Twedt, and R. R. Wilson. 2006. The role of the Wetland Reserve Program in conservation efforts in the Mississippi River Alluvial Valley. The Wildlife Society Bulletin 34:914–920.

LMRCC (Lower Mississippi River Conservation Committee). 2012. Fishing the lower Mississippi River. LMRCC, Vicksburg, Mississippi. Available: www.lmrcc.org/wp-content/

uploads/2013/09/LMR-Fishing-Guide-Version2.pdf. (May 2015).

LMRCC (Lower Mississippi River Conservation Committee). 2015. Restoring America's greatest river: a habitat restoration plan for the lower Mississippi River. LMRCC, Vicksburg, Mississippi. Available: www.lmrcc.org/wp-content/uploads/2015/04/RARG-FINAL-4-21-2015.pdf. (May 2015).

Nissen, R. A., editor. 2014. Habitat enhancement using water level management on the upper Mississippi River. Water Level Management Task Force - River Resources Forum. Available: www.mvp.usace.army.mil/Portals/57/docs/Navigation/WLM_summary-May%20_2014.pdf. (June 2015).

Palmer, M. A., E. S. Bernhardt, J. D. Allan, P. S. Lake, G. Alexander, S. Brooks, J. Carr, S. Clayton, C. N. Dahm, J. Follstad Shah, D. L. Galat, S. G. Loss, P. Goodwin, D. D. Hart, B. Hassett, R. Jenkinson, G. M. Kondolf, R. Lave, J. L., Mayer, T. K. O'Donnell, L. Pagano, and E. Sudduth. 2005. Standards for ecologically successful river restoration. The Journal of Applied Ecology 42:208–217.

Pokrefke, T. J., editor. 2012. Inland navigation: channel training works. American Society of Civil Engineers, ASCE Manual of Practice No. 124, Reston, Virginia.

Rivers and Harbors Act of 1888. U.S. Code, volume 33, section 601.

Rohweder, J., J. T. Rogala, B. L. Johnson, D. Anderson, S. Clark, F. Chamberlin, D. Potter, and K. Runyon. 2012. Application of wind fetch and wave models for habitat rehabilitation and enhancement projects – 2012 update. Available: www.umesc.usgs.gov/management/dss/wind_fetch_wave_models_2012update.html. (June 2015).

Saucier, R. T. 1994. Geomorphology and quaternary geologic history of the lower Mississippi valley. U. S. Army Engineer Waterways Experiment Station, Vicksburg, Mississippi.

Soar, P. J., C. R. Thorne, and O. P. Harmer. 2005. Hydraulic geometry analysis of the lower Mississippi River. Final Report to U.S. Army European Research Office, Edison House, London.

UMRBA (Upper Mississippi River Basin Association). 2007. Integrating NESP and EMP: a UMBRA vision for the future of Ecosystem restoration on the upper Mississippi River system. UMRBA, St. Paul, Minnesota. Available: www.umrba.org/publications/usacenav/nesp-emp-vision11-13-07.pdf. (June 2015).

UMRBC (Upper Mississippi River Basin Commission). 1982. Comprehensive master plan for the management of the upper Mississippi River system. UMRBC, St. Paul, Minnesota.

UMRCC (Upper Mississippi River Conservation Committee). 1993. Facing the threat: an ecosystem management strategy for the upper Mississippi River. UMRCC, Rock Island, Illinois.

USACE (U.S. Army Corps of Engineers). 1988. Record of decision for Melvin Price Locks and Dam second lock (Locks and Dam No. 26, replacement, second lock), Mississippi River, Alton, Illinois and Missouri, 23 November 1988. USACE, St. Louis District, St. Louis.

USACE (U.S. Army Corps of Engineers). 1992. Avoid and Minimize measures design memorandum no. 24 progress report. USACE, St. Louis District, St. Louis.

USACE (U.S. Army Corps of Engineers). 1992–2010/2011. Avoid and Minimize measures design memorandum no. 24 progress report. USACE, St. Louis District, St. Louis.

USACE (U.S. Army Corps of Engineers). 1996. Problem appraisal report for water level management, upper Mississippi River. USACE, Water Level Management Task Force, St. Paul District, St. Paul, Minnesota.

USACE (U.S. Army Corps of Engineers). 1997. Channel maintenance management plan. 1997 original updated annually. USACE, St. Paul District, St. Paul, Minnesota.

USACE (U.S. Army Corps of Engineers - MVP). 1999. Pilot pool drawdown: Pool 8, upper Mississippi River Wisconsin and Minnesota. Definite project report, environmental assessment. USACE, St. Paul District, St. Paul.

USACE (U.S. Army Corps of Engineers). 2000. Upper Mississippi River habitat needs assessment: summary report 2000. U.S. Army Corps of Engineers, St. Louis District, St. Louis.

USACE (U.S. Army Corps of Engineers). 2003. Habitat rehabilitation and enhancement project (HREP) planning and sequencing framework, 11/10/03 final. Available: www.mvr.usace.army.mil/Portals/48/docs/Environmental/EMP/HREPSequencingFramework-FinalNov03.pdf. (June 2015).

USACE (U.S. Army Corps of Engineers) 2004. Integrated feasibility report and programmatic environmental impact statement for the UMR-IWW system navigation feasibility study. USACE, Mississippi Valley Division, Vicksburg, Mississippi.

USACE (U. S. Army Corps of Engineers). 2007. Upper Mississippi River Pool 8 drawdown results. Water Level Management Task Force. Available: www.mvp.usace.army.mil/Portals/57/docs/Navigation/River%20Resource%20Forum/Pool%208%20WLM%20Results%20Report.pdf. (June 2015).

USACE (U.S. Army Corps of Engineers). 2010. Upper Mississippi River Restoration Environmental Management Program 2010 report to Congress. USACE, St. Paul District, St. Paul, Minnesota; Rock Island District, Rock Island, Illinois; and St. Louis Districts, St. Louis.

USACE (U.S. Army Corps of Engineers). 2011. Upper Mississippi River navigation chart. Available: www.mvr.usace.army.mil/Missions/Navigation/NavigationCharts/UpperMississippiRiver.aspx. (May 2015).

USACE (U.S. Army Corps of Engineers). 2012. Upper Mississippi River Restoration Environmental Management Program environmental design handbook. USACE, Rock Island District, Rock Island, Illinois.

USFWS (U.S. Fish and Wildlife Service) 2004. Final biological opinion for the Upper Mississippi River-Illinois Waterway System navigation feasibility study. Prepared by U.S. Fish and Wildlife Service, Rock Island Field Office, Rock Island, Illinois. Available: www.fws.gov/midwest/mussel/documents/final_biological_opinion_upper_mississippi_river-illinois_waterway_system_aug_2004.pdf. (May 2015).

USFWS (U.S. Fish and Wildlife Service). 2006. Upper Mississippi River National Wildlife and Fish Refuge comprehensive conservation plan. USFWS, Winona, Minnesota.

USFWS (U.S. Fish and Wildlife Service) 2013. Biological Opinion: Channel Improvement Program, Mississippi River and Tributaries Project, lower Mississippi River. USFWS, Mississippi Field Office, Jackson, Mississippi. Available: www.fws.gov/mississippiES/_pdf/LMRBiologicalOpinion.pdf. (May 2015).

USGS (U.S. Geological Survey). 2014. Water data for the nation. Available: http://waterdata.usgs.gov/nwis/. (May 2015).

USOFR (U.S. Office of the Federal Register). 2011. Reservoirs at headwaters of the Mississippi River; use and administration. Code of Federal Regulations, Title 33, Part 207, Section 340. U.S. Government Printing Office, Washington, D.C.

Ward, J. V., K. Tockner, and F. Schiemer. 1999. Biodiversity of floodplain river ecosystems: ecotones and connectivity. Regulated Rivers Research and Management 15:125–139

Water Resources Development Act of 1976. Public Law 94–587, 94th Congress, 2nd Session (22 October 1976).

Water Resources Development Act of 1986. Public Law 99–662, 99th Congress, 2nd session (17 November 1986).

Water Resources Development Act of 1990. Public Law 101–640, 101st Congress, 2nd Session. (28 November 1990).

Water Resources Development Act of 1992. Public Law 102–580, 102nd Congress, 2nd Session (31 October 1992).

Water Resources Development Act of 1999. Public Law 106–53, 105th Congress, 2nd Session (17 August 1999).

Water Resources Development Act of 2007. Public Law 110–114. 107th Congress 2nd Session (8 November 2007).

Way, C. M., A. J. Burky, C. R. Bingham, and A. C. Miller. 1995. Substrate roughness, velocity refuges, and macroinvertebrate abundance on artificial substrates in the lower Mississippi River. Journal of the North American Benthological Society 1995:510–518.

Wisconsin v. U.S. Army Corps of Engineers et al. 1973. Opinion and order 73-C-183, U.S. District Court for the Western District of Wisconsin.